RESEARCH AND DEVELOPMENT
OF HIGH TEMPERATURE MATERIALS
FOR INDUSTRY

RESEARCH AND DEVELOPMENT OF HIGH TEMPERATURE MATERIALS FOR INDUSTRY

CO-ORDINATING EDITOR
E. BULLOCK

CO-EDITORS
R. BRUNETAUD
J.F. CONDÉ
S.R. KEOWN
S.F. PUGH

COMMISSION OF THE EUROPEAN COMMUNITIES
INSTITUTE OF ADVANCED MATERIALS
JOINT RESEARCH CENTRE
PETTEN ESTABLISHMENT, THE NETHERLANDS

ELSEVIER APPLIED SCIENCE
LONDON and NEW YORK

ELSEVIER SCIENCE PUBLISHERS LTD
Crown House, Linton Road, Barking, Essex IG11 8JU, England

Sole Distributor in the USA and Canada
ELSEVIER SCIENCE PUBLISHING CO., INC.
655 Avenue of the Americas, New York, NY 10010, USA

WITH 40 TABLES AND 156 ILLUSTRATIONS

© 1989 ECSC, EEC, EAEC, BRUSSELS AND LUXEMBOURG
Softcover reprint of the hardcover 1st edition 1989

British Library Cataloguing in Publication Data
Research and development of high temperature materials
for industry
1. High temperature materials
I. Bullock, E.
620.1'1217

ISBN-13: 978-94-010-7008-9   e-ISBN-13: 978-94-009-1145-1
DOI: 10.1007/978-94-009-1145-1

Library of Congress CIP data applied for

Publication arrangements by Commission of the European Communities, Directorate-General Telecommunications, Information Industries and Innovation, Scientific and Technical Communication Unit, Luxembourg

EUR 12374

LEGAL NOTICE
Neither the Commission of the European Communities nor any person acting on behalf of the Commission is responsible for the use which might be made of the following information

EDITORIAL PREFACE

The first CEC review of the "Technological Requirements for High Temperature Materials R&D", published in November 1976 was an active expression by the European Commission to encourage the spirit of co-ordination and interaction across the breadth of European high-temperature materials research actions, and to identify general areas of common interest which were appropriate for centralised European research effort in support of the development of industry and technology within the Member States.

The study was envisaged to provide a forward look of some ten years. The decade following has seen a revolution in European materials research effort with the establishment of several major co-operative research exercices both under the auspices of the European Commission and externally within the Community and worldwide. Research expenditure for materials development throughout Europe has increased ten-fold.

The period has also witnessed significant changes in European R&D philosophy, particularly in the high temperature field, which have been motivated generally by changes in the technological/ economic climate of European industry, and a redefinition of industrial needs with respect to materials performance and development. Concurrently the emergence of innovative technology (new materials, processes, applications has led to considerable advances in a number of fields, beyond the scope of the original review. In retrospect the selection of ten years as an outlook, period seems to have been correct, and the occasion is now timely for a further forward look into high temperature materials research needs for industry.

The present review has been extended to respond to the increased international awareness that industrial technology and economic competitivity should provide the general "steer" in defining research and development needs. This mood of international competitivity in the materials field has been heightened primarily by the emergence of Japan as a major producer of high technology materials supported by superb marketing and publicity skills, and from the early movements within Europe to unify the independent efforts of the individual nations into a European co-operative force to meet this competitive challenge. The earlier rather idealistic concepts of a co-ordinated "European Research Effort" are now increasingly presented as an essential strategy for survival.

In response to this trend, the current review has been structured to include two approaches:-
i)   a general appreciation of the prospects of the particular materials classes relevant to high temperature technology, considering their potential for future development and exploitation. This approach reflects the "development push" of the developers of materials and of materials production and processing technologies to achieve improved properties and performance.
ii)   the viewpoint of the major industrial users of structural materials, to indicate the service demands imposed upon structural materials and components in the high temperature industrial and energy technologies and to identify crucial areas where the shortfall of materials capability impedes technological progress. In addition to the major industrial user sectors, a number of technological areas associated with the design performance and application of materials in plant components have been included in this approach under the generic title "Optimisation of Components".

The individual materials or technology topics covered by the book are represented by original contributions by specialist authors who have been selected on the basis of their association with the particular field of materials or engineering. It is inevitable that authors contributions from the two approaches will overlap to some extent. In order to allow the optimum expression of both viewpoints, without incurring too much duplication of subject matter, a number of "general orientations" for the main sections of the study have been drawn up.

Authors in all sections were invited to address the current state of art with respect to materials capability for their targetted end-use, to identify areas of particular shortfall with an indication of the realistic prospects for overcoming materials problems, and to itemise specific areas of research and development need to which priority should be given over the next decade. The authors recommendations for future research priorities are presented within the individual contributions, and have also been gathered into a separate chapter (7. - Conclusions) but still referenced by organisation and number to the orginal contribution. Editorial conclusions relevant to the separate topics, which have been drawn both from other contributions within the book and from personal knowledge, have also been included in Chapter 7 - Conclusions.

The editors have consciously made the decision not to draw generic conclusions from the authors recommendations. The temptation to identify a series of general research areas for the purpose of directing future research funding has been resisted, for two principal reasons:

i) the conclusions and recommendations of the authors are often very well defined and address specific and real research needs. Any attempt to collate these findings into more diffusely defined general areas, would reduce the impact of the individual contributions and of the book as a whole and, detract from its principal objective, to identify materials research and development needs for industries.

ii) the mechanisms for funding of research in Europe are highly diverse and are often of very different character. The tailoring of research programmes, particularly on a European scale, in which political, economic and sectorial interests must be satisfied in addition to industrial and technological need, is most effectively achieved by defining research programmes within the operating rules of the funding authority.

The expressed objective of the book is to identify the priorities for research and development in the short term future (ca. ten years) of structural materials operating in the major high temperature technologies.

The selection of topic areas and of contributing authors was confirmed by a panel of expert editors who were chosen to represent the major sectors of the book. The editors were further responsible for harmonising the authors contributions to their area of responsibility to promote the overall coherence of the book, to minimise duplication and to draw conclusions for individual topics from the field of interest as a whole, and from their own expertise.

The review has been restricted to materials/industrial processes, operating at high service temperatures. No strict guideline on lower temperature limit has been made, so that topics such as, aluminium-matrix composites, and titanium are included. In general, however, a limit of 550-600°C has been selected by the authors themselves.

The book has attempted to cover all of the principal areas of high temperature industrial use of structural materials but inevitably a number of areas are not represented, where authors could not be found, or where factors such as commercial sensitivity would not allow an authoratative statement to be made. The viewpoints expressed in the book are in general those of the individual contributing authors themselves rather than of their affiliated organisations.

Content of the Review

The review has been structured into 3 major groups.
* Materials Potential
* Materials Production and Processing
to provide a developers outlook

* Materials Constraints in the High Temperature Industrial
  Technologies
* Optimisation of Components
giving the viewpoint of materials users, and of plant and component designers.

* Research Trends in Industrial Development
* Conclusions
to focus upon future orientations and research needs.

The force driving materials development is the market pressure for improvement of process economics by increasing plant efficiency, implying higher temperatures, lower costs, less maintenance, and therefore greater component and material capability.

Over the last two to three decades the major efforts to enhance high temperature materials capability have been devoted to metals and alloys, where the leading technology demanding highest performance requirements has been the gas turbine industry. Much of the original research and development of the most advanced materials therefore relates to turbine applications although experience gained in this area has been spun-off into other technologies.

Inevitably any particular materials class will have a final development limit. As a system approaches its potential limit, the return in improved capability becomes economically less and less attractive until the only opportunity to advance material capability is by making a quantum step to a new material base of higher intrinsic capability.

Historically, the broad sweep of materials development has reached just such a discontinuity at the present time. The range of high temperature steels and nickel/cobalt base alloys, conceived in the 1940s and so spectacularly developed over the past forty years, are approaching their theoretical limits of exploitation. The "high tech" advances of the last decade have been more in structural than in chemical engineering raising the "added value" of the materials themselves by for example, processing and production technology to achieve improved microstructures, by designing and engineering component shape and form to alleviate process loading, by fabrication of components made of more than one material e.g. with coatings, composites, bonded materials, etc..

The next decade will look increasingly to materials innovation to provide the next step-out in capability. The complexity of requirements for materials working in modern energy and process environments makes the task formidable, but the current indications of innovation and ingenuity in research suggest that the developers and producers are rising to the challenge.

Faced with this background, the authors contributing to the two sections covering "Materials Potential" and "Materials Production" were invited to consider the prospects of the materials currently in service or in near-service development, and of new and innovative materials still at experimental status but considered as promising candidates for future exploitation. Authors have been asked to include in their contributions, aspects covering:

* the state of the art and current capability of materials or the production and processing of materials, for service application
* critical property shortfall vis à vis technical application, properties, degradation modes, etc.
* future development trends, realistic outlook for economic and development potential, technology transfer, cost effectiveness, availability and management of resources
* new materials or production routes, innovative developments, prospects and scope of exploitation
* indication of significant R&D efforts currently being undertaken e.g. in Europe, U.S.A. and Japan
* future research needs and development needs; across the full spectrum from fundamental research to demonstrator projects/ pilot plants/in-plant investigations; ranking of R&D actions with respect to time and needs.

The viewpoint of the materials user industries leans necessarily to economic pragmatism. The requirement to operate plant or machinery efficiently over protracted periods of time places considerable emphasis on reliability, safety, and component endurance as well as on performance and initial cost. Economic use of resources, high security of operation, minimisation of research, development and accreditation costs, all influence profitability so that factors such as process monitoring, design, life prediction and modelling assume considerable importance.

The prodigious costs of investment in new processes and of accrediting new materials for plant/component service creates a natural reluctance to take up novel materials and design until performance and behaviour are well established. Moreover, operating conditions are generally becoming more demanding following the increasing pressure for energy economy for example by optimising use of heat, by combining process cycles, by raising temperatures for thermal efficiency, by use of less pure fuels and cheaper materials, and more recently by the increasing awareness of the need to protect the environment. All of these factors restrict the freedom of the industrialist to make optimum use of innovation in materials and design.

A practical review of research needs should take account of the constraints of the user technologies. Authors for these contributions were selected on the basis of their close association with the relevant industry or branch of engineering. Specifically, they were requested to address the industrial view of the current material/component shortfall and its constraint on further process development, and to provide a prognosis of future process developments taking into account the economic situation, and the relative importance of materials problems such as;

* materials cost, availability
* reliability, safety, performance
* essential materials development needs and priorities
* fulfillment of industrial needs by current research and development profiles
* re-orientation of research philosophy future/needs e.g.
    * industrial development
    * demonstrator projects
    * inspection procedures
    * component security/safety.

Authors contributing to the section 5 (Optimisation of Components) were asked to identify the contribution made by these methodologies to the exploitation of materials and components in industry. The current state of art and future evolution, and the prospects of benefit to industry.

Principal aspects should cover
* the current state of art
* the importance in relation to materials role in industry
* the extent to which lack of development holds back industrial progress

* the future orientations, major areas for development, for optimum industrial benefit
* the research needs and priorities.

The nature of industrial research also shifts emphasis on a more generic scale. The higher temperature technologies are by nature highly sensitive to the supply of fuel and materials. The somewhat frenzied period of activity worldwide to develop synthetic fuel processes in the years following the 1973 fuel crisis, have all but disappeared in the succeeding decade of oil surplus. Similarly the alarm of resource limitations of certain minerals of the 1960s has relaxed into an economic rather than a strategic factor.

These broad fluctuations in development philosophy have a major influence on research funding patterns and an understanding of the factors influencing general industrial strategy in the mid-term future is a major asset in the efficient deployment of research resources. The current attention to raising materials capability in terms of performance, properties and production suggests a buoyant and confident outlook to materials development and exploitation.

Increasingly, however, the scale of the major technological sectors is expanding, so that economic viability acquires a European, rather than a national dimension. The high costs of accreditation, standardisation, duplicating research and development and of investment in manufacturing, testing and development plant all point to the economic attractions of a larger technology base and internal market.

Following this trend, the major progress in materials research and development in the next decade may well depend upon advances in the political and economic domain, without which the wealth of European scientific innovation will not be realised.

E. Bullock
Joint Research Centre
Petten

LIST OF CONTENTS

1. Advanced High Temperature Materials

M.H. Van de Voorde

Joint Research Centre,

Petten, NL

1. INTRODUCTION

Materials are recognized to be the building blocks of advanced energy technology and industrial innovation. Many industries are intensively engaged in the development of new and better materials. Their efforts are crucial to innovations that will render the companies more energy efficient and more capable of meeting international competition in the future.

Processes involving high temperatures form an important part of modern industrial activity. Fundamental thermo dynamics predicts that an increase in temperature leads to a higher overall efficiency e.g. in the conversion of thermal energy to mechanical and subsequently to other forms of energy. For example if the gas turbine inlet temperature is increased from 900°C to 1250°C (for the same specific fuel consumption) the specific power increases by about 30%. Other processes which become more efficient as the temperature is increased are those in which the limiting factors involve microstructural or molecular kinetic processes; in such cases higher efficiency corresponds to a shorter reaction time and therefore higher productivity for the installed equipment.

Currently many materials in HT service are working at the limit of their capability. The demand for strength, toughness and corrosion resistance means that materials must now be designed for very specific application needs. For this reason the field of high temperature materials embraces a wide range of metals, alloys, engineering ceramics and composites. Some of the most prominent and exciting applications are for propulsion units (turbine and rocket), for high speed aircraft and spacecraft, new and improved methods for energy production, conversion and utilization including nuclear energy, metals processing, the chemical and automotive industry. It has become apparent that materials are one of the major obstacles to efficient utilization of high temperatures and consequently higher efficiencies and to optimized plant design. Without adequate high temperature materials and a knowledge of their "personality" we cannot exploit the new theories of science or implement the advanced designs of new technologies. Gas turbine technology, which was the most impressive HT development area of the 1950-80s, progresses at the rate of availability of new materials. Other areas in which material high temperature limitations are holding back industrial progress include for example space technology magneto-hydrodynamics, nuclear fusion, etc.. So, in the advanced technological situation of today the search for new and improved high temperature materials is essential to maintain progress.

2. MATERIALS DEVELOPMENT

Before evaluating new materials we shall review the potential of conventional materials systems. With ferrous alloys the temperature capability, strength and corrosion increases with increasing chromium content. Once temperatures in the region of 900°C are required the oxidation rate and strength combination of these alloys becomes critical.

Superalloys

At higher temperatures, alloys based on nickel and cobalt, "superalloys", are required. These alloys demonstrate higher strength at higher temperatures than the ferrous alloys, which is achieved only by reducing their chromium content. The lower chromium is necessary to guarantee microstructural stability and allow increased additions of alloying elements such as Al, Ti and Nb for optimum creep properties. However, as the level of strengthening alloying elements increases the alloys have to be cast, rather than wrought. These high quality alloys have been developed primarily for gas turbine applications. Latest advances to avoid problems of high temperature creep by grain boundary sliding, have been the development of directionally solidified (DS) and single crystal (SX) turbine blade alloys. Typically the temperature capability of these alloys has risen by 300°C over the last 30 years.

Ultimately the nickel-based systems are, intrinsically temperature limited since the alloys rely for strength at high temperatures upon a dispersion of coherent precipitation based on the intermetallic γ' (Ni_3Al) phase and this phase dissolves above 1000°C.

Newer developments are towards Oxide Dispersion Alloys (ODS), strengthened by oxide particles which are not soluble in the matrix and thus are stable to higher temperatures than DS or SX alloys. ODS alloys are however generally anisotropic, are not weldable, have limited tensile strength up to 1000°C and poor thermal fatigue characteristics.

While little is to be expected in improving materials for higher temperature capability, the major further developments are likely to be through manufacturing technology that allows the design of improved cooling systems, and by short term improvements (50 to 100°C) coming from thermal barrier coatings (TBC). The major step forward is however envisaged through the use of ceramic systems operating uncooled at temperatures approaching 2000°C by the year 2000.

Today turbine discs require high tensile strength, coupled with good low cycle fatigue life to withstand the frequent change in stress accompanying each cycle, defect tolerance and long life. Future requirements are for even higher usable strengths and temperature capability. Likely candidates for such an arduous task are the nickel aluminides, nickel metal matrix composites and in the longer term, reinforced ceramics. All three areas have associated with them a number of technically challenging problems. For metal matrix composites, the availability of suitable refractory fibres, combined with the thermal/chemical stability of the fibre/matrix combination and resistance to thermal cycling are likely to be the major problems. Limited ductility will cause concern with the intermetallics (nickel - aluminides), whilst fibre availability, fibre/matrix control, life and integrity requirements will seriously hinder the introduction of reinforced ceramics into turbine disc applications.

In other applications including the chemical, metallurgical and petrochemical industries it is easier to achieve a balance between high temperature corrosion resistance and strength since strength requirements appear to be more moderate. Since the latter industries require materials in considerable quantities, the material cost factor is an important element in the selection criteria.

Refractory Metals

Refractory metals, possess attractive high temperature strength but are handicapped by their very poor oxidation resistance, mainly due to volatile or molten oxides.

Attempts to provide adequate protective coatings have been successful but should the coating become damaged, the consequences of allowing the environment to access the underlying metallic component would be disastrous. Precious metal coatings are possible for certain specialized applications as the melting point of platinum is 1772°C, but the price is prohibitive confining their use to aircraft applications.

Ceramics

Monolithic ceramics, and in particular silicon nitride (Si_3N_4) and silicon carbide (SiC) show potential for application as HT structural materials. These materials are stronger than the nickel superalloys above 1000°C, have superior creep strength and oxidation resistance and are potentially cheaper. In addition their density is less than half that of the superalloys (typically 3,2 gm/cm^3 compared to 7,9 gm/cm^3). The Achilles heel of these materials, however, is their intrinsic flaw sensitivity, brittleness and consequent lack of reliability. The main source of mechanical brittleness in ceramics derives from their crystalline structure which does not permit the existence or the movement of dislocations (as with ductile materials) and hence local plastic flow, under an applied stress. For operational acceptance ceramic components must demonstrate a reliability in operation at least as good as the metal components they replace. This can be achieved in one of the three ways. The first approach is to learn to live with the brittleness (low K_{1C}) and develop a fundamental understanding of the micromechanics of failure (i.e. flaws and their relation to strength). In this way statistical methods, non destructive evaluation or proof test methodologies can be used to specify design parameters such as strength or component life. The second approach is to identify the sources of strength degrading flaws and develop improved processing methods to totally eliminate these strength limiting defects. As the critical defect size for ceramics is ca. 100μm i.e. some two orders of magnitude less than that for metallic materials at typical operating stresses, this solution is not as easy as it may first appear. Substantial advances have however been made with both of these approaches over the last 10 years with proof of concept and component demonstrations having been carried out successfully by several companies in the industrialised world. The two approaches discussed so far rely heavily on the design stress being kept, at all times, below the failure stress. If for any reason the failure stress is exceeded, catastrophic failure will ensue. A basic change in failure mechanism from this flaw sensitive brittle, catastrophic failure to a more forgiving failure mechanism is seen to be essential for aero engine operation. This can only be achieved by the third approach: to design ceramic microstructures with improved resistance to fracture, and hence some tolerance to defects.

Fibre reinforced ceramic matrix composites have recently received a great deal of attention for use in high temperature structural applications. The reason for this interest lies in the assumption that strong ceramic fibres can prevent catastrophic brittle failure in ceramics by providing various energy absorbing process during a crack advance and thus provide some defect tolerance. The critical issues for the efficient utilization of reinforcements are high strength and stiffness, low density, low diameter and most importantly thermal stability, both during fabrication and under component operation. It is unlikely that the ceramic fibres which have been available for some time will provide adequate long term reinforcement at temperatures greater than 1400°C, and most will experience problems above 1200°C. Susceptibility to decomposition, oxidation, accelerated grain growth and creep are the major technical hurdles to be overcome. Recognising the need for more refractory ceramic fibres, suppliers are working towards improved products and some new fibres are available.

In addition fibre coatings will be required to prevent fibre-matrix chemical reaction and serve as mechanically weak boundaries for toughening. The long term compatibility of these fibre coatings with both matrix and fibres will also be an important factor in maintaining as-fabricated composite properties.

Coatings

With increasingly greater demands imposed by advanced processes, it is becoming more and more difficult to combine in a single material the required structural properties and resistance to attack by hostile environments. Coatings and claddings hold out the best hope for achieving a degree of resistance beyond that of any viable engineering alloy candidate. Application of specifically tailored coatings and claddings to substrate chosen for excellent mechanical properties can produce cost effective composite systems which optimize both corrosion resistance and strength.

The early coatings were diffusion aluminide type and which have provided excellent service particularly in aircraft gas turbines. The recent overlay type coatings, based upon MeCrAlY systems also perform well, finding application in the energy technologies and petrochemical industries.

Although room is available for coating alloy development, the future for new high quality coatings has to be found in coating processing techniques. However, no matter, how much more successful developments may be realized, the prospects are limited by the melting point of the superalloys.

Coating the more refractory metals with conventional overlay coatings to provide protection is not possible because, if they are to be used appreciably above the temperatures at which current superalloys operate, the oxidation degradation rate of the coating would be excessive.

Ceramic coatings have the advantage of being resistant to oxidative environments with low coefficients of thermal conductivity and these ideal combinations of properties make them potential for thermal barrier coatings.

The problem is the mismatch of the coefficients of thermal expansion between the ceramic coating and the metallic substrate and hence produce stresses during temperature cycling resulting in coating/substrate failure. The significant differences in coefficient of expansions between popular engineering ceramics o.a. silicon nitrides, silicon carbides, zirconia and the base alloy make thermal barrier coating unsuitable. To accommodate the changes in thermal expansion coefficients, porous coatings have been successfully developed, with increased thermal barrier efficiency and acceptable generation of stresses upon thermal cycling e.g. $ZrO_2.MgO$. In addition experiences have shown that ceramic coatings accommodate better to the iron based alloys than to superalloys. Thermal barrier coatings could be more prosperous being developed when parallel efforts are devoted to the investigation of substrate alloys with controlled coefficients of expansion.

The favoured thermal barrier coating e.g. $ZrO_2.MgO$ have the disadvantage to be hot corrosion sensitive; the MgO being attacked by sulphur and vanadic contaminants. Therefore, these ceramic coatings have no great value for applications in high temperature corrosive environments.

In addition, the aggressive contaminants react with the stabilizing additions MgO, CaO, Y_2O_3 in the ZrO_2 coating systems. This effect results in loss of stabilization of the zirconia coating and during thermal cycling leads to premature coating failure by stresses induced by phase instability.

Al_2O_3 and ThO_2 are quite resistant to most corrosive species but these materials cannot be used as thermal barrier coatings because they are not available in porous form. An open porous structure is ideal as a thermal barrier but not for resisting corrosive attack.

Ceramic thermal barrier coatings are in their infancy of development and particularly for applications in corrosive environments.

3. ECONOMICS AND SUPPLY

Many common materials, or the precursers from which they are produced, are widely and readily available, e.g. sand, limestone and bauxite for cement, glass etc. No difficulties are foreseen in meeting demand for such materials and progress can only be anticipated in economic terms to combat the increasing costs of fuel for processing and for transport.

The position of metals may be very different. All the major industrial metals except iron, aluminium, magnesium and calcium occur only in restricted locations and at levels of concentration often well below one per cent. The mining, concentration and refining of metals therefore involves the handling of large amounts of material and the consumption of much energy. While in most cases the established reserves of ore are sufficient to guarantee supplies well into the next century, the cost of production will, in most cases, increase because of the need progressively to exploit decreasing grades of ore. Additionally, costs will increase because of rising fuel costs since minerals extraction technology is generally a fuel-intensive process. It is the prospect that we may reach a stage in which we can no longer afford the cost, particularly in energy requirements for extraction from very low grades ores.

Since metallic ores are recovered from specific locations which we relatively few in number where natural processes have concentrated the deposits, recovery must take account of the availability of labour, power and transport. Some metals may be particularly influenced by national political policies. For example, cobalt production was severly disrupted in 1978 by rebel action in Zaire - the source of about half the world's requirements - leading to a dramatic cobalt shortage and to a market price increase from about 6 US/lb to a peak of 25 US/lb in less than a year. The price has never fully recovered its former position in relation to that of other metals.

Other element of strategic importance e.g. chromium, essential for modern industry, are also recovered in the main from third world countries of uncertain political stability. Hence the need for contingency plans to involve the establishment of stockpiles, the exploration of substitution programmes and the recycling of materials. Procedures of recovery and recycling of the precious metals are, of course, well established and for almost all metals some scrap recovery is operated for the bulker forms, but extension of these to include smaller and more dispersed forms of the metals should be encouraged.

Exploration for new deposits of metallic ores continues world-wide using the modern geographical techniques based on aerial surveys by visual, magnetic and electroconductive methods, as well as the time-honoured land surveys following geological indications. The sea remains a potential source for a number of metals. Magnesium has been extracted from sea-water, although the wide occurrence of land deposits of magnesia and dolomite render this generally uneconomic. The marine nodules widely scattered on the sea bed, particularly in the Pacific, provide a vast reserve of manganese with useful levels of copper, nickel and cobalt. A typical composition is manganese 35%, copper 2.5%, nickel 2%, cobalt 0.4%. If suitable methods of recovery can be developed and if a reasonably balanced demand for the constituent metals can be established they could provide valuable contributions to the world's requirements.

Importance of raw materials

The industrial countries are alarmed concerning their dependent position on the supply of raw materials from third world countries, particularly because of the vital role of these primary materials for industrial trading, defence and standard of living. Japan imports almost all of its raw materials. This applies also for Western Europe for high temperature alloy based materials as chromium, cobalt, nickel, titanium and tungsten ores. The position of the U.S. is more optimistic, but concern exists with regard to the more vulnerable metals particularly with respect to political-economic movements in third world countries. Another important factor is that mineral producers in advanced countries such as Canada and Australia are keen to export costly upgraded ores.

It may be useful to recall the vital uses of critical metals as nickel, manganese, antimony, tantalum, titanium, tungsten, platinum, aluminum in stainless steels, heat resistant steels, superalloys etc. in the chemical industry, aeronautics and space, energy technology, etc.

4. FUTURE STRATEGY

A future materials strategy aimed at improving the value content of manufactured products and in a broader context combating the wasteful consumption of materials will fall in the broad categories of materials substitution, conservation and performance. Opportunities lie in the development of near net shape manufacturing techniques, which minimise metal removed during processing and the associated wasted scrap. Computerised methods will improve product design as well as better control during manufacturing. We may design lighter components, thereby saving energy and material costs. We may reduce the quantity of scarce materials used. A good example is the specific function of chromium in a corrosion resistant application where a steel component may be coated with a layer of a chromium alloy rather than having a large volume content of chromium within the alloy. We may design complex engineering structures to operate over a longer and more reliable life span, and to avoid over-design.

The development of high quality materials and structures introduces a "New concept of design", which implies:-
Improved performance, i.e. enhanced reliability and resistance to failure;
Improved durability of materials and their life extension. Problems of losses through corrosion, erosion and wear will be addressed through the introduction of new protective technologies and the introduction of advanced ceramics.

For ensuring safe materials supply for the future, we must develop a materials conservation ethic and recyclability concepts. An optimum utilisation of materials shall yield benefits in terms of performance, materials conservation as well as energy consumption and a clearer environment.

5. ENVIRONMENTAL PROTECTION

With the growth of the world's population and the consequent increase in the demand for the products of industry, attention is necessarily being paid to preservation of the natural environment, including the prevention of contamination of the air, sea and land by waste or discarded materials of all kinds, and the avoidance of extinction of species of animal and vegetable life. The rejection of some products of industrial processes as waste is currently based solely on economic grounds and if suitable uses for them could be established, contamination of the environment could be avoided. For example, much of the sulphur dioxide generated in the roasting of sulphide ores for the recovery of metals is trapped as sulphuric acid or sulphates, but the total amount produced exceeds the present demand for the acid and the excess is therefore emitted to the atmosphere. It contributes in some degree to the formation of acid rain, and the consequent deleterious effects on agriculture and forestry.

Within the materials area it is apparent that all developments of new products should involve parallel studies of the influence of the suggested processes on the environment and should include the establishment of safe methods of waste disposal. In the same context the collection and recovery of material after having served its useful life should be considered as contributing to environmental protection as well as to being economically and strategically desirable.

2. MATERIALS POTENTIAL

2.1. Alloys

2.1.1. High Temperature Steels for Power Plant

R.D. Townsend

C.E.R.L.

Leatherhead, U.K.

1. High Temperature Steels

Ferritic and austenitic steels have been used for many different high temperature applications in oil and chemical plants and in power generation equipment. Such applications have been generally successful, and there are many examples of successful operation at high temperatures for times well in excess of 150,000 h. However, there are particular characteristics of these two classes of material which severely restrict their range of application in terms of permissible operating parameters and these, if infringed, can lead to failures in service. The problem essentially is that for some characteristics, whether it be toughness, strength, or environmental interaction, these materials show too little tolerance to minor variations in chemistry, heat treatment, fabricating process, or operating parameters.

The two types of steel do, in fact, show an interesting and contrasting spectrum of physical parameters and properties which, in terms of engineering applications, expands the total range over which both can be used, but at the same time, severely restricts the application of each of them. Thus, ferritic steels are inexpensive, austenitic steels generally expensive. Both types can be easily fabricated into a variety of cast, wrought, or forged products, but the austenitics are more difficult to machine due to work hardening. Ferritic steels can be hardenend by heat treatment and variations in this allows a wide range of useful engineering properties. Austenitic steels usually enter service in the fully solution treated condition. Welding is generally satisfactory, but the ferritics require postweld heat treatment and can be susceptible to reheat cracking. The austenitics do not require postweld heat treatment, but in comparison to the ferritics, austenitic welds are much more difficult to inspect with ultrasonic and MPI. The general corrosion resistance of the austenitics is much better than the ferritics, but in aqueous solutions they are much more susceptible to stress corrosion cracking and if sensitised, can be susceptible to intergranular attack. For high temperature applications, the austenitics are generally stronger in creep and rupture, but because they have higher coefficients of expansion and lower thermal conductivity can be weaker in resistance to creep/thermal fatigue interactions. The properties of both types of material do, however, change with high temperature exposures. With the ferritic steels, this can take the form of a degradation in the structure as (alloy) carbides precipitate and coarsen, giving rise to a general softening characterised by lower hardness, tensile strengths, and reduced creep resistance. In contrast austenitic steels can significantly strengthen at high temperatures due to the precipitation of fine dispersions of alloy carbides and/or intermetallics. High temperature exposure can also cause significant embrittlement in both classes of material; in the ferritics this arises due to segregation of group IV elements to grain boundaries, which can impair low temperature toughness and cracking resistance at high temperatures. In the austenitic steels, similar embrittlement can occur if there are significant quantities of delta ferrite in the structure, as for example in a weld or casting, but more frequently, embrittlement in the austenitic steels is associated with the precipitation of intermetallics such as sigma phase or chi phase.

Given this wide range of contrasting properties, it is hardly surprising that in the main these two classes of steel are each used in quite specific regimes of operating parameters, and there are relatively few instances in which an austenitic steel and a ferritic steel offer equally viable options to the engineering requirement. The properties which chiefly dictate this situation are on the one hand the comparative weakness of ferritics at high temperatures and on the other the susceptibility to stress corrosion cracking displayed by austenitics in aqueous environments at low temperatures. As a consequence, ferritic steels are rarely used at operating temperatures much above 570°C and austenitics are generally not in situations where they would be exposed to water. In a steam boiler, for example, this would restrict their use to temperatures above ca 450°C when superheated steam can be considered to be fully unsaturated.

2. Ferritic Steels for High Temperatures

Ferritic steels are used at temperatures up to about 570°C. The applications include those at high temperatures (essentially above 400/450°C) at which the design is based on the creep properties and those at lower temperatures where the design is based on the time independent tensile properties. For engineering purposes, we can broadly divide them into two categories: those which are weldable and as a consequence have C contents usually <0.15%, and the higher strength varieties in which the C contents are up to 0.25-0.3%. In both types, the ferritic steels can be subject to various forms of embrittlement which in many cases can be traced to the presence of residual elements S, P, Sn, As, and Sb and even to alloying elements such as Mn and Si. The most important forms of embrittlement are those associated with temper embrittlement and creep embrittlement, the latter frequently associated with welds. It is fortunate, therefore, that almost concurrent with the realisation of the damaging role of trace elements has come the technological means of removing them by means of scrap selection and secondary refining methods of steel making. These have recently been reviewed by Jaffee (1), who predicts that secondary refinements of steel may achieve 50% of all steel products by the year 2000. Examples of the importance of this new technology are given below.

3. Nonwelded Turbine Materials--Rotors and Bolting

Conventionally melted low pressure NiCrMoV rotors are subject to temper embrittlement and to avoid this, the operating temperature must be restricted to 350°C. Many laboratory studies of temper embrittlement in NiCrMoV steels (2, 3) have shown that high purity base steel, free of Mn and Si and low in P, As, Sb, and Sn, which segregate to prior austenite grain boundaries, do not temper embrittle. In commercial production, ladle furnace refining now permits the production of similar high purity super clean steels for full size LP turbine rotors on a cost-effective basis (4), which are also essentially free of inclusions (<0.005 volume%). Fig. 1 shows that the FATT at the centre of a trial rotor actually decreased about 10°C after aging for 17,000 h in the temper embrittlement range 350-480°C. A potential application is supercritical plant with steam conditions such as 4500 psi/600°C, in which a super clean LP rotor could operate at temperatures up to 460°C without the need for rotor cooling which would be necessary with conventional material.

Ladle refining steel making has also been used for 1CrMoV HP and IP rotors. The TVA Gallatin rotor failure, which burst in 1976, had been air melted in the 1950s, was high in P and S, and

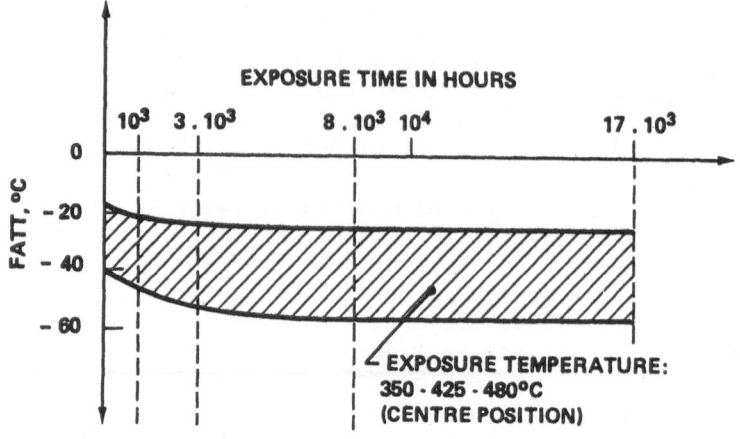

**FIG. 1 FATT AT CENTRE OF TRIAL 3.5NiCrMoV ROTOR AFTER ISOTHERMAL
AGEING FOR TIMES UP TO 1700 h AT 350°C, 425°C AND 480°C**

the critical flaw was found to localised in an area of brittle
"A" type segregation containing MnS inclusions along which the
crack had propagated (4). Since then, several investigations
have been made of secondary steel making techniques such as ladle
furnace refining and electroslag remelting (ESR) (5), and vacuum
carbon decarburisation (VCD) (6, 7) all of which produced rotors
of greatly improved quality. The clean 1CrMoV HP IP rotor steel
has been found to have slightly higher creep rupture strength
than conventional material because fewer MnS particles are
present on grain boundaries to act as nuclei for cavities.

An application in which adequate high temperature ductility
is a prerequisite for successful operation is with high
temperature bolting. Traditionally, steels selected for these
applications were of the 1-2% CrMoV type and relied on high C
(0.2-0.6%) contents to give fine dispersions of alloy carbides
and thus increase the stress rupture strength and, in particular,
resistance to stress relaxation. The service conditions of bolts
are, however, particularly onerous and situations can arise,
often caused by differential temperatures along the bolt, which
cause both high stresses and strain concentrations in the hottest
part of the stud. Since these high strength materials have very
low ductilities, a number of serious failures have been recorded
(8). Extensive work on this problem has indicated that to obtain
good ductility in these materials, it is important to obtain fine
grained material by careful control of heat treatment (9) and
also to reduce the residual element content by reducing the "R"

parameter, which is an empirical factor devised by King (<u>10</u>) to describe the potency of residual elements in controlling high temperature ductility, such that R=(P) + 2As + 3.6(Sn) + 8.2(Sb), where the elements contents are in weight percent. Commercial application of the importance of reducing the residual element contents has now been realised and much cleaner and higher purity alloys have been produced using careful scrap and ore selection and modern steelmaking practise using basic electric furnaces for primary melting followed by VAD (vacuum arc degassing) in the ladle and secondary refining by VAR. By these methods, the "R" factor can be controlled well below 0.15 giving much improved high temperature strength and ductility (<u>11</u>, <u>12</u>).

An alternative approach to residual element control as a means of improving high temperature ductility is to increase the Cr content of the steel. Clearly this increases the cost of the material, but the pioneering work of Pickering and Irvine (<u>13</u>) has shown that the high alloyed ferritic steel based on 9-12% Cr have an excellent spectrum of high temperature properties and, in particular, display good resistance to creep cavitation. Middleton's work (<u>14</u>) indicates that this increased resistance is associated with the increased chromium content which eliminates weak precipitate free zones close to the boundary and reduces shear strains along the prior austenitic grain boundaries (Fig. 2). The resistance to cavitation appears to be maintained

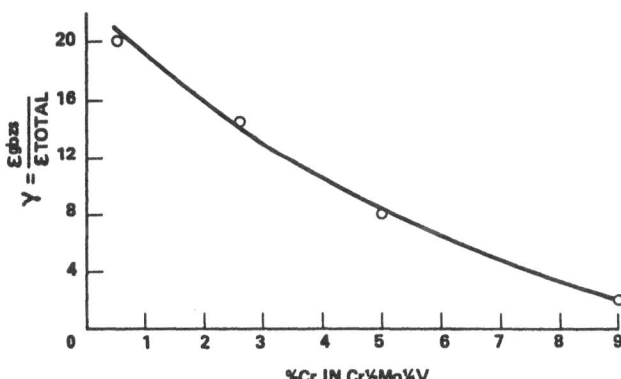

**FIG. 2 EFFECT OF Cr CONTENT ON GRAIN BOUNDARY SHEAR STRAIN IN
SLOW TENSILE TESTS AT 700°C AT RATES OF 5 x 10^{-6} min (14)**

even in steels containing fine dispersions of vanadium and niobium carbides. This is an important observation and underwrites the recent development of high strength 12CrMoV steels modified and strengthened by additions of Nb and W. These materials have been extensively studied in Japan (<u>15</u>) and are now

under active investigation in Europe (Cost 501). By varying the
Nb and W contents, it would seem possible to produce material
with rupture strength at 600°C and 650°C equivalent to 1CrMoV at
568°C. The new 12Cr steels would, therefore, have potential as
rotor material in advanced high efficiency fossil plants
operating at temperatures up to 650°C (16). One essential
feature of this type of development, particularly in the high W
steels, is the need to avoid abnormal segregation in the ingots
by adapting ESR steelmaking. Further developments with the 12Cr
steels may come about by introducing nitrogen into the steel as a
partial replacement for carbon. Nitrogen, like carbon, is an
interstitial element but has low solubility in molten steel;
recent developments in Germany (17), using the pressurized
electroslag remelting process PESR, have enables N contents to be
increased from a norm of 0.05% to (0.12-0.33)%. Provisional data
indicate that steels with increased nitrogen contents show
improvements in impact properties and short-term rupture
strengths over similar carbon strengthened steel with
conventional nitrogen levels. Further development of these
materials will require building bigger steel making capability
for the PESR process and investigating the weldability of the
material.

4.   Weldable Boiler Piping, Header, and Tubing Materials

The role of MnS particles as nuclei for cavitation has been
studied extensively by Middleton (18, 19) in 1/2CrMoV steels and
by Middleton and Cane (20) with 2-1/4Cr1Mo steel. In both steels
MnS particles were preferential sites for cavities in weld HAZ,
both during postweld heat treatment (reheat cracking) and during
creep testing at 565°C. Reducing the MnS content by lowering S
to 0.005 eliminated cavitational damage in the material and
significantly improved high temperature ductility. This
improvement was achieved even in steels deliberately doped with
high levels of residual elements, and it is clear that the prime
cause of cavitation is the presence of incoherent MnS as nuclei.
The role of the temper embrittling elements Sb, Sn, P, and As
appears secondary and is associated primarily with the way in
which they modify the size and distribution of MnS particles.
Increasing the Sn content, for example, from 0.01 to 0.06 wt%
reduced the cavity diameter from (1-5) μ to (0.5-1) μ and
significantly impaired ductility. Other factors were more
important in controlling cavitation, and B additives are
particularly so in that increasing B from 0.1 ppm to 15 ppm
increases propensity for grain boundary sulphides dramatically,
and causes a catastrophic loss of high temperature ductility. As
in the case of the rotor steels, there is clearly the prospect of
reducing the susceptibility of these steels to cavitational
damage during postweld heat treatment and service by using
secondary refining techniques during steel making but it may

require some expansion of this type of capability before it becomes economic for these types of material.

The low alloy ferritic steels of the type 1Cr1/2Mo, 2-1/4CrMo, and 1/2CrMoV have, in fact, been used as weldable steels for pressure containing parts such as tubing, pipework, steam chests, and castings. They have been operated in the creep regime at temperatures in the range 500-570°C and in most cases this service has been successful for operating times well in excess of 150,000 h, far exceeding the original design requirements. Problems have occurred in short-term operations, but these have generally been associated not with the base material, but with weldments which have shown considerable susceptibility to creep cavitational damage during fabrication (reheat cracking) or early service. This problem was particularly acute in the CrMoV type steels in which the dispersion of fine vanadium carbides within grain interiors tends to direct creep strain towards grain boundaries. It is now known that this problem can be corrected by proper control of postweld heat treatment, which coarsens the VC distribution and by control of Mn, S, and residual element distributions (18, 19, 20).

As the operating times have increased, other forms of cavitational creep damage have become apparent and these have taken the form of catastrophic failures in seam welded pipework in the United States, in which creep cracking has been observed along the weld fusion line (21), extensive cavitational damage to the extrados of bends in high temperature pipework in Germany (22), and extensive cracking in the intercritically annealed/over tempered region of weldments, the so-called type IV cracking phenomenon which has been observed in Denmark, Germany, and the U.K. (23, 16). A further problem which has come to the fore over the last few years has been associated with creep-fatigue interaction. This was always a damage mechanism in very thick section components such as turbine casings (8) where the problem was often associated with weld repairs but more recently a spate of thermal fatigue cracking has been found associated with the inter-tube ligament region of thick section steam headers in the United States and Europe.

These problems are nearly all manifestations of prolonged service and a concomitant material degradation, but in nearly all cases, there have been additional factors which have initiated the failure. Thus, in the case of seam-welded pipes, defects from the original welding process are suspected to have initiated the cracking, the damage to bends, and type IV cracking in weldments was probably enhanced by the presence of high system stresses, and the ligament cracking in headers by temperatures and rates of temperature change beyond the original design concept.

In this decade and beyond, the chances of obtaining failures of this type are increased significantly because of current trends in major plant construction/utilisation. The first concerns a worldwide trend to extend the life of fossil-fired plants, and the second the trend to enhance the steam conditions and hence thermal efficiency of any new fossil plant to be constructed. The first trend comes about by the realisation by major utilities that plant life extension, by a policy of major component repair/replacement can be achieved at a much lower cost than new stations (<u>25</u>) and also avoids the public debate and legislation associated with new construction. The latter recognizes that if new plants are to be constructed, they need to be as cost-effective and flexible as possible to meet the changing needs of the future. In the context of flexibility, this means that both the life extended plant and the new high efficiency plant need to have load following capabilities. This requirement is likely to subject the materials within the plant to operating parameters, in terms of stresses, temperatures, and rates of change in temperature and stres, which potentially could encroach upon the original design constraints. There is a need, therefore, to develop and use new steels which have sufficient tolerance to meet these changing conditions. In order to ensure good resistance to thermal fatigue interaction, it would be desirable that these materials should be ferritic rather than austenitic. It is also important that the ferritic steels chosen should have strengths equivalent to say the 300 series austenitics so that the improved thermal properties of the ferritic are not offset by the need for thicker sections. As discussed previously, the 9-12% Cr steels have a spectrum of properties which encompass the new requirements.

Until fairly recently, the difficulty with the 9-12% Cr steels has been to optimise the composition and heat treatment to give the required blend of good strength with adequate fabricability and ductility. Thus, of the well-established steels in this range, the British steel 9Cr1Mo has good fabricability and ductility, but its creep strength is only about the same as 2-1/4Cr1Mo (Fig. 3); the French steel EM12 (9Cr2MoNbV) is strong, but tends to be brittle, whereas the German steel X20 (12CrMoV) is also fairly strong but because of its high C content (0.2%), tends to be difficult to weld and is also relatively brittle. Significant attempts to improve these steels have been made in the last ten years; these include the development of 9Cr2Mo steel HCM9M in Japan and T91 in the United States, which is low C 9Cr1Mo alloy (with additions of N, Nb, and V). On provisional data, both materials have good toughness, fabricability, and weldability, but T91 would seem to have more potential on account of its high rupture strength. Indeed, because of its excellent combination of properties, there is tremendous interest world-wide in developing T91 for high temperature applications.

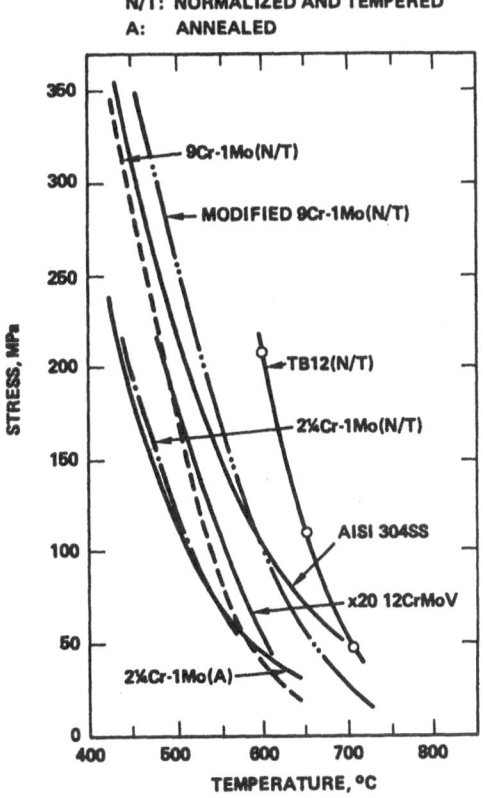

N/T: NORMALIZED AND TEMPERED
A: ANNEALED

9Cr-1Mo(N/T)

MODIFIED 9Cr-1Mo(N/T)

TB12(N/T)

2¼Cr-1Mo(N/T)

AISI 304SS

x20 12CrMoV

2¼Cr-1Mo(A)

STRESS, MPa

TEMPERATURE, °C

FIG. 3 10⁵ h CREEP RUPTURE STRENGTH OF
SOME HEAT RESISTANT STEELS

Detailed investigations of the properties of T91 in terms of fabricability, welding technology, and properties are in progress. The range of candidate components include thin section tubing for steam generator units of fossil boilers and thick section components such as headers, pipework, casings, steam chests, tube plates, and shell bodies. Given this impetus, it is clear that T91 is destined to be established as a significant and widely used material for high temperature applications over the next decade. On a word of caution, however, it should be noted that T91 can be susceptible to the phenomenon of type IV cracking of weldments. This type of cracking arises from the concentration of creep strain in the narrow over-tempered (softened) region of weld heat affected zones adjacent to the unaffected parent material. Typically, the strength of this zone is some 20-30% below that of the normalised and tempered base material. It must be emphasised that type IV cracking is not restricted to T91 but is, in fact, a generic problem to all ferritic steels (16) and arises as a natural consequence of the transformation behaviour of these materials. For the continued development of T91 and other weldable ferritic steels, however, it is clear that the phenomenon of type IV cracking should be studied in detail in order to minimise the effect if at all possible, but in any case, so that allowances for it can be accommodated in design.

Although T91 has attained considerable momentum as a commercial material, there are already indications that the development of a much stronger ferritic steel would be of great benefit. As always, the driving force is to reduce wall thicknesses of thick section components such as headers and pipework and thus ease the problems of fabrication and welding, and in service, to lower the restraints on terminal components and difficulties associated with thermal cycling.

These goals are by no means unattainable. In Japan, developments (15) have been taking place on a whole series of high chromium ferritic steels, many of which have 10^5 h rupture strengths, up to 1-1/2 x to 2 x that of T91. These alloys containing 9-12% Cr, 0.5-2% Mo and strengthened by V, Nb, and sometimes W are usually austenitised at the comparatively high temperature of 1050°C, Table 1. An example in HCM12 a 12% CrlMolWVNb alloy developed as a duplex material (martensite plus delta ferrite) to improve weldability and increased chromium over T91 for better corrosion resistance (26). A major feature of the alloy is its high tempering temperature (800°C) made possible by a high A_{c1} temperature which is claimed to give stable creep behaviour at 600°C and 650°C. The result is an alloy which has similar strength to T91 but improved corrosion resistance and weldability, and because of its duplex structure reduced susceptibility to type IV cracking. If these claims are substantiated, the material has attractive properties as superheater and reheater tubing in advanced designs. Another example is 9Cr0.5Mol.8W NbV (27) in which the partial replacement of Mo by W is claimed to improve the rupture strength significantly, giving a predicted 10^5 h rupture strength of 196 MPa at 600°C. The main problem with the commercial development of these steels is the need to extend the short-term database on the stress rupture properties and to demonstrate the capability to fabricate and weld the variety of product forms required for high temperature applications and to investigate the effects of fabrication and welding on properties.

5. **Austenitic Steels for High Temperature Applications**

Given the potential to increase the strength of ferritic steels well within the range of the 300 series austenitics, there appears to be very little necessity on the grounds of strength alone to use austenitics for thick section components. The justification for using austenitics in thick section must therefore, rest in their superior general corrosion resistance compared to ferritics and possibly with fabrication considerations; for example, with welded structures whereby problems of distortion arising from postweld heat treatments can be avoided by using austenitic steel rather than a ferritic. Such considerations were clearly of importance when selecting 316 as the structural material for the European Fast Reactor Primary Vessel and of 308 for BWR main steam pipe work. In both applications, attention had to be given to the possibility of stress corrosion cracking. Experience has since shown that in

Table 1

STRONG 9Cr AND 12Cr BOILER TUBING MATERIALS

Wt%	C	Si	Mn	Ni	Cr	Mo	W	V	Nb	N	B	10^5h Creep rupture strength (MPa) 600°C	650°C
9CR-1Mo	0.10	0.5	0.4	--	9	1.0	--	--	--	0.02	--	39	20
Mod.9Cr-1Mo	0.10	0.35	0.45	<0.2	8.75	0.95	--	0.21	0.08	0.05	--	98	49
Mod.NSCR 9	0.08	0.05	0.5	0.1	9	1.6	--	0.16	0.05	0.03	0.003	128	69
HCM 12	≤0.14	0.5	0.7	0.1	12	1.0	1.0	0.25	0.2	0.02	--	98	50
TB 9	0:08	0.05	0.5	0.1	9	0.5	1.80	0.20	0.05	0.05	--	196	98
TB 12	0.08	0.05	0.5	0.1	12	0.5	1.80	0.20	0.05	0.05	0.003	206	108

the Fast Reactor Primary Vessel case, the most significant risk
of SCC is during fabrication, in service SCC being unlikely in a
sodium cooled vessel; the situation with BWR pipework is,
however, essentially that of an in-service problem of SCC of
sensitised pipe butt welds. In the European Fast Reactor
Programme, the risk of SCC has been further minimised by the
adoption of the low C grade—316L (N), this decision has the
additional advantage of improving creep rupture strength and
ductility (28), and there is the possibility of using this
material for high temperature (>600°C) thick section fossil plant
applications if a ferritic steel with adequate strength cannot be
developed. In existing plants, the normal 316 grade has given
problems of distortion at the Drakelow C supercritical plant in
the UK and creep fatigue cracking in the Eddystone 1
supercritical unit in the United States (8, 16). The 316L (N)
grade may have sufficient ductility to avoid creep-fatigue
interactions, but still not have adequate tensile strength to
avoid problems of thermal ratchetting. It is possible,
therefore, that in those limited situations requiring a high
strength austenitic material in thick section, the only viable
steel with an adequate database and proven service history is the
British Steel Esshete 1250 (16Cr10Ni8MoNbMoV), which has given
good performance as pipework at Drakelow C for over 80,000 h at
600°C.

The main future potential use for austenitic steels at high
temperatures would seem to be in thin section applications which
require good corrosion resistance in nonaqueous environments and
high strength. The most obvious application would be as
superheater and reheater tubing in advanced fossil plants with
steam temperatures up to 600/650°C and metal temperatures up to
700/750°C. Unfortunately, this temperature range also
corresponds to the maximum fireside corrosion rates due to molten
salt attack and the available data indicate that the 300 series
steels and strong alloys such as Esshete 1250 and 17-14 CuMo have
inadequate corrosion resistance. One solution to this problem
has been the development of coextruded tubing which uses a strong
creep resistant, ductile, and weldable material as the inner core
and high chromium steels as the outer casing for improved
resistance to molten salt attack. In the UK extensive use has
been made of coextruded tubing comprising a core of Esshete 1250
clad with AISI 310 (24Cr20Ni) and corrosion benefit factors of
over 2.5x have been achieved in corrosion rates (29). More
limited use of an alternative composite tube comprising Alloy
800H overlaid with a cladding IN 671 (50Cr50Ni) has been gained
by the American Electric Power Company and by the CEGB. This
tubing gives corrosion benefit factors of up to 10x, but the
IN 671 is a duplex alloy and is difficult and expensive to
fabricate.

Furthermore, coextrusion adds considerably to the cost of tubing and a more elegant solution to the problem would be the development of an alloy which can be used as monotubing with strength and corrosion resistance equivalent to the coextruded tubing currently available. For boiler tubing, such a material needs to have good fabricability and weldability. Considerable progress has been made with the development of 20-25% Cr class of austenitics such as HRC3 (modified 310 with strengthening additions of Nb and N) and mod. 800H with added Ti. At still higher chromium contents, attractive and potentially useful alloys have been developed in the range 30-35 Cr. Examples are NKK's Cr35A, which has potential as an advanced cladding material and also Cr30A, which is strengthened by Mo, Ti and Al and has potential as an advanced monotube material for use in superheaters with 650°C steam temperature and 5000 psi pressure. An important point to note with these alloys is that the gain in strength and corrosion resistance and hence increase in operating temperature and concomitant efficiency can only be achieved at considerable price premium. Jaffee (4) indicates that the relative price of alloys compared to carbon steel = 1 is 3 times for low alloy steels 1Cr1/2Mo, 2-1/4CrMo, 10 times for 9-12% Cr steels, 15 times for the 18-8 stainless steels, 25 times for teh 20-25 Cr steels, and up to 40 times for the high Cr high Ni alloys. Over this range, the permissible operating temperature can be increased from 450°C to 650°C. Although this comparison is made on the basis of raw material costs and the price differential will be somewhat lower after fabrication and installation in a boiler, it is clear that considerable confidence in the quoted properties of the exotic high strength/ corrosion resistant alloy will be required before they are used extensively. There is an urgent need, therefore, to extend the experience with the new alloys with regard to all characteristics required such as fabricability, weldability, and confirmed long-term strength and corrosion resistance.

6.   Transition Joints

No review of steels for high temperature applications would be complete without some brief discussion of welded transition joints between low alloy (ferritic) and austenitic steels in superheaters and reheaters and occasionally on pipework. These joints must accommodate significant differences in chemical, mechanical, and physical properties and often operate at temperatures well within the creep range of the ferritic steel. Experience indicates that the service lives of transition joints can be significantly shorter than those achieved between similar metals (30). The most common type of dissimilar metal welds are those between a low alloy ferritic steel such as 2-1/4Cr1Mo and an austenitic steel such as 316. Typically, failures occur due to low ductility cracking in the heat affected zone of ferritic

steel close to the weld fusion line and welds made with
austenitic metal are often weaker than similar welds made with a
nickel base filler. However, it must be emphasised with
transition joints that marked changes in the mechanism and
location of failure can occur depending on the precise details of
weld chemistry, heat treatment, and microstructure and the exact
conditions of operating stresses and temperatures.

Depending on the particular combination of these variables,
the location of failure can change from along the weld fusion
line, in the coarse grained HAZ of the ferritic, in the fine
grained overtempered region of the ferritic HAZ, to within the
weld metal, or in some rare instances, even within the HAZ of the
austenitic (31). A typical example of this type of dilemma is
the comparison between low alloy (2-1/4Cr1Mo)-austenitic joints
and high alloy (9Cr1Mo)-austenitic joints in that welds made with
nickel base fillers give better performance than austenitic
fillers for the former but the reverse situation pertains for the
latter. It is extremely important, therefore, with transition
joints to regard each partial or combination of parent and filler
metals as a unique entity requiring detailed investigation over
the complete spectrum of likely service conditions. This
statement is particularly valid for combinations of the advanced
ferritic and austenitic steels discussed in this chapter with
which current knowledge of transition joint behaviour is
particularly lacking.

7. Future Research and Development Needs

1. For low alloy ferritic steels there is still a need to
   continue fundamental work on the complex interactions between
   residual element and inclusion content/distribution and the
   various forms of embrittlement. The work should be targeted
   at identifying the maximum levels of tramp elements
   permissible for successful use in service. In parallel with
   this work is the need to evaluate and develop the modern
   techniques of secondary refining in steel making to assess
   which processes are most technically/economically viable for
   long-term development for bulk steel production. An important
   feature of all this work is the demonstration of successful
   in-service performance such as arranged trials of LP rotors
   made from the high purity super clean NiCrMoV steels and for
   high purity bolting materials.

2. Continued research and development is also required on the
   stronger 9Cr and 12Cr materials. The strategy to be adopted
   for these materials should be:

   (a) to establish T91 as a viable commercial material for high
        temperature applications. This requires continued

exploration of the long-term properties, an evaluation of weld properties, particularly the susceptibility to type IV cracking, the development of fabrication techniques for forged, wrought, and cast products, and the successful demonstration of T91 components such as headers, pipework, and steam chests in service.

(b) To continue to support the European Cost.501 programme on modified 12CrMoV rotor materials and the recognition of the need for a major plant demonstration of in-service performance within the next five years.

(c) to initiate long-term research and development on the weldable grades of 9Cr and 12Cr known to be significantly stronger than T91. The scope of this work should essentially be similar to that described for T91 with the recognition at the outset that successful development of such a material will require a 10-year programme and eventually major plant trials.

3. Future R&D on austenitic materials should:

(a) establish the long-term properties of 316L for fast breeder reactor components.

(b) comprise a limited development and property evaluation of strong austenitics such as Esshete 1250 and possibly modified Alloy 800 for thick section components—this work would be particularly necessary if the the strong ferritic steel developments prove unsatisfactory.

(c) concentrate on the development of high Cr austenitic steels for superheater and reheater applications requiring high corrosion resistance and strength. A particular requirement in this development is the need for in-service trials by means of corrosion probes and inserted superheater sections.

4. Concomitant with all the developments on ferritic and austenitic steels is the need to evaluate the properties of weldments for the ferritic steels, particularly the stronger varieties. This should concentrate on the generic problem of Type IV cracking and, for dissimilar metal welds, the properties and filler metals required for joints between the new strong ferritic and austenitic steels.

8. REFERENCES

1. Materials & Electricity, Jaffee, R.I., Metallurgical Transactions A, Volume 17A, 755 pp., May (1986)

2. Gould, G.C., Temper Embrittlement in Steel ASTM STP407, 59 pp. (1967)

3. McMahon, C.J., Jr., et al., Part 1, Impurity Segregation and Temper Embrittlement, EPRI Report NP-1501, September (1980)

4. Jaffee, R.I., International Conference, Chicago, ASM, 1 pp. (1987)

5. Swaminathan, V.P., Jaffee, R.I. and Steine, J.E., Steel Forgings ASTM STP 903, 126 pp (1986)

6. Ewald, J., Berger, C., Heiburg, K.H. and Wiemann, W., Steel Research 57 Part I, 83 pp., Part II, 172 pp. (1986)

7. Sasaki, R., Kaneko, R. and Kuriyama, M., Hitachi Special Report, 16 pp., February (1986)

8. Advances in Materials for Fossil Power Plants, Hall, D. and Whitley, G.H., Proceeding International Conference, Chicago, ASM, 21 pp., September (1987)

9. CERL Report TPRD/L/2604/R83, Kimmins, S.T., and Gooch, D.J., CEGB

10. King, B.L., Middleton, C.J. and Townsend, R.D., CEGB/CERL Report RD/L/R1919, June (1975)

11. Everson, H., Orr, J. and Dulieu, D., Proceedings International Conference, Chicago, ASM, 375 pp., September (1987)

12. Oakes, G., Bridge, M.R. and Judge, S., Proceedings International Conference, Chicago, ASM, 411 pp., September (1987)

13. Pickering, F.B. and Irvine, K.J., Published originally in ISI Special Report 86, 34 pp. (1964)

14. Middleton, C.J., CEGB, CERL Report RD/L/N/56/79

15. Funita, T., Metal Progress Magazine, August (1987)

16. Townsend, R.D., CEGB Experience and UK Development in Materials for Advanced Plant, Proceedings International Conference, Chicago, ASM, 11 pp., September (1987)

17. Uggowitzer, P.J., Anthamatten, B., Spiedel, M.O. and Stein, G., Development of Nitrogen Alloyed 12% Cr Steels, Proceedings International Conference, Chicago, ASM, 181 pp., September (1987)

18. Middleton, C.J., CEGB CERL Reports Part 1, RD/L/N37/79, Part 2, RD/L/N38/79, and Part 3, RD/L/N47/79, August (1979)

19. Metal Science, Middleton, C.J., Vol. 15, 154 pp. (1981)

20. Cane, B.J., and Middleton, C.J., Metal Science, Vol. 15, 295 pp. (1981)

21. Viswanathan, R., Dooley, R..B and Saxena, A., Paper submitted to Journal of Pressure Vessel Technology (1987)

22. Bendick, W., Müsch, H. and Weber, H., 14th Inter-Ram Conference for the Electric Power Industry, Toronto, 391 pp. (1987)

23. Answald, W., Blum, R., Newbauer, B. and Paulsen, K.E., Proceedings International Creep Conference, Tokyo (1986)

24. Nitta, A., The Second EPRO-CRIEPI Workshop on Fossil Power Plants, EPRI (1987)

25. Townsend, R.D., Australian Materials Forum, Volume 9, No. 1, 90 pp. (1986)

26. Marsuyama, F., Haneda, H., Yoshikawa, K. and Iseda, A., Chicago, ASM, 259 pp., September (1987)

27. Masumoto, H., Sakakibara, M., Takahashi, T., Sakurai, H. and Fujita, T., First International Conference on Improved Coal-Fired Power Plants, Electric Power Research Institute, Palo Alto, California, November (1986)

28. Optimisation of 316 Steel for Strength and Ductility, Lai, J.K., CERL Report RD/L/N/210/80, CEGB

29. Flatley, T., Latham, E.P. and Morris, C.W., ASM, 219 pp., September (1987)

30. CEGB service data, Price, A.T.

31. Creep Strain Distribution in 9Cr1Mo-Alloy 600 Transition Weld, Soo, J.N., Confidential CEGB Report

2.1.2. Developments in Heat-Resisting Alloys

for Petrochemical Plant

J.J. Jones

Lake & Elliot Paramount Ltd.

Braintree, U.K.

1. INTRODUCTION

Modern industrial processes, such as those of the
Petrochemical, Mining, Nuclear and Steelmaking industries, can
operate today at high levels of output efficiency by using heat
resisting alloy steel castings.
This section briefly reviews the CAST HEAT RESISTING ALLOYS and
cast forms developed over the past thirty years for the
petrochemical industry, in particular for continuous high
temperature process plant for the bulk production of organics
such as hydrogen, ethylene, methanol etc. However, the cast
alloys primarily developed for these processes find many other
applications throughout industry. Some interesting current
developments and future needs particularly in modification are
also discussed.

2. PETROCHEMICAL PROCESSES

Petrochemical "cracking" processes are carried out in large
furnaces where pressurised mixtures of hydrocarbon feedstock,
steam and air are passed through long reaction tubes or coils.
Since the desired reactions are endothermic (absorbing heat), the
tubes/coils have to be maintained at high temperatures (800 to
1150°C). They are continuous stream processes and the tubes have
design lives of 11.4 years (100,000 hours). However, in service,
the tubes and coils are subjected to a range of damaging
mechanisms not included in the design computations which
generally shorten the lives to 2 to 8 years.
Two particular processes which illustrate these problems are

REFORMING ("catalytic cracking")
and PYROLYSIS ("steam cracking")

Whilst there is an "overlap" of conditions and alloys for both processes, there are also important differences which have demanded the development of alloys with special properties and these will be described later.

In REFORMING, the reactions take place inside banks of vertically suspended catalyst-filled tubes which 'feed' into a collector manifold or header. The tubes are about 10 metres long, with bore sizes in the region of 100mm and wall thicknesses between 12 and 25 mm, depending on the alloy and service design conditions.
Tube metal skin temperatures are usually in the range 800°C-1050°C and pressures are in the region of 3-4 N/mm². The main life-controlling conditions are usually creep "rupture" damage and thermal shock.

FIGURE 1: Coil assemblies for a cracking furnace in 25/20 CrNi steel assembled by welding centrifugally cast tube and statically cast return bends

In PYROLYSIS, the reactions occur inside long 'serpentine' coils fabricated by welding 10 metre tube 'legs' to 180° and 90° bends; (see Fig. 1).

Bore sizes are in the range 50 - 125 mm and the wall thicknesses between 7 and 15 mm.

The coils are usually suspended vertically and, during the passage through the coils, the feedstock deposits a coke layer in the bores which has to be burned away by a steam/air mixture every 30 to 60 days.

Tube metal skin temperatures can reach 1150°C during the final passes and pressures are in the region of 0.5 N/mm^2.

In addition to creep damage and the effects of the higher temperatures, the life of the coil alloys is also dependent on carburisation, oxidation, thermal stresses (due to de-coking cycles) and stresses caused by the strong coke layer (which also impairs heat transfer thus demanding even higher tube skin temperatures). Hot erosion (from carbon solids during de-coking) and thermal fatigue can also occur, both inside the bends, and in the tube walls due to differential expansion of carburised layers.

TABLE 1

PARALLOY GRADE	NOMINAL COMPn.				A.C.I ref	other ref	Typical Applications *			
	Cr	Ni	C	other			(a)	(b)	(c)	(d)
H20	25	20	0.4	-	HK	HK40	●	●		●
H39	25	35	0.4	-	HP	HP40	●			●
H24W	24	24	0.4	1%Nb	-	IN519	●			●
H39W	25	35	0.4	1%Nb	-	HPmod	●	●		●
H34CT	25	35	0.4	5%W	-	-			●	●
H48T	25	48	0.4	5%W	-	22H	●			●
CR32W	20	32	0.1	1%Nb	-	800H			●	●
CR39W	25	35	0.1	1%Nb	-	-			●	●
SH24T	24	24	0.4	1%Nb +Ti	-	-	●			●
SH39TZ	25	35	0.4	1%Nb +Ti,Zr	-	-	●	●		●

* KEY: (a) = Reformer Tubes (b) = Pyrolysis Tubes/Bends
        (c) = Manifolds   (d) = Other Fittings

(Note: A group of about 10 different compositions represents the alloys available to the Petrochemical Industry today. Since most are not yet included in standards or design codes, they are identified in this paper by their NOMINAL compositions, along with any common reference name, and the proprietary alloy code used by the authors' company. See Table 1)

3. CENTRIFUGAL CASTINGS

Prior to 1960, the majority of furnaces were equipped with wrought tubes, usually stainless steels or high nickel alloys. Because of hot-working limitations, the alloys were low carbon and tube walls relatively thin. Thus the processes could only operate at temperatures, pressures and output efficiencies limited by these wrought tubes.

A major breakthrough was the foundry development of the centrifugal casting process which permitted tubes up to about 5 metres length to be produced virtually in any required alloy composition.

The centrifugal casting process used can be described as follows: - Molten alloy is poured into the end of a horizontal tubular steel mould spinning at about 1000 RPM. The liquid alloy is quickly distributed as a uniform "annulus/lining" along the length of the mould bore, and held there by a force of 100g or higher. The combination of high "g" force and directional solidification ensures a clean, dense casting.
Any slight shrinkage porosity is confined to a narrow band in the tube bore which is removed by bore machining. This has the additional advantage of ensuring a smooth bore for improved corrosion resistance, and also improving thermal efficiency and tube weight considerations.
The outer surface of the tube is left unmachined; the characteristic 'textured' cast finish assisting the heat flux efficiency in the furnace.

The grain size of centrifugally cast tube appears relatively coarse when compared with wrought material but this is desirable as it contributes to the higher strength of the cast heat resisting alloys.
During production, tubes are subjected to many inspection stages. For example, samples are machined from actual tubes for such tests as tensile or stress/rupture testing when required.
Tube lengths are welded together to fabricate the full service lengths.

4. STATIC CASTINGS

In addition to centrifugally cast tubing, there is also a need for statically-cast components such as 90° and 180° bends, complex manifolds, tube sheets, tube supports, and tube hangers. These are required also in heat resisting alloy grades.

Modern petrochemical standards demand levels of cast surface smoothness almost unattainable by traditional foundry techniques; especially in the bore of cast bends where grinding is not always possible. Similarly, casting must achieve high standards of "soundness" (freedom from porosity) and "cleanliness" (freedom from non-metallic inclusions).

This has been achieved by modernising foundry practices as follows:-

1. Computerised manufacturing methods.
2. Special furnace and ladle practice to ensure cleanliness.
3. Special moulding techniques to ensure smooth sound castings.
4. Powerful radiographic equipment to permit high sensitivity radiography of sections up to 200mm thick.
5. Sophisticated inspection and finishing techniques.

4.1. Weld Fabrication

The centrifugally cast tubes are welded to other components such as bends made by static casting (or made by forming centrifugally cast tubes to shape), manifolds or inlet connections to produce the required service 'geometry'. The welding technology has also been developed to keep-pace with the newer alloys. In most cases "matching composition" welds are made using automatic "T.I.G." and "M.I.G." processes.

5. IMPORTANT ALLOY PROPERTIES

With the adoption of centrifugal casting furnace designers were given more freedom of alloy choice. A whole new range of alloys was opened-up for development i.e. the high carbon alloys with increased creep strengths due to the alloy carbides present in the microstructure and a grain-size larger than wrought materials.

A number of these alloys had been available and used for years as "static" (or conventional) castings but the arrival of the centrifugal process accelerated their study and use.

Initially, there were reasonable doubts expressed about other properties such as tensile ductility (compared with wrought) but it was quickly recognized that these cast alloys displayed a range of properties suited to the petrochemical processes which were adjusted to accommodate these different types of tubes.

(Table 2 gives a summary of mechanical properties of the alloys considered).

The two post important alloy properties which make major contributions to the service lives of cast heat resisting alloys are creep strength and carburisation resistance.

TABLE 2

NOMINAL COMPOSITION				PARALLOY GRADE	ROOM TEMPERATURE TENSILE 'AS-CAST' Condition. MINIMUM						100,000 HOUR STRESS to RUPTURE (MEAN - EXTRAPOLATED)					
					U.T.S		0.2% PS		ELONGn.		850C	1550F	950C	1750F	1050C	1950F
Cr	Ni	C	Other		N/mm2	psi	N/mm2	psi	%		N/mm2	psi	N/mm2	psi	N/mm2	psi
25	20	0.4	-	H2O	425	62000	240	35000	10		20.89	3213	8.14	1130	2.81	341
25	35	0.4	-	H39	430	62500	235	34000	4.5		27.59	4216	10.78	1490	3.14	365
24	24	0.4	1%Nb	H24W	440	64000	220	32000	10(tube) 8(static)		27.98	4283	11.22	1556	3.69	370
25	35	0.4	1%Nb	H39W	440	64000	220	32000	8(tube) 6(static)		35.93	5495	16.43	2303	7.70	995
25	35	0.4	5%W	H34CT	430	62500	235	34000	7(tube) 3.5(static)		-	-	11.17	1553	4.17	513
25	48	0.4	5%W	H48T	430	62500	235	34000	5(tube) 3.5(static)		26.88	4105	11.59	1614	4.39	541
20	32	0.1	1%Nb	CR32W	420	61000	175	25500	20		24.76	3741	13.43	1896	-	-
25	35	0.1	1%Nb	CR39W	420	61000	175	25500	20		26.06	4000	10.10	1398	3.24	387
24	24	0.4	1%Nb +Ti	SH24T	440	64000	220	32000	10(tube) 8(static)		39.18	5986	16.59	2309	6.06	743
25	35	0.4	1%Nb +Ti,Zr	SH39TZ	440	64000	220	32000	8(tube) 6(static)		(50)*	(7700)*	(23)*	(3200)*	(11)*	(1400)*

(Note: * Interim Values)

MECHANICAL PROPERTIES OF CAST HEAT RESISTING ALLOYS

5.1. Creep Strength

Two **different** forms of Creep Strength can be considered:-

(a) Creep stress to "rupture"
Grain boundary creep voids accumulate to become "creep
fissures" or cracks. Eventually, failure by creep rupture would
occur if the alloy remained in service much beyond this stage.
The stress which an alloy can sustain at temperature for a given
time (usually 100,000 hours) before failure by creep rupture is
termed the: "creep stress to rupture value."

(b) Creep rate
In the conditions of pyrolysis type furnaces, it has been
observed that coils can sometimes suffer from excessive creep
strain during service to a stage where the coils distort severely
and have to be replaced.
This property is not directly related to the creep "stress to
rupture" strength, and occurs at high temperature (1050°C and
above) with relatively low stress. It is reported to be more
common in the low tungsten, niobium-bearing grades than in the
4-5% tungsten, niobium-free grades and this difference is
probably related to the carbide type of morphology.
In this instance it is actual service performance which is
providing alloy data and not laboratory testing. This is because
until recently, laboratory techniques have not included
"meaningful" creep rate data for long times at high temperature
owing to difficulties in measurement and cost.

More reliable quantitative data are needed on this
particular creep property.

ALLOY DEVELOPMENT STAGES:

It may be helpful to consider that the development of alloys
can be classified as belonging to different "generations":-

The first generation were the original wrought low carbon
tubes.

The second generation started about 1950 and represents the
initial Cr/Ni/Fe high carbon cast grades the study, development
and wider acceptance of which was greatly accelerated by the
centrifugal casting process.

The third generation started about 1970 and the increasing use
of cast alloys, coupled with the increasing demands of the
flourishing oil industry led to the search for stronger and
better alloys. The target was achieved by additions of further
elements such as niobium and tungsten to some "second generation"
grades.

The fourth generation started about 1980, and even stronger alloys were produced by small additons of Ti, Zr and sometimes Rare Earths to "third generation" grades. These alloys are referred to as "Synergistic" or "Microalloys".

Whilst this grading of alloys into "generations" may be considered a little simplistic, it permits the main development features of creep strength to be explained. Examples of alloys from the different "generations are given in the table below.

Generation	Alloy Common Name	Composition wt %				
		Cr	Ni	C	Nb	W
second	HK series	25	20	0.4		
	HP series	25	35	0.4		
third	IN 519 type	24	24	0.4	1	
	HP-Nb Mod	25	35	0.4	1	
	HP-W Mod	25	35	0.4		5
	22H	28	48	0.5		5
	800H type	20	32	0.5	1	
		25	35	0.1	1	
fourth	Paralloy SH24T	24	24	0.4	1	+ Ti
	Paralloy SH39TZ	25	35	0.4	1	+ Ti/Zr

The "second generation" (about 1950), exploited the cast-carbide morphology of the alloys. The "as-cast" structures of cast heat resisting alloys are all basically similar. The macrostructure consists of an outer layer of columnar grains with equiaxed grains near the bore. The microstructures also have much in common. Alloy "HK" (25/20/0.4 - Cr/Ni/C) for example, displays a network of primary carbides ("M_7C_3") surrounding the austenite grains. The primary carbides remain throughout service, only changing in form ("rounding off") with increased time and temperature. They are, therefore, creep-inhibiting barriers.
There is however, a second form of "defence". During even a short time of exposure at service temperatures, fine "secondary" carbides ("$M_{23}C_6$") precipitate within the austenite matrix where much of the carbon was held in solid-solution after casting. This mechanism of secondary carbide precipitation is called "aging" and these secondary carbides also act to inhibit dislocation movement. With increasing service time/temperature, the secondary carbides both re-dissolve and agglomerate thus losing their "barrier" effect.

The selection of an alloy is based on a 'balance' between alloy cost and life expectation. Of the "second generation" alloys available, the HK series (25Cr.20Ni) eventually became the most widely used for fired tubes.

The "third generation" (about 1970) saw dramatic improvements in the creep stress to rupture strength by additions of critical amounts of other elements including niobium and tungsten. An increase was also made in the nickel content to compensate for the 'side-effect' of elements like Nb which could render the material more suceptible to embrittlement by "Sigma phase" during long-term exposure in the range 650°C to 850°C.

Both the primary and secondary carbides are altered by these additions. Particularly the secondary carbides which form more slowly and remain finely dispersed for longer than in the more simple "second generation" alloys.

Development was also carried-out during this period of lower carbon cast alloys where superior ductility was also required. This need arises, for example, in "Manifolds" where complex designs and multi-branch joints impose additional stresses when thermal cycling and tube distortion occur.

The "fourth generation" (starting about 1980) continues the theme of ever-increasing creep stress-to-rupture strength which will permit furnaces to operate hotter and with faster throughput than before. This is being achieved by exploitation of the phenomenon know as "synergism". In alloy terms, this means that where two additions are each known to have a strengthening effect, when added together in only small amounts, the combined effect is not only greater than their individual effects but also greater then the summation of their effects: In other words, there is a "magnification" of strengthening characteristics. Small additions (less than 0.5 %) of titanium, zirconium and certain rare earth elements have proved to be effective.

There is no doubt that these alloys (also termed "micro-alloys") are stronger than the third generation series. Though they are not yet widely used, the development and service information to date is very encouraging. There is evidence that their service ductility is higher than the "third generation" series and it is interesting to note that the 24/24 Cr/Ni synergistic alloy achieves similar creep strength to the 25/35 Cr/Ni "third generation" 'HP modified' with 10% less nickel.

5.2. Carburisation resistance

Because of the higher temperatures and the nature of the chemistry of the actual process, pyrolysis coils suffer the

additional damage of carburisation and oxidation.
The more serious of these, carburisation, occurs because the
inside surface of the tubes and bends absorb carbon from both the
gas stream itself and the layer of coke which gradually deposits.
As the inner tube wall becomes increasingly contaminated by
carbon, the mechanical properties and corrosion resistance are
impaired and therefore, carburisation is a major life-controlling
feature.

The whole science of alloy development for pyrolysis is
complicated by the fact that there are several operational
factors which each play a part in tube damage. Operating
conditions vary widely from plant to plant (more so than in
reforming) and it is difficult to simulate actual service by
means of laboratory testing.
In earlier years much valuable work was done by "pack
carburising" alloy samples to enable the effects of alloy
additions to be evaluated and alloys to be "ranked" in
order or their ability to resist carbon absorbtion.
This type of testing enabled a small "family" of alloys to be
created which had increased resistance to carburisation. The
principle elements most effective in this respect are nickel,
chromium, silicon and tungsten.

An effective and relatively simple improvement is obtained
simply by increasing the silicon content of the "second
generation" "HK" (25/20) alloy to about 2%, and the alloy is
often used for the inlet end of coils where temperatures are not
too high (800°C to 940°C). For the hotter outlet legs, higher
alloy grades are required and the 25/35 Cr/Ni-base types are most
commonly used. Tungsten levels over 4% are particularly helpful
in the outlet legs, and the alloy 25/35/0.4/5 - Cr/Ni/C/W is a
good example.

There are mixed opinions about the effect of niobium in the
alloy. Whatever laboratory results may indicate regarding
carburisation resistance there is growing service evidence that a
25/35/0.4 - Cr/Ni/C with just a little tungsten (1-2%) in
combination with niobium can suffer from excessive stretching at
high temperature leading to coil distortion. This does not
appear to be the case with the niobium-free, higher tungsten
grades.

Study methods now introduced use synthetic process gas
mixtures as the carburising medium, to simulate actual service to
the extent that carbon is deposited on the specimens and de-coke
cycles can be introduced into the test. In this way alloys have
been graded more accurately with regard to their resistance to
carburisation. Also, the more complex aspects of the carburising
reactions have been studied in detail. As a result of this work,

the effects of "physical" factors such as; tube bore surface
finish, grain structure, purity levels and cyclic de-coking on
carburisation rates have now been explored.

Amongst the latest aspects to be considered is the matter of
"coke deposition rate" (as opposed to "carburisation rate"). For
economic reasons, plant operators would like to reduce the number
of "de-cokes" necessary in each year, as de-coking causes loss of
production time. Furthermore, the coils can be subjected to
damaging thermal stresses during de-coke. Thus, any technique to
inhibit or even slow down the coke deposition rate would be
helpful. Apart from any changes to the process chemistry itself,
there are two approaches which are being evaluated with the cast
tubes themselves:
(i) "Bi-metallic" (or "co-axial") tubes, where the outer cast
     layer is the main stress supporter and the inner cast layer
     is an alloy which does not readily catalyse coke deposition.
(ii) Coated tubes where the bore surface is coated or impregnated
     with a substance which also inhibits coke deposition.

6. CHOICE OF ALLOY AND TUBE DESIGN

For Reformer Furnaces, from the point of view of creep strength,
the selection of a particular alloy for service in reforming
furnaces is often a matter of selecting an alloy which in the
opinion of the design engineer offers the best combination of
creep strength at temperature coupled with the thinnest wall to
improve thermal efficiency.
The main design property is the "100,000 hour creep stress to
rupture value". In practice, it can be obtained only by
"extrapolation" from available shorter term laboratory test data.
The reliability of such a predicted value is obviously influenced
by the "quality" of the test data which is studied.

For example, data banks should contain as many actual tests
of 10,000 hours or more duration as possible, and any results of
20,000 or 30,000 hours enhance the data bank further.
The reliability of the predicted "10,000 hour" value is also
influenced by the statistical method used to make the
extrapolation. The method widely accepted for these high
temperature alloys is the "LARSON-MILLER" technique which, as
longer-term test data and service data become available, is
proving to give sensible predicted values providing computer
aided statistical data analysis is carried out. The relationship
between alloy grade selected and the design temperature and tube
wall thickness is shown in Fig. 2. The design tube wall
thickness is calculated from a relatively simple equation where a
higher "allowable stress" value results in a thinner wall for a
particular temperature and pressure service environment. The
advantage to be gained in higher temperatures or thinner tubes by
using "third generation" alloys can be clearly seen.

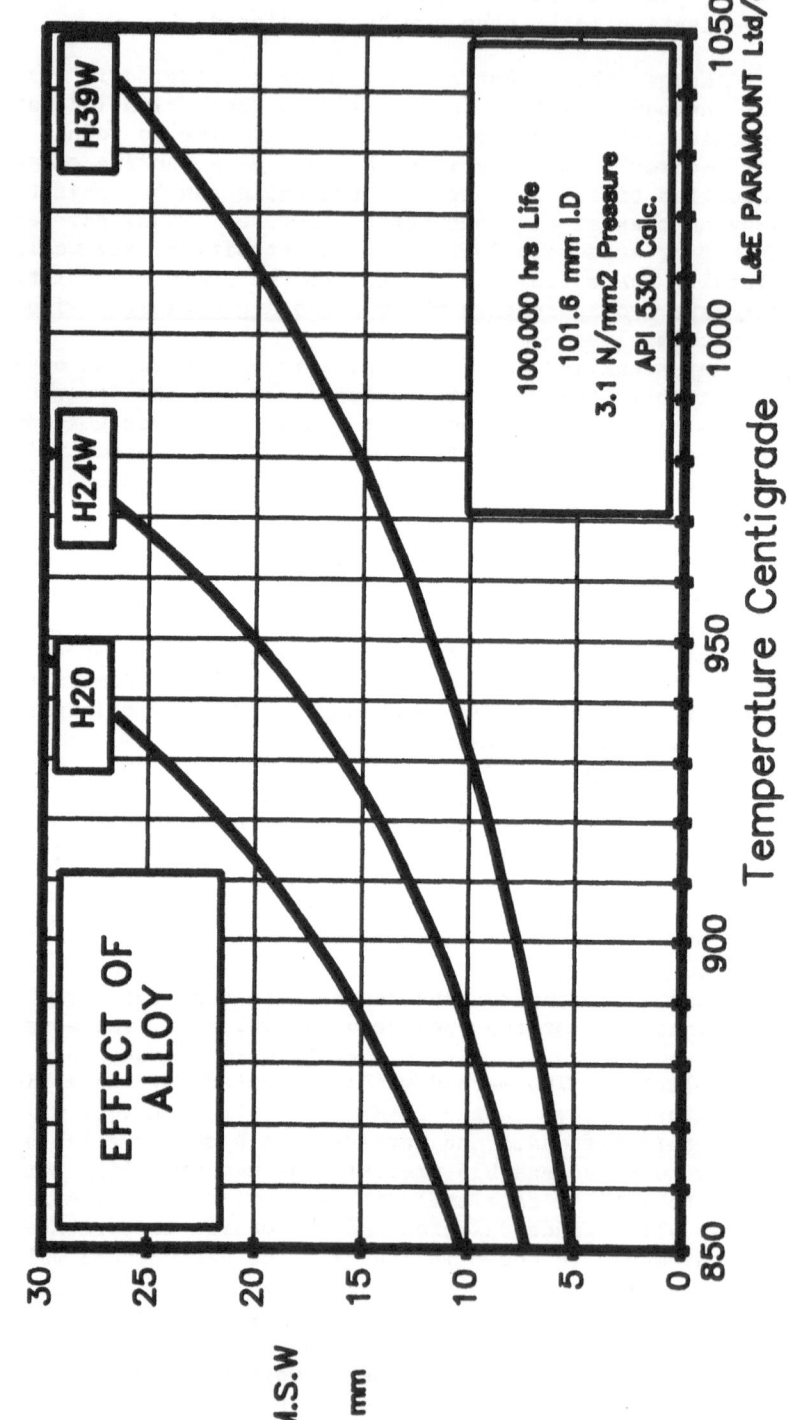

Figure 2 TUBE WALL THICKNESS —vs— DESIGN TEMPERATURE

Today, thirty years after centrifugally cast tubes first became available, it is not uncommon to find older furnaces still using thick-walled "second generation" "HK" (25/20) type tubes. However, there is an increasing trend for plant designers to take the opportunity offered by a maintenance shut-down to replace these by the stronger "third generation" types ("HP modified" 25/35/Nb or "IN519" (24/24/Nb)). This permits the designers to "modernize" the furnaces and, with the majority of reformer furnace operators, enjoy the production and operational advantages of these stronger alloys.

For Pyrolysis Furnaces, with the lower pressures involved, the minimum wall to resist creep rupture is less than in a reformer furnace therefore attention is concentrated on selecting an alloy with sufficient resistance to the relevant life-controlling features including:

(i)   Carburisation
(ii)   Stress caused by a coke layer
(iii)   Stress caused during de-coking cycles
(iv)   Creep strain

Because of the complexity of pyrolysis processes and the variation from plant-to-plant only the first of these (carburisation resistance) has been studied in detail and quantitative alloy data made available. The other three features are more difficult to measure and predict in service and the selection of any particular alloy is often made in the light of previous experience or pilot-scale field trials.

7. FUTURE NEEDS

Cast heat resisting alloys have made a major contribution to the petrochemical industry and are now essential for its continued efficiency. However, even after more than twenty five years of experience with these cast materials, several points still remain today which tend to delay a wider appreciation of the alloys and the advantages which they offer:

Most of the cast alloys which the petrochemical industry find to be essential for their operations are not even mentioned by existing design codes, standards, or specifications.

Most alloy purchasers write their own specifications for an alloy (both the metallurgical and quality aspects). This lack of "unification" can lead to confusion regarding properties and applications for any particular alloy.

In the absence of a "unified" standard which could guide both the foundry and user, the major source of data is that published by the foundries themselves, and one could say that

some of the claims made in some brochures may not be as
'up-to-date' as one would wish.

The "time scale" involved in a new alloy being accepted for
wide- spread use can be as long as 15 years:
first 5 years for a foundry to complete a sensible test data bank
of properties. (especially stress to rupture values)
plus 5 years for the industry to be able to "accept" a new
alloy.
plus 5 years before sufficient "field-service" data becomes
available to permit a reasonable judgement to be made
regarding the alloys suitability.

8. ACKNOWLEDGEMENTS:

The author would like to express his appreciation to the
management of L&E PARAMOUNT Ltd for permission to present this
paper, and also to acknowledge his gratitude for the help and
encouragement given by Ray Atkinson, Basil Hall, Cliff Hughes and
Ray King.

9. REFERENCES:

1. Steam Hydrocarbon Reformer Furnace Design", Krauss, M.,
    Petroleum Mechanical Engineering, ASME conf. (1969)

2. "Improvements in Making Hydrogen", Voogd, J. and Tielrooy,
    J., Hydrocarbon Processing, No. 9, Vol. 46, 115-120 pp.
    (1967)

3. "Materials for Steam Reforming", Report by the Battelle
    Inst. Group Sponsored Research Project. (1970-1974)

4. "Material Selection Consideration for Petrochemical Furnace
    Tubes", Swales, G.L., Rev. int. Htes Temp et Refract, t. 13
    (1976)

5. "Metallurgical Problems in Steam Reforming Plant", Edeleanu,
    C., Metals and Materials, March (1967)

6. "The Strengthening of Centrifugally Cast Heat Resistant HK40
    Tubes", Zaghloul, M.B., (Doctor of Eng. Thesis; Tokyo
    University 1976)

7. "A Time-temperature Relationship for Rupture and Creep
    Stresses", Larson, F.R. and Miller, J., Trans. ASME, Vol. 74
    (1952)

8. "The Long-Time Creep Rupture Properties of the HP-45
    Alloys", Van Echo, J.A. and Roach, D.B., Battelle Report
    SFSA (1982)

9. "A New Generation of Heat Resisting Alloys", Jones, J.J. and Steiner, J.L.D., NACE "Corrosion 85"

10. "Properties of Cast Chromium Nickel Alloy Steel Tube", Jackson, J.F.B., Slater, D. and Dawson, D.W.O., Materials Technoly Symposium (1964)

11. "Recommended Practice for Calculation of Heater Tube Thickness in Petrochemical Refineries", American Petroleum Institute A.P.I. RP 530

12. "Data Sheets on the Elevated Temperature Properties of Centrifugally CAST 25Cr-20Ni-0.40 Steel Tube for use in Reformer Furnaces" (SCH-SSCF), National Research Institute for Metals, Japan, Creep Data Sheet No. 16A

13. "Understanding the Larson-Miller Parameter", Furillo, F.T., Puroshothaman, S. and Tien, J.K., Scriptam Metallurgica, Vol. 11 (1977)

14. "High Temperature Creep Rupture Properties of Cast Austenitic Steels", Jones, J.J., ASME Conf. (1986)

15. "Design and Operating Experience with High Temperature Fired Tubes in Petrochemical Plants", Jones, W.G., IMechE, Jan. (1976)

16. "The Structure of Centrifugally Cast Stainless Steel", Hou, W.T. and Prof. Honeycombe, R.W.K., Cambridge University

17. "An Improved Cast Austenitic Steel for Heat Resisting Applications", Cox, G.J. and Jordan, D.E., AFS Transactions

18. "Design Features, Material Selection and Performance of Steam Reforming Furnaces", Kawai, T., Takemura, K. and Zaghloul, M.B., INPEC, Bombay (1984)

19. "Tungsten and Niobium Modified HP alloys for High Temperature Service" Steel, C., Pidgen, R.E. and Engel, W., NACE (1983)

20. "Contribution factors to the Unusual Creep Growth of Furnace Tubing in Ethylene Pyrolysis Service", Hendrix, D.E. and Clark, M.W., NACE (1985)

21. "Cast Reformer Headers and Manifolds", Blackburn, J., Ammonia Plant Safety Symposium (1976)

2.1.3. High Temperature, Oxidation-Resisting

FeCrAl Steels

S.R. Keown

Metallurgical Consultant

Sheffield, England

1.   INTRODUCTION

In his paper on high temperature steels, Townsend (1) has comprehensively described creep-resisting steels for application in the temperature range 450° to 650°C for a wide range of components in power generation and in oil and chemical plant.

Flatley and Morris (2) have extended this approach to consider high temperature corrosion environments at similar operating temperatures and they describe the cladding of creep-resisting steels with more corrosion-resistant, high chromium alloys.

Another group of high temperature steels for application in the temperature range up to 1400°C are the FeCrAl steels which are used primarily for oxidation-resistance in applications typified by electrical resistance heating elements. Davidson (3) has reviewed heating element materials ranging from NiCr, FeNiCr, FeCrAl metallic alloys to non-metallics materials which can operate up to 1800°C.(4) FeCrAl steels offer a useful temperature applications range of 700°C to 1400°C at the most economical cost. Unfortunately FeCrAl steels are not creep-resisting and are sometimes brittle; consequently component design has to allow for these shortcomings but, as will be described later, these are prime areas for further research and development.

1.1. FeCrAl Steels

There are essentially 3 groups of alloys defined by the

chromium and aluminium contents. In the absence of carbon, the steels are fully ferritic since all the constituent alloying elements are ferrite-forming elements.

Type	Typical Alloy	Cr	Al	Si	
1. Low Al	Sicromal 10	18	1	0.9	
2. Low Cr	F.A.L.	13	4	0.7	
3. High Cr. High Al	Kanthal Al	22	5.5	0.3	
" " " "	Fecralloy A	20	5	0.3	0.3Y

The Sicromal alloys are essentially ferritic stainless steels with oxidation resistance enhanced by relatively small additions of approx. 1% Al and 1% Si. Typical applications are in heat treatment furnaces, pyrometer tubes etc.

With higher aluminium contents of about 4%, the F.A.L. type of FeCrAl steels (equivalent grades are Resistalloy 134 and Kanthal Alkrothal) were developed primarily for electrical resistance components, rheostats and resistors. The high aluminium content imparts a high specific electrical resistivity whilst the 13% Cr confers corrosion resistance, allowing these alloys to be used for heating elements up to about 1050°C.

The most important group of FeCrAl steels are those with high aluminium and high chromium contents with 20-23% Cr and 4-5.5% Al. An addition of 0.05-0.5 Y or Ce can improve the oxidation characteristics and the ductility of the steels. Typical alloys are Fecralloy, Kanthal AF and Sandvik OC404. The most important application of FeCrAl steels is for electrical resistance heating elements but the steels are also candidate material for a wide range of high technology applications (6).

1.2. FeCrAlY Steels

The FeCrAlY steels were jointly developed in the UK and in the USA during the 1960's (5). Previously the FeCrAl alloys were first developed in Sweden in 1930s and introduced for resistance heating elements a few years later (6).

The physical metallurgy of FeCrAlY steels has recently been reviewed (5) to show that the excellent oxidation resistance of the steels is attributable to the formation of an α-Al_2O_3 film, on pre-oxidation above 1000°C, and the influence of yttrium which provides adhesion of the alumina film to the steel by grain boundary pegging and/or by a vacancy sink mechanism at Al_2O_3

particles. The yttrium addition also prevents grain growth of the steel during long exposure at elevated temperatures leading to improved ductility.

FeCrAlY steels are superior to the austenitic NiCr of FeNiCr alloys for resistance heating elements for a variety of reasons:-

a)   higher operating temperature (up to 1400°C)

b)   lower density and higher resistivity to give more heating elements per unit weight of material

c)   cost is independent of the erratic and high price of nickel

d)   better resistance to sulphur

FeCrAlY steels can be readily hot and cold worked to give a wide range of product forms, from 0.025mm diameter wire for resistance heating elements to thick plate for cladding furnace linings. The steels can be fabricated by welding, deposited by plasma spraying, hot extruded as composite clad tubing and used in powder form for hot isostatic pressing. FeCrAlY and FeCrAlCe steels are being considered for major application as metal support systems for car exhaust catalysts (autocatalysts) at temperatures up to 900°C. Recent EC legislation will necessitate the widespread use of autocatalysts in Europe by the early 1990's (7) and metal supports are vying with ceramic substrates for this important market valued at $800m by 1990 (7). Other potential applications for FeCrAlY steels include electronic printed circuit board substrates, tubes for fluidised bed components and chopped or woven fibres for aerospace and insulation systems.

In all these applications the unique oxidation resistance of the steels at temperatures above 650°C is a crucial factor, together with the benefit of ready formability to almost any component specification of shape and size.

1.3. Problems and Research Requirements

As with all ferritic stainless steels, the FeCrAl steels experience poor hot strength, low impact resistance and 475°C embrittlement. Research is required to improve the creep-resistance of FeCrAl steels by the controlled addition of ferrite-forming elements such as niobium and molybdenum to give precipitation or solute strengthening effects. Oxide dispersions-strengthened, metal matrix composites utilising Al_2O_3 dispersions in FeCrAlY alloys are already available but the prohibitive cost of these alloys has so far prevented their widespread

application. Toughness should be enhanced by greater attention
to grain control during processing or by alloying with grain-
refining additions. 475°C embrittlement can be tolerated by
careful control of heat treatment cycles, although more
fundamental work on the α'-phase might lead to the elimination or
diminution of the embrittlement effect by alloying or processing.

There is also the possibility of further enhancement of the
superior oxidation characteristics of FeCrAlY steels by detailed
research into other combinations of rare-earth additions and by
further consideration of the basic alloy constitution.

As pointed out by Davidson (3), 'the FeCrAl alloys represent
the highest inherent temperature capability of the common
metallic materials'. Due to the relatively low cost of FeCrAl
steels, further alloy development is to be encouraged not only
for heating elements but for a wide range of high temperature
applications where oxidation resistance is the major
consideration.

2. REFERENCES

(1) Townsend, R.D., "High Temperature Steels", (this
     publication)

(2) Flatley, T. and Morris, C.W., "Claddings", (this
     publication)

(3) Davidson, J.H., "Electrical Heating Elements", Review of
     Technological Requirements for High Temperature Materials
     R&D, EUR 5623EN CEC (1979)

(4) "Kanthal Super Handbook", Kanthal Furnace Products,
     Hallstahammar, Sweden (1986)

(5) Keown, S.R., "The Physical Metallurgy of Fecralloy Steels",
     Stainless Steels '87, Institute of Metals, London, Book 426,
     pp. 345

(6) "Kanthal Handbook", Kanthal Furnace Products, Hallstahammar,
     Sweden, (1987)

(7) "Autocatalysis" Materials Edge, No. 6, July/August, (1988),
     pp. 17

2.1.4. Superalloys base : Nickel, Cobalt, Iron, Chromium

C. White

Inco Alloys Ltd.

Holmer Road, Hereford

Superalloys have been defined as "alloys developed for elevated temperature service, usually based on Group VIII A elements, where relatively severe mechanical stressing is encountered and where high surface stability is frequently required" (1).

In addition to aircraft, industrial and marine gas turbines, superalloys are now used in nuclear reactors, power generation, petrochemical applications, rocket engines, space vehicles, etc. The development of superalloys has been well documented over the years and is very usefully summarised in ref. (2).

The performance and efficiency of gas turbine engines is a direct function of the maximum cycle temperature and it is this which has provided the motivation for the continuous development of materials capable of operating at higher temperatures in the turbine section of the engine. Turbine entry temperatures have risen from around 700°C in the Whittle W1 engine in 1941 to around 1350°C in today's advanced engines.

Three main groups of superalloys are commonly identified, nickel base, cobalt base and iron base.

Although not normally considered as a superalloy, chromium base alloys will also be briefly discussed.

1. NICKEL BASE ALLOYS

Nickel base superalloys are the predominant high temperature materials used in the turbine section of the engine. They have outstanding strength and oxidation resistance over the temperature range encountered. The alloys have a fully austenitic FCC structure and they are capable of maintaining good tensile, rupture and creep properties to high temperatures. The superalloys can be used to 0.8 Tm (melting point) and for times up to 100,000 hours at somewhat lower temperatures.

Nickel alloys have little inherent resistance to high temperature oxidation and their remarkable high temperature endurance is due in part to the formation with the addition of chromium, of Cr_2O_3 rich protective scales. Because these scales have a low cation vacancy content, they restrict the diffusion rate of metallic elements outwards and oxygen, nitrogen, sulphur and other aggressive atmosphere elements inwards (3). The strength of the alloys is mainly due to the formation, on alloying with aluminium and titanium, of the intermetallic gamma prime (γ') an ordered form of the FCC matrix based on Ni_3TiAl. Other elements are added to effect solid solution strengthening and, in the case of carbon and boron, to modify or strengthen the grain boundaries, etc. The role to the various elements added to nickel base superalloys is summarised in Table 1. It is not

TABLE 1:   ELEMENTS USED IN NICKEL BASE SUPERALLOYS

Purpose	Cr	Al	Co	Fe	Mo	W	Ti	Ta	Nb	Hf	C	B	Zr	Mg
Matrix Strengtheners	X		X	X	X	X								
Gamma prime formers		X					X	X	X					
Carbide formers	X				X	X	X	X	X	X				
Oxide scale formers	X	X												
Grain boundary strenghteners										X	X	X	X	X

intended to review the physical metallurgy of the alloys, this is discussed comprehensively in the literature (3, 4, 5 and 6).

The development of nickel base alloys can be broadly divided into two phases. Initially, i.e. in the period 1940 to around 1970 alloy development alone was sufficient to meet design needs. In recent years, process dominated developments have emerged as the major contributor to improved performance. The process development chronology for the three major turbine components is summarised in Fig. 1. (7).

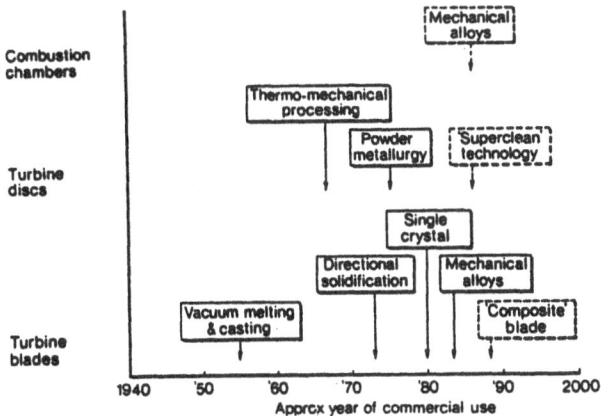

Fig. 1   The chronology of process dominated
         developments (after Ref. 7)

1.1. Turbine Blade Materials

The range of wrought blading alloys in use today was
developed from the basic 80 % nickel, 20 % chromium used for the
past 60 years for electrical resistance applications. It was
found that considerable improvements to the creep rupture
strength of this composition could be effected by additions of
titanium and aluminium and associated other compositional
modifications. At a given temperature the volume fraction of γ'
is directly proportional to the amount of total hardening
elements present i.e. aluminium, titanium and niobium. An
increase from 14-60 % can quadruple the strength of nickel base
superalloys. It is this effect which formed the basis of nickel
base alloy development in the period 1940-1960.

The γ' phase, in common with many other ordered structures,
has the unusual property of showing an increase in strength with
increase in temperature. In the superalloys it is extremely
finely dispersed as sub-micron particles. It closely matches the
matrix phase in structure, composition and lattice parameter.
The interface between the phases, therefore, has very low free
energy and there is little driving force for particle coarsening.
The fine structure is thus stable to high temperatures.

A major breakthrough occurred in the 1950's with the
introduction of vacuum melting and refining. Vacuum melting
prevented the variable losses of the reactive hardening elements
and it therefore allowed more titanium and aluminium to be added
for γ' precipitation. Vacuum refining removed some of the
volatile trace elements, present in the raw materials used, which

adversely affected creep strength and ductility (8, 9) and hot
workability (10). The principal effect of these impurities was
to weaken the grain boundaries. Levels of a few p.p.m. of
elements such as Bi, As, Sb, Sn, Tl, In, Ga, Zn, Pb segregating
to the grain boundaries are sufficient to have a deleterious
effect. The higher concentrations of aluminium and titanium
necessary for increased γ' precipitation also had the advantage
of raising the γ' solvus thereby maintaining the precipitate
stability to higher temperatures. However, high titanium and
aluminium contents decrease the melting point of the alloy
thereby narrowing the forging range to the point where it becomes
impossible to forge without the danger of incipient
melting (4, 11, 12). Thus alloys such as NIMONIC alloy 115 and
Udimet 700 represent about the limiting composition which could
be forged to turbine blades, particularly complex blades with
intricate cooling passages.

Material development has only partly accounted for the marked
increase in turbine entry temperature that has been achieved
since the Whittle engine. As can be seen from Fig. 2 (13), the

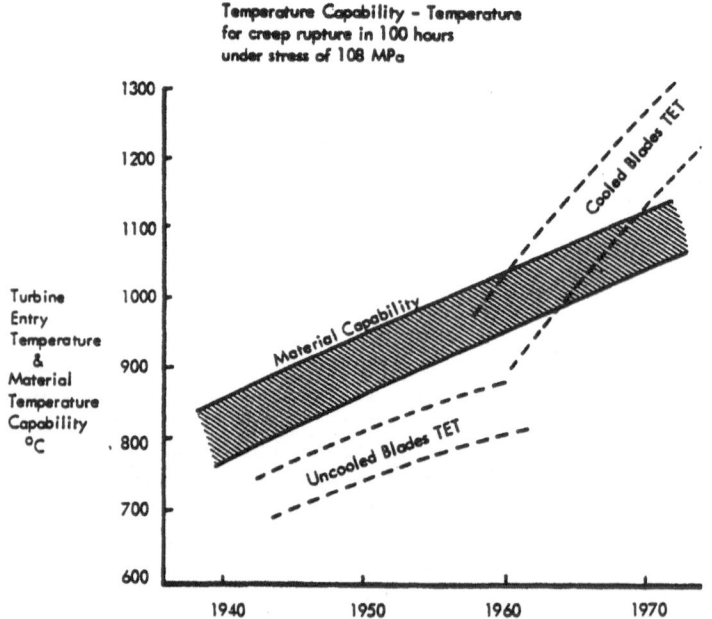

Fig. 2 The contributions of turbine blade cooling and
material development to increases in turbine
entry temperature. (After Ref. 13).

development of blade cooling technology has made a significant
contribution.

Investment casting allowed much greater freedom of design of complex turbine blades and as forgeability was not a relevant factor, even greater additions of aluminium and titanium could be made. IN100, for instance, which contained about 10.5 % Ti + Al has a γ' volume fraction of about 70 % and combined high strength with relatively low density. However, early casts of IN100 were prone to embrittlement after exposure to intermediate temperatures resulting in the formation of sigma phase an other intermettalic compounds. These undesirable, hard phases generally referred to as topologically close packed (TCP) phases (14) are characterised as composed of close packed layers of atoms forming in "kagome" (basket weave) nets aligned with the octahedral planes of the FCC matrix. These generally detrimental phases appear as thin plates, often nucleating on grain boundary carbides. To avoid their presence, phase computation (PHACOMP) techniques were developed (15, 16) to predict the likelihood of phase formation of sigma and other TCP phases. PHACOMP control is now standard on most of the highly alloyed superalloys is use today (17, 18).

Since it was impractical to increase the titanium and aluminium content still further to increase strength other approaches had to be adopted. Lowering the titanium content was found to improve castability and this led to alloys such as B1900 with a low Ti:Al ratio but with additions of tantalum and molybdenum to maintain strength. The replacement of molybdenum with tungsten gave rise to the alloy Mar M200 which, in its original form, gave inconsistent creep life and limited intermediate temperature ductility. It was found that the addition of about 2 % hafnium improved ductility and also minimised porosity. The beneficial effect of hafnium was found to apply to a whole range of cast alloys, e.g. Mar M 004 = IN 713 + Hf and also led to the development of a number of new alloys based on this concept e.g. Mar M002 and Mar M 247.

The next major change in turbine blading capability was the introduction of directional solidification (DS). It can be said that alloy development gave way to process dominated technology. DS allowed the production of castings with grains aligned in the direction of maximum tensile stress with no grain boundaries normal to this direction. This avoided the intermediate temperature ductility problem which is associated with grain boundary weakness. The crystal growth direction {100} in DS is the direction of lowest Youngs Modulus and this produces a major reduction in the thermal stresses originating from the temperature gradient in the blade during service (7).

Single crystal blades are in effect an extension of DS technology. The complete absence of grain boundaries allowed the removal of those elements primarily introduced for grain boundary strength, i.e. carbon, boron, zirconium, hafnium. This minimised the danger of incipient melting and allowed higher solution treatment temperatures to be safely used resulting in less chemical segregation, a more uniform distribution of γ' and an improved temperature capability. Initially the standard compositions used for directional solidification were used for single crystal blades, e.g. Mar M002, but as single crystals are free of some of the constraints of conventional materials (no grain boundaries), alloys have been specifically developed to meet the engineering requirements of particular engine components.

Modified directional solidification techniques have been used to produce aligned structures in superalloy eutectics in which the main strengthening phases are carbides or inter-metallic compounds. For example the system $\gamma/\gamma'/\delta$ contains aligned lamellae of Ni_3Al (γ') and Ni_3Nb (δ) in a γ matrix and has a 70°C advantage over DS Mar M200. Other examples are $\gamma/\gamma'/\alpha$ in which the strengthening phases are Ni_3Al (γ') and α Mo, and the NiTaC and CoTaC series of alloys. DS eutectics have some major property disadvantages, however. The structures are relatively coarse and properties are markedly anisotropic. For transverse fracture toughness at lower temperatures they rely on the crack stopping properties of a soft, ductile second phase. Under combined high stress and high temperature the continuity of the hard phase can be broken by the formation of grain boundaries which are penetrated by diffusion into them of the soft phase, with consequent weakening of the structure. In addition, any mismatch in thermal expansion coefficients of the two phases can cause sufficient plastic flow in the soft phase to break up the single crystals. Processing is also inherently difficult and although from Fig. 3 it is apparent that the strongest DS eutectics have a higher temperature capability than single crystal materials at the lower stress levels, to date no eutectic system has been exploited commercially. Until some new discovery is made in this area there will be no "technological push" from DS eutectics for turbine blade applications.

The development of the blading alloys over the past 45 years is summarised in Fig. 4.

Dispersion strengthened materials obtained by "mechanical" alloying" are discussed in section 2.2.5.

1.2. Turbine Discs

In the 1950's the alloys used for turbine discs were extensions of the austenitic alloys used in industrial gas turbines. These were superceded in the 1960's by the nickel iron base alloys such as Alloy 901 and Inconel alloy 718. Also used

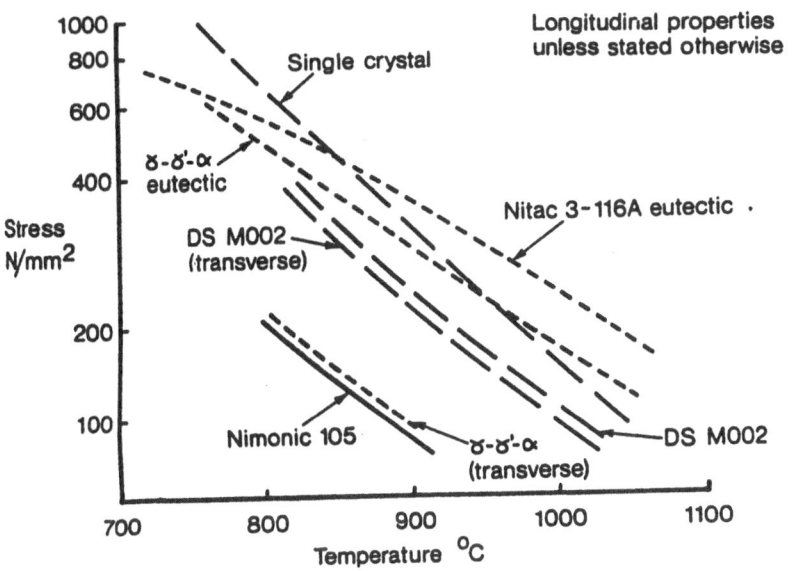

Fig. 3   1000 h creep rupture properties of advanced high
temperature superalloys (After Ref. 7).

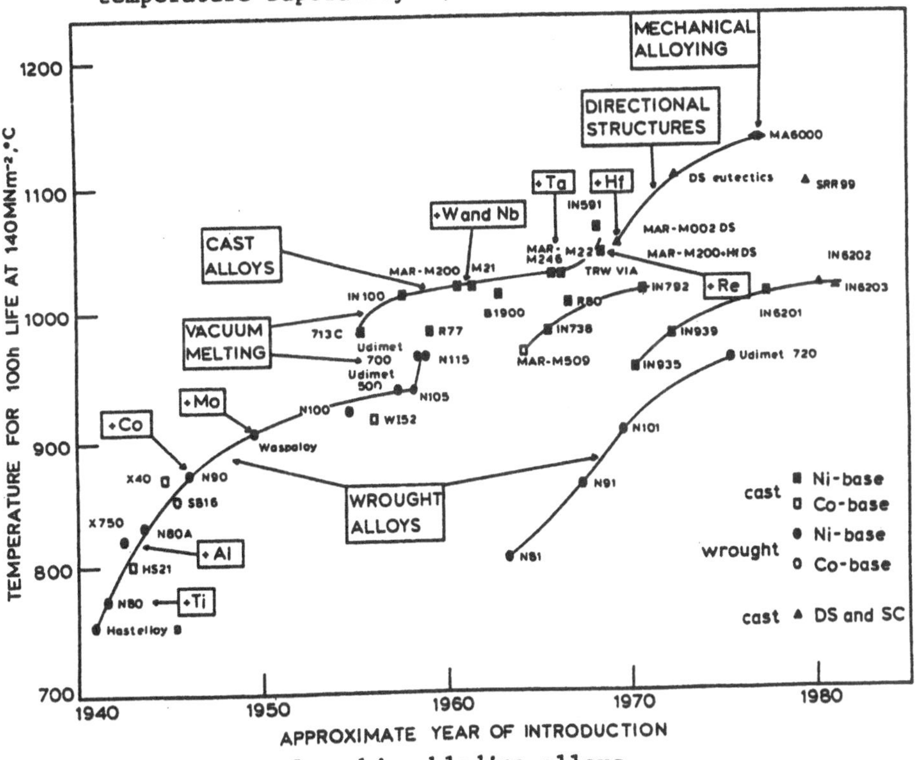

Fig. 4   Development of turbine blading alloys
DS = directional solidified   SC = single crystal
(After Ref. 2)

were the nickel base alloys Nimonic alloy 90, Waspaloy and Astroloy, developed originally for turbine blade applications. However, the requirements for blading and discs are different, turbine blading tending to be creep dependent whereas discs require good tensile strength to prevent bursting in the event of an overspeed. It also became recognised in the late 1960's that low cycle fatigue (LCF), to withstand the frequent changes in stress accompanying each flight cycle, was the main life limiting parameter.

In order to achieve improved LCF properties, "Thermo-mechanical processing (TMP) technology" was developed. In TMP, the deformation is confined to specific temperature ranges in relation to the γ' and carbide solvus temperatures (19). Size of the forging stock, the deformation rate and design of the forging sequence are other critical parameters. Waspaloy, for example, is "warm worked" at slightly below the solvus and also below that at which thermally induced recrystallisation and grain growth occurs. Thus the grain size is controlled during the forging process and the correct amount of residual strain is retained after deformation. Tensile properties are significantly increased over the blade version of the alloy, with corresponding increases in LCF but at the expense of creep life (19).

Because the elements involved in the formation of carbides, γ' and the undesirable intermetallics segregate during solidification it is necessary to minimise segregation relative to conventional ingot casting. To achieve this, electroslag remelting (ESR) or vacuum arc remelting (VAR) of previously vacuum induction melted or vacuum refined material are now universally used to produce disc alloys. To reduce the segregation effects still further, powder processing involving inert gas atomisation, in which the molten metal stream is broken up by jets of argon, followed by consolidation using hot isostatic pressing (HIP) was introduced in the mid 1970's. The extent of any segregation in this case is restricted to the powder particle size. Discs can be produced in a number of ways;

a. conventionally forged from the HIP powder billet,
b. forged from extruded powder billet using the the Gatorising (R) process (20), i.e. final forming in the superplastic conditions or
c. for the most attractive cost benefit, by direct HIP to near final shape.

Although powder disc superalloys have higher strength than conventionally melted disc materials, their full commercial potential has not yet been realised.. The reason for this relates to defect sensitivity. It has been realised that most materials contain defects/inclusions/inhomogeneities. At low stress levels premative fatigue crack nucleation from these defects is relatively rare and fatigue life predictions based on model materials are valid. However, at high stress levels, material defects can behave as fatigue crack nucleators and under these circumstances severe fatigue life reductions compared to safe-life predictions based on model materials have been noted. It is, therefore, the presence of small scale inclusions inherent in current powder materials which limits the maximum stresses at which powder products can operate, since defects which are undetecteble using current NDT techniques could cause premature failure (21).

Stringent control of the power production process can reduce the incidence of such defects but their total elimination cannot be guaranteed.

It is important therefore that for highly stressed "critical" parts (a turbine disc is such a component as catastrophic failure could cause the loss of the aircraft) full scale component tests are carried out to provide lifing information and to validate fracture mechanics based calculations (22). This is expensive and time consuming.

The past few years has seen developments, not only in powder alloys but also in conventionally melted material to minimise their defect sensitivity by microstructural control but perhaps more importantly by the production of "superclean" material. The object is to refine and cast the alloy without it contacting deleterious refractory products. The VADER (R) process, Osprey (R) process and Electron Beam (EB) melting are examples of these developments. A more detailed exposé of developments in disc materials is given in ref. (23).

1.3. Combustion Chambers

These are fabricated from sheet, chamber walls can experience peak temperatures in excess of 1000°C and thus good oxidation and thermal fatigue resistance, reasonable strength and good fabrication and welding characteristics are required.

The simple solid solution alloys such as NIMONIC alloy 75 and Hastelloy X were used in early gas turbines. The need for higher strength resulted in the development of the precipitation strengthened alloy C 263. The ever present requirement to operate with higher metal temperatures resulted in the development of alloys with improved oxidation resistance and strength such as NIMONIC alloy 86. These new alloys have included small additions of rare earth elements to contribute to oxidation resistance.

Dispersion strengthened sheet alloys are discussed in section 2.2.5.

2. COBALT BASE ALLOYS

Cobalt base superalloys were used in early aircraft gas turbine engines for both the turbine blades and vanes. In modern gas turbines, because of their relatively low strength, their use is now restricted to some nozzle guide vanes (investment cast) and combustion chambers (sheet).

There is no phase equivalent to γ' in cobalt base alloys. The chromium level required to form a Cr_2O_3 scale is higher in a cobalt base than in a nickel base alloy and they therefore contain 25-35 % chromium. There is always a solid solution strengthener, normally a refractory metal, such as tungsten and an element to stabilise the FCC phase and increase the stacking fault energy, usually about 10 % nickel. Additional strengthening is provided by a carbide distribution so that the carbon content is significantly higher than in nickel base alloys. The metallurgy of cobalt base alloys is discussed in detail in ref. (1).

The cobalt crisis in the 1970's has caused an apparently permanent drop in the use of cobalt base alloys as the gas turbine designers and users learnt to manage without them for many applications. Typical alloys still used are Mar M509 and Haines Stellite 31 (a modification of X40) for nozzle guide vanes, and Haynes 188 for combustion chambers.

3. IRON BASE ALLOYS

For the purpose of this review, these are defined as alloys with an austenitic matrix hardened by an intermetallic or carbide precipitate and containing $\geq$ 40 % iron. Thus using this definition many of the nickel-iron base alloys such as Alloy 901 and Inconel alloy 718 are considered as nickel base alloys.

Iron base alloys are therefore essentially limited to the early disc alloys such as A286 which is still seeing significant use particularly in industrial gas turbines. During the past years there have been developments of new iron base alloys for applications other than discs e.g. XF-818 which is a cast alloy containing 8 % Mo developed as an alternative to the cast cobalt base superalloys.

An interesting new series of alloys has been introduced during the past few years, e.g. Incoloy alloys 904, 907 and 909 and Pyromet CTX-3. These are nickel-iron-cobalt alloys whose outstanding characteristics are a constant low coefficient of thermal expansion, a constant modulus of elasticity and high strength. Their use is for gas turbine casings, shafts and shrouds. The low expansion enables close control of clearances and tolerances e.g. for shafts, for greater power output and fuel efficiency. The high strength increases the strength-to-weight ratios for lower weight in aircraft engines.

4. CHROMIUM BASE ALLOYS

Aircraft engine designers are continually striving to operate the gas turbines at higher temperatures because of increased efficiency. As the melting point of chromium is 1850°C compared to 1455°C for cobalt, the use of chromium as a base alloy would appear to be attractive. In addition to a higher melting point, chromium also has a higher elastic modulus and a lower density, all three properties being highly desirable for aircraft gas turbine components. Unfortunately the disadvantages outweigh the advantages. Chromium undergoes a transition from ductile to brittle behaviour usually above room temperature, and there is further embrittlement resulting from nitrogen pick-up during exposure in air.

Activity in the development of chromium base alloys reached its peak in the 1960's and is usefully summarised in ref. 24.

The inherent brittleness of chromium base alloys, has, however, not been overcome and this has limited their usefulness and to date precluded their use.

5. FUTURE PROSPECTS

Superalloys have played a significant role in the development of the gas turbine because of their unique combination of high temperature properties. As mentioned earlier, the past 10 years has seen alloy development gradually giving way to process dominated development and it is envisaged that this trend will continue in the next decade.

5.1. Turbine Blading

Modern turbine entry temperatures can reach 1370°C and much higher temperatures are being considered for future engines. These temperatures are above the melting point of some of the strongest superalloys and it would be easy to infer from this that superalloys have passed their zenith and "new" materials e.g. ceramics (discussed in Chapter 2.5.) will take over. However, these "new" materials are not without their problems and it is envisaged that superalloys will retain their prime position for at least the next decade.

Because of their melting point limitations, considerable amounts (> 15 %) of compressor delivery air is required to provide cooling air to present day turbine blades. Also, superalloys have a relatively high specific gravity not helped in recent years by the need to add refractory elements for solid solution strengthening. This adds weight to the engine and imposes very high centrifugal loading on the rotating parts (21). Thus lighter blades and more efficient use of cooling air are desirable characteristics.

The use of coatings, to improve oxidation/sulphidation resistance and to act as thermal barriers, can enhance superalloy blade performance. Ceramic thermal barrier coatings are being used successfully in combustion chambers and nozzle guide vanes, and they also have potential for turbine blading providing a sufficiently strong bond can be maintained between the coating and the parent metal. Coatings are discussed in more detail in Chapter 2.3.

Dramatic increases in the melting point of superalloys are unlikely but there is no doubt that some modest increases in high temperature capability are possible by further improvements to single crystal, directional solidification, eutectic and mechanical alloying technology. Unfortunately even more so than was the case in the past, it is unlikely that advances in any one alloy system will cover the whole range of properties required. This suggests that the ultimate superalloys turbine blade could be a "composite" blade made from different materials/alloys each having optimum properties for its specific region in the blade. The complexity of the manufacturing problems involved should not be underestimated.

The "wafer blade" (26) and "spar and shell" concepts are examples of this type of approach. In the wafer blade concept, complex cooling passages are accurately etched into the wafers which are diffusion bonded together and the blade is then machined from the stock. The advantage of this system is in the very accurate positioning of the cooling passages which allow higher turbine entry temperatures to be used without increasing the actual blade temperature. In addition its use of cooling air is more efficient. At present, the cost effectiveness is questionable but some form of "composite" blade could prove a real benefit. In the "spar and shell" technique a central creep resistant spar, maintained at a relatively "low" temperature is surrounded with a shell, conforming to the blade profile, of a highly oxidation/corrosion resistant alloy.

5.2. Turbine Discs

Over the past years it has been established that the maximum allowable operating stresses are dependent on the defects inherent in the material and which cannot be detected by non-destructive testing techniques, e.g. ultrasonics. Developments are being and should continue to be directed at removing "defects" altogether.

However, if defects cannot be detected by non-destructive testing techniques it will be necessary for any new material and process to be evaluated, at least initially, using destructive techniques, e.g. billet and disc cut-up assessment, rig test, etc., in order to set the processing criteria. The "nil" defect concept is not possible using current manufacturing techniques and for components made in high strength alloys an initial life will need to be calculated based on the initial estimates of the maximum inhomogeneity size likely to be present. This involves the use of fracture mechanics based lifing philosophies.

Once a satisfactory set of processing procedures has been established, strict process control (statistical process control is becoming increasingly relevant) and audited quality assurance procedures are then required to ensure the very highest product reliability.

Improvements to disc alloys are therefore related in the first instance to developments in melting techniques to obtain super clean material. Double vacuum melting i.e. VIM + VAR is currently the norm for disc materials and VIM + ESR are currently at or near product qualification. However, even with these established techniques considerable improvements are possible e.g. automated control of melt rate, arcgap, etc. A vacuum eletroslag remelting furnace has recently been built and the potential of this technique has still to be established. Future developments will be ceramic free powder atomisation, vacuum arc double electrode remelting (VADER) and electron beam cold hearth refining.

Having obtained clean material it is necessary to establish the microstructure required for optimum disc life. This may mean new and further developments of the processing techniques, particularly thermo-mechanical processing. As in the case of turbine blade materials, the technical requirements for discs are often conflicting e.g. fine grain size for resistance to crack initiation versus coarse grain size for resistance to crack propagation. Also different property characteristics are required for the hub in relation to the rim. Therefore a future possibility is the bi-metal disc where the rim is made from a different alloy to the hub. Atomised powders are typically used, fed into a shaped can and HIPed. At present the technique is restricted to small discs and extending the technique to larger discs will present a considerable challenge.

6. AVAILABILITY OF RESOURCES

Shortages in alloying additions for superalloys were experienced in 1973-74 and in 1979. Both these crises were considered to be due to a lack of investment rather than a basic depletion of resources. This probably still applies today in that the recession over the past years has not encouraged new mining projects. However, more worrying than overall scarcity is the prospect of sudden disruption of supply of critical elements such as experienced with cobalt in 1979. Due to the cobalt crisis much work on substitution was done and this has had a permanent effect on the use of cobalt base alloys even though the price of cobalt is currently back to realistic levels. There is little evidence of cobalt replacement as an addition element in nickel base alloys but new alloys will probably be designed with reduced cobalt levels.

The element which is perhaps causing some concern at present is chromium. There really is no substitute and its ore and production are highly localised in regions of potential disruption. The supply of tungsten, hafnium, yttria could also present problems.

It should also not be overlooked that superalloys have voracious appetites for energy from the extraction of the elements used to the melting, fabrication, etc. of the alloys and the ultimate components. Energy crises lead to stagnation as witnessed 10 years ago but it has also stimulated the development of more efficient gas turbines.

7. DEVELOPMENT NEEDS

Superalloy technology is principally directed, at least initially, at the aircraft/aerospace industry. It is an area of science and technology in which Europe and the USA have been dominant over the past 40 years and it is imperative that this position is maintained in future years in view of the total commercial value of the aerospace industries.

Much of the technology developed initially for the super-alloys used in the aircraft turbine industry e.g. vacuum melting, double melting, etc. is now also specified for nuclear, oil and chemical industry applications. The spin-off factor from further work on superalloys must therefore not be underestimated.

7.1. Possible Areas of Research in the Next Ten Years

7.1.1. Blading

The development of conventional new alloys is unlikely to be

very fruitful. Single crystal technology can however be further
improved by both alloy and process development.

The instability of gamma prime in nickel base alloys is a
fundamental limitation to higher temperature strength and a more
stable strengthening mechanism will be required to affect
temperature capability. Such mechanisms operate in mechanical
alloying and DS eutectics. Mechanical alloying is still in its
infancy and its full potential has yet to be realised. However,
current problems of poor strength and ductility at intermediate
temperatures will need to be overcome. DS eutectics have not to
date demonstrated the total inventory of properties necessary for
commercial application. Further exploitation would seem to
depend on the discovery of new eutectic alloys. This assumes
that this type of alloy is technically acceptable and that the
pronounced property anisotropy can be accommodated.

Probably the most likely new development will come from the
composite blade technology, e.g. the wafer blade concept, complex
coatings, etc. and considerable activity is required in this
area.

Ceramics have a fundamental limitation of a lack of
ductility and low defect tolerance. Developments are therefore
needed to overcome these limitations and to increase our
understanding of the micro structural control necessary to allow
more accurate prediction of component behaviour.

7.1.2. Discs

The main development thrust for improved disc materials is
likely to come from the following areas and extensive R&D
activity will be required :

a. the development of superclean materials;
b. achieving a thorough understanding of fracture mechanics based
   lifing philosophies;
c. the development of composite discs.

Having achieved some understanding of the first two criteria
could then lead to the development of tougher materials.

7.1.3. Combustion Chambers, Nozzle Guide Vanes

High temperature capability, thermal fatigue resistance and
fabricability are the important criteria. Conventional alloys
are restricted in their high temperature capability and therefore
further developments and use of mechanically alloyed materials
and ceramics for these applications are required.

8. REFERENCES

(1) Preface, "The Superalloys", Sims, C.T. and Hagel, W.C., (Eds.), John Wiley and Sons (1972).

(2) Betteridge, W. and Shaw, S.W.K.S., Mat. Sci. & Tech., 3 (9), Sept., pp. 682 (1987).

(3) "Strengthening Mechanisms in Nickel-base Superalloys", Decker, R.F., Climax Molybdenum Company Symposium, Zurich, May 5-6, (1969).

(4) "The Nimonic Alloys", Betteridge, W. and Heslop, J., (Ed.), Edward Arnold, (1974).

(5) "The Superalloys", Sims, C.T. and Hagel, W.C., (Eds.), John Wiley and Sons (1972).

(6) "Phase Stability in Superalloys", Stickler, R., in Sahm P.R., and Speidel, M.P., (Eds.), High Temperature Materials in Gas Turbines, Elsevier Scientific Publishing Company, (1974).

(7) Meetham, G.W., The Metallurgist and Materials Technologist, pp. 387, (1982) Sept.

(8) Thomas, G.B. and Gibbons, T.B., Metals Technology, pp. 95, (1979), March.

(9) Holt, R.T. and Wallace, W., Int. Metals Rev., 203, March (1976).

(10) Heslop, J., and Knott, A.R., Metals and Materials, 5, pp. 59 (1971).

(11) The Development of Gas Turbine Materials, Chapter 4, White, C.H., Applied Science Publishers, London, (1981).

(12) Beeley, P.R. and Driver, D., Metals Forum, 7, pp. 146 (1984).

(13) "The Development of Gas Turbine Materials", Chapter 1, Driver, D., Hall, D.W. and Meetham, G.W., Applied Science Publishers, London, 1981, Ed. Meetham, G.W.

(14) Beattie, H.J. and Hagel, W.C., Trans. AIME 233, pp. 277, (1965).

(15) Boesch, W.J. and Slaney, J.S., Metal Progress, 86, pp. 109 (1964).

(16) Woodyatt, L.R., Sims, C.T. and Beattie, H.J., Trans. AIME, <u>236</u>, pp. 529 (1966).

(17) Wallace, W., Metal Sci., <u>9</u>, pp. 547 (1975).

(18) Ashdown, C.P. and Grey, D.A., Metal Sci., <u>13</u>, pp. 627 (1979).

(19) The Development of Gas Turbine Materials, Chapter 7, Turner, F., Applied Science Publishers, London, (1981).

(20) Moore, J.P. and Astley, R.L., United Aircraft Corporation, US Patten, 3,519,503 (1970).

(21) Alexander, J.D., Proc. Inst. Mech. Engrs., <u>197</u>, pp. 75 (1983).

(22) Pickard, A.C., Mat. Sci & Tech, <u>3</u> (9), pp. 743 (1987).

(23) Sczerzenie, F. and Maurer, G.E., Mat. Sci & Tech., <u>3</u>, (9), pp. 733 (1987).

(24) "The Superalloys", Chapter 6, Klopp, W.D., Ed. Sims, C.T. and Hagel, W.C., John Wiley & Sons (1987).

(25) Weaver, M.J., Mat. Sci. and Tech. <u>3</u>, (9), pp. 695 (1987).

(26) George, D.G., AIAA/SAE/ASME 15th Joint Propulsion Conference, Las Vegas, USA 18-21 June 1979.

Compositions of some typical superalloys

	Ni	Cr	Co	Ti	Al	Mo	C	Zr	B	Others
Wrought turbine blades										
Nimonic 80A	Balance	19.5	-	2.4	1.4	-	0.07	0.07	0.003	
Nimonic 90	Balance	19.5	16.5	2.4	1.4	-	0.08	0.07	0.003	
Nimonic 105	Balance	15.0	20.0	4.7	1.3	5.0	0.13	0.10	0.006	
Nimonic 115	Balance	14.5	13.3	3.8	5.0	3.3	0.15	0.045	0.016	
Udimet 700	Balance	15.0	18.0	3.5	4.0	5.0	0.10			
Cast turbine blades										
IN713LC	Balance	12.0	-	0.6	5.9	4.5	0.05	0.10	0.010	2.0Nb
IN100 (Nimocast PK24)	Balance	10.0	15.0	4.7	5.5	3.0	0.18	0.06	0.014	1.0V
B1900	Balance	8.0	10.0	1.0	6.0	6.0	0.10	0.10	0.015	4.0Ta
MarM200	Balance	9.0	10.0	2.0	5.0	-	0.15	0.05	0.015	12.0W;1.0Nb
MarM002	Balance	9.0	10.0	1.5	5.5	-	0.15	0.05	0.015	10.0W;2.5Ta;1.5Hf
MarM247	Balance	8.3	10.0	1.0	5.5	0.7	0.15	0.05	0.015	10.0W;3.0Ta;1.5Hf
Rene77	Balance	14.5	15.0	3.3	4.3	4.2	0.07	0.04	0.016	
Rene80	Balance	14.0	9.5	5.0	3.0	4.0	0.17	0.03	0.015	4.0W
MarM509	10.0	23.5	Bal	0.2	-	-	0.60	0.50	-	7.0W;3.5Ta
$\gamma/\gamma'/\alpha$	Balance	-	-	-	6.0	32.0	-	-	-	
Nilac 3 116A	Balance	1.9	3.7	-	6.5	-	0.25	-	-	4.0V;8.2Ta;6.3Re
Turbine disc										
Waspaloy	Balance	19.5	13.5	3.0	1.3	-	0.08	0.06	0.006	
Astroloy	Balance	15.0	17.0	3.5	4.0	5.25	0.06	-	0.030	
Nimonic AP1	Balance	15.0	17.0	3.5	4.0	5.0	0.025	0.04	0.025	
Alloy 901	42.0	13.0		3.0	0.30	5.7	0.04	-	-	Bal Fe
Inconel 718	Balance	19.0		0.9	0.6	3.0	0.04	-	-	20Fe;5.2Nb
Sheet										
Nimonic 75	Balance	20.0		0.4			0.10			
C263	Balance	19.0	20.0	2.2	0.5	6.0	0.06	-	-	
HS188	22.0	22.0	Bal	-	-	-	0.10	-	-	14.0W
Nimonic 86	Balance	25.0		-	-	10.0	0.05	-	-	

Single crystal alloy compositions

Alloy	Composition (wt%)								
	Ni	Cr	Co	Ti	Al	Mo	W	Ta	Others
Alloy 444	Balance	8.6	-	1.98	5.1	-	11.1	-	
Alloy 454 (P & W 1480)	Balance	10.0	5.0	1.5	5.0	-	4.0	12.0	
Nasair 100	Balance	9.0	-	1.2	5.75	1.5	10.5	3.3	<0.01C
SRR99	Balance	8.5	5.0	2.2	5.5	-	9.5	2.8	0.02C
RR2000	Balance	10.0	15.0	4.0	5.5	3.0	-	-	1.0V 0.02C
RR2060	Balance	15.0	5.0	2.0	5.0	2.0	2.0	5.0	0.02C

Composition of High-strength, Low-expansion Superalloys, wt-%

Alloy	Ni	Co	Nb	Ti	Al	Si	Fe
Incoloy 903	38.0	15.0	3.0	1.4	0.9	0.1	Bal
Incoloy 904	33.2	14.5	-	1.6	0.1	-	Bal
Incoloy 907	38.4	13.0	4.7	1.5	0.03	0.1	Bal
Incoloy 909	38.2	13.0	4.7	1.5	0.03	0.4	Bal

TRADEMARKS

NIMONIC, NIMOCAST, INCONEL and INCOLOY are registered trademarks of Inco Alloys Limited.

Udimet	-	Registered trademark of Special Metals Inc.
Waspaloy	-	Registered trademark of United Technologies Corporation
Hastelloy } Haines (HS)}	-	Registered trademark of Cabot Corporation
Gatorising	-	Registered trademark of United Technologies Corporation
Pyromet	-	Registered trademark of Carpenter Technology.

2.1.5. Oxide Dispersion Strengthened (ODS) Alloys

C. White

Inco Alloys Ltd.

Holmer Road, Hereford

One of the factors controlling the maximum service temperature of the nickel base superalloys is the instability of the γ' phase. When solutioning of the γ' occurs, properties revert to that of the solid solution and high temperature creep strength decreases markedly. One way to overcome the problem and maintain creep strength is to introduce a stable dispersoid. These are usually chemically inert oxide particles.

Dispersion strengthening of tungsten (1) by thoria particles was recognised in 1919 and sintered aluminium product (2) (SAP), and alumina dispersion in aluminium was developed in 1949. A number of researchers (3, 4, 5) then applied this idea to other systems including nickel. They demonstrated that the dispersion strengthening of nickel with various oxides was a viable concept but few of these techniques achieved significant production importance.

The first practical application of dispersion strengthening was the development by Anders et al. (6) of TD nickel (nickel with 2 % thoria) and TD nichrome (Ni–20Cr–2ThO$_2$). The earlier workers had used mechanical mixing of the various powders but for the production of TD nickel a chemical co-deposition process was used. The technique basically involves the precipitation of nickel hydroxide from a nickel salt solution, by ammonium hydroxide and ammonium carbonate, under controlled conditions of pH and agitation. The hydroxy-carbonate is precipitated on to thoria particles of 50 to 100°A which are present as an aquasol or colloidal suspension. The precipitate is dried, baked at 300°C and pulverised to give nickel oxide powder surrounding thoria particles. This powder is reduced to nickel surrounding thoria by the action of dry hydrogen at 700°C. The resulting powder is cold compacted, sintered extruded and thermo-mechanically processed to sheet or bar.

Relatively simple solid solution strengthened alloys containing thoria can also be produced via this technique, i.e. TDNiCr and TDNiMo. These are produced by selectively reducing the oxides of the various elements at different temperatures. However for the chromium containing alloy difficulties in reducing the oxide led to an alternative production method. Chromium powder is added to the $NiThO_2$ and homogenised for prolonged periods at high temperatures in a reducing atmosphere.

Although these alloys possessed excellent high temperature creep resistance, their low temperature properties were poor in comparison with conventional γ' strengthened materials. The addition of reactive elements such as aluminium and titanium required for γ' precipitation strengthening was not possible in these chemical methods as their oxides could not be reduced by hydrogen.

The disadvantages of the above process and its variants were neatly overcome in the late 1960's by the invention of the Mechanical Alloying process (7). Basically the process embodies the high energy ball milling or attrition of powder materials. The use of mechanical alloying to produce ODS nickel and iron base superalloys has a number of advantages over competitive processes such as chemical co-deposition, the main ones being that

a) no fine metal powders are required and
b) it allows reactive elements such as titanium and aluminium to be included in the alloy and thus for γ' strengthening to be achieved in addition to oxide dispersion strengthening.

The process is carried out in attritors and ball mills, the "attritive elements" being steel balls. These are maintained in a highly energetic state of random motion by a rotating system of paddles in the case of the attritors, or their available kinetic energy is used in the case of the ball mills. Elemental or master alloy powders together with the dispersoid, currently yttrium oxide, are charged into the attritor or ball mill and are then processed for a specific length of time. A repetitive sequence of particle welding and fracturing occurs in the mill at a rate determined by the process kinetic energy, temperature and atmosphere. During milling, an equilibrium powder particle size distribution is established and each individual particle acquires the alloying constituents in the correct proportions.

The powder produced can be compacted either by hot extrusion or hot isostatic pressing (HIP) but subsequent hot working operations must be modified to compensate for the different levels of working involved with these consolidation methods.

The development of optimium high temperature properties in ODS materials, whether produced by mechanical alloying or not, is critically dependent on the development in the final product of a coarse, elongated grain structure appropriate to the nature of the product. This in turn is critically dependent on the thermo-mechanical history of the material. Temperature, strain, strain rate and direction of working all have to be controlled within fairly tight limits.

The aim of thermomechanical processing is therefore two-fold, to produce the required product form, i.e. sheet, plate, tube, etc., and to optimise the stored energy content of the product. This provides the driving force necessary for secondary recrystallisation which develops the coarse elongated grain structure and which in turn greatly enhances the elevated temperature mechanical properties. It has been shown that the secondary recrystallisation response is dependent on the entire processing history including the mechanical alloying stage and the powder consolidation variables.

Studies on bar have shown that creep rupture life at elevated temperatures is proportional to the grain aspect ratio up to about 6:1, above which it is relatively independent (8). In practice a minimum grain aspect ratio (GAR) of 8:1 is considered acceptable. As with many other aligned structures, the properties of ODS alloys can be markedly anisotropic.

The history and development of the mechanical alloying process is detailed in references (9, 10, 11, 12).

A large number of alloys have been developed over the past 15 years and the following are the main ones currently in use or under evaluation. The compositions are given in Table 1.

1. INCONEL ALLOY MA754

This is essentially a nickel-chromium alloy (with minor titanium and aluminium additions) and containing 0.6 % yttria. Isothermal annealing produces a stable secondary recrystallised grain structure that is coarse and highly elongated in the direction of working with a GAR as high as 10:1. The alloy is used for brazed nozzle guide vane and band assemblies in US military aero engines and is under evaluation for vanes in industrial gas turbines. The creep rupture behaviour of the alloy in comparison with that of a conventional nozzle guide vane alloy (Mar-M 509) is shown in Fig. 1 (12). The flatness of the

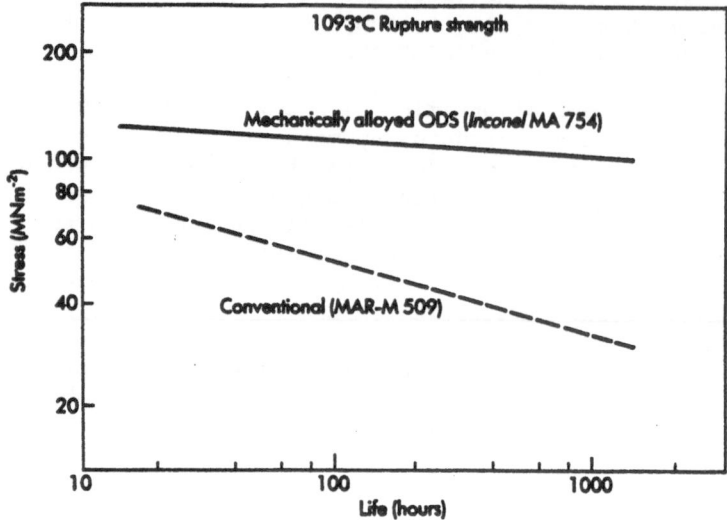

Figure 1
Stress-log time plot of INCONEL alloy MA754
versus a conventional nozzle guide vane alloy,
Mar-M 509. (After ref 6).

INCONEL alloy MA754 curve means that by appropriate selection of
design stresses, tenfold and greater increases in the component's
life can be obtained.

Other variations of the ODS Ni-Cr alloys have been developed
such as INCONEL alloy MA758 containing 30 % Cr for enhanced
resistance to oxidation at about 1150°C in air and attack by
molten glass.

2. INCONEL ALLOY MA6000

Benjamin and Cairns (13) reported in 1970 on the development
of an yttriated version of NIMONIC alloy 80A, designated INCONEL
alloy MA753. This was the first alloy to combine the benefits of
γ' precipitation at intermediate temperatures with dispersion
strengthening at higher temperatures where γ' precipitates
becomes ineffective. However, the evaluation by the gas turbine
engine manufacturers suggested that the alloy did not have
sufficient intermediate temperature strength (i.e. beyond
760°C/100 hr/276 MPa) for advanced turbine blade applications. A
number of alloys were subsequently developed based on ODS
versions of existing alloys, e.g. IN 738, B 1900, etc., but the
coarse elongated grain structures required to make the oxide
dispersion strengthening effective at high temperatures could not

be obtained. In 1977 further work led to the development
(14, 15) of INCONEL alloy MA6000, which achieved the inventory of
properties necessary to compete successfully with the most
advanced superalloys (see Fig. 4 of section on Superalloys).

The alloy composition is complex, being strengthened with
several solid solution elements, for intermediate temperature
strength and yttria for elevated temperature creep resistance (16). Secondary recrystallisation to a coarse elongated
structure is more difficult to achieve by static annealing,
possible because the relatively low solidus of approximately
1295°C prevents annealing at higher temperatures. A high aspect
ratio grain structure can be produced over short distances in a
sufficiently steep stationary temperature gradient. Alternatively, extremely high aspect ratio structures can be developed
over commercial lengths of 1 metre or more by the application of
a moving hot zone annealing technique (17, 18).

The creep rupture strength of INCONEL alloy MA6000 as
compared to that of conventional and single crystal superalloys
is shown in Fig. 2 (19). The alloy is stronger than the single
crystal alloy at high temperatures/low stresses but weaker at low intermediate temperatures/ high stresses. The reason for the inferior inter-mediate temperature strength is that it is not yet possible to produce ODS single crystal structures. Instead columnar grain structures are produced and these make it impossible to obtain the same γ' strengthening contribution as in single cyrstals. Also the texture displayed is <110> rather than the more favourable <100>.

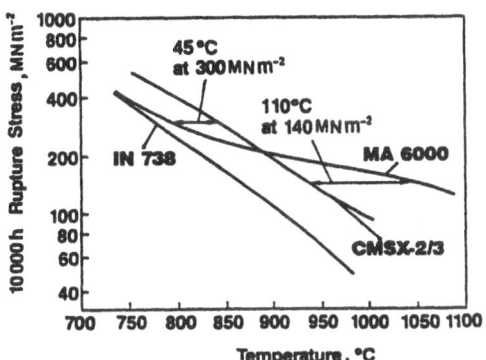

Fig. 2. Comparison of 10,000 hr
rupture strength of
INCONEL alloy MA6000 with
a conventionally cast
(IN 738) and single
crystal (CMSX - 2/3)
superalloy. (After
ref. 19).

The immediate applications for the alloy are for first and
second stage turbine vanes and blades machined from solid bar.
Forged airfoil components have also been developed. The
characteristics of the alloy are such that blade cooling can be
reduced or eliminated altogether as the metal temperature can be

increased by 100°K or more in engines where the stresses are medium or low.

3. INCOLOY ALLOY MA956

This is a ferritic, Fe-Cr-Al alloy having improved oxidation resistance and high temperature strength over conventional nickel base sheet alloys at temperatures in excess of 1000°C. The alloy is used in the form of sheet, plate, spinnings, rings and forgings for combustion chambers, afterburner and turbine casing section, etc.

4. MECHANICAL ALLOYED ALUMINIUM ALLOYS (12)

The mechanical alloying process has also been applied to aluminium alloys. The dispersoid is now no longer Y_2O_3 but a mixture of aluminium oxide and carbide. Because of its high affinity for oxygen there is always some aluminium oxide present either on the powder particles at the start of processing or formed during milling. Aluminium carbides are formed by the decomposition of an organic compound added to assist processing. Both types of dispersoid are finally reduced to 30-50 nm in the alloyed powder.

Two alloys are currently commercially available. Inco MAP alloy AL-9052 (composition is given in Table 1) is similar to the

TABLE 1

Nominal compositions in wt% of some mechanically alloyed dispersion strengthened alloys

	Ni	Fe	Al	Cr	Mg	Li	C	O	Y_2O_3	Ti	W	Mo	Ta	B	Zr
Superalloys															
INCOLOY alloy MA956		Bal	4.5	20					0.5	0.5					
INCONEL alloy MA754	Bal	1.0	0.3	20			0.005		0.6	0.5					
INCONEL alloy MA6000	Bal		4.5	15			0.005		1.1	2.5	4.0	2.0	2.0	0.01	0.15
Aluminium alloys															
IncoMAP alloy AL-9052			Bal		4.0		1.1	0.6							
IncoMAP alloy AL-905XL			Bal		4.0	1.3	1.1	0.6							

conventional alloy 5083 but with an ultra fine grain size. This gives 50 % better strength, higher fracture toughness and improved resistance to stress corrosion cracking. The alloy is currently being evaluated in military aerospace applications where marine corrosion is also a factor.

Inco MAP alloy AL-905XL is an Al-Mg-Li alloy with lower density and greater stiffness than the age hardenable conventional alloy 7075-T73, widely used for airframe applications. Because the alloy does not require age hardening, it makes it possible to produce forgings and heavy sections with homogeneous metallurgical structures. Such forged airframe components are currently being evaluated in service.

5. FUTURE PROSPECTS

It has been demonstrated over recent years that ODS alloys, particularly since the advent of the mechanical alloying process, have moved on from being a laboratory curiosity to a commercial reality. The full potential of the process has not yet been realised and further developments are clearly possible.

In the case of nickel base alloys, improvements to INCONEL alloy MA6000 are under evaluation (20). Compositions similar to those of single crystal cast alloys and the Ni_3Al (γ') and NiAl (β) intermetallics are being studied. The inherent low ductility of these intermetallics has led to the adoption of powder metallurgy as the preferred processing route.

Fabrication and joining can present problems e.g. most fusion welding processes are unsuitable because of dispersoid agglomeration in the fusion zone. Diffusion bonding and brazing techniques are normally used for ODS materials. This may mean some modifications to plant design in order to take full advantage which present and future ODS alloys offer over conventional materials.

The mechanical alloying technique has to date been successfully applied to nickel, iron and aluminium base alloys. Titanium base alloys remain a challenge for the future.

6. REFERENCES

(1) Jeffries, A., Trans. AIME, 1919, 60, pp. 588.

(2) Irmann, R., Techn. Rundschau, 41, pp. 19, (1949).

(3) Cremens, W.S. and Grant, N.J., Proc. ASTM, 58, pp. 714, (1948).

(4) Tracy, V.A. and Worm, D.K., Powder Metall., 10, (1962).

(5) Murphy, R. and Grant, N.J., Ibid, 10, (1962).

(6) Anders, F.J., Alexander, G.B. and Wartel, W.S., Met. Progr., pp. 88, (1962), Dec.

(7) Benjamin, J.S., Metall. Trans., 1, pp. 2943, (1970).

(8) Benjamin, J.S. and Bomford, M.J., Ibid., 5, pp. 615, (1974).

(9) Benjamin, J.S. and Volin, T.E., Ibid, 5, pp. 1929, (1974).

(10) Benjamin, J.S. and Larson, J.M., Proc. AIAA Conf. 1976, Dallas, Texas. American Institute of Aeronautics and Astronautics.

(11) Weber, J.H., Proc. 25th Nat. SAMPE Symp., 25, pp. 752, (1980), San Diego Ca., SAMPE.

(12) Hack, G.A.J., Met. & Materials, 3, Aug., pp. 457, (1987).

(13) Benjamin, J.S. and Cairns, R.L., New P/M Materials, Proc. Int. Powder Metallurgy Conf., July, pp. 47-71, (1970).

(14) Merrick, H.F., Curwick, L.R., and Kim, Y.G., May, (1977), NASA CR-135150.

(15) Benn, R.C., July 1980, NASA Contract NA S3-21448.

(16) Hack, G.A.J., Met. Powder Rep., 36, pp. 425, (1981).

(17) Cairns, R.L., Curwick, L.R. and Benjamin, J.S., Metall. Trans., 6A, pp. 191, (1975).

(18) Benn, R.C., Curwick, L.R. and Hack, G.A.J., Powder Metall., 24, pp. 191, (1981).

(19) Singer, R.F., Met. Sci. Tech., 3, pp. 726 (1987).

(20) Benn, R.C., May 1981, NADC-79106-60.

2.1.6. Refractory Metals and Alloys

R. Eck

Metallwerk Plansee GmbH

A-6600 Reutte

1. INTRODUCTION

It is widely agreed to classify "refractory metals" as those with a melting point above 2000°C. From these ten metals the elements W, Mo, Ta, Nb and Re in metallic form have gained considerable technical significance, while the metals Hf, Os, Ir, Ru and Tc are up to now only important for some special fields of application.

Availability of Mo, W and Nb is good with abundant reserves. The expensive metals Ta and Re are much less available.

As indicated by their high melting point the most significant properties of all refractory metals are high strength at high temperatures, high temperature for loss of work hardening (recrystallisation temperature), high ductile-brittle transition-temperature and accordingly difficult working and forming. With the exception of Re they have a b.c.c. crystal structure and form very stable carbides. All five elements have poor oxidation resistance. Mo W and Re form volatile oxides at very low temperatures, Ta and Nb have a high solubility not only for O but also for N and H. All five elements exhibit excellent corrosion resistance against a specific groups of agents, as described later.

Of the less significant metals hafnium is very resistant to high temperature oxidation and corrosion by water. It occurs in combination with zirconium and resembles it so closely chemically that it is separated from it only with difficulty. Hafnium can be used as a control material in pressurized water nuclear reactors (neutron absorber). It is also used as a minor alloying addition in certain titanium alloys, refractory metals and superalloys. Technetium is formed only by nuclear fission or neutron activation, it is radioactive and does not find any application as a high temperature material.

The most significant properties of the technically important refractory metals are summarized in Table 1. Detailed standard information on chemical and physical properties and description of applications of W, Mo, Ta, Nb and Re are contained in references (1-13).

These refractory metals are essential in many industries for very specific applications, but at the same time dominate special applications where there is no suitable substitute.

Considering productional capabilities there are no limitations on either for size or production capacities at present. Maximum sizes for powder metallurgically produced molybdenum alloys for instance reach 5000 kgs ingots and production capacities for semifinished products of this metal are not restricted at the moment.

Table 1 Properties of Refractory Metals

	Melting point (°C)	density (g/cm²)	Thermal conductivity (W/m.°K)	Electrical conductivity (% IACS)	Temperature of beginning catastrophic oxidation in air (°C)	Modulus of elasticity at 20°C (kN/mm²)	Tensile strength (recrystallized/worked) (N/mm²)		Ductile above temp. (°C)	Special properties
							20°C	1000°C		
Tungsten	3410	19.3	166	31	600/900	400	700/3400	200/2000	-50 +300 dependent on working	forms volatile oxides
Molybdenum	2620	10.2	146	34	400/800	320	500/1300	100/500	-100 -100 dependent on working	forms volatile oxides
Rhenium	3170	21.0	71	9.3	200/600	460	800/4000	300/2500	-270	forms volatile oxides
Tantalum	3000	16.6	54	13.9	400	190	300/1500	100	-200	high solubility for H.N.O.
Niobium	2470	8.6	52	13.2	300	100	200/1000	70	-100	high solubility for H.N.O.

2. TUNGSTEN

Tungsten metal being 2.5 times heavier than steel melts at 3410°C which is the highest melting point of all metals, surpassed only by graphite, diamond and the carbides of Zr, Nb, Ta and Hf. The most important property of W is high temperature strength which makes W and its alloys suitable for applications up to 3000°C in hydrogen or inert gas atmosphere and in vacuum. The disadvantage of high densities is compensated by formability of W and W alloys which is one advantage against high temperature materials like ceramic compounds based on oxides, nitride, borides or others. This makes it the first choice for many very high temperature applications. Poor oxidation resistance causes the same problems as for molybdenum, but beginning at a higher temperature. Loss of metal becomes unacceptable in air above approximately 700°C.

The b.c.c. crystal structure is responsible for lack of ductility at low temperature and difficult forming. Alloying a maximum of 26 w/o Re makes this alloy ductile at room temperature dependent on heat treatment possible. One of the most important alloys of tungsten are the ODS (oxide dispersion strengthened) alloys adding 1 to 2 % of thoria, zirconia, ceria or lanthania. Oxide additions improve machinability, but also make it suitable for application as TIG welding electrodes.

Only 15 % of the W produced worldwide is produced as pure W or W alloys. More than half of the W is needed for production of hard metals and heavy metals, which are liquid phase sintered. Heavy metal is used as armour piercing projectiles in large quantities.

Production of tungsten and refractory tungsten alloys is exclusively performed applying powder metallurgy methods using furnaces equipped with tungsten heating elements, operating up to 2700°C. The high melting W or WRe alloy are the target surface material to produce X-rays using stationary or rotating anodes. These materials are also used for applications where high heat fluxes have to be controlled like divertors for fusion reactor experiments or exit nozzles of jet stream motors burning various fuels. An old conventional application at and above 3000°C are lamp filaments applying a special overlapping recristallized structure for excellent creep resistance.

Specific properties that impede a wide use of tungsten and refractory alloys are high oxidation rate even at low temperatures and the fact that tungsten can neither be melted nor fusion welded without causing unacceptable embrittlement.

Recent research programs have tried to overcome low

temperature brittleness of tungsten by alloying molybdenum and trying to develop lighter alloys by adding similar carbidic constituents to tungsten-molybdenum alloys as to molybdenum. Advanced equipment and modern tools should enable research and development programmes to improve brittleness of tungsten and tungsten alloys. Development of coatings that improve oxidation resistance of tungsten could be a profitable research programme if successful.

3. MOLYBDENUM

Molybdenum melts at 2.610°C, has a high modulus of elasticity and due to its low density of 10.2 g/cm³ has an excellent strength-to-density ratio at high temperatures. Poor low temperature ductility-compares badly with Ta and Nb but is superior compared to W. Oxidation resistance or Mo in air is poor. Above 500°C MoO_3 forms and reaches a vapour pressure of 1 bar at 700°C. Mo can be used at high temperatures in hydrogen or inert gases or vacuum only.

The state of the art in the protection of Mo against oxidation is that for small parts $MoSi_2$ coatings by PVD processes have been successful. In the glass industry protection of molybdenum is common by coating it with glass. Also alumina and MCrAlY coatings can solve oxygen protection problems. In these applications temperatures are not usually higher than approximately 1400°C and exposure times are limited to hours or days dependent on actual temperature.

Only 3 to 4 mio kgs of molybdenum are produced as pure molybdenum metal or Mo-based alloys. More than 90 % of this volume is processed by powder metallurgical methods starting from powders. Arc melting is the second method of consolidation but is less economic than the PM method. Arc cast molybdenum has a lower gas content, so weldments exhibit a lower pore content. Electron beam welding is the preferred method and is applied primarily to molybdenum alloys.

Molybdenum alloys are based on Ti-Zr-Hf-C dispersions for TZM, ZHM and MHC alloys produced by PM methods primarily for dimensional reasons. TZM contains 0,5 % Ti, 0,08 % Zr and 0,01 - 0,04 % C and is the most important Mo alloy with improved high temperature strength characteristics due to higher recrystallization temperature. Zr-Hf-C and Hf-C alloys, exhibit the best high temperature strength remaining work hardened up to 1400°C. These alloys can be welded but the quality of the seam cannot be compared with that of steel weldments. The only alloy of molybdenum which achieves ductile welds is Mo41Re. Fastening techniques for molybdenum and molybdenum alloys still rely on riveting and bolting for all fabricated structures.

Pure molybdenum is resistant to most molten glasses and ceramics, many fused salts and some liquid metals at high temperatures. Molybdenum is used in many glass, quartz and ceramic melting processes. Fabricated parts for high temperature furnaces and furnace equipment like heating elements, radiation shields and containments are very often combined with tungsten. Molybdenum has many uses in the electronic industry. Applications in aerospace and the nuclear industry are also of long standing. A quite recent application for advanced Mo alloy is as dies for isothermal forging or superalloys and titanium alloys.

The catastrophic oxidation of molybdenum, beginning between 600 and 700°C is its greatest drawback. R & D programmes to develop protective coatings have been partly successful for specific problems but would be worth selection for a joint programme of potential users and producers of molybdenum. To develop a coating that matches the thermal expansion of Mo and at the same time withstands the reaction with oxygen at reasonable costs needs new ideas. Much effort has been directed to testing molybdenum as an inner wall material for fusion reactors already. This effort should be continued including possible metal and nonmetal composites and coating programmes.

4. RHENIUM

Rhenium is one of the heaviest metals. A density of 21,0 g/cm^3 is only surpassed by Ir and Os. With a melting point of 3170°C, exceeded only by tungsten, the high temperature strength of pure Re is greater than for any other metal up to 2000°C. This metal is ductile at room temperature even in the recrystallized condition. Oxidation resistance of Re is poor, volatile oxides are formed at even lower temperatures than molybdenum.

Production of rhenium follows the powder metallurgical steps. Mostly wire, ribbon or highly worked products are produced. Rhenium needs a recrystallizing anneal after every 5 % reduction because work hardening is the highest of all metals known. This is a decisive disadvantage for fabrication of larger parts.

One of the possible ways of overcoming the difficulty of working is by alloying with molybdenum or tungsten. MoRe alloys are weldable and WRe is ductile at room temperature if the Re content is low enough not to form a brittle σ-phase.

The main high temperature application of WRe is for X-ray anodes and for thermocouples. MoRe is used for high temperature heat pipes, thermocouples and for constructions where ductility

is necessary over the full temperature range. All high
temperature applications need inert or hydrogen gas atmosphere.
About 80 % of the rhenium produced is used for PtRe or pure
catalysts.

5. TANTALUM

For quite a number of years the annual production of
tantalum metal has remained at ca. 1000 tonnes distributed
between applications as; 450 t for production of capacitors,
300 t are alloyed to cemented carbides and special alloys in form
of carbides, 100 t of Ta are added to superalloys as alloying
element and 150 t are used in the chemical industry for anti-
corrosion reasons.

Two properties of tantalum are application determining.
a. The chemical industry takes advantage of the unique corrosion
   resistance of this metal. Ta shows an excellent corrosion-
   resistance against hot acids and a series of other chemical
   compounds that are usually processed at temperatures up to
   150°C.
b. The melting point of nearly exactly 3000°C leads to high
   temperature applications in vacuum and the very stable carbide
   determines another high temperature use. The density of
   16,6 g/cm³ is expecially in applications needing a high
   strength to weight ratio, where Nb may show more promise. As
   distinct from molybdenum and tungsten, tantalum is ductile at
   room temperature if not embrittled by gases which can be taken
   into solution into the Ta-matrix up to several percent at
   comparatively low temperatures.

For corrosion applications the expensive metal Ta is used to
clad low cost metals such as steel for chemical reactors. Heat
exchangers in the chemical industry and furnace parts for high
temperature use in vacuum are constructed using Ta or Ta alloys
only.

Tantalum semifinished and final products can be fabricated
after arc casting, EB melting or by powder metallurgy methods by
working at room temperature and annealing in vacuum. The maximum
available Ta-ingot is reported to have a diameter of 400 mm.

Tantalum can be joined by TIG welding techniques and all
other methods in vacuum. This property is valuable for
structures that have to be in service at temperature near 2500°C
in vacuum.

Corrosion resistance of pure Ta can be improved by adding
tungsten, usually 2,5 to 10 w/o. Grain structures can be
controlled by addition of elements that inhibit grain growth or

control grain configuration at recrystallization temperatures and above. During the periods of high priced tantalum, development of TaNb alloys for corrosion and pure Nb for certain capacitor applications was successful.

As for most refractory bcc alloys, the property of tantalum which places the biggest restriction on its exploitation is low oxidation resistance together with embrittlement in standard gas environments above room temperature. A protective coating could help to take advantage of good ductility characteristics at all temperatures.

6. NIOBIUM

Supplies of niobium minerals are relatively abundant but processing to metal is expensive. Its principal applications are as an alloying element and as a carbide stabiliser in steels.

Niobium has a number of properties which make it potentially suitable for high temperatures applications: High melting point of 2470°C, a low density of 8,60 g/cm^3, good low temperature ductility, and high strength at high temperature. It also has properties of special interest for nuclear applications. Its low modulus of elasticity is a disadvantage. Resistance to oxidation is poor. The high rate of pick up of O, N and H even at low temperatures causes severe embrittlement.

Niobium like tantalum can be melted by arc casting or by EB-melting depending on the purity required. Powder metallurgy preparation of niobium and its alloys is the standard production route and is preferred if thin sheets of excellent deep drawing qualities is required. Welding of niobium is easy in vacuum or inert gas conditions.

Standard Nb alloys are the following: Carbide dispersed alloys containing ZrC, HfC and W, Mo and/or Ta for high strength and only Zr, Hf and/or V alone for high ductility. They can be formed into wire using standard swaging and drawing techniques. Well known alloys are Nb46Ti for superconductor application and Nb1Zr. Both exhibit a relative ease of fabrication. Nb$_3$Sn and intermetallics of this group have superconducting properties but are brittle. They must be formed in situ by reaction of Nb with tin after the forming processes.

Application of niobium and niobium alloys as a high temperature material is very limited to very specific needs: Weldable structures that have to be lightweight is a typical example. They can be used at temperatures up to only 1700°C maximum in vacuum due to very low creep strength and high vapour pressure of the metal. This metal is still considered to be exotic and not widely enough known throughout industry.

The application of niobium based materials for superconductors has left the stage of laboratory and prototype scale production, but fundamental and developmental research for these materials has been carried on and has been stimulated recently. Of course an effective protection of niobium as for all refractory metals against oxidation and against solution of nitrogen and hydrogen is a worthwhile programme which should pay within a short period if really successful.

7. SUMMARY OF RESOURCES, YEARLY PRODUCTION, PRICES AND FUTURE R & D NEEDS

Table 2 summarizes world resources, production volume and cost data for the five most important refractory metals. Only production of Re is limited because concentration of Re in Cu (Mo) ores is known to be as low as 20 µg/g.

Table 2

Element	W	Mo	Re	Ta	Nb
Known world resources tonnes	ca. 1 M	abundant	Cu by product	abundant	abundant
Production/y. tonnes	43,000	91,000	15	1,000	10,000
Cost $ per pound	6 (1975)	3 (1976)	600 (variable)	27 (1975)	60 (1987)

Future research and development needs to be summarized here are protection of Mo, W (Re) and Ta, Nb surfaces against oxygen bearing gases and melts by known methods such as chemical, physical or electrochemical coating, plating, cladding or by processes yet to be developed.

Improvement of strength properties at high temperatures and ductility properties at low temperatures is an obvious target for future development together with research on improvement of weldability of base metals and alloys. The big step forward in the application of the bcc refractory metals would come from the development of an economic coating to protect against high temperature oxidation and embrittlement by pick up of H, N and O

from air. The chance of finding a universal solution to this
problem is very small but by concentrating on one specific
application the likelihood of success is greatly increased.

The potential for further development of the refractory
metals as alloying elements in other base metals seems to be only
marginal not only for steels and superalloys but also for metals
melting at medium or low temperatures.

LITERATURE

1. "Powder Metallurgy of Refractory Metals and Applications",
    Eck, R., The Intern. J. of Powder Metallurgy & Powder
    Technology, Vol. 17, No.e, 2 pp. (1987)

2. "Pulvermetallurgie and Sinterwerkstoffe", Benesovsky, F.,
    ed., Metallwerk Plansee Ag & Co. KG, Reutte 1973.

3. "Sondermetalle", Kieffer, R., Jangg, G. and Ettmayer, P.",
    Springer-Verlag, Wien/New York (1971)

4. "Vanadium, Niob, Tantal", Kieffer, R. and Braun, H.,
    Springer-Verlag, Berlin, Gollingen, Heidelberg (1963)

5. "Tantalum and Niobium", Miller, G.L., Scientific
    Publications, London (1959)

6. "Behaviour and Properties of Refractory Metals", Tietz,
    T.E., Stanford University Press, Stanford, California (1965)

7. "The Science and Technology of W, Ta, Mo and Nb and their
    alloys", Promisel, N.E., ed., Pergamon Press, London (1964)

8. "Arc-Cast Mo-Base TZM alloy. Properties and Applications",
    Briggs, J.L. and Barr, R.Q., High temperature - High
    Pressure, Vol. 3, 363-409 pp. (1971)

9. Climax Mo.Co.: Molybdenum metal (1960)

10. "Mechanical Properties of Advanced Molybdenum Based Ti-Zr-
    Hf-C Alloys", Eck, R. and Tinzl, J., Modern Development in
    Powder Metallurgy, Vol. 15 - 17, 129-143 pp. (1985)

11. "Molybdän-Rhenium Legierungen als Scheissbare
    Hochtemperatur-Konstruktionswerkstoffe", Eck, R.,
    Proceedings Vol. 2., 11. Plansee-Seminar, Reutte,
    39-57 pp. (1985)

12. "Pulvermetallurgie der hochschmeltzenden Metalle", Eck, R.,
    Metall, 690-695 pp. (1978)

13. "Niobium", Stuart, H., Proceedings of the International
Symposium San Francisco Cal. Nov. 8 - 11 (1981)

2.1.7. Titanium

S.F. Pugh

Consultant Metallurgist

Abingdon, Oxon., U.K.

INTRODUCTION

Titanium MP 1670°C cf Ni 1453°, Fe 1537° is clearly a potential base for refractory alloys. Its density is 4.5 g/cm^3 cf Ni 8.9, Fe 7.87 makes it particularly suitable for aerospace applications and for rotating components. Although very high in the electromotive series it forms a very protective oxide film which readily heals in water and more oxidizing environments giving protection up to about 550°C. Above that temperature the metal dissolves oxygen slowly. Titanium has much in common chemically and physically with zirconium but the two metals have found entirely different applications. The very low neutron capture cross section of zirconium has led to its exploitation as fuel element cladding and pressure tubes operating at 280-350°C in pressurised water nuclear reactors. About 3/4 of the titanium produced each year goes to aerospace applications, as skin material for Mach 3 military aircraft, and compressor discs and blades for jet engines. The amount consumed in this way has remained constant over the past decade. Low temperature applications particularly as condenser tubes in sea water or brackish water cooled power stations is gradually increasing. The metal titanium is a late arrival on the industrial scene and it has considerable development potential.

Fabrication

The special features of the production of titanium alloys are dominated by its chemical reactivity and ability to dissolve oxygen. The metal is produced by reduction of the chloride by magnesium or sodium to give a sponge, which on consumable arc vacuum melting gives ingots with about 1000 ppm oxygen in solution.

High temperature alloys for gas turbine rotating componenets are double or triple melted to achieve homogeneity with electromagnetic stirring of the melt. A steady rate of melting is achieved by temperature sensors and feed back of feed rates and/or current. The consumable arc process avoids contact of the melt with refractories but the vacuum system must be very good to keep the oxygen level in final product below 2000 ppm. This is now so well done routinely that some in-house scrap can be recycled while still maintaining a low oxygen level.

Titanium is expensive and the low utilisation in heavily machined components adds considerably to product costs. Processes for making near final shape products in one operation are under development. In particular some progress is being made in the production of high quality castings. The cast hollow cooled turbine blade is tempting goal. So far titanium alloys in jet engines run at near gas temperatures and they could be used at higher gas temperatures if cooled.

Secondary scrap from machining is likely to be contaminated with tungsten (etc.) carbide cutting tips which do not dissolve or "sink-out" during consumable arc melting. Such material can be reclaimed by "skull" melting on a cooled copper health, with electron beam melting.

Alloying

Up to 885°C titanium has a close packed hexagonal structured and above 885°C to the melting point it is body centred cubic β. Zirconium follows the same pattern so that the Ti-Zr phase diagram shows continuous α and β phases right across the composition range. Most other elements dissolve preferentially in either α or β. Elements such as Mo and V stabilise the β phase and when more than about 20 wt % is added the formation of alpha phase on cooling is suppressed completely. Aluminium stabilises α Ti and raises the β change point. α+β alloys are much used for high temperature applications and it essential that both phases should be solid solution strengthened by incorporation of both α and β stabilisers. The phase change can be used for controlling the structure. Since the hcp phase is anisotropic and subject to cleavage, thermomechanical working has been a necessary development particularly to avoid alignment of the orientations of the grains.
Titanium alloys are unusual among high strength materials in that dispersions of hard second phase particles play little or no part in alloy structure or strengthening methods.

For the highest temperatures of service the α phase is most suitable since it has a high creep strength. Thus have evolved the near alpha compositions which may contain a small proportion of β and derive a few percent increase in strength from particles of second phase.
One such alloy IMI834 currently undergoing engine tests has sufficient creep strength to give full engine life service at 650°C.

Titanium alloys are currently in regular service at temperatures up to 550°C. Intermetallic compounds such as Ti_3AZ are treated in a separate section.

Oxidation resistance and protection

The oxide film which forms on titanium in air or water resembles that on aluminium or chromium in that while still very thin it confers complete protection on the metal and its alloys. Above 550°C, however, oxygen dissolves too rapidly for very long exposures to be envisaged. Thus IMI834 which survives for 20 000 flights running at 650°C on the basis of its mechanical properties, is restricted to 650°C for a few minutes per flight under hot day take off conditions with a cruise temperature of 500 to 550°C.

Development of high temperature strength has therefore gone beyond the improvement in oxidation resistance. It seems unlikely that bulk alloying could effect an improvement. There has been considerable work on coatings, but as yet no viable technique of protection has been found. This is clearly an area needing further R & D.

Powder Metallurgy - Oxide Dispersions

Much effort has gone into the development of the powder route for preparation of titanium alloys, but in straight competition with the ingot method it is unlikely to succeed economically. The advantage is now seen to lie in obtaining fine grain structures by rapid solidification. Provided that a sufficiently high rate of cooling can be achieved, the particles can be quite coarse (shot), produced for example by the spinning electrode technique. One line of experiment has been to alloy with erbium, neodymium or yttrium and form their oxides in situ relying here on the ability of titanium to dissolve oxygen. With 1 to 2 % addition quite spectacular increases in strength up to 700°C have been achieved.

These experiments are all at a very early stage and it is likely to be a long time before a commercial materials is evolved.

For shot consolidation dynamic compacting is being investigated using explosives or projectiles driven by compressed air. The particles are deformed and friction welded together, so avoiding the need to heat and risk oxidation. This technique again is at an early experimental stage.

Quality Control and Specifications

Since the purchasers of high temperature titanium alloys are the builders of air frames and aero engines then the same well established high standards of production control and product testing are required. Thus the licensing procedures and property tests of engine components are similar to those applied to the nickel base superalloys.

Titanium technology therefore shares many of the needs for NDT techniques particularly on-line methods to detect structural, composition variation at the earliest stage in manufacture. A good example is the recent use of IR spectroscope to detect impurities in $TiCl_4$ feed material. Ther is also the need as for many other metals to have on-line real time flaw detection. A particular problem with titanium is to detect segregation of oxygen and nitrogen.

For routine control of batches of material the mechanical properties measured include room temperature and high temperature tensile tests, creep and post creep tensile tests (to detect embrittlement due to ageing, or creep damage) and for components subject to "flutter" a high cycle fatigue test.

An alloy must get CAA and FAA approval for critical applications, and then a long series of engine tests on well characterised material must be done before general use, on a commercial basis can begin.

Future R & D requirements

Because titanium is a relatively newly exploited metal there is still plenty of scope for further development. In terms of annual tonnage utilisation the biggest increases are likely to be in "low" temperature applications. For HTMs the following specific developments are likely to repay R & D in Europe.

1. Development of high quality castings for aero engine components, particularly hollow turbine blades. The choice of mould and core materials, and mould dressings and preparation are here key factors.

2. In alloy development there remains the very long term problem of employing rapid solidification powder (shot)

metallurgy and possibly oxide dispersions to make a new and improved structure for high temperature service.

3. For long term service at temperatures higher than can be tolerated by existing alloys, it is essential to discover and develop an effective coating to protect against oxidation and pick up of oxygen, nitrogen and perhaps hydrogen. This is probably a more urgent need than the development of further increases in creep strength.

In view of the importance of aero engine developments in Europe there is a parallel need to keep at the forefront of materials development for that application. Because development costs of a new product or process are very high, collaboration on a European scale could be of great benefit in allowing cost sharing and in speeding up the rate of progress.

2.1.8. The Platinum Group Metals

S.F. Pugh [*] and R.J. Seymour [+]

1. OCCURRENCE, ANNUAL PRODUCTION AND MAJOR APPLICATIONS

The platinum-group metals (PGMs) comprise six elements in Group VIII of the Periodic Table - ruthenium, rhodium and palladium (the "light" group) and osmium, iridium and platinum (the "heavy" group). The PGMs are often considered collectively in view of their position in the Periodic Table and their close association in nature. They are also very rare, compared with other metals, the annual production in tons of new metals averaged over the years 1975-1984 (1) is as follows :

Pt 96	Gold	1250	Uranium	25,000	Copper	6 million	
Pd 96	Silver	8000	Nickel	550,000	Iron	300 million	
					(Aluminium	15 million)	

In view of their high intrinsic value, there is good reason to recycle and re-use the PGMs - secondary sources currently account for about an extra 12 % over the world's supply of primary platinum metal.

Their comparative scarcity also confers a strategic resource sensitivity. Production statistics and estimated composition of PGMs from the major sources are shown in Table 1.

[*] Consultant Metallurgist; Abingdon, Oxon (U.K.)

[+] Johnson Matthey Technology Centre, Reading (U.K.)

Table 1 :   Production and Estimated Composition of PGMs from the
            Major Sources (2) (3).

Metal	% by Weight		
	South Africa	USSR	Canada
Pt	64.02	30	43.4
Pd	25.61	60	42.9
Ir	0.64	2	2.2
Rh	3.20	2	3.0
Ru	6.40	6	8.5
Os	0.13	0	0

Industrial usage is widespread as shown in Table 2.

Table 2 :   Industrial Use of Platinum and Palladium.

Application	Consumption (tonnes in 1987)	
	Platinum	Palladium
Autocatalyst	36.5	6.6
Jewellery	31.7	5.3
Chemical/petrochemical industry	8.0	--
Glass industry	3.8	--
Investment	15.7	--
Electrical/electronic	5.8	51.7
Dental	--	31.8
Other	3.8	8.7
Total	105.3	104.1

{Data from Platinum 1988}

2. PHYSICAL PROPERTIES AND RESISTANCE TO OXIDATION

The relevant physical properties are listed in Table 3
below.

Table 3

	Ru	Rh	Pd	Os	Ir	Pt
Density g/cm³ at 20°C	12.2	12.4	12.02	22.5	22.65	21.45
Melting point °C, T_m	(2500)	1960	1554	3050	2443	1769
Electrical resistivity microhm. cm. at 20°C	7.6 (0°C)	4.7	10.8	9.5	5.3	10.6

In most cases the density has little technological significance since PGMs are used in applications where weight is not a critical factor, or where their contributions to the overall weight is low.

While PGMs have adequate T_m for use as HTMs their cost restricts their use to applications where their refractoriness is combined with other properties not possessed by other refractory metals. The PGMs compete also with non metallic refractories and here again they are used in applications demanding their particular chemical properties and/or their superior mechanical toughness.

Resistance to oxidation in air at very high temperatures is a property of some PGMs and their alloys, not possessed by other refractory metals (11, 12). In the case of platinum itself, at temperatures up to 400-500°C a thin film of transparent solid oxide (PtO_2) forms on the surface. At higher temperatures, instead of continuing to thicken (as is the case with many base metals) the oxide enters the vapour phase and the film completely disappears. If the metal is then quenched rapidly, it is seen to be bright and oxide-free. It is also resistant to sulphur-bearing atmospheres. Of the other platinum metals, rhodium and palladium most closely resemble platinum in their behaviour at high temperature. The thin solid oxide film on rhodium thickens up to 1040°C and is visible as a tarnish, but this disappears at higher temperatures. On palladium, a dark brown oxide film develops up to 850°C which suddenly vanishes when the transition temperature is exceeded. Rhodium and iridium lose weight at temperatures above 1000-1200°C due to the relatively high vapour pressures of the oxides. Usually, iridium and rhodium are not used per se as a basis for HTM's but are usefully alloyed with platinum, although pure iridium is used for some reactive oxide containment applications where temperatures in excess of 2000°C are encountered. With the more reactive platinum metals, ruthenium and osmium, their reactions with oxygen are more complicated. On heating ruthenium in air, a thin film of RuO_2 forms on the surface which becomes visible as a brown coloration above 400°C. It is the only solid oxide that exists in equilibrium with the metal and oxygen, and does not transform until a temperature of 1540°C is reached. The formation of RuO_2 is only an intermediate stage in the oxidation of Ru metal, and the observed weight losses are due to the formation of new volatile species, RuO_3 and RuO_4. Osmium is much more reactive, being covered with a layer of OsO_2 at room temperature which reacts with air to form the poisonous OsO_4. This has a very high vapour pressure so that finely divided osmium powder loses weight at a measurable rate.

Among other HT properties of technological importance (4) (5) is that under oxidising conditions Pt and Ir are not attacked by a wide range of molten oxides and glasses. Under reducing conditions, platinum (and PGM's) can reduce 'stable' oxides such as Al_2O_3, forming platinum intermetallics which have a higher thermodynamic stability. Air weakening effects, which have been observed in the temperature regime above 1200°C, are attributed to the presence of oxidisable impurities, and appropriate impurity control is therefore necessary to maintain stress-rupture properties (6).

All the PGMs are attacked at high temperatures by fluorine and chlorine.

3. METALS, ALLOYS AND COMPOSITES FOR HIGH TEMPERATURE SERVICE

The PGM's can be treated in sub-groups of pairs of elements, each pair occupying the same vertical columns of the periodic table. The mechanical properties of individual pairs are similar and reflect the crystal structures. Platinum and palladium, both face-centred cubic when pure and annealed are readily worked both hot and cold. At the other extreme, osmium and ruthenium are very hard and cannot be cold worked. This reflects their heaxagonalclose-packed structure. Rhodium and iridium have intermediate properties and are also face-centred cubic (general Ref. 2). In all cases working is strongly affected by the presence of impurities. Recent demands for large pure iridium crucibles have led to improvements in purity and hot fabricability, so that a wide range of semi-finished forms can be supplied.

Platinum, with its excellent corrosion and oxidation resistance, good mechanical processing characteristics and relatively high abundance, is too weak and soft for many engineering applications. As most of the high temperature

applications for platinum involve its freedom from corrosion, it
is important that any alloying additions are chosen to preserve
this characteristic. Pt can be strengthened by alloying with
rhodium, iridium or ruthenium, a 10 % addition increasing the
tensile strength of annealed wire from 12.8 kg.mm^{-2} to 31.5, 38.7
and 58.3 kg.mm^{-2} respectively. The maximum practical limits for
workability are alloys containing 40 % Ir and 14 % Ru. Although
alloys of platinum and rhodium in all proportions may be
fabricated, 20-25 % Rh is the maximum practical limit for high
temperature work. For service at elevated temperatures under
oxidising conditions, rhodium is the preferred alloying addition.
The high temperature strength increases with the rhodium content
(see Fig. 1, Ref. 2) while the excellent oxidation resistance of
platinum is maintained, with up to 25 % Rh.

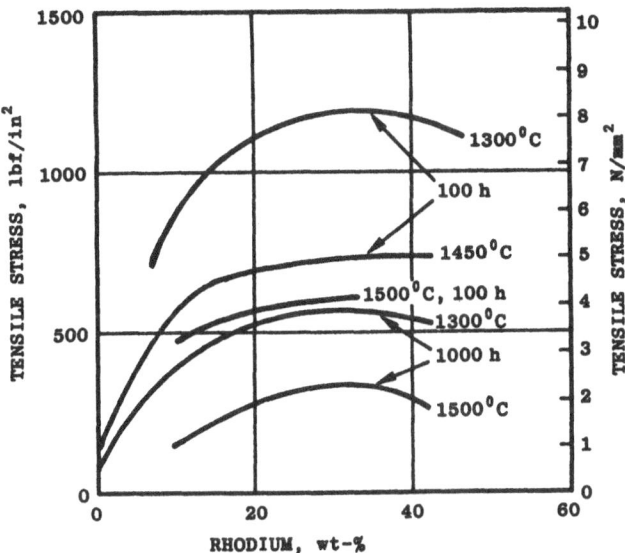

FIGURE 1.   Stress rupture data[13] on alloys
            of high rhodium content.

Some of the high melting PGM intermetallic compounds - such
as $HfIr_3$, $TaIr_3$ and $ZrPt_3$ - are extremely resistant to corrosion
and oxidation and could possibly extent upwards the range of
temperature currently appropriate to containment of reactive
oxides, but considerable difficulties remain as regards their
preparation and processing (7).

Oxide dispersion strengthening of platinum gives a strong
material that has been widely accepted by industry (8). A highly
reproducible fine dispersion of zirconia (ZrO_2) can be made by an
internal oxidation process, using dispersoid concentrations less

than 0.5 volume per cent (about 0.06 weight per cent). ZGS (zirconia grain stabilised) platinum can be processed to sheet or wire, is significantly stronger and more creep resistant than the conventional Rh-Pt alloys (Fig. 1), yet retains the inherently good high temperature corrosion/oxidation properties of pure platinum. Further improvements in strength are under development. 10 % Rh-Pt and Au-Pt versions of ZGS platinum are available, having improved high-temperature strength and non-wettability by molten glass respectively.

As with all dispersion-strengthened alloys, attempts to fusion weld result in a substantial weakening of the alloy due to coarsening of the dispersoid. With platinum, the density differences between platinum and the zirconia dispersoid result in the latter floating to the surface of the molten weld pool.

This has been partially overcome by use of a high rhodium content filler rod during the welding (9) producing glass tight welds, but the differential diffusion of rhodium and platinum after prolonged use at the service temperature produces Kirkendall diffusion voids at the weld/parent metal interface. Solutions to these problems were arrived at either by the development of unique and proprietary joining techniques, or for applications where rhodium could be tolerated, the development of ZGS 10 % Rh-Pt alloys. The use of these materials greatly reduces the effects of preferential rhodium diffusion and associated Kirkendall porosity.

ZGS 10 % Rh-Pt is significantly stronger at high temperatures than ZGS Pt (10), and is used for some critical components in the glass industry. A gold containing material ZGS 5 % Au-Pt exhibits non-wetting characteristics with molten glass and is also becoming an important material of construction in the glass processing industry.

The cost of platinum alloys and their dispersion strengthened counterparts inevitably leads to attempts to find cheaper alternative materials with similar properties. Platinum coated base metals and palladium-rich alloys are cheaper but both of these approaches have limitations with regard to strength, corrosion resistance and reliability. A new commercial composite material consisting of two ZGS platinum outer layers diffusion bonded to a palladium core, combines these approaches (11). The outer layers provide high temperature strength and resistance to both grain growth and corrosion while the cheaper palladium core, which makes up the bulk of the composites, gives good rigidity and electrical conductivity.

The performance of the composite is limited by diffusion of palladium into the ZGS platinium layers, its significance depending on the thickness of the layers, the temperature and the period of operation. In practice, while some Kirkendall diffusion effects are encountered at temperatures higher than 1100°C-1200°C, these can be minimised with correctly designed components. Fig. 2 compares the stress rupture properties of the composite with other platinum containing materials. Applications include analytical apparatus, glass industry equipment and bursting disc foils for use at high temperature and/or in corrosive environments.

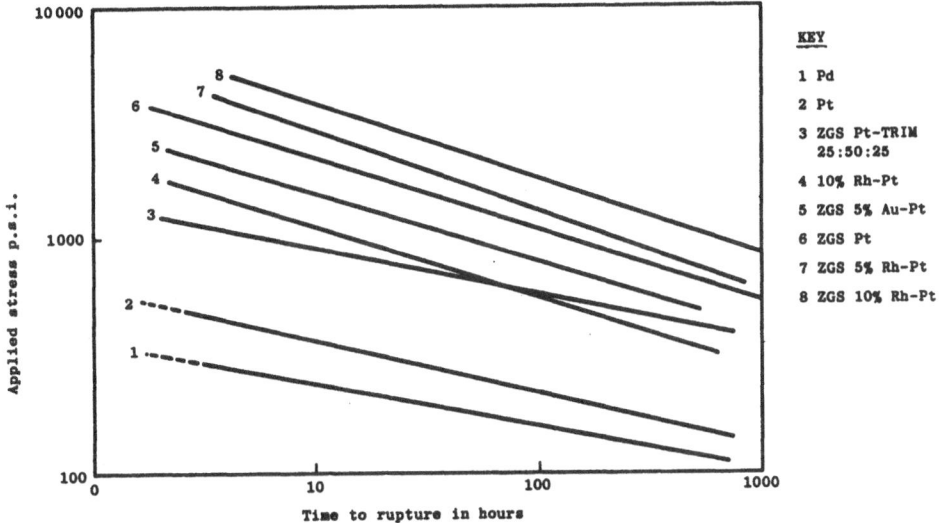

FIGURE 2. Stress rupture properties of ZGS platinum compared with comparative Pt-alloys.

4. HIGH TEMPERATURE PRESENT AND POTENTIAL FUTURE APPLICATIONS OF PGMs

Of the major applications of platinum mentioned in section 2 about half are for uses at high temperatures and of these half are for catalysis. Successful competition with other refractory metals depends on the ability of PGMs to catalyse specific chemical reactions. For example recent work has shown that a ruthenium catalyst used in the Fischer-Tropsch reaction converts methane to paraffins of chain length C_{12} to C_{18} suitable for diesel and jet engine fuels (12), while platinum produces gasoline C_5 to C_{11}.

Since the late 1970's, the largest use of platinum has resulted from the introduction of the motor car exhaust catalyst (the catalytic converter) to reduce emission of noxious gases into the atmosphere. Significant amounts of palladium and

rhodium are involved in this application. Although initially
this application was confined to motor vehicles sold in the USA,
the market will continue to grow as legislative requirements are
introduced in other parts of the world including Eurpe, Australia
and the Far East. Lean-burn engines will still require oxidation
catalysts to meet the legislative requirements. Recycling from
catalytic converters, while technically feasible, is not yet a
major business.

A very promising potentially major new application of
platinum catalysts is in the phosphoric acid fuel cell using
platinum to catalyse both anodic and cathodic reactions with
oxygen and hydrogen respectively (13). Pilot plant has already
been installed for combined heating and electricity supply. One
projection has estimated that by the end of the century about one
third of the PGM output could be used in this application alone.

Traditional uses of PGMs in industry continue in providing
equipment for chemical and physical laboratories. The platinum-
platinum rhodium thermocouple introduced in 1886 and the iridium-
iridium rhodium thermocouple for use up to 2000°C introduced in
1933 are finding more applications with the increase in control
and automation of industrial processes (14). In the glass
industry the use of platinum alloys as crucibles for special
glass including optical glass, and for spinnerets in fibre glass
production has been established for over 25 years. High
temperature strength, oxidation resistance, and freedom from
contamination of the glass are critical properties here, and
have become the material of choice in production of optical glass
fibres. Use of optical fibres in transmission, and fibre glass
in polymer composites in transport equipment and boat building
have maintained a growing demand in this industry. Recently
improved strength has been provided in zirconia dispersions.

Platinum and iridium have been used extensively as crucibles
for growth of single crystals from oxide melts at temperatures up
to 1500°C using the Czockralski technique. The products are
required for solid state devices and laser applications. More
recently iridium crucibles of 4 mm wall thickness and over 6
litres capacity have been fabricated for use up to 2100°C for
production of single crystals of gadolinium gallium garnet (GGG)
for memory chips and yttrium aluminium garnet for lasers.

Other established uses of PGMs involving high temperatures
and potentially aggressive environments include heavy duty
electrical make and break contacts, tube liners for certain
chemical reaction tubes such as those used in fluorocarbon
polymer production, and electrical heating elements.

From 1977 onwards the Voyager space vehicles have had an

onboard electrical power supply obtained from thermocouples attacked to a heat source. The latter consists of $^{238}PuO_2$ encapsulated in iridium (14). Decay of the unstable plutonium isotope provides the heat. Oxide dispersion strengthened platinum tubular heaters with dispersed yttria or zirconia have been tested as possibly providing space station auxiliary propulsion jets. A 10^5 hour life at 1300°C in CO_2 has been demonstrated and in methane at 600°C (16).

As with most materials development, ultimate success in most applications is dependent on the viability and marketability of the project that they are intended to serve. Ultimate development must therefore be alongside the relevant engineering project. The phosphoric acid fuel cell if it reaches the 10 Gigawatt p.a. installation rate by 2005 could provide a major new market for PGMs.

The use of platinum electrodes in spark plugs improves their performance by enhancing durability. Another possible new application is the use of a thick platinum sheath on copper MHD electrodes. Resistance to oxidation and attack by hot slags from high temperature coal flames is required here (17). The overall objective is to extract power from burning coal more efficiently.

5. GENERIC PROBLEMS AND NEW DEVELOPMENTS

There are several new engineering developments where use of PGMs with some technical optimisation could supply critical components. Such developments would be mainly motivated by the technical needs of the project. There remain, however, a number of generic problems, and possibilities of speculative new materials developments based on PGMs as described below.

Purity and Susceptibility to Contamination

The presence of certain impurities in platinum can have a deleterious effect on it's high temperature stress rupture properties and e.m.f. characteristics, the latter being important in thermocouple applications. Metal batches from different sources contain different impurities and/or varying impurity levels. Other impurity elements may be introduced during extraction, refining and processing operations. Control of purity of metal from different sources would therefore be desirable. Additionally, work is needed to determine how impurities from the service environment diffuse into the grain boundaries – and which are the important impurity species.

A very important and insidious impurity is silicon. Traces of silicon are readily introduced into the PGMs from silica and silica-bearing refractories if the surroundings become reducing

in character even momentarily. Even small quantities of platinum
silicide can form intergranular films of low melting platinum-
platinum silicide eutectics with catastrophic effects on the high
temperature properties.

Specific impurities can sometimes cause order of magnitude
differences in the room temperature working characteristics of
the metals. Ruthenium, normally very hard and difficult to work,
is fairly ductile if highly-pure forms are made by techniques
such as electron-beam zone refining.

The susceptibility of PGMs to contamination is one reason
why handleability and containment of molten PGMs is a problem;
the normal practice is to use special refractories containing
zirconia or magnesia. However further understanding of the
interface chemistry of molten platinum-solid refractory systems
would be useful, and extremely important if direct melt processing techniques (such as melt spinning) were adapted for the
fabrication of PGMs in semi-finished form (tape, wire, etc.).
The availability of such an approach would be particularly useful
for the difficult-to-work metals such as ruthenium and rhodium.
It would also be attractive from the point of view of energy
savings in the manufacture of platinum and palladium. Amorphous
PGM alloys made by melt spinning (e.g. Pd-Cu-Si) might have
useful high temperature structural and catalytic applications if
it were possible to increase their devitrification (glass-crys-
talline transition) temperature.

High Temperature Diffusion

Although PGM coatings offer a cost-effective means of using
these metals, diffusion processes limit the usefulness of
platinum-clad base or refractory metals at high temperature.
Migration of these metals into the platinum and their subsequent
oxidation results in embrittlement and eventual failure of the
cladding. Barrier systems are needed to prevent high temperature
diffusion. Possible approaches include the development of
interface barrier materials or in-situ development of secondary
structures (e.g. intermetallics). There are three particularly
important interface systems requiring study :

- PGM/refractory metal (notably tungsten)
- PGM/base metal (notably nickel)
- PGM/ceramic (for applications in the range 1200-1600°C).

Although platinum aluminide coatings on base metal
superalloys provide a protective layer capable of resisting both
high temperature oxidation and hot corrosion, some problems have
been encountered with regard to ductility of the aluminide
coatings at low temperature. This limits their use in
applications where temperature cycles occur.

Solid state bonding is normally used to join ZGS alloys. However, on occasions when welding is employed, the Kirkendall effect can reduce the strength of completed joins. This effect is only a problem in welds of ZGS platinum or ZGS rhodium-platinum with higher Rh-content rhodium-platinum alloys, where voiding and porosity due to self-diffusion of rhodium and platinum leads to reduces strength. Matching of alloys to be joined alleviates the problem.

Fabricability

Various aspects of the fabrication of PGM structural components require improvement. The machining of platinum - often accompanied by rapid tool wear - could be improved by a tribological study of tool-platinium interactions. Other means of bonding the metals - other than normal welding practice - such as laser welding might produce structures with improved strength and reliability (particularly with the dispersion strengthened materials).

The commercial acceptability of the PGM intermetallics - particularly those containing refractory metals - is limited by the difficulty of fabricating these materials in bulk form. If ductility could be induced in these materials, this might lead to new high temperature structural and catalytic applications.

6. REFERENCES

(1) Platinum Review, Johnson Matthey, (1988).

(2) "Supply and Use Patterns for the Platinum-Group Metals", U.S. National Materials Advisory Board Publication, NMAB-359, (1980).

(3) "Platinum-Group Metals (Mineral Commodity Profile)", Loebenstein, J.R., U.S. Bureau of Mines Publication, June 1983.

(4) Edelmetall-Taschenbuch, publ. Degussa, (1967).

(5) Corrosion Guide, 2nd Edition, Ed. Rabald, E., Elsevier Publishing Co., Amsterdam, (1968).

(6) Boswell, P.G., Platinum Metals Review, (1982), 26(1), pp. 16.

(7) McGill, I.R., Platinum Metals Review, (1977), 21(3), pp. 85.

(8) Selman, G.L., Day, J.G. and Bourne, A.A., ibid, (1974) 18(2), pp. 46.

(9) Heywood, A.E., Conference on Less Common Metals, Hongzkou PRC (1982), paper 44. Chine Society of Metals and The Metals Society London.

(10) Selman, G.L. and Bourne, A.A., Platinum Metals Review, (1976), 20(3), pp. 86.

(11) Rowe, M.S. and Heywood, A.E., ibid, (1984), 28(1), pp. 7.

(12) King, F., et al., ibid, (1985), 29(4), pp. 146-154.

(13) Robson, G.G., Platinum (1986), Johnson Matthey Plc.

(14) Platinum 1986.

(15) Platinum Metals Review, 30(1), Jan. 1986.

(16) Ibid, 29(4), (1985) pp. 167.

(17) Ibid, 30(1), (1986), pp. 2-11.

(18) Knapton, A.G., Platinum Metals Review, 23 (1), 1979, pp. 2.

(19) Chaston J.C., Platinum Metals Review, 19 (4), 1975, pp. 135.

GENERAL READING

1. INCO Publications on Platinum, Palladium, Rhodium and Ruthenium (1965).

2. Darling, A.S., "Some Properties and Applications of the Platinum-Group Metals", Int. Met. Rev., (1973), Review No. 175.

3. Dowsing, R.J., "Spotlight on the PGM's", Parts 1 and 2, Metals and Materials, May 80 and June 80.

4. Seymour, R.J., "The Platinum-Group Metals - Present and Future Technology", The Metallurgist and Materials Technologist, Nov. 82, pp. 505.

5. "Supply and Use Patterns for the Platinum-Group Metals", NMAB-359, (1980).

6. Platinum Metals Review, published quarterly by Johnson Matthey PLC, 43 Hatton Garden, London EC1N 8EE.

7. Hunt, L.B., "A History of Iridium", Platinum Metals Review, 31, (1), pp. 32-41, (1987).

2.1.9. Intermetallic Compounds for High Temperature Use

P. Costa

ONERA

29, avenue Division Leclerc, F-92320 Châtillon

The particular intermetallic compounds which are of interest for applications at elevated temperatures while heavily stressed are alloys of high chemical stability, with high melting temperatures and high elastic moduli. This stability is the result of very strong interactions between unlike metal atoms. In most cases, bonding retains its metallic character; even when the electro-negativity difference between metallic elements is large, charge transfers and the ionic contribution to bonding remain very limited. As a consequence, at least when the sizes of unlike atoms are not too different, the memory of usual metallic arrangements is preserved, and the compounds are ordered derivatives of usual

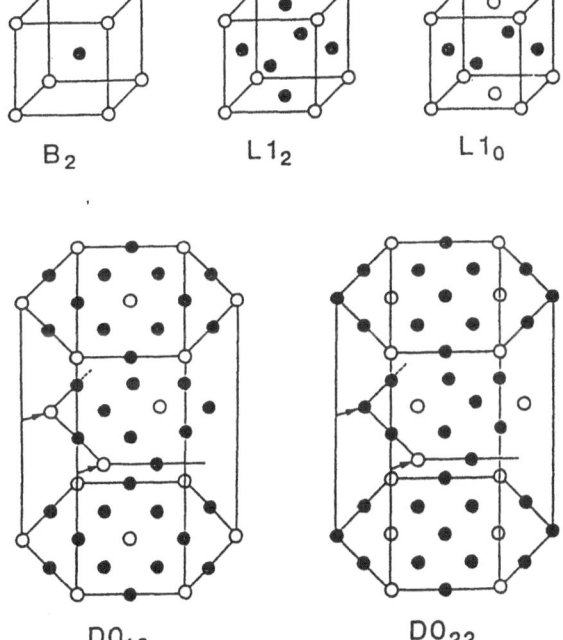

B_2 $L1_2$ $L1_0$

DO_{19} DO_{22}

Fig. 1   The main ordered structures derived from cc, fcc and hcp Bravais lattices

metallic close-packed structures (simple face-centred cubic, body-centred cubic or hexagonal) (Fig. 1). The occurrence of a particular ordering arrangement is governed by electronic - structure and statistical - thermodynamics laws which are now fairly well understood, at least qualitatively (1). The nature of these arrangements is often more complex than it was first assumed; this is apparent from recent high resolution electron microscopy work, on the Ti-Al system for instance (1 bis). When the difference in size between atoms is large, different structures occur, such as Laves phases, in order to achieve close-packing of atoms.

The main intermetallic compounds have high strength and good oxidation and corrosion resistance, and relatively low specific gravity. Unfortunately, they are in general brittle, at least at low temperatures.

The mechanical properties can be explained in terms of complex superdislocation structures, which inhibit slip by for example formation of extended dislocation core configurations leading to high friction forces, or dissociations leading either to planar slip behaviour or to sessile configurations, which in turn increase elastic limits, strain-hardening rates and brittleness. Another cause for limited toughness and ductility is a reduced cohesion of grain boundaries, which leads to brittleness in polycrystalline materials. Also, some intermetallic compounds do not possess the five independent shear modes necessary for general shear (von Mises criterion).

A large proportion of the intermetallics suitable for high temperature applications are aluminides; we shall review in the next pages possible developments and applications for nickel and titanium compounds. Other families of alloys might be of interest for high temperature use; the highest values of melting temperature and the reduced (i.e. divided by the specific gravity) elastic moduli for instance are not obtained for the ordered structures which derive from fcc, bcc or hcp structures, but from structures which are more specific to the multi-atomic systems like the A-15 or the Laves phases. All of them are brittle. These might in the long run turn out to be a better compromise between refractories and toughness than are ceramics. A special mention should be made of silicides, which might give a happy compromise between high temperature mechanical properties and oxidation resistance.

1. THE Ll$_2$ COMPOUNDS

The Ll$_2$ structure is the principal and the simplest ordered derivative of the fcc Bravais lattice. It corresponds to a A$_3$B type of composition, and it is adopted by several compounds of interest, the main one being the γ' (Ni₃Al) phase of nickel base superalloys. A large number of similar compounds exists with B being an element belonging to the 3B or 4B columns of the periodic table, or a transition element, and A a transition element. The structure will also tolerate considerable solid solutioning, in which ternary elements may be substituted into both the A or the B atom. For Ni₃Al, most of the elements of the 3B, 4B and 3A to 6A columns can be substituted into aluminium, and Co, Fe, Mn and Cu into nickel.

1.1. Ni₃Al and Related Compounds

1.1.1. Yield Strength

Ni₃Al, like several other interstitial alloys, shows a peculiar behaviour of the temperature dependence of the elastic limit, with a steady increase up to about 600°C, and then a decrease (Fig. 3). An extensive discussion of the physical origin of this phenomenon based upon dislocation locking processes has been given by Pope & Ezz (2) and others (3-10) (Fig. 2).

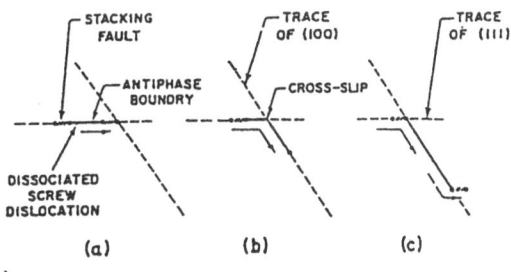

a)

Fig. 2a Mechanism of cross-slip pinning, from a
to c, proposed by Kear and Wilsdorf (3)

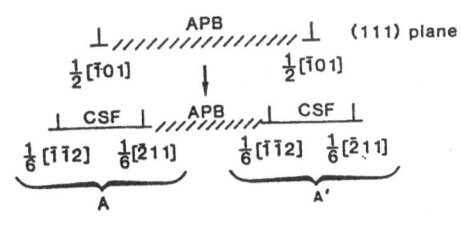

b)

Fig. 2b Dissociation of (101) perfect
dislocations. A and A' should be
considered as extended cores

These effects are also responsible for the analagous behaviour of γ - γ' superalloys. However, the yield strength of γ' alloys is much lower than that of the γ - γ'. It can be

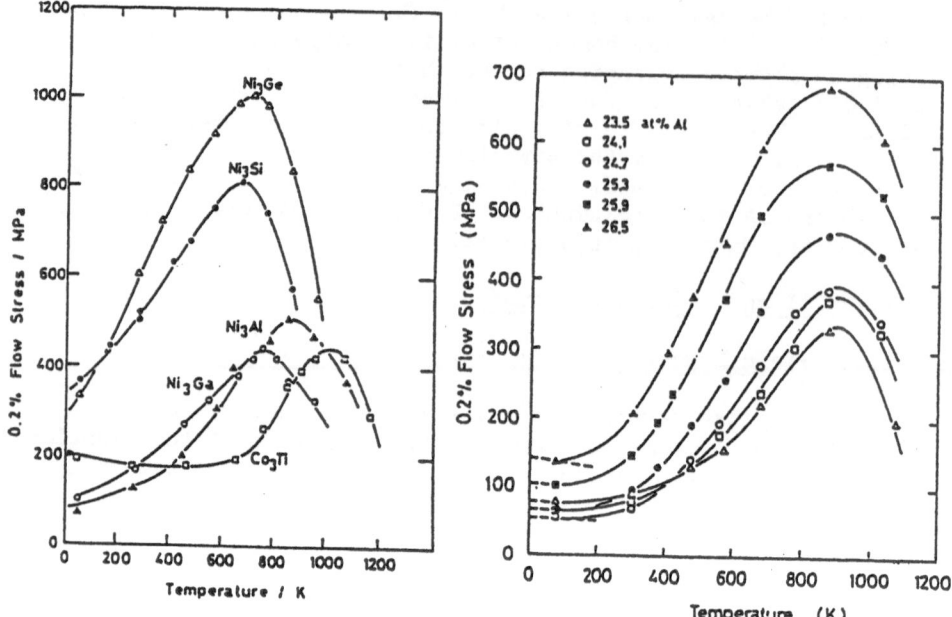

Fig. 3   Yield stress of Ll$_2$
compounds as a function
of temperature (4)

Fig. 4   Yield stress of Ni$_3$Al
compounds as a
function of stoechio-
metry and temperature
(11)

increased either by the addition of an excess of aluminium (the
elastic limit is for 26.5 at % Al approximately double that for
25 at % in the whole temperature range) (11), Fig. 4, or by the
substitution of ternary elements such as Si, Ge, Ti, Zr, Hf, Nb
and Ta. The alloys retain the same behaviour for the temperature
dependence of the flow stress. The extent of the increase of the
critical resolved shear stress (CRSS) is a function of a
parameter which includes misfits of both lattice parameter and
modulus (12). Hf is the most effective; 4 % of this element led
to an 0.2 % flow stress of 800 MPa at room temperature, and
1200 MPa at 500°C (13). Rapid solidification techniques can give
even higher values, up to 1000 MPa at room temperature for
complex alloys (14).

1.1.2. Ductility

As for most intermetallics, ductility is, for the Ni$_3$Al type
alloys, the poorest property. It is very limited at room
temperature - below 1 % elongation - except for a few examples
that will be given below. This behaviour is due to the incidence
of elastic grain boundary fracture. It is not correlated with

grain boundary precipitation, since rapidly solidified alloys and alloys solidified using classical techniques are equally brittle. More precisely, the correlation is proven to be better in terms of electronegativity, rather than valency. The concept of average electronegativity has to be used for ternary compounds (17).

In fact, better ductility might be achieved :

1. By recrystallization to a very large grain size although this is a point of contention at present (18), (16);

2. Or by the introduction of a second phase (19);

3. Or by a reduction of the degree of order of the alloy (15, 20).

One should however be careful about the exact nature of the effect, which might be a direct consequence of the small size of domains, and of the fact that γ phase films which are expected to increase the strength of grain boundaries are often present on APB's. In this line, Cahn (21) has shown that a strong correlation exists between brittleness and the fact that order is still retained at the melting temperature, a situation which upon cooling leads to grains which are single domains. He suggests that APB's and the γ films which decorate them, acting as obstacles to the dislocation passage, enhance the ductility by decreasing the stresses exerted by dislocation pile-ups on grain boundaries.

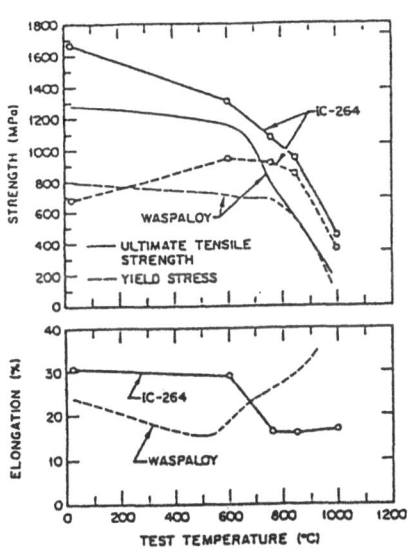

4. Or by reducing the oxygen content of the alloy, to which high temperature ductility seems to be very sensitive.

5. Or by the introduction of oligo-elements, which decrease the grain boundary cohesion. Impressive results have been obtained by Liu and his co-workers of the Oak-Ridge team and others (22-25, 29) (Fig. 5).

Fig. 5 Tensile properties of IC-264 (NI-16.7A1-1.0Zr-8Cr-0.1B at%); test performed in air (25)

At this stage, the situation can be summarized by saying that there is little hope of having Ll$_2$ nickel base alloys with acceptable ductility at high temperature, unless γ disordered phase is present at grain boundaries.

1.1.3. Creep Properties

The activation energy for self-diffusion of ordered structures is higher than in disordered solid solutions. This effect should be reflected in improved creep strength. However, in spite of a lower specific gravity, the improvement is not sufficient to give a creep strength that can compete with that of the multiphase precipitation-strengthened structures of the alloys, which are already on use for high temperature applications (26) (Fig. 6). The addition of ternary elements such as Zr, Hf, Ta, which increase the elastic limit gives a dramatic improvement of the creep behaviour; but the rupture lives are well below those of the superalloys used for turbine discs or blades (28, 29).

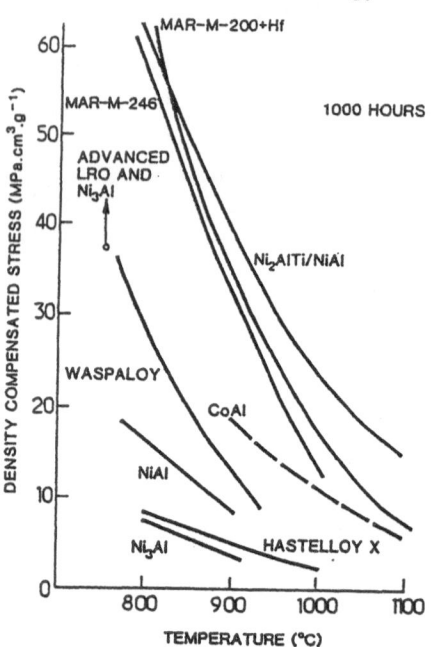

Fig. 6 Creep resistance of ordered compounds (26)

1.1.4. Fatigue Properties

Available work on fatigue properties is very limited (30, 32). Fatigue crack growth rates for boron-doped alloys are, at room temperature, lower than for commercial superalloys. They rapidly increase with temperature, with a high sensitivity to environment, mainly oxygen; these variations are parallel to those which have already been described for the ductility. High cycle fatigue behaviour seems also to be excellent; this is probably a consequence of a very high cyclic work-hardening rate, similar to the monotonic work-hardening rate.

1.1.5. Conclusions

To summarize the preceding results, it can be said that the Ni$_3$Al type alloys have high elastic moduli, high melting temperature, high yield strengths and high rupture strengths at

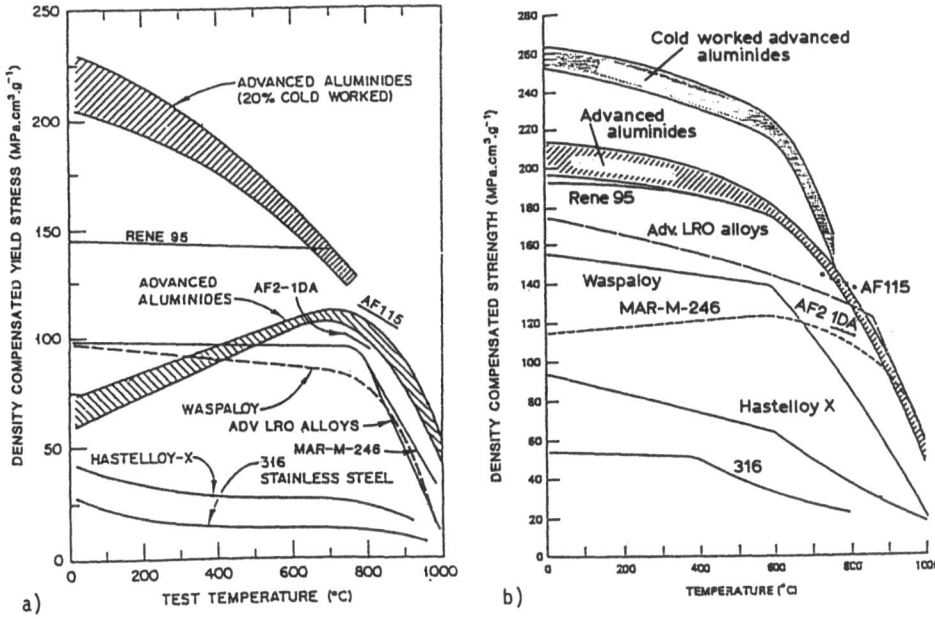

Fig. 7 Yield stress (a) and rupture stress (b) of "advanced"
Ni_3AL compounds (27)

average temperatures (600°C), large strain hardening rates and,
for the best alloys, large ductilities and rupture strengths at
room temperature (Fig. 7). Conversely, creep behaviour and
ductility at medium temperatures are poor or very poor.

- For use a lower temperatures, "advanced" cold worked nickel
  aluminides could be developed with a specific (i.e. density
  corrected) value of the yield stress of 240 $MPa.cm^3.g^{-1}$ and
  specific Young's modulus of 24 $GPa.cm^3.g^{-1}$, which equals or
  even surpasses titanium alloys. However, due to the cold work
  requirement, they cannot be used for hot and die-forged parts
  such as compressor or fan blades; and, for the same reason,
  welding would be forbidden. In the author's opinion, there is
  no real advantage over super plastically formed, diffusion
  bonded (SPFDB) titanium alloys. The only well defined application is for bolts, that would fully use the very high elastic
  limit of these compounds.

- At medium temperatures, even the best alloys, such as IC 264,
  which are ductile, cannot compete with superalloys because of
  their low creep rupture strength.

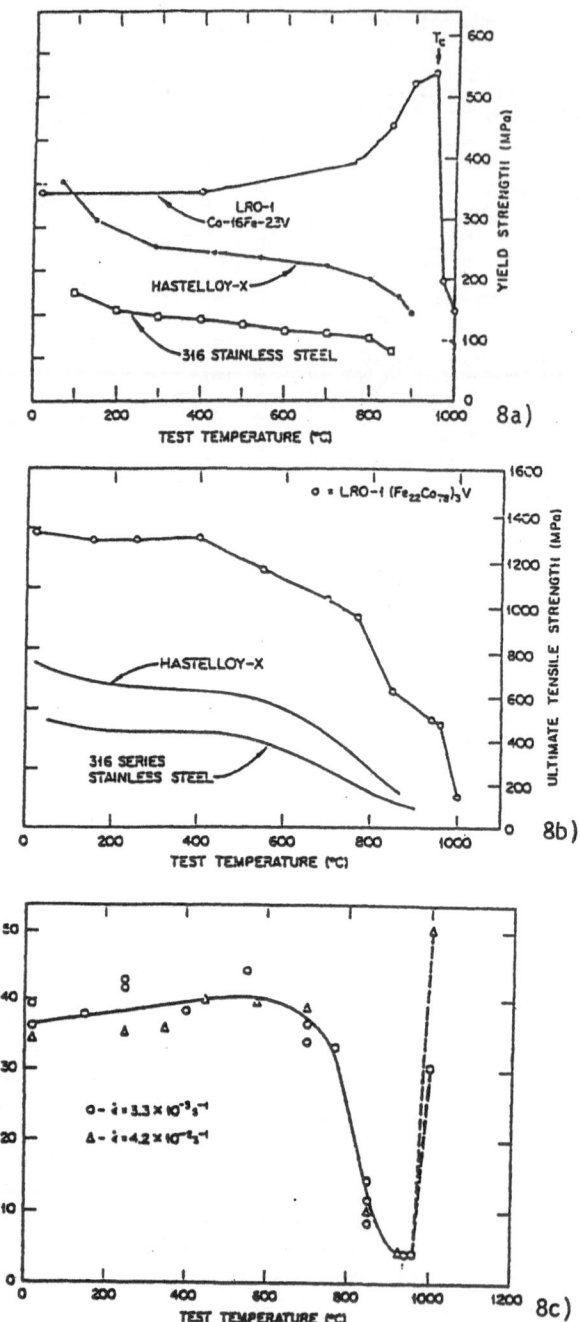

Fig. 8   Cobalt-base LRO alloys
          a) Elastic limit; b) Rupture strength;
          c) Elongation (LRO 1: $(Co._{78}Fe._{22})_3V$) (22)

- At high temperatures, single crystals could be used for very
hot non rotating parts exposed to an oxidising environment
(distributors) provided that the thermal fatigue resistance is
adequate.

1.1.6. The $(Fe,Co,Ni)_3V$ Ll_2 Alloys "LRO" (Low Range Order) Alloys

Another family of Ll_2 alloys has been extensively studied in
recent years by Liu & co-workers (22, Fig. 8 and 9) : the
$(Fe,Co,Ni)_3V$ Ll_2 alloys. Fe_3V, Co_3V and Ni_3V have non cubic
structures and are
brittle. But for
electron to atom
ratios between 7.5
and 8, ternary or
quaternary alloys
follow the Ll_2
structure, with
ordering temperatures
in the 650–950°C
interval far below
the melting point.
Iron base alloys have
lower ordering
temperatures; high P
values are obtained
for cobalt and nickel
base alloys. The
elastic limit of
these alloys is about
400 MPa at room
temperature, and
increases slightly up
to the ordering
transition
temperature. It is
very low above the
ordering temperature.

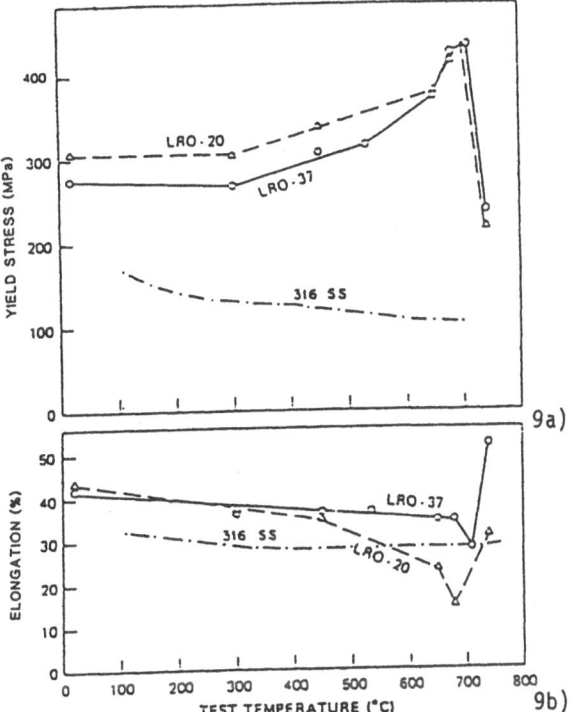

9a)

9b)

Fig. 9   Iron-base LRO alloys
a) Elastic limit; b) Elongation
(22) (LRO 20: $Fe._{50}Ni._{50})_3V$;
LRO 37: $(Fe._{50}Ni._{50})3(V._{98}Ti._{02}))$

The strain
hardening rate is
very high, leading to
ultimate tensile strengths of approximately 1000–1200 MPa at room
temperature. This property decreases with temperature, mainly in
a 100–200°C interval below the ordering temperature, down to
values close to the yield strength; ductility is then reduced in
most cases to a very low level. This ductility drop is more
moderate for Fe-base than for Ni- or Co-base alloys.

Since fairly small amounts of Ti, Zr or Hf substituted for

vanadium increase the elongation to rupture, it can be reasonably assumed that the ductility properties are related to modifications of the grain boundary strength.

As for Ni$_3$Al, and for the same reasons, the fatigue properties of these alloys are excellent, and probably even better than for Ni$_3$Al (33, 34), but the creep properties are rather limited below the ordering temperature, and very poor above. Liu claims that some of the cobalt base alloys have mechanical properties comparable with those of superalloys such as Waspaloy, and iron base alloys have better properties than stainless steels; but, due to the absence of chromium, the oxidation properties are very poor, which limits the possible uses of these products. As a consequence, no high temperature application can be defined, except for steam turbines (the resistance to steam corrosion in the 500-600°C temperature range is better than that of stainless steel) or in nuclear reactors (limited swelling under irradiation).

2. THE B$_2$ COMPOUNDS

Several intermetallic compounds of possible industrial interest such as NiAl, CoAl and FeAl have the B$_2$ structure, Fig. 1. This is the simplest ordered structure derived from the

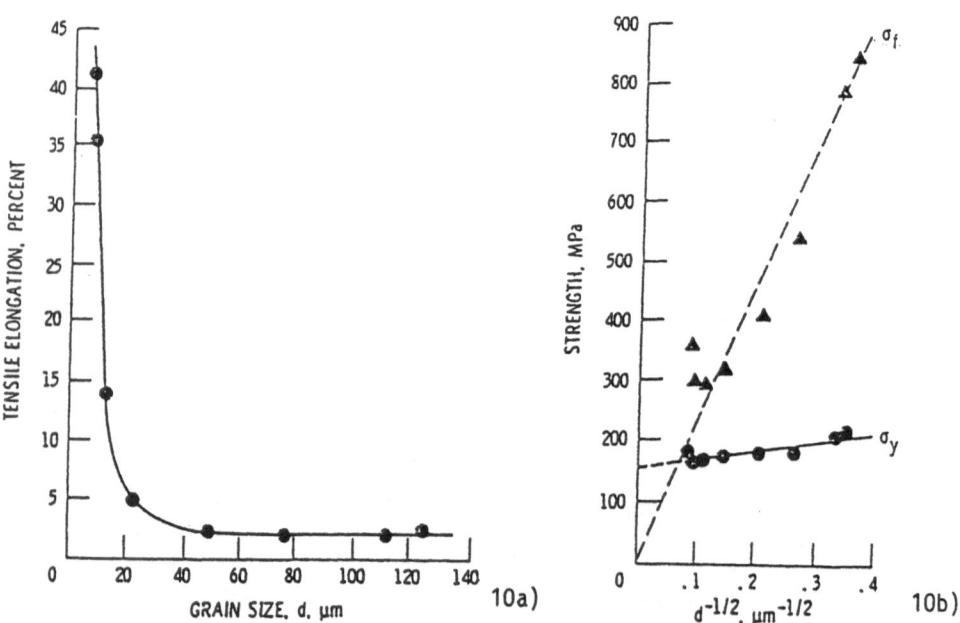

Fig. 10 a) Relationship between grain size and ductility for NiAl at 400°C (35)
    b) Hall-Petch law for NiAl at 400°C

body-centered-cubic structure. These compounds have high melting temperatures and the highest values of the reduced modulus of any of the cubic intermetallics. NiAl and CoAl retain their order up to melting temperatures. They also have excellent resistance to oxidation, a consequence of the high aluminium content. As a result, NiAl is used for the coatings of high temperature parts; FeAl also has excellent properties which have been reviewed by Tomaszewicz (57), and are illustrated by their use for electric heater elements (Kanthal alloys).

These compounds, when polycrystalline, are brittle at room temperature. Several authors have tried to increase the ductility by reducing the grain size. It has been shown that for 40 μm grain size for NiAl the transition could be lowered from 600 down to 400°C (35), Fig. 10, and even to room temperature for rapidly solidified alloys, with a ductility of about 1 % in flexural tests (36). This figure could be further increased by an anneal at 1000°C, which indicates that grain boundary segregates further enhance the brittleness of the alloy.

The elastic limit of stoichiometric binary B2 compounds is low (37, 38) (Fig 11). It can be increased by the partial substitution of another transition element. But departure from stoichiometry has a much bigger hardening effect. Excess of nickel is accommodated by substitution for aluminium, excess of aluminium by the creation of nickel vacancies. In both cases, point defects are clustered, which explains the magnitude of the observed increase (38, 39).

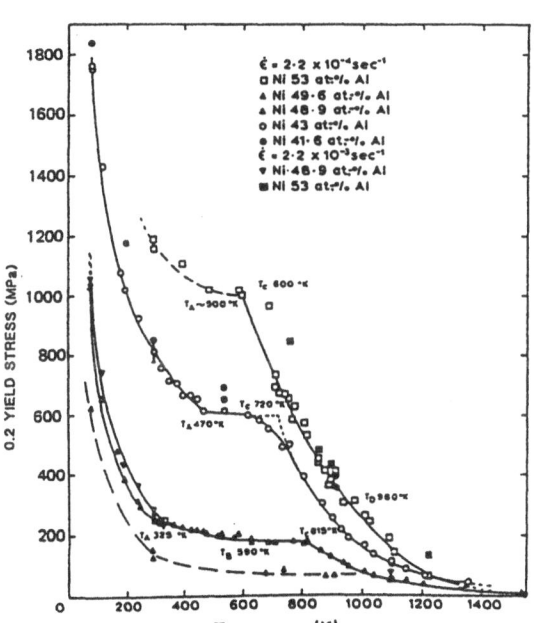

Fig. 11 Elastic limit of NiAl single crystals, as a function of stoichiometry and temperature (37)

Although fine grained examples have high ultimate tensile strengths at intermediate temperatures, the creep properties of the

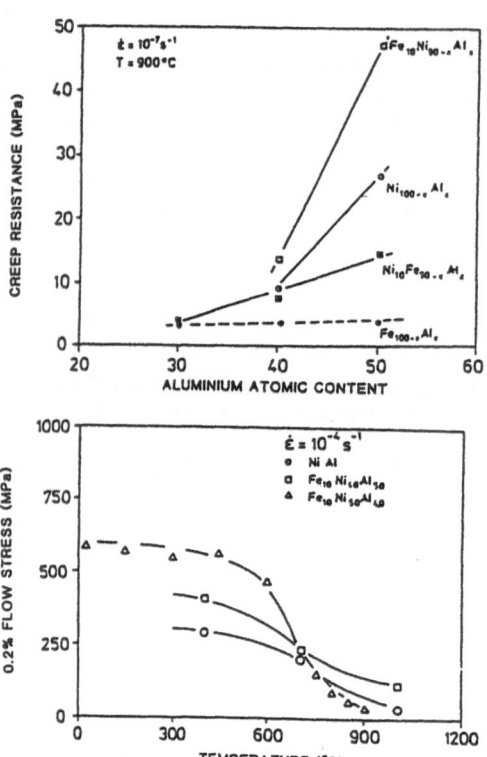

Fig. 12 High temperature
properties of several
(Fe,Ni)Al alloys (40)

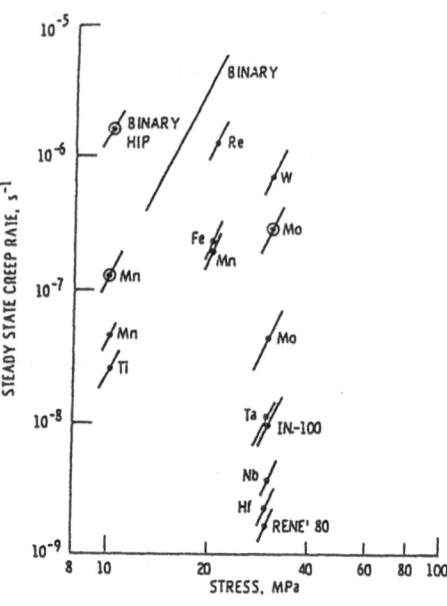

Fig. 13 Creep properties of
NiAl type alloys (42)

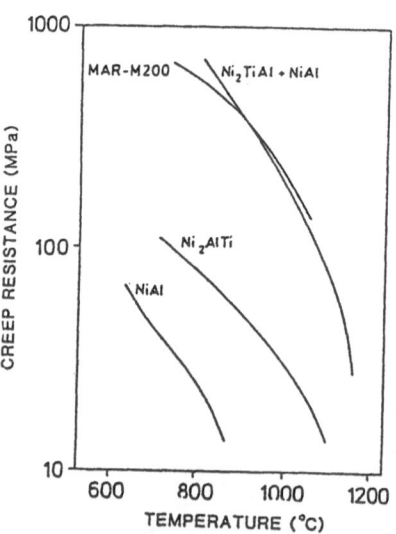

Fig. 14 Creep resistance at
the secondary creep
rate of $10^{-7}s^{-1}$ of
Ni_2AlTi and
$Ni_2AlTi+NiAl$,
compared with
MAR-M 200 super-
alloy (43)

stoichiometric compounds are
very poor (Fig. 12). The
creep behaviour may be
enhanced by substitution of
limited amounts of elements of
the second or the third series
of transition elements such as
Ta, Nb or Hf (41, 42). The
values of the steady state
creep rate compare with those
of good superalloys such as
IN 100 or René 80 (Fig. 13).
As far as creep is concerned,
the paramount values for this
family are reached for Heusler
Ni_2AlTi alloys, obtained by
further ordering the structure
(atomic diffusion is made more
difficult); the alloy has a
fine distribution of

softer NiAl particles (43), Fig. 14.

Unfortunately, due to the onset of grain boundary sliding, creep strength diminishes with reduction of grain size, which is required to increase the ductility for any of these alloys.

3. TITANIUM ALUMINIDES

The titanium-aluminium phase diagram includes a very large number of phases (1). The main ones are $TiAl_3$, a DO_{22} type structure resulting from the ordering of the fcc aluminium solid solution; Ti_3Al at the other end of the diagram, a hexagonal DO_{19} phase resulting from the ordering of the α-titanium solid solution, and in between, TiAl, a cubic phase derived like $TiAl_3$ from the fcc. Due to their high melting temperatures, high elastic moduli and high elastic limit, they are materials of great interest for use in the 400-600°C temperature interval (and even higher if the oxidation problems are mastered), with the reservation that the ductility still needs to be improved.

3.1. Ti_3Al

Pure Ti_3Al is very brittle. The reduced elastic limit is higher than those of modern Ti-α alloys or Ni-base superalloys

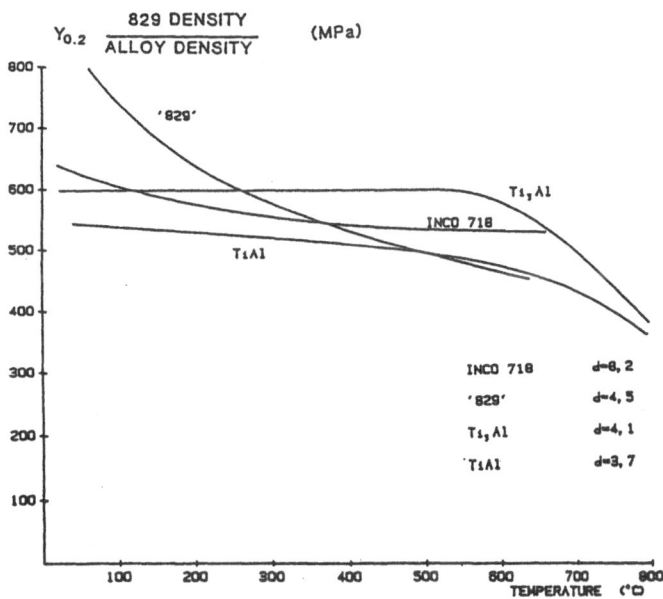

Fig. 15 Reduced elastic limit of TiAl and Ti_3Al, compared with Inco 718 superalloy and IMI 829 titanium alloy (by courtesy of A. Vassel)

which are in use for temperatures in the 300-700°C interval (44)
(Fig. 15). The elastic moduli are very high (E = 150 GPa) and
decrease very slowly with increase in temperature. Little work
has been performed to elucidate the deformation mechanisms; some
evidence has been given for climb dissociation (45), which
suggests that the strength of the alloy might be related, as for
Ni$_3$Al, to dislocations becoming sessile at intermediate and high
temperatures. Deformation rates at high temperatures are very
low, but the creep life is limited by lack of ductility.

To overcome this obstacle, ternary alloys including niobium
have been developed (46), (Fig. 16); other elements such as

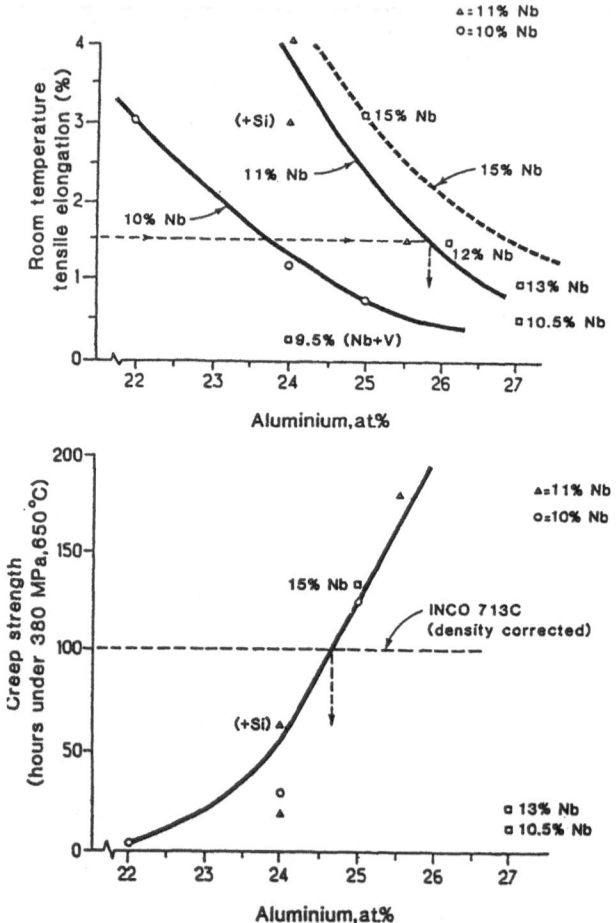

Fig. 16 Properties of (Ti$_3$Al,Nb) alloys (46)

vanadium could also be tested. The best compromise between
ductility and creep properties is achieved for alloys containing

an excess of aluminium (e.g. Ti-25.5 Al-13Nb, atomic %; rupture life at 650°C under 380 MPa ≃ 100 hours; elongation to rupture at room temperature ≃ 3 %).

The presence of a fine dispersion that cannot be sheared by dislocations is expected to improve the ductility of the alloys by causing a more uniform distribution of glide. Tests have been made of alloys containing 20 nm diameter non-coherent erbia or yttria precipitates : the results were inconclusive (47, 48).

3.2. TiAl

The concentration range for TiAl is restricted to 52-54 at. % aluminium. Above 54 at. %, very small Ti_3Al_5 particles occur (49, 50); below 52 at. % fine platelets of a Ti_2AlN phase are observed, even for very low nitrogen content. The reduced modulus of TiAl is very high, since E = 178_3 GPa and $\rho = 3.9$ g.cm^{-3} (51, 52).

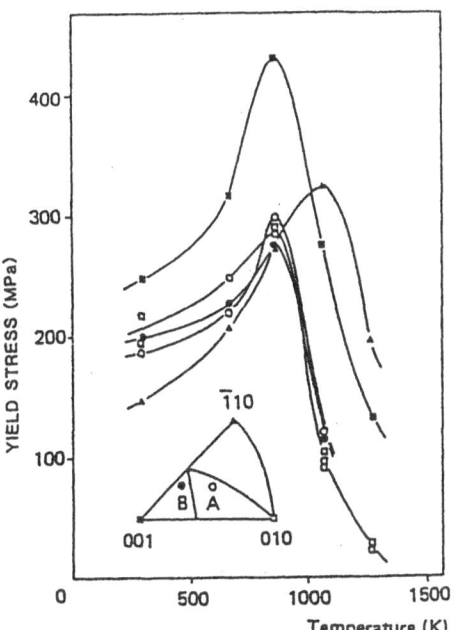

Fig. 17 Elastic limit of TiAl single crystals, as a function of orientation and temperature (53)

The elastic limit for single crystals undergoes an anomaly which is very similar to that of Ni_3Al; it increases steadily up to 700°C and then decreases with the onset of (101) glide (53) (Fig. 17).

For polycrystals (55), Fig. 18, the elastic limit is constant up to the brittle-ductile transition (700°C), and deformation occurs mainly by twinning; above 700°C glide is dominant.

Little information is available on the results of attempts to improve this alloy. Additions of Nb or W have been tried, with limited results. It might be profitable to investigate the effects on mechanical properties of extensive Ti_2AlN or Ti_3Al_5 dispersions (56).

3.3. Applications

Due to their brittle behaviour, the application of titanium aluminides is restricted in aero engines to static parts,

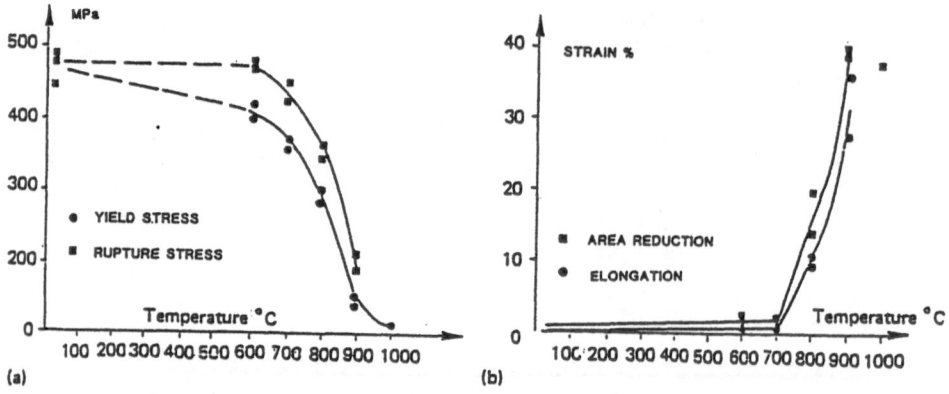

Fig. 18 Mechanical properties of polycrystalline TiAl (55)

subjected to intermediate temperatures (300-700°C), and low
stresses, such as casings or compressor vanes. For casings they
compete with medium or high temperature composites. Critical
development of these alloys is the ability to superplastic
forming.

4. CORROSION MECHANISMS

Intermetallics are extensively used as protective coatings
against high temperature corrosion and oxidation. This property
is conferred by high concentrations of aluminium at high
temperature, by chromium at more moderate temperatures, i.e. up
to 950°C (the high vapour pressure of Cr_2O_3 being a limiting
factor) or by silicon. NiAl base, or NiCrAl alloys which can in
some respects be considered as basically nickel aluminides,
although γ phase is present, are described in section 2.3.

Silicides could be an interesting alternative route to
compounds already studied. Si_3Ti_5 has a melting point of 2130°C,
and is present in eutectic rapidly solidified Ti-Si systems (58)
that could be used as protective coatings for titanium.

5. FABRICATION OF INTERMETALLICS

The limited ductility of intermetallics is the main factor
determining the choice of fabrication process. It must be
optimized both to produce the best metallurgical structure - a
fine grain size is one of the keys to better ductility - and to
allow the fabrication of components. The possibility of
superplastic forming is being investigated. Since that process
requires an ultra fine grained material it should produce tough
components. Similarly, isothermal forging is a possible way to

form intermetallics, such as Ti_3Al for example.

Powder metallurgy is extensively used. The technique is particularly well adapted if the compound undergoes the order-disorder transformation before melting. Compaction can then be performed in the temperature window where the solid is disordered. Hot isostatic pressing (HIP) can be used in this case, and also for compounds which retain order up to the melting temperature. The main drawback of HIP is that the chemically modified zone at the surface of the powder particles is not sufficiently deformed. Hot extrusion is better in this respect.

One of the limitations of powder metallurgy is the high oxygen content of the resulting alloy. The same applies also to most of the RSR routes (either atomization or melt-spinning). The Osprey techniques could be a good compromise; for Ni_3Al Baker et al. (59) have reached an oxygen content of 25 ppm; 100 ppm is more usual for atomized powders. Achieving better gas purity by improving the powder production technologies (both gas atomization and centrifugal methods such as PREP) is one of the major goals for intermetallics development.

6. CONCLUSIONS AND PROSPECTS

Ni_3Al, NiAl, their more complex derivatives and the long-range order (LRO) alloys are the intermetallic compounds that have been most extensively studied. The range of their present and potential future applications is clearly rather limited, even assuming that improvements in properties that have already been judged to be feasible are realized. They certainly cannot compete with the latest range of superalloys at least in their more demanding applications. For other intermetallics which are beginning to attract attention, the prospects are more hopeful. In particular, titanium aluminides are of interest for applications at medium temperatures (400-700°C) and low stresses; in this prospect, research leading to an improvement of the ductility and the fracture toughness is required. The development of multi-phase structures seems the most promising line to follow, to achieve this goal. New developments should avoid sacrifice of the attractive features of the pure compound, namely low density, high elastic moduli and adequate oxidation resistance. The development of multiphase alloys which combine a strong ordered intermetallic phase with a more ductile phase (which could also be an intermetallic compounds) is worth more extensive study; this combination, which has proved to be the most efficient for the conventional superalloys, has also been successful for some aluminides (Fig. 14).

Although the rewards of success would be very high, the cost of development of a commercially viable "new" material would also be high, and the chance of success is not great. It is necessary to achieve a major improvement leading to an oxidation resistant, high strength, ductile, refractory intermetallic compound (research so far indicates that this is going to be difficult). The field of research could be extended to other intermetallic systems with even higher melting temperatures, with a preference for cubic systems such as A15 or C15 alloys - acceptable ductility is more likely to be found for high symmetry structures - and also for silicides or beryllides which combine a high thermal stability and an excellent oxidation resistance. This would imply following a different track, i.e. exploiting materials which are expected to be rather brittle, at least at low temperature.

Here it is necessary to face competition from ceramics, including oxides, carbides and nitrides, some of which have both lower density and better resistance to oxidation and creep. Some ceramics also have good thermal shock resistance, but perhaps intermetallics are better in this respect. For both ceramics and intermetallics the main problem, even in components not subject to high stresses, is in avoiding catastrophic damage during the low temperature regimes of the working cycle, from ingested debris or water and/or high thermal stresses. More specifically, in the case of turbine blades, the ability of superalloys to sustain high thermal stresses has permitted the successful use of cooled blades, when the maximum metal temperature is only 950-1000°C, while the gas temperature is 1450°C. These designs are not likely to be applicable to intermetallics; they would then need to retain mechanical properties up to 1450°C to be worth consideration.

7. ACKNOWLEDGEMENTS

The author wishes to thank Drs. T. Khan, A. Lasalmonie, A. Loiseau, S. Naka, D. Pope, J.F. Stohr, A. Vassel and P. Veyssière for valuable discussions and comments.

8. REFERENCES

(A more extensive bibliography could be obtained using Ref. 2, 34 and the various contributions to the Boston Symposium on High Temperature Ordered Intermetallic Alloys (MRS Symp. Proc. Vol. 39, Koch, Stoloff & Liu eds. (1985)).

(1) Loiseau, A., Lasalmonie, A., Van Tendeloo, G., Van Landuyt, J. & Amelinckx, S., Int. Conf.on Titanium, Munich, pp. 1467 (1984).

(2) Pope, D.P. & Ezz, S.S., International Metal Reviews, 29, pp. 136 (1984).

(3) Kear, B.H. & Wilsdorf, H.G.F., Trans. AIME, 224, pp. 382 (1962).

(4) Mishima, Y., Oya, Y. & Susuki, T., Trans ISIJ, 25, pp. 1171 (1986).

(5) Beauchamp, P., Douin, J. & Veyssière, P., Phil. Mag. A 55, pp. 565 (1987).

(6) Lasalmonie, A., Chenal, B., Hug, G. & Beauchamp, P., Phil. Mag. A 56 (1988), under press.

(7) Yoo, M.H., Scripta Met., 20, pp. 915 (1986).

(8) Douin, J., Veyssière, P. & Beauchamp, P., Phil. Mag. A 54, pp. 375 (1986).

(9) Veyssière, P., Phil. Mag. A 50, pp. 189 (1984).

(10) Veyssière, P., Horton, J.A., Yoo, M.H. & Liu, C.T., Phil. Mag. Letters, to be published (Jan. 1988).

(11) Noguchi, O., Oya, Y. & Susuki, T., Met. Trans. A, 12A, pp. 1647 (1981).

(12) Mishima, Y., Ochiai, S., Hamao, N., Yodogawa, M., & Susuki, T., Trans. JIM, 27, pp. 648 (1986).

(13) Mishima, Y., Ochiai, S., Yodogawa, M., & Susuki, T., Trans. JIM, 27, pp. 41 (1986).

(14) General Electric Co., US Patents 4606888, 4609528, 4613368 and 4613480.

(15) Inoue, A., Tomioka, H., Masumoto, T., Met. Trans. 14A, pp. 1367 (1983).

(16) Weihs, T.P., Zinoviev, V., Viens, D.V. & Schulson, E.M., Acta Met., 35, pp. 1109 (1987).

(17) Taub, A.I. & Briant, C., Ordered Intermetallic Alloys, Eds. Stoloff, N.S., Koch, C.L., Liu, C.T. and Izumi, O., Mat. Res. Soc., Pittsburgh (1987).

(18) Hanada, S., Watanabe, S. & Izumi, O., J. Mat. Sc., 21, pp. 203 (1986).

(19) Takasugi, T., Izumi, O. & Masahashi, N., Acta Met., 33, pp. 1259 (1985).

(20) Lasalmonie, A., Private communication.

(21) Cahn, R.W., Mat. Res. Soc. Symp. Proc., 81, pp. 27 (1987).

(22) Liu, C.T., International Metal Reviews, 29, pp. 168 (1984).

(23) Lasalmonie, A., Private communication.

(24) Miller, M.K. & Horton, J.A., Scripta Met., 20, pp. 79 (1986).

(25) Liu, C.T. & Sikka, V.L., Journal of Metals, 38, pp. 5, 19 (1986).

(26) Structural uses for ductile ordered alloys, NMAB-419, Nat. Acad. Press, pp. 74 (1984).

(27) "High Temperature Alloys Theory and Design", Liu, C.T., Ed. Stiegler, J.O., AIME, pp. 289 (1984).

(28) "High Temperature Ordered Intermetallic Alloys", Liu, C.T. & White, C.L., Symp. MRS, pp. 365 (1985).

(29) "Creep Properties of Ni_3Al-based Intermetallic Alloys", Sikka, V.K. & Weir, J.R., Jr., ORNL internal report, Feb. 1987.

(30) Fuchs, G.E. & Stoloff, N.S., Scripta Met., 21, pp. 863 (1987).

(31) Kuruvilla, A.K. & Stoloff, N.S., Scripta Met., 21, pp. 873 (1987).

(32) Bonda, N.R., Laird, C. & Pope, D.P., Acta Met., 35, pp. 2371 (1987).

(33) Ashok, S., Kain, K., Tartaglia, J.M. & Stoloff, N.S., Met. Trans., 14A, pp. 1997 (1983).

(34) Stoloff, N.S., Int. Met. Reviews, 29, pp. 123 (1984).

(35) Schulson, E.M. & Parker, D.M., Scripta Met., 17, pp. 519 (1983) and 19, pp. 1497 (1985).

(36) Gaydosh, D.J., Jech, R.W. & Titsan, R.H., J. Mat. Sc. Lett., 4, pp. 138 (1985).

(37) Pascoe, R.T. & Newey, C.W.A., Met. Sci. Jour., 5, pp. 50 (1971).

(38) Lequeux, M.J., Thesis, ONERA publication 1980-3.

(39) West, G., Phil. Mag., 9, pp. 979 (1964).

(40) Sauthoff, G., Z. Metallkunde, 77, pp. 1171 (1986).

(41) Vedula, K. & Stephens, J.R., M.P.R. 2, pp. 84 (1987).

(42) Vedula, K. Anderson, G., Pathase, V. & Aslanicks, I., Proc. Int. Powder Met. Conf., Toronto (1984).

(43) Polvani, R.S., Tzeng, W.S. & Strutt, P.R., Metall. Trans. 7A, pp. 33 (1976).

(44) Sastry, S.M.L. & Lipsitt, H.A., Metall. Trans. 8A, pp. 1543 (1977).

(45) Thomas, M., Vassel, A. & Vevssière, P., Scripta Met., 31, pp. 501 (1987).

(46) Blackburn, M.J. & Smith, M.P., Patent nr. 80/13485.

(47) Rowe, R.G. & Sutliff, J.A., Proc. Mat. Res. Soc. Symp., Boston (1985).

(48) "Designing with Titanium", Naka, S., Bristol (1986).

(49) Miida, R., Hashimoto, S., Watanabe, D., Jap. J. Ap. Phys. 21, pp. 159 (1982).

(50) Loiseau, A., Lasalmonie, A., Van Tendeloo, G., Van Landuyt, J. & Amelinckx, S., Acta Cryst. B41, pp. 411 (1985).

(51) Loiseau, A., Lasalmonie, Mat. Sc. and Eng., 67, pp. 163 (1984).

(52) Kaufman, M.J., Konitzer, D.G., Shull, R.D. & Fraser, H.L., Scripta Met. 20, pp. 103 (1986).

(53) Kawabata, T., Kanai, T. & Izumi, O., Acta Met. 33, pp. 1355 (1985).

(54) Hug. G., Loiseau, A. & Veyssière, P., Phil. Mag., under press.

(55) Sastry, S.M.L., Lipsitt, H.A., Proc. 4th Int. Conf. on Titanium, pp. 1231 (1980).

(56) Lasalmonie, A., Private communication.

(57) Tomaszewicz, P. & Wallwork, G.R., Rev. High Temp. Mat. 4, pp. 75 (1984).

(58) "Titanium Rapid Solidification Technology", Chumbley, L.S., Ohls, M.A. & Fraser, H.L., Eds. Froes, F.H. & Eylon, D., The Metallurgical Society Pennsylvania, pp. 211 (1986).

(59) Baker, I., Ichishita, F.S., Surprenant, V.A. & Schulson, E.M., Metallography, 17, pp. 299 (1984).

2.2. Coatings

2.2.1. Claddings and Co-Extruded Tubes

T. Flatley & C.W. Morris

Central Electricity Generating Board

Leeds, England

1.   INTRODUCTION

Claddings are generally used to provide corrosion protection
and are widely available for sheet, or plate and tube products.
In the case of sheet or plate products, the major advantage is
one of economics where a cheap substrate of adequate strength is
substituted for the generally more expensive corrosion (or in
some cases erosion) resistant material. Such plate products are
produced by either hot or cold rolling processes. These products
are generally utilised at relatively low temperature and where
there is no pressure containment requirement. Mechanically
bonded bi-metallic (or multiplex) tubes can be produced by cold
drawing two or more separate but concentric tubes over a die to
form a good mechanical fit. Such cold clad tubes have a tendency
for layer separation if high temperatures are met, due to stress
relaxation and are generally of limited application at
temperatures below 250°C where good heat transfer is not
necessary for the process requirement.

High temperature corrosion (e.g. fuel ash corrosion, hot
corrosion, oxidation or combinations thereof) can lead to
premature failures of boiler tubes and expensive plant shutdowns.
The long term solution to such problems usually requires the use
of more corrosion resistant materials as the most cost effective
solution, often in the form of claddings. It is in the area of
high temperature components, particularly when there is a
pressure containment requirement, that the technology and
technical requirements for corrosion resistant claddings become
far more complex. The tube products in some applications operate
at elevated temperatures in the proof stress regime (i.e. above

400°C) and in other cases the creep regime at temperatures up to
650°C. In many cases anisothermal operation is the norm, with
heat fluxes approaching 500kW m^{-2} being encountered. Such
adverse operating conditions have led to the development of a
small range of metallurgically bonded co-extruded (King, Robinson
1973; Flatley, Latham, Morris 1981) and multi-plexed tubes
(Bennet et al 1986) (Table 1) which can meet the exacting
requirements of high temperature and pressure operation. Whilst
the range of materials has been relatively limited (i.e.
predominently ferritic/austenitic steels and Ni/Cr alloys), some
combinations have had extensive usage, particularly in the power
generation industry e.g. TP310 (modified) on either an
Esshette 1250 steel inner or a carbon steel inner. (Figure 1)

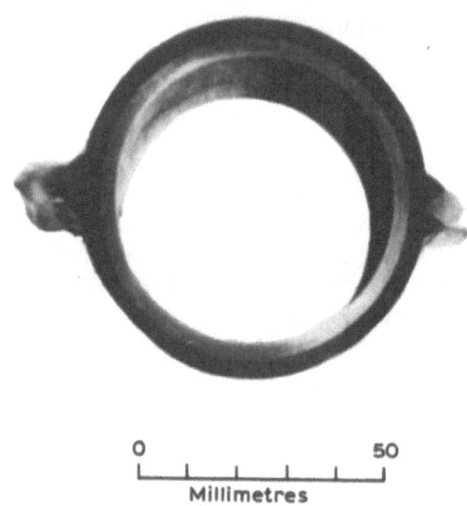

FIGURE 1: Furnace wall tube: Inner carbon steel 440
Outer TP310, Weld TP309

2. STATUS OF CO-EXTRUDED TUBE MANUFACTURE

The use of a double wall co-extruded tube has been developed
for the power generation (Laxton et al 1988) and chemical process
industries to protect tube steels from increasingly aggressive
corrosion environments (Flatley, Thursfield 1984). The bi-
metallic tubes are formed by hot extrusion of a dual composite
billet to produce a strong metallurgically bonded tube having all
the mechanical characteristics of a simple integral tube. Such
bi-metallic tubes therefore permit the joint optimisation of
corrosion and mechanical properties not previously obtainable
from a single alloy monobloc tube.

Table 1: Summary of Co-extruded Tube Experience

Application	Location	Installation Period	Outer Material	Inner Material	Overall od Size Range (mm)	Overall Wall Size Range (mm)	Quantity (m)
Coal Fired Power Utility Boilers	Superheater/ Reheater	1969-1986	Tp310 or Tp310M+Nb	Steel 1250	33-65	3.2-9	300,000
	Superheater	1979	12RE10	12R72	44-54	6.4-8.3	1,300
	Superheater	1985	50Cr50Ni	Steel 1250	42	5.4	16 (Trial)
	Reheater	1974-1982	50Cr50Ni	In 800H	54	5.95-9.2	3,000+? in USA
	Furnace Walls	1974-1986	TP304 or TP310M or TP310+Nb	Carbon Steel 440	50-76	6-8.3	30,000
	Furnace Walls	1982	13Cr4Al	Carbon Steel 440	76.2	7	30 (Trial)
	Furnace Walls	1981	TP446	Carbon Steel 440	68.5	7	38 (Trial)
	Furnace Walls	1981	50Cr 50Ni	Carbon Steel 440	76.2	8.3 (Trial)	20
	Furnace Walls (Wallblower ports)	1982	12CrMoV	Carbon Steel 440	76.2	8.3	20 (Trial)
Black Liquor Recovery Boilers	Furnace Walls	1969 on	Tp304	Carbon Steel SA210 Grade A1	50.8-76	5.6-7.2	Estimated as 250,000
Ammonia Reforming Plant	Quench Coolers	1973/4	15Mo3 Carbon Alloy	Alloy 600	38-42	4.9-6.5	8,000
PWR Nuclear Plant	Outlet Nozzles	1980-1981	Tp304L	Carbon Steel	111	21	300
Combined Cycle Power Plant	Gasifiers	1984	NAC-CR35A Tp310+Nb	ASTM A213 T11	-	-	-
Supercritical Coal Fired Units	Superheaters	1984	SUS310S	17/14CRMo	-	-	-
	Superheaters	1984	NAC-CR35A	17/14CRMo	-	-	-
Laboratory Trials	-	1986	Feralloy -2 1/4Cr- Fecralloy Triplex		28	5	Trial 3m lengths
			Feralloy -Tp 321- Fecralloy Triplex				
			Feralloy -800H- Fecralloy Triplex				

2.1. Metallurgical Bonding

The achievement of a fully bonded bi-metallic tube is assured by the direct hot extrusion of a large diameter (~200mm) composite billet at temperatures of around 1200°C. As the extruding billet is grossly reduced in section (e.g. factors of 10:1 on cross sectional area reduction), the reduction pressures and temperatures create a "pressure forged" welding type effect at the bi-metallic interface, promoting rapid interdiffusion and extensive bonding between the two layers.

Typical relative tube layer thicknesses of 1:1 to 1:4 are commonly produced but ratios greater than 1:5 are feasible. The thinner component of these co-extruded tubes can be either the inner or outer layer dependent on the industrial application required.

2.2. Co-extrusion Process

Direct extrusion is by far the most satisfactory process for producing seamless tubes in stainless steel and high nickel alloys. Unlike rotary piercing where the billet at some stage is subject to severe tensile forces, the forces involved in extrusion are compressive throughout. This aspect can be particularly advantageous with higher alloyed materials that normally have limited hot workability and low ductility. Bi-metallic tubes are produced by the co-extrusion of a duplex billet, i.e. the solid (inner layer) forged billet is placed centrally inside a tight fitting cylinder of the outer layer material. The well established Sejournet hot extrusion process and Pilger cold reduction plant are used, identical to that used for cold finished stainless steel tubes.

2.3. Composite Billet Production

Production of the outer layer billet component has resulted in a choice of three different manufacturing routes, all with the same purpose of producing a duplex composite billet prior to extrusion. A prime objective is to produce a large cylindrical billet outer with the minimum of metal wastage. Two of the three routes make use of a solid metal outer billet cylinder whilst the other involves a compacted metallic powder outer layer composite. The solid outer layer cylinders are produced either by hot piercing and up-extrusion (pre-lifting) of a forged billet to form an enlarged hole, or by direct centrifugal spin casting to the outer layer dimensions.

Irrespective of the method chosen to produce the pre-extrusion composite billet, all three manufacturing routes provide a sound metallurgically bonded co-extruded product. The

majority of industrial experience, however, is with co-extruded
tube produced by the forged inner/solid wrought outer billet
method. The powder outer route requires an additional
compaction/sintering process and control of concentricity to
ensure uniform outer layer thickness. This is more demanding
than with outer layers from a forged/machined route.

2.4. Quality Control

Quality control has been rigidly applied in terms of 100%
ultrasonic testing of the bond between the inner and outer layer
components. Techniques have been developed for routinely
detecting lack of bond defects, currently as small as 12mm^2 on
all co-extruded tubes that are manufactured for the power
generation industry in the UK (CEGB Standard 680710, 1987). Co-
extruded tubes meet the requirements of mechanical testing given
in BS3059, BS1113 and ASTM A450.

2.5. Welding

The welding requirements for co-extruded tubes have been
embodied in CEGB Standard 680710 (1987) which covers Weld
Procedure and Welder Approval, Attachment Welds, Welding
Electrodes and Filler Wire, Weld Acceptance Standard, Weld
Repairs, Post-weld Heat Treatment and Welding Equipment. Only
metal arc and inert gas shielded welding are used. These
procedures have been adopted for combinations of ferritic,
austenitic and nickel alloy. In general standard vee-
preparations have been used, though in some instances a stepped
profile (outer layer machined back to the interface) has been
employed. For co-extruded to co-extruded butt welds, power
industry experience has generally been with TIG root and MMA
matching filler for the inner layer. The outer layer has been
welded with MMA fillers of the matching corrosion resistant out
layer, e.g. for TP310 outer layers a TP309 filler has been used
and for IN 671 an IN 72 filler (Flatley, Latham, Morris 1987).
For co-extruded to monobloc butt welds the outer corrosion
resistant outer fill has been omitted, the fill being with the
same matching inner layer material. One minor welding problem
which has occurred (in-works) was hot shortness associated with
TIG welding of an Nb stabilised 310 outer layer; however, this
was overcome by specifying an MMA weld with a controlled, lower
heat in-put. Power generation industry practice has been to
subject all shop and site welds to non-destructive examination.
In general butt welds have been inspected by 100 % radiography,
no doubt contributing to the very low incidence of failures (5 in
about 70,000 welds in the CEGB plant, some of these not being
attributable to welding defects). In some cases tubes have been
butter-weld repaired but this has always followed weld procedure
trails. No problems have occurred.

The general philosophy for welding co-extruded tubes has been to adopt existing practices and weld consumables. The only change has been to adopt a composite two stage welding process to provide both the load bearing inner component and corrosion resistant outer layer. This conservative approach to welding has proved to be highly successful.

2.6. State of the Art

The stage has now been reached whereby some combinations of clad metallurgically bonded co-extruded tubes can be manufactured routinely in large quantities (e.g. 25,000 metres), with demonstrable cost benefits to the user. (Figures 2, 3) For co-extruded combinations of TP310 - mild steel, TP310 - Esshette 1250 and 50 Cr 50 Ni - Alloy 800, standards which cover design, tube manufacture and quality control, and tube element fabrication, site installation and quality control, have been developed (Fellows 1976; CEGB Standard 680710, 1987).

3. CRITICAL PROPERTIES

All high temperature materials selected for co-extruded tube combinations must be thermally compatible in terms of their heat treatment characteristics, heat transfer and thermal expansion behaviour. Although in theory most combinations of transition metal alloys of similar melting range and hot flow strength, can be co-extruded, in practice it is important to consider the requirements for achieving a sound metallurgical bond without melting occurring at the interface during the extrusion process. Control of composite billet interfacial fit, clean contamination free conditions and minimisation of interfacial oxidation during billet heating are important but achievable factors. Good bonding is critically dependent on the very short term high temperature/high pressure solid state inter-diffusion period available during the rapid co-extrusion process. A continuous metallurgical bond is essential particularly with high heat fluxes as any defects may result in layer separation by thermal fatigue.

3.1. Material Property Requirements

Specific physical property attributes required by materials selected for co-extruded tubes include:
* Corrosion (or erosion) resistance for internal and external environments.

* Proven high temperature strength for either the proof or creep regimes of at least one of the layers.

* Thermal conductivity of the composite to be unimpaired by any

FIGURE 2: Wrapper tubes, lugs and attachment on TP310 – Esshete 1250 superheater pendant. Lug attachment: TP310. Welds: TP309 for attachment.

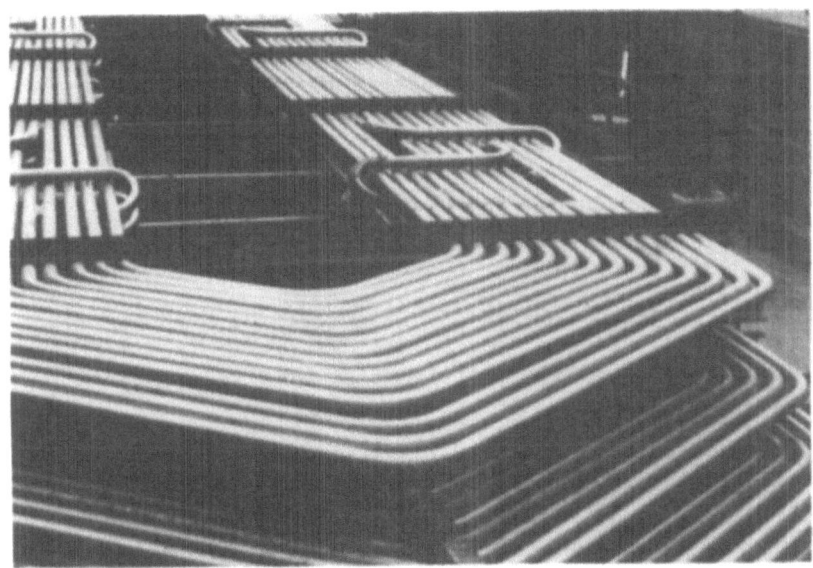

FIGURE 3: Superheater pendant assembled for installation. Material: TP 310 – Esshete 1250. Tube butt welds: Essweld (equivalent to E 1250) and TP309.

small interfacial bond defects, and should meet the industrial process heat transfer requirements.

* Thermodynamic phase stability or compatibility for manufacture, heat treatment and plant service.

* Comparable thermal expansion characteristics to minimise the need for stress relaxation and yield in service.

* Adequate hot or cold ductility for tube manufacture and fabrication.

* Load bearing components should be weldable using established technology.

3.2. Materials Property Shortfalls

During the 20 year development and service application of co-extruded tubes, a number of materials property shortfalls have been identified, some being of more or less importance, depending upon the industrial application, i.e. power generation as opposed to chemical process industries (e.g. black liquor recovery boilers in Scandinavia and North America). The more important limitations are listed below.

* Inadequate longer term corrosion performance of the more ductile higher chromium austenitic stainless steels (e.g. 25 Cr) when operating in the more extreme environments (temperature, corrosivity).

* Inability to maintain and/or regrow a stable and protective Cr_2O_3 - type oxide at temperatures below 600°C with 25 Cr austenitic steels.

* Low ductility and limited fabricability of high chromium nickel alloys (e.g. 50 Cr 50 Ni) despite their exceptional and established corrosion resistance at high temperatures.

* Poor product yield (i.e. length) of the hot extruded composite hollow because of increased end cropping to ensure removal of the characteristic lack of bond zone at both ends of the tube hollow.

* Lack of design criteria for load bearing capability of components welded to the tube outer layer (i.e. lug attachments, membrane fins to form tube panels): Thermal fatigue of stainless membranes welded to stainless TP304-carbon steel tubes in Scandinavian black liquor recovery boilers has previously caused problems (Egnell, Tornblom 1973; Odelstam 1983).

* There will be creep limitations to the new designs of advanced steam generating plant operating at higher steam temperatures of greater than 565°C and high pressures (greater than 17 MPa).

  Higher creep strengths are required to obviate impractical tube dimensions (i.e. excessive wall thickness) in terms of manufacture, heat transfer and weight loadings.

* Limited data on models to predict longterm (e.g. > 100kh service) ageing effects, to quantify design and service life assessments (e.g. interdiffusion across bond, bulk micro-structural changes, ageing of the less ductile welds).

4. FUTURE DEVELOPMENTS

Co-extruded tubes are now a well established commercial product, about 500 kilometres being installed in industrial plant mostly in the UK power industry and in Scandinavian black liquor recovery boilers. Some installations have now seen almost 20 years service (Flatley, Latham, Morris 1987). The next stages in the development of this product can be grouped into four main categories:

a) Billet manufacture & optimisation of composite billet route.

b) Further alloy development for either corrosion or creep improvement for advanced plant cycles.

c) Mechanical property characterisation and long term ageing assessments.

d) Assessment of the need for interfacial layers to prevent inter-diffusion.

4.1. Billet Manufacture

Product yields should be optimised by considering changes in the composite billet manufacturing route using the available technologies listed below:

* Forged billet outer/forged billet inner - most popular method to date.

* Centricast outer/forged billet inner - second most popular method.

* Power outer/forged billet inner - attractive alternative for metallurgically difficult alloys but can result in problems with concentricity and outer layer thickness control.

* Hot metal sprayed outer/forged billet inner – commercially available process which may be applicable to tube hollow as well as billet.

* Centricast outer/hot isostatic press/forged billet inner – may improve bonding and yield but the long high temperature dwell (of normal manufacturing) can bring about unwanted phase changes and inter-diffusion.

* Powder outer/hot isostatic press/forged billet inner – may improve bonding and yield, and overcome concentricity problems.

It is also possible to consider the same technologies to produce the inner layer billet but the formation of an initial cast structure may significantly alter mechanical properties compared to normal wrought tubes and hence not meet ISO standards for design purposes.

4.2. Alloy Development

a) Corrosion Performance

High temperature corrosion performance in coal and oil fired combustion environments requires the presence of chromium at concentrations greater than 25 weight % (Brook, Meadowcroft, Morris, Flatley 1982; Latham, Flatley, Morris 1982). In stainless steels and Cr-Ni alloys, this is generally accompanied by a decrease in ductility, particularly if alpha-chromium is a favoured phase. Practical experience shows that 50 Cr 50 Ni is an excellent high temperature corrosion resistant alloy, particularly for the power industry but suffers from problems of low ductility and limited fabricability during manufacture. An ideal compromise is to aim for a compositional specification which maximises chromium content but minimises the potential for alpha-chromium formation. Alloys with 30 % – 40 % Cr and 30 % – 50 % Ni, balance iron are single phase (Nash 1986) and worthy of further development and investigation to:

i) Assess corrosion performance in plant, e.g. on air/ steam cooled corrosion probes or as boiler tube inserts.
ii) Assess the methods and economics of manufacture in the form of a co-extruded tube.

For less aggressive environments, TP 310 is well established and lives now exceed the design requirement; however, TP 309 or TP 446 may be economically more attractive alternatives if production routes can be developed and their corrosion performance established. Of the single (or near single) phase alloys currently being investigated in Japan, NAC CR35A (35 Cr 45 Ni Fe Nb) (Tamura et al 1985) is receiving considerable interest.

b) Load Bearing Capability/Creep Properties

The move to higher temperature and pressure process cycles (e.g. 590-620°C steam at 31 MPa) to achieve improved plant efficiencies during the next decade, will require materials with increased creep strength, significantly better than the 300 series austenitic steels (Townsend 1987) Potential candidates include an improved Esshette 1250 (either heat treatment, micro alloying additions, or cold work to aid precipitation). Sandvik 12R72 is potentially a useful load bearing tube alloy steel having an extensive long term creep rupture database (and 20 year plant trial data as co-extruded tubes) which demonstrates long term strength and stability. However, there appears to be much less data on like to like butt welds at desirous stresses applicable to steam raising plant. Such additional data needs establishing in view of the promising 90kh of this material as trial straight lengths in UK plant which started in 1969 (King, Robinson 1973).

c) Experience Resistance

Experience within the power generation industry shows that sootblower ash erosion is one of the major causes of tube failures. Alloys can be selected with a combination of erosion/corrosion resistance suitable for development into co-extruded tubes if necessary. It is worthy that consideration is given to this important problem area. The same concepts for inner/outer layer materials property selection will apply to both corrosion and erosion resistant materials.

4.3. Long Term Mechanical Property & Ageing Characterisation

To underwrite long term installations and maximise economic benefits, it is necessary to understand and quantify long term thermal degradation behaviour of the co-extruded tubes. Existing plants, particularly power generation are now expected to operate to between 250 kh and 300 kh. New plants are likely to be designed for >200 kh. Thermal ageing characteristics are therefore required for:

* The bulk alloys

* The welds

* The inner - outer layer metallurgical bond

* This will require mechanical testing of new and aged materials (e.g. bond shear tests, creep rupture testing, thermal fatigue assessments) and the microstructural characterisation of the processes involved. These findings will assist in revision of the established design and assessment codes such as CEGB Standard 680710 (1987) (also described by Asbury and Brooks 1987).

4.4. Interfacial Layers

European practice has not found the need to employ a thin nickel interlayer at the co-extruded tube inner-outer layer interface (Faulkner, Bridges 1973). This contrasts to Russion and Japanese thinking which incorporates this layer. It needs to be established whether a nickel layer would aid bonding during the composite billet extrusion process and thereby improve product yield. There may also be some benefit in minimising the interfacial transfer of carbon and other mobile elements during manufacture and long term service.

5. FUTURE RESEARCH AND DEVELOPMENT

Current R&D on co-extruded tubing is now centred in Japan where there is extensive research into materials for the boilers of advanced design in the power generation industry. Particular attention is being paid to developing improved creep resistant materials for inner layers and to assessing high chromium nickel type alloys for outer layer corrosion protection. These developments appear to be based on the successful usage of co-extruded tubes within the UK power generation industry. Particular attention in Japan is being given to the development and assessment of high chromium nickel alloys which are single phase. In the United States there would appear to be little interest currently in the development of new combinations of co-extruded tubing despite their successful large scale production of IN671/Alloy 800 in the 1970's (Rahoi 1974). Materials which are currently on trial in the USA have been obtained from either the UK or from Japan. To date Japanese experience with co-extruded tubes has been limited to trial sections within boilers but no doubt within the next few years they will have extensive service experience.

6. CONCLUSION

The next major steps in the successful development of co-extruded tube materials in Europe will be driven by the requirement of advanced cycle plant and will require:

* The optimisation of the composite billet production route for selected combinations of alloys to give better value for money by improving yield and thereby unit costs.

* The corrosion assessment, including plant trials, of improved alloys with chromium contents in the range 30 % - 40 % and nickel of 30 % - 50 % balance iron.

* The determination of long term degradation modes of various materials combinations and the validation of two stage composite welding processes selected from monobloc tube welding technology.

* An additional useful policy development would be the establishment of potential new applications of co-extruded tubes over a wider range of industrial processes. A wider study would identify further materials combinations with potential for commercial exploitation but requiring further R&D prior to large scale applications.

7. REFERENCES

Asbury F.E., Brooks, E.,
"Superheater Tubes for Advanced Coal-fired power plant in the UK"
ASM International Conference on Advances in Materials and
Technology for Fossil Power Plants, Chicago - September 1987

Bennett, M.J., Hudson, Robert G.C., Scott, K.T.,
Shrimpton, G.R.D., Symonds, A.E., Tuson, A.T.
"The protection of Conventional Substrate Alloy in Aggressive
Environments by Fe Cr AlY Steel Co-extruded Overlayers"
UKAE Harwell Report, ARE R 12395, October 1986

CEGB
"Co-extruded Tubes"
CEGB Standard 680710 (1987)
Published by Reprographic Services, CEGB, Sudbury House, London

Egnell, L., Tornblom, H.
"Thermal Fatigue in Composite Steel Tubes"
Industrial Conference on Creep and Fatigue in Elevated
Temperature Applications - September 1973, Philladelphia

Faulkner, R.G., Bridges, P.J.
"Compatability Studies of Alloys for Use in Composite Tubes at
High Temperature"
J. Inst. Metals 101, pp. 103-107, 1973

Fellows, K.G.
"The Design Properties of Co-extruded Tube"
Proceedings of Seminar by Institution of Corrosion Scient and
Technology & CEGB at Harrogate, November 1976

Flatley, T., Morris, C.W., Latham E.P.
"Co-extruded tubes improve resistance to fuel ash corrosion in UK
Utility Boilers"
Materials Performance 20(5), 1981

Flatley, T., Thursfield, T.
"Review of Corrosion Resistant Co-extruded Tube Developments for
Power Boilers"
1984 ASM Conference on Coatings and Bi-metallics, South Caroline
- November 1984

Flatley, T., Latham, E.P., Morris, C.W.
"CEGB Experience with Co-extruded Tubes for Superheated &
Evaporative Sections of PF Fired Boilers"
ASM International Conference on Advances in Materials Technology
for Fossil Power Plants, Chicago - September 1987

King, C.W., Robinson, M.T.
"Development and use of composite tubes for coal fired boilers"
5th European Congress on Corrosion, Paris (1973)

Latham, E.P., Flatley, T., Morris, C.W.
"Comparative Performance of Superheated Steam Tube Materials in
Pulverised Fuel Fired Plant Environments - Corrosion Resistant
Materials for Coal Conversion Systems - Edited by
Meadowcroft, D.B. and Manning, M.I.

Laxton, J.W., Meadowcroft, D.B., Clarke, F., Flatley, T.,
King, C.W., Morris, C.W.
"The Control of Fireside Corrosion in Power Station Boilers"
Published by CEGB 1988

Nash, P.
Binary Alloy Phase Diagrams Volume 1
Publisher Willam W. Scott Jr
ASM 1986

Odelstam, T.
"Performance of Composite Furnace Tubes in Black Liquor Recovery Boilers"
Presented at 4th International Symposium on Corrosion in the Paper Pulp Industry - Stockholm June 1983

Tamure, M., Yamanouchi, N., Tanimura, M., Murase, S.
"Promising Alloys for Heat Exchangers of Advanced Coal Fired Boilers"
Industrial Heat Exchanger Technology Conference ASM Pittsburgh 1985

Townsend, R.D.
"CEGB Experience and UK Developments in Materials for Advanced Plant"
ASM International Conference on Advances in Materials Technology for Fossil Power Plants, Chicago - September 1987

Rahoi, D.W.
"Controlling Fuel Ash Corrosion with Nickel Alloys"
Metals Engineering Quarterly - February 1974

8 ACKNOWLEDGEMENTS

This paper is published by permission of the Central Electricity Generating Board. The views expressed are those of the authors and do not necessarily reflect the views of the CEGB.

2.2.2. Coatings, Metallic and Ceramic (Thermal Barrier)

for Turbine Applications

R. Mevrel

ONERA

B.P. 72, F-92322 Châtillon Cedex

1. INTRODUCTION

High temperature alloys used in hot sections of gas turbines are generally designed for specific mechanical and thermal properties : creep, mechanical or thermal fatigue, etc. Their use in the aggressive environment of combustion gases often requires the presence of a protective coating, as their chemical compositions do not offer a sufficient intrinsic resistance against high temperature corrosion.

A protective coating must meet several criteria :
- give adequate environmental resistance,
- be chemically and mechanically compatible with the substrate,
- be applicable.

2. MAIN TYPES OF COATINGS (1)

Two types of coating systems are currently used and developed : metallic and ceramic coatings (thermal barrier coatings).

2.1. Metallic Coatings

They form a protective oxide scale by interaction with the oxidizing environment. The actual protection is conferred by this scale which isolates the metallic component from the gas. This scale must therefore fulfill certain conditions :
- be stable, dense and compact,
- be adherent in thermal cycling conditions,
- grow slowly.

Only few oxides can be envisaged : Cr_2O_3, SiO_2, Al_2O_3. Cr_2O_3 and SiO_2 are limited to service temperatures below 900°C, the former because it forms volatile suboxides at higher temperatures, the latter because potentially deleterious diffusional effects can occur between coating and substrate. Al_2O_3 remains the most adequate oxide and its adherence can be enhanced by the incorporation of "active elements" such as yttrium in the coating alloy.

These considerations explain the development of the main types of metallic coatings :
- aluminides
- modified aluminides
- MCrAlY (M = Ni and/or Co and/or Fe).

Aluminide coatings (2) are still the most widely used to protect nickel- and cobalt-based superalloys. They are generally formed by pack cementation which is a simple technique : the components are immersed in a powder mixture in a semi-sealed retort. The retort is then placed inside a furnace and heated at high temperature (between 750°C and 1150°C) under a protective atmosphere. In the case of an aluminizing process, aluminium is transferred from the pack to the substrate surface where it reacts and an aluminide layer forms by diffusion with substrate. It is therefore a simple process, easily controlled, non-line-of-sight and inexpensive to run. These advantages justify its wide use for protecting superalloys. But due to its mode of formation, it suffers from several limitations : the range of accessible coating compositions and structures is rather narrow - only one element in practice can be transferred in gas phase from the pack - and in some applications it is necessary to enrich the substrate surface not only with aluminium but also with chromium, yttrium, etc. Moreover, the composition and the structure of the substrate alloy can seriously affect the performances of the coating.

Modified aluminides have been developed to overcome some of these limitations. Prior to the aluminizing process, the substrate surface is modified by a thermochemical treatment or a deposit. The most widely used duplex processes consist or chromizing + aluminizing and platinum deposit + aluminizing. The coatings thus obtained are significantly more resistant against hot corrosion in presence of sodium sulphate deposits.

MCrAlY (M = Ni and/or Co and/or Fe) alloys whose compositions and structures are entirely independent from the substrate can now be deposited due to the development of techniques such as EB-PVD (electron beam evaporation) and plasma spraying under controlled atmosphere. As a consequence, specifically tailored coatings can now be applied. The industrial development of EB-PVD for MCrAlY coatings has been restricted mainly to the United States. Plasma spraying processes under reduced pressure or inert atmosphere have been developed for the same application throughout the world in the last ten years; they are now commercially available in Europe. MCrAlY overlay coatings provide a significant improvement in high temperature corrosion resistance compared with conventional aluminides. However, despite the flexibility they permit, these techniques remain line-of-sight processes and, in addition to their relatively high cost, this can be a real drawback for coating components having complex shapes.

2.2. Thermal Barrier Coatings

Thermal barrier coatings are generally constituted of an oxide layer (thickness : about 300 µm in turbine applications, up to about 2 mm in diesel engines) plasma sprayed on top of an MCrAlY bond layer deposited on the substrate. Most often the oxide is zirconia-based, whether partially or completely, stabilized with proper additions of Y_2O_3, CaO or MgO. The benefits of thermal barrier coatings are multiple : higher inlet gas temperatures with the same metal temperature, lower cooling flow within the metallic component, reduction of temperature transients on the metal surface. All these advantages can result in improved efficiency, extended lifetimes and simplified designs. Such ceramic coatings have been applied on sheet metal combustor components for more than 15 years. Only recently (3) have they been used in the turbine section of commercial gas turbine engines (certification in the United States of thermal barrier coatings over the entire airfoil of PW4000 turbine vanes in 1986). This progress could be achieved by a proper control of porosity, microcrack distribution, and residual stresses in the coating.

3. RECENT TRENDS AND SHORTCOMINGS

The recent trends in the coating field are mainly related to the search for higher temperatures and, in the case of industrial and marine gas turbines, to the use of low grade fuels.

3.1. High Temperature Corrosion Degradation Mechanisms

About ten years ago, an apparently new form of hot corrosion attack (named type II hot corrosion) was identified in land-based and marine gas turbines. Extensive studies in the U.K. and the U.S.A. showed that this degradation occurs at rather low temperatures (between 650 and 750°C) as a result of the interaction between sodium sulphate deposits, the SO_3 present in the gas phase and the substrate alloy. New coating compositions rich in chromium have been defined with claimed increased resistance against this type of corrosion (4).

Active elements such as yttrium or rare earth increase the adherence of alumina layers formed on MCrAlY alloys, particularly in thermal cycling conditions. Recent microstructural and analytical studies have shed new insights into the role of these elements. The beneficial effect of other elements, for example silicon, platinum, which are known to improve hot corrosion resistance, remain unclear.

3.2. Mechanical Properties of Coated Superalloys

The search for higher performances in gas turbines has led to the development of single crystal superalloys. Several studies (in the U.K. (5), in France (6), and in the U.S.A. (7)) have shown that the presence of a protective coating (aluminide or MCrAlY overlay) can deteriorate the mechanical and thermomechanical fatigue behaviour of these alloys. As a consequence of thermal mechanical fatigue cracking, the use of overlay coatings in military aircraft (in the U.S.A.) has been limited, despite their high oxidation resistance (8). This effect is not properly described because the intrinsic mechanical properties of the coatings are still poorly understood. In particular, the ductile-brittle transition temperature is a concept of very limited use. It is to be noted that protective coatings have no deleterious effects on creep resistance of superalloys provided the thermal treatments associated with the coating cycle are compatible with the standard heat treatments of the substrate alloy and that interdiffusion phenomena remain limited in service.

3.3. Diffusional Stability

Extended interdiffusion phenomena have been reported between conventional coatings and superalloys having high temperature capability, such as O.D.S. (oxide dispersion strengthened) and D.S.E. (directionally solidified eutectics). These effects can result in lower environmental resistance and reduced creep life. "Diffusion barriers" consisting of a thin ceramic (nitride, carbide or oxide) layer or a dispersion of oxide particles have been proposed with mixed prospects : a continuous layer is likely to be incompatible mechanically; moreover, a nitride or carbide layer is a preferential path for oxidation propagation, and as a result spalling of the coating is a major risk. A more viable alternative would be to define an optimum coating composition adapted to the substrate and exploit the flexibility of the plasma spraying process for example.

3.4. Alternative Processes

The coating techniques presently in production for depositing MCrAlY alloy (plasma spraying under controlled atmosphere, EB-PVD) have in common to be line-of-sight processes. This can be a serious drawback in some applications, for example to coat components having complex shapes. In addition, protecting internal cooling passages may be a necessity in the near future. Several alternative coating techniques which are non line-of-sight (and economical) are being developed in Europe to deposit MCrAlY coatings : electrolytic codeposition (9), electrophoretic deposition (10) prior a vapour phase aluminizing treatment, etc. Other techniques exist for protecting internal cavities or passages : low-pressure aluminizing (11), pulse-aluminizing (12).

3.5. Ceramic Coatings

Several degradation mechanisms of thermal barrier coatings have been identified. Failure generally occurs by propagation of a crack within the ceramic layer near the bond coat, and subsequent spalling. Several factors may influence this failure : thermal expansion mismatch, bond coat oxidation and plasticity, ceramic sintering and structural evolution, residual stresses, etc. No satisfactory modelling of the thermomechanical behaviour of ceramic coatings has yet been proposed. Progress in the performance of thermal barrier coatings has been achieved mainly through empirical developments : selection of optimum composition (ZrO_2- 6 to 8 wt % Y_2O_3) to obtain a tetragonal metastable structure by plasma spraying, control of porosity/ microcrack distribution and residual stresses by adjusting spraying conditions, and selecting oxidation resistant MCrAlY bond coats.

More recent developments have concentrated on modifying the ceramic composition in order to improve spallation resistance and corrosion resistance in low grade fuels. Systems investigated are for example ytterbium-stabilized zirconia (13), ceria-yttria stabilized zirconia and other ternary systems (14).

Zirconia-based coatings have also been produced by EB-PVD. The columnar structure obtained provides an excellent thermal shock resistance. Reproducibility problems have limited the development of this technique.

Two challenges remain :
- produce more performant and reliable coatings to exploit the full potential of thermal barrier coatings on turbine blades;
- improve the hot corrosion resistance of ceramic coatings in presence of molten salts and in the case of contaminated fuels.

4. INDICATION OF SIGNIFICANT R & D EFFORTS

Under NASA sponsorship (15) several studies are carried out at Garret Turbine Engine Co., General Electric Aircraft Engine Business Group, Pratt & Whitney Aircraft, in order to develop engine-life models for thermal barrier coatings. Another programme (at Pratt & Whitney Aircraft) concerns the development of life prediction and constitutive models for two coated single crystal alloys in gas turbine airfoils (16).

In Europe, within COST 501 Round 2, a work programme concerning "Improved coatings for aero gas turbines" has been recently proposed.

5. FUTURE RESEARCH AND DEVELOPMENT NEEDS : INDICATIONS

5.1. Metallic Coatings

Fundamental studies on the oxidation behaviour of metallic coating alloys, in particular on the alumina scale growth mechanisms, on the role of addition elements (Pt, Si, Y, etc.), the scale/alloy interface, are necessary to understand the oxide/alloy adherence which is essential for the integrity of the protection.

Coatings generally degrade the (thermo)mechanical fatigue properties of advanced superalloys. Work is needed to understand the mechanical properties of the coating material and the degradation mechanisms, and model the mechanical behaviour of the composite system.

Industrial development of techniques already existing on a laboratory scale (electrolytic codeposition, electrophoresis, etc.) can provide economical and technically advantageous alternative processes for depositing MCrAlY coatings.

Programmes have already started to develop titanium alloys having temperature capabilities such that they would need protective coatings in the oxidizing environment of the compressor stage. Moreover future temperature requirements for turbine airfoils and combustor may be beyond the conceivable nickel superalloy/cooling technology and new materials, inter-metallic compounds, ceramics, etc., will have to be envisaged. Formulating new coating composition and processes is likely be necessary to ensure the protection of these materials.

5.2. Ceramic Coatings

Thermal barrier coatings should provide the largest near-future engine performance and/or durability improvement in gas turbine engines. In the past, the approach to develop thermal barrier coatings has been essentially empirical. In order to define more performant systems, fundamental studies are needed to understand the degradation mechanisms (role of ceramic microstructure, microcracking, porosity, residual stresses, oxidation of bond layer, ceramic/metal interface, etc.).

This should be accompanied by an effort to model the thermomechanical behaviour of the composite substrate/bond layer/ceramic coating and thus optimize the coating formulation and deposition process. There is a need in Europe for a rig test which could permit to evaluate the performances of thermal barrier coatings under simulated service conditions (controlled corrosive environment, cyclic testing, internal cooling, etc.).

New ceramic compositions are required to define coatings resistant in corrosive environments (molten salts, contaminated fuels and atmosphere,) and, in the long run, coatings having higher temperature capabilities than partially stabilized zirconia.

New processes for depositing ceramic coatings will have to be developed in order to protect components having complex geometry.

6. CONCLUSION

A major step in the evolution of high temperature protective coatings in the foreseeable future is the development of thermal barrier coatings. In more and more applications, the selection of a coating will have to be considered at the component design stage, with an effort to model the thermomechanical behaviour of the coating/substrate composite system.

7. REFERENCES

(1) "A guide to the control of high temperature corrosion and protection of gas turbine materials", Duret-Thual, C., Morbioli, R., Steinmetz, P., EUR 10682 EN (1986).

(2) "Pack cementation processes", Mevrel, R., Duret, C., Pichoir, R., Mat. Sci. Technol., 2, pp. 201-206, (1986).

(3) "Current status of thermal barrier coatings – An overview", Miller, R.A., Surf. Coat. & Technol., 30, pp. 1-11, (1987).

(4) "Low temperature hot corrosion in gas turbines. A review of causes and coatings therefore", Goward, G.W., Turbomachinery Inter., pp. 24-28, (May-June 1985).

(5) "The mechanical properties of coated nickel based superalloy single crystals", Wood, M.I., Restall, J.E., in High Temperature Alloys for Gas Turbines and other Applications 1986, Proc. Conf. Liège, 6-9 Oct. 1986, pp. 1-18.

(6) "Influence of protective coatings on the mechanical properties of CMSX-2 and Cotac 784", Veys, J.M., Mevrel, R., Mat. Sci. & Engng, 88 (1-2), pp. 253-260, (1987).

(7) "Creep-fatigue behaviour of NiCoCrAlY coated PWA 1480 superalloy single crystals", Miner, R.V., Gayda, J., Hebsur, M.G., NASA TM 87110.

(8) "The durability and performance of coatings in gas turbine and diesel engines", Fairbanks, J.W., Hecht, R.J., Mat. Sci. & Engng, 88, pp. 321-330, (1987).

(9) U.K. Patent Appl. 2 167 446 (1985). Electrode deposited composite coatings (BAJ Ltd.).

(10) "Procédé et dispositif pour la réalisation de revêtements protecteurs métalliques", Brevet F 2 259 911 (1982), (SNECMA).

(11) "Procédé en caisse de formation de revêtements protecteurs sur des pièces en alliages réfractaires et dispositif pour sa mise en oeuvre", Brevet F 85 01454 (1985) (CNRS).

(12) "Alternative processes and treatments", Restall, J.E., Wood, M.I., 2, pp. 225-231 (1986).

(13) Stecura, S., NASA TM-87062 (1985).

(14) "Advanced thermal barrier coating systems", Reardon, J.D., Dorfman, M.R., J. Mat. for Energy Systems, 8(4), pp. 414-419, (1987).

(15) "Thermal barrier coatings", Stearns, C.A., Aerospace America (May 1987), pp. 27-29.

(16) "Life prediction and constitutive models for engine hot section anisotropic materials program", Swanson, G.A. et al., NASA CR-174952 (1986).

2.3. Composites

2.3.1. Metal Matrix Composites

R. Warren

Chalmers University of Technology,

S-40296 Göteborg

1. INTRODUCTION

The main advantages expected from reinforcement of metals are increased yield strength, fracture strength, fatigue strength, creep resistance and elastic stiffness, most often with a reduction in density. Increased wear resistance and reduced thermal expansion are also predicted in most cases. Compared with plastic matrix composites, superior transverse and shear strengths and higher temperature capability are important benefits. The main drawbacks of fibre reinforced metals (FRM) at present are difficulties of preparation (in contrast to unreinforced metals and reinforced plastic) and high cost.

The two major areas of progress in the last five years have been the re-inforcement of super-alloys by tungsten wire and the strengthening of aluminium by particles or whiskers. While re-inforced aluminium cannot, in general, be considered for high temperature applications; some alloy composites are pushing toward 400°C capability and so warrant mention.

2. Fibres for Reinforcement of Metals

New fibres are constantly being developed, Table 1 lists significant examples of fibres representing most fibres currently available. Property values are typical rather than exact. Fibres can be produced by a variety of techniques which to a large extent determine the form, microstructure and properties of the fibre.

Metal fibres (wires) are normally prepared by conventional

154

Table 1. Principal Fibre Types and their Properties at Room Temperature.

Fibre and form	Method of pre- paration	Diameter (μm)	Specific gravity	Mean fracture stress (MPa)	Axial Youngs modulus (GPa)	Coeff. of thermal expansion $(K^{-1}x10^6)$	Approx. price ($/kg)
Tungsten	drawn	10-500^3	19.2	2500^3	400	5	2000-100^3
Steel (M or Y/C)	drawn	10-250^3	7.8	2500^3	210	15	60-2^3
Boron (M/C)	CVD	150	2.6	3500	400	8	500
SiC (M/C)	CVD	150	3.4	3800	450	4.5	1000
SiC (Y/C)	PP	12±3	2.6	2500	200	4.5	200
SiC whisker (R/S)	PP	0.1-2	3.2	10000	700	4.5	300
α-Al$_2$O$_3$(Y/C)	PS	20±5	3.9	1500	380	7	200
δ-Al$_2$O$_3$(R/S)	PS	3±1	3.5	2000	300	7	50
Carbon (Y/C):							
high modulus	PP	10	2	3000	600	0	2000
med. strength	PP	8	1.9	4200	300	0	200
Al$_2$O$_3$/27%SiO$_2$ (R/S)	PS	3	3.0	850	150	-	40
Al$_2$O$_3$/47%SiO$_2$ (R/S)	melt	3	2.7	1750	105	-	2
S-glass (Y/C)	melt	3-20^3	2.5	4000	90	3	5

Notes

1. C - continuous; S - short; M - monofilament; Y - multifilament yarn; R - random

2. CVD - chemical vapour deposition; PP - pyrolysis of polymer precursor fibre; PS - pyrolysis or sintering of a salt and/or oxide suspension or gel in fibre form.

3. Fibre diameters can be chosen in this range. For metal fibres the strength and price depend on diameter (extent of wire drawing).

drawing techniques. The diameter can be chosen freely; the strength and cost increase with decreasing diameter. The wires are drawn singly i.e. as monofilaments but can also be readily gathered into multifilament strands. Boron and SiC monofilaments are produced by chemical vapour deposition (i.e. the material is the solid product of a vapour phase chemical reaction) at around 1000°C onto a tungsten or carbon fibre (1,2).

Certain inorganic fibres, notably, carbon and SiC can be produced by first preparing a precursor fibre of a suitable polymer and then burning this under carefully controlled conditions (pyrolysis) (1,2). By this method the fibres are produced, as many hundred parallel fibres with diameters around 10 μm in a continuous yarn. Similar techniques are used to produce aluminium oxide fibres from a precursor fibre consisting of a suspension of salts or oxide particles in a suitable liquid binder (1). As well as being produced as continuous, multifilament yarn, alumina fibres can be produced in the form of a collection of less oriented and shorter fibres. In the latter form the fibres can be processed further to give blankets, rope, paper, board etc. Mixed oxides with sufficiently low melting points, e.g. glasses, can be produced more cheaply in these forms by drawing or blowing the fibres directly from the melt. Adding SiO_2 to alumina reduces the melting point and therefore the cost of fibre production but also reduces the elastic modulus of the fibre.

Most inorganic materials can be produced in the form of single crystal fibres (whiskers) by growth from a suitable vapour reaction. Generally the yield is low and the production expensive. However, competitive methods have recently been developed for the preparation of SiC whiskers e.g. from rice hulls (1,2). These are supplied in the form of loose random fibres with diameters from about 0.1 to 1 μm and lengths up to about 100 μm.

The development of FRM depends very much on fibre development. As well as improvement in the mechanical properties of fibres, efforts are being made to improve high temperature structural stability and compatibility with metal matrices. For large volume applications such as in automobile components, the cost of the fibres must be reduced; a price level of below 10$/kg is a reasonable target for many applications. Fibres currently being developed include multifilament forms of boron nitrides, silicon-based ceramics and alumina-based mixed oxides (4) as well as whiskers of various refractory compounds. A particularly interesting new development is a semi-continuous, aligned -alumina fibre with average fibre diameter of about 3 μm (5).

Included in this survey are SiC-particle reinforced

Al-alloys. The SiC particles are readily available in the form of abrasives and therefore at a very competitive price. Moreover the processing of their composites is more straightforward than with fibres (see section 3.3.1.).

3. Principle Composite Types

A large number of metal matrix composites based on various combinations of fibre and matrix have been produced on a laboratory scale and a few have found applications or show considerable promise for specific applications. It is convenient to arrange them into three main categories (Table 2). Included in the first group are composites consisting of SiC particles in Al-base matrices. These are here classed as composites rather than dispersion hardened alloys because of the relatively large SiC particle size, usually around 10-20 μm, and because they exhibit properties similar of those of short, random fibre composites.

4. Expected Properties

Assuming that no fibre degradation occurs during manufacture and subsequent service, it can generally be assumed that the properties of fibre reinforced metals will be given by relatively straightforward rules of mixtures based on the as-received fibre properties. Broadly; this implies that substantial improvements in strength and stiffness are expected in the fibre direction in aligned fibre composites while only moderate improvement in stiffness and little or no improvements in strength are expected in short, random fibre composites or in off-axis directions in aligned fibre composites. However, even in the latter cases, improvements in strength might be expected through particle and dispersion hardening mechanisms in the matrix; such effects can be expected to increase with decreasing fibre diameter.

The potential of many metal matrix composites is limited by degradation of the fibre or matrix through their interaction with each other. Responsible phenomena include fibre dissolution, chemical reaction, poisoning or structural changes in the fibre caused by matrix atoms, deposition of matrix precipitates at the fibre/matrix interface and thermal expansion mismatch leading to thermal cycling damage. Fibre/matrix compatibility can sometimes be improved by the application of a surface layer of another material on the fibre; these naturally add significantly to the cost of the fibre. Since, to be effective they must have a significant thickness (2 μm) they are generally only applicable to larger-diameter fibres.

Table 2. Main category of fibre reinforced metals.

Composite category	Examples	Main features
Short or long, random or partially-oriented fibres or ceramic particles in light metal matrices - prepared by melt infiltration or powder metallurgy	SiC whisker/Al alloys Al_2O_3/Al alloys Al_2O_3-SiO_2/Al alloys SiC-particles/Al alloys	o Moderate strength and stiffness to around 300°C o low density o wear resistance
Long, parallel fibres in light metals:		
- prepared by melt infiltration	Carbon/Al & Mg alloys SiC/Al alloys Al_2O_3/Al & Mg alloys Steel/Al alloys B/Al alloys	o High axial stiffness and strength to 200°C (Mg), 300°C (Al), 600°C (Ti) o Low axial thermal expansion
- prepared by melt infiltration or diffusion bonding	B/Al alloys SiC/Al alloys SiC/Ti alloys	o Low density
Long, parallel fibres in superalloys - prepared from PM or foil matrix and hot pressing	W/Ni, Fe, Co alloys	o High UTS and creep strength to about 1100°C

5. Particulate and Short-Fibre Reinforced Al-Alloys

Both particles and whiskers confer improved elastic modulus and strength at higher temperature at the expense of ductility and fracture toughness, apparently by dispersion hardening of the matrix rather than load transfer reinforcement. Fig. 1. shows the effect of temperature on the strength of selected composites. Reinforcement is retained up to at least 400°C; the relative effectiveness of α-alumina fibres appears to improve with increasing temperature. The steeper fall in strength with temperature of the SiC composites is probably due to interaction between the whiskers and matrix alloying additions, notably Mg (21,22). The relative benefits of alumina fibres at higher temperatures should therefore become even more apparent for long term exposures.

Several studies have shown that short-fibre reinforcement improves the high-cycle, unnotched fatigue strength of the matrix. Thus in rotating bend the 10^7 cycle limit is improved from around 100 MPa for unreinforced alloys to 150-300 MPa for 20 vol.% reinforcement. Because of their brittleness, however, the composites are sensitive to notch fatigue and exhibit very poor crack tolerance (23). Their static fracture toughness, K_{Ic}, lies in the range 6-25 $MNm^{-3/2}$ depending on the details of the microstructure (24).

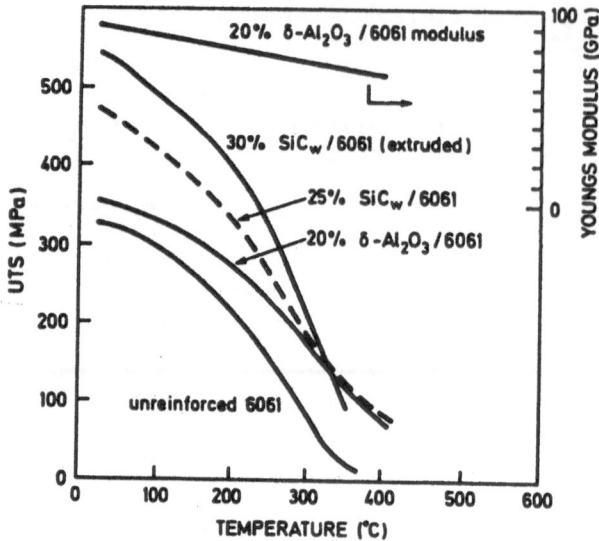

Fig. 1: Effect of temperature on strength of short-fibre
reinforced squeeze-cast Al alloys, refs. (18,19,20).

Alloy 6061: Al - 1Mg, 0.6 Si, 0.25 Cu, 0.25 Cr (wt%).

The thermal conductivity of Al alloys is reduced only
slightly by incorporation of SiC whiskers and somewhat more by
alumina fibres. It remains significantly higher than that of
Fe-base alloys. Short fibre MMC seem to be relatively
insensitive to thermal cycling (25,26). A number of
investigations confirm a marked improvement in wear resistance of
Al-alloys by reinforcement.

Donomoto et al (27) have made a thorough analysis of the
potential of short fibre reinforcement of the ring groove in a
diesel piston. The wear properties of a 5 vol.% alumina in 6061
were comparable to those of the alloy, Ni-resist. Improvements
over Ni-resist included improved seizure resistance, lower
weight, improved bonding to the Al-alloy substrate, lower
sensitivity to thermal cycling and improved thermal conductivity.
The improved conductivity leads to a 20°C fall in the piston
working temperature in the region of the groove.

The excellent specific stiffness of short fibre reinforced
AL-alloys makes them attractive for use in moving structural
parts such as connecting rods in automobile engines and as
compressor vanes in turbines. However, for these applications
improvements in their toughness will be necessary. It will also
be necessary to gain knowledge of the corrosion properties of
these composites.

6. Long, Parallel Fibre Reinforced Al-Alloys

The principle properties of three Al-alloy matrix composites are listed in Table 5 and the effects of temperature on their strength are shown in Fig. 2. The behaviour of these three systems is representative of other composites of this type. Tensile properties in the fibre direction are in good agreement with the simple rule of mixtures allowing for some degradation of the fibres during composite preparation. The transverse properties are dominated by the matrix. The simple uniaxial geometry means that the role of the matrix is more clearly discerned than in short fibre composites. Thus, for α-alumina composites, Li additions to Al are necessary to produce adequate fibre/matrix bonding. The precipitates in conventional Al-alloys seem to reduce the fibre strength (29).

More generally, it is found that in composites based on a pure Al matrix the toughness and axial tensile strength are improved but transverse and shear strengths suffer. Thus for the 50 vol.% Al_2O_3-15%SiO_2 composite in Table 5, replacing the Al-5%Cu matrix with pure Al leads to an increase of UTS from 590 to 860 MPa, an increase of impact energy from 2 to 9 J/cm^2 but a decrease of transverse strength from 230 to 100 MPa. A very beneficial property of these composites is their very high compressive strength. In compression the ceramic fibres are extremely strong while the relatively high shear strength of the matrix suppresses buckling instability of the microstructure.

Like the short fibre composites these composites exhibit significantly improved fatigue strength in crack free specimens. They suffer, however, from relatively poor toughness. Attempts to improve fracture energy by modifying the fibre/matrix bond in -alumina fibre composites met limited success (30) although controlled fracture was obtianed in braided fibre composites which was attributed to the inhomogeneity of fibre distribution (31).

The composites retain their strength properties to at least 400°C. Composites of yarn oxide and SiC fibres also exhibit good long-term thermal stability; any interaction between fibre and matrix constituents occurs very slowly in the solid state.

The application of α-alumina fibre composites in connecting rods has been analysed in some detail (28,32). Producing rods with the same dimensions as a steel counterpart results in about 25% weight reduction, which reduces inertial loads on the rod making possible a redesign with even greater weight saving. Composite rods with α-alumina and stainless steel reinforcement have been successfully tested in motorcycle engines (33).

Table 5.   Properties of selected continuous, unidirectional
fibre re-inforced Al-alloys (taken from suppliers
data).

Composite Properties *	50% yarn $-Al_2O_3/$ Al-3% Li 0°		50% yarn $Al_2O_3-15\%SiO_2/$ Al-5% Cu 0°		35% yarn SiC 6061 0°	
	0°	90°	0°	90°	0°	90°
Specific gravity	3.25		2.9		2.6	
Tensile modulus (GPa)	210	140	150	110	110	95
Tensile strength (MPa)	560	175	590	230	700	100
Compressive strength (MPa)	2900	350	2200	540	1300	220
Fatigue strenth rotating bend, 10^7 cycles	450	110	250	80	280	--
Charpy impact energy (J/cm^2)	--	--	2.2	2.0	--	--
Coeff. of thermal exp $(10^6 K^{-1})$	7.2	20	7.6	14	--	--
Thermal conductivity W/mK	38	--	--	42	--	--

* 0° and 90° refer to the angle of test direction w.r.t. the fibre direction

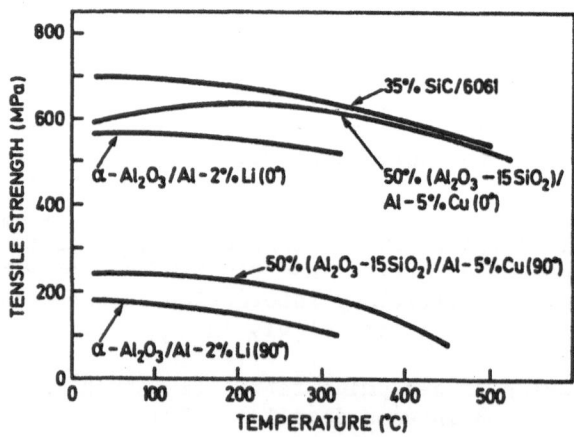

Fig. 2. Tensile strength of continuous fibre reinforced Al-alloys
as a function of temperature (suppliers data except
$\alpha-Al_2O_3$ fibre data taken from (28).

Composites of Al-alloys reinforced with CVD SiC and B (SiC coated) fibres are to be included in this group. They are characterised by exceptional specific strength and stiffness, but they are expensive and methods available for their fabrication are very restricted.

7. Long Parallel Fibre Reinforced Ti-Alloys

The idea of reinforcing Ti-alloys with fibres is attractive since this would extend the potential of Ti, first by increasing its stiffness and secondly by improving its high-temperature creep strength. However, all fibres currently available react chemically with Ti. Composites must therefore be prepared using solid-state methods. Even the most resistant fibres, namely expensive, large diameter CVD-SiC fibres with a specially developed carbon enriched surface layer exhibit significant reaction around 900°C (34). Thus the potential for Ti-based composites is restricted at present, to the low to intermediate temperature range, where increased stiffness can be exploited.

TUNGSTEN WIRE REINFORCED SUPERALLOYS

Tungsten fibre reinforced superalloys (WRS) have been reviewed recently (14,35) while their application in turbine blades specifically is treated in refs. (36-39). It is predicted that their use could permit increases in blade temperature of up to 150°C over current values.

The short-term tensile strengths and ductilities of typical unidirectional WRS as a function of temperature are shown in Fig. 3. The composites generally exhibit good agreement with ROM strengthening allowing for some loss of strength during fabrication. One advantage of metallic fibres is their ductility. However, in some matrices W wires can be embrittled below 200°C by taking up carbon or oxygen from the matrix (41). The tensile creep behaviour of WRS loaded in fibre direction is summarised in Fig. 4, compared with that of other blade materials, in terms of Larson-Miller plots. The creep strength of Fe-base, Ni-free matrix composites follows a rule of mixtures while that of Ni-alloy matrix composite falls increasingly below this with increasing exposure, as nickel activates creep processes in tungsten. Nevertheless, it is seen that WRS are potentially superior to existing blade materials. This advantage is partly offset by their higher density, a factor that would be of less significance in static engine components.

Other properties of significance in blade applications are high- and low-cycle fatigue, thermal conductivity (36,38) and gas corrosion resistance. The high-cycle fatigue strength (10^6 cycles, tension/tension ratio 0.1) for a 35 vol.% WRS is

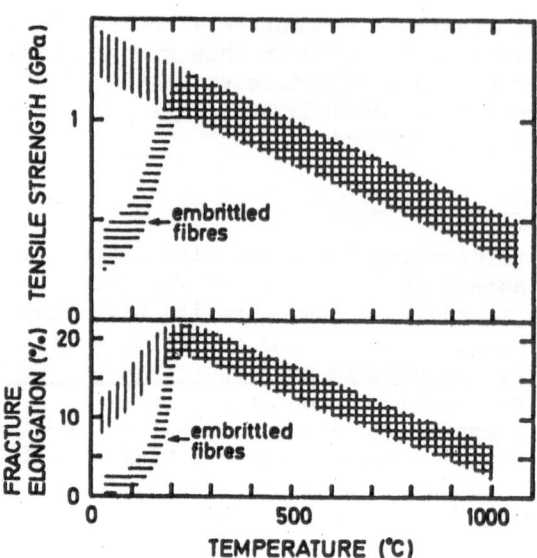

Fig. 3.:   Tensile strengths and ductilities of typical
            unidirectional tungsten reinforced superalloys with 40
            vol.% fibres. Property levels depend on the matrix
            alloy and fibre alloy composition and fibre diameter.
            (Based on refs. (14, 35-41).)

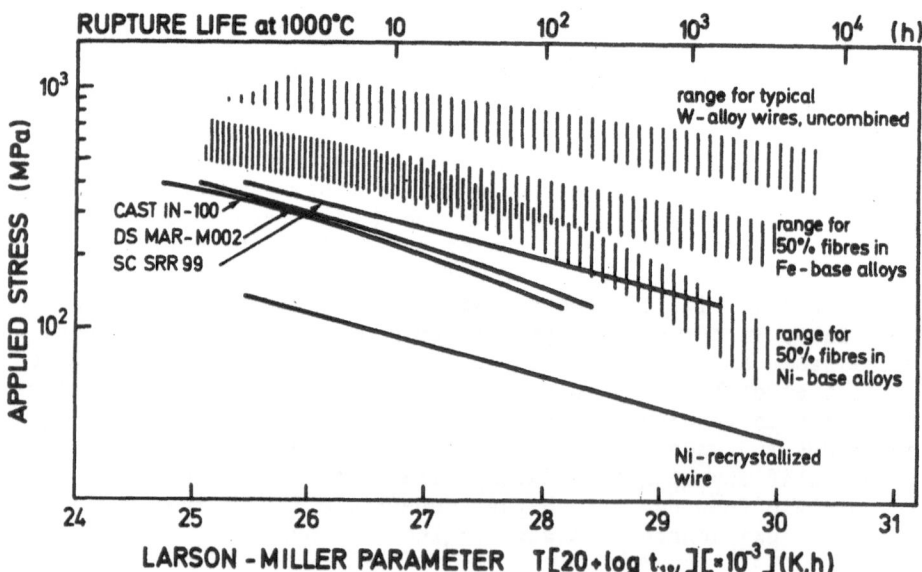

Fig. 4.:   Larson-Miller plot of creep behaviour of tungsten
            reinforced superalloys and unreinforced superalloys.
            t_R is time to rupture or 2% secondary creep strain.

reported as 490 and 250 MPa at room temperature and 900°C
respectively (38). Depending on the direction relative to the
fibre axis and temperature, the thermal conductivity of a 50
vol.% WRS is between 3 and 6 times greater than unreinforced
superalloys (36). This results in a considerably more effective
blade cooling and a reduction in thermal stresses during
operation. Tungsten is extremely sensitive to oxidation and
consequently all fibres must be well covered by matrix in
oxidizing conditions.

In practice, the high-temperature performance of WRS is
limited by degradation caused by fibre matrix interactions (14).
One problem is recrystallization of the fibre activated by matrix
atoms, in particular Ni. Thus, alloys with Ni-content greater
than a few percent begin to lose their properties rapidly above
about 1000°C. In ferritic matrices the most important effect is
dissolution of the fibre which becomes significant above 1000°C
causing loss of fibre and embrittlement of the matrix. Thermal
mismatch stresses leading to component distortion, matrix failure
or fibre debonding during thermal cycling are most serious in
austenitic alloys which have the highest coefficients of thermal
expansion (15-20 x 10^{-6} C^{-1}) but are also significant in ferritic
alloys (10-15 x 10^{-6} C^{-1}).

Two families of alloys have emerged as very promising
matrices for WRS, first the Incoloy 903, 907 and 909 series, and
second Fe-Cr-Al alloys with between 15 and 25% Cr and 5-10% Al.
The Incoloy alloys are Fe-Ni-Co, low expansion alloys with
expansion coefficients close to that of tungsten (4-6 x 10^{-6} C^{-1})
up to around 500°C; this matching reduces thermal cycling effects
significantly. The Fe-Cr-Al alloys, being Ni-free, do not
stimulate premature recrystallization of the tungsten and creep
activation; they have the added benefit of excellent hot
corrosion resistance. Nevertheless, Ni-activated fibre
recrystallization in the Incoloy alloys and fibre dissolution in
the Fe-Cr-Al alloys will necessitate the application of a
diffusion barrier layer between fibre and matrix if these
composites are to be used for extended service around 1100°C.
The barrier must resist diffusion of Ni and/or W and at the same
time be sufficently thin to resist mechanical damage. Hitherto
efforts to develop such layers have met with limited success
(14). The addition of a layer will of course add to the cost of
the material. The cost of WRS turbine blade is estimated to be
roughly the same as that of a single crystal superalloy blade.

8. Future Development

SiC-particle reinforced Al-alloys have already proved to be a competitive engineering material; current production is estimated to be around 100 tons per annum. An important development in this composite type is expected to be optimisation of the particle morphology i.e. shape, size distribution and uniformity of spatial distribution, leading to improved ductility and toughness. There is considerable potential for the use of alternative materials as reinforcement particles, in particular low cost ceramics. Particle reinforced Mg-alloys have already been prepared and their further development can be expected.

A promising area of application for short-fibre composites appears at present to be as local reinforcement in Al-alloy castings to impart wear resistance and dimensional stability. However, if they are to compete more generally with the particulate composites, reductions in price and further improvement in mechanical properties of the composite must be achieved. Property improvement can be expected through improved understanding and control of fibre morphology and through the development of new fibre-compatible Al-alloy matrices. Sufficient cost reduction may be achieved by the development of cheaper fibres and the development of processing techniques such as continuous casting and liquid phase forging (see section 3.3.1.).

Long fibre composites will probably always suffer from a higher price and restriction in methods of fabrication. However, improvements in fibres and processes will make them competitive for special applications. Development can be expected in the preparation of long fibre preforms e.g. multidirectional woven structures and laminates. A rewarding avenue of study would be the development of hybrids consisting of more than one fibre or composite type.

In the development of high-temperature composites, the search for structurally stabler fibres and more compatible matrices will continue. Matrices of interest include Cu, which is inert to most fibres but is not a barrier to the access of oxygen, and intermetallic compounds such as Ni_3Al and $TiAl_3$ in which the reactive individual metals exhibit lower activity.

REFERENCES

1. "Inorganic Fibres and Composite Materials", Bracke, P., Schurmans, H., and Verhoest, J., Pergamon, Oxford, UK (1984)

2. Andersson, C.H. and Warren, R., Composites 15, 16-24 pp. (1984)

3. Faraed, A., Fang, P., Kocjak, M.J. and Ko, F.M., Am. Ceram. Soc. Bull. 66 pp.

4. Mah, T., Mendiratta, M., Katz, A. and Mazdiyasni, S., Am. Ceram. Soc. Bull., 66, 304-308 pp. (1987)

5. Staceyy, M.H., Taylor, M.D. and Walker, A.M. Safimax, ICI New Science Group, P.O. Box 11, Runcorn, U.K. WA7 4QE

6. Maruyama, B. and Rabenberg, L., in Interfaces in Metal Matrix Composites, Eds. Dhingra and Fishman, 233-238 pp. (1986)

7. Warren, R., Andersson, C.H. and Carlsson, M., J. Matls. Sci. 13, 178-188 pp. (1978)

8. Dicarlo, J.A., as in ref. 7, 1-14 pp.

9. Metcalf, A.G., in Interfaces in Metal Matrix Composites, ed. Metcalfe, Academic Press, New York, 65-123 pp. (1974)

10. Warren, R. and Andersson, C.H., Composites 15, 101-11 pp. (1984)

11. "Constitutions of Binary Alloys", Hansen, M. and Anderko, K., McGraw-Hill, New York, 122 pp. (1958)

12. Tressler, R.E., as in ref. 30, 285-328 pp.

13. "Constitution of Binary Alloys", Hansen, M. and Anderko, K., Mc-Graw-Hill, New York, 649 pp. (1958)

14. "The Mechanical Properties of Fibre Reinforced Superalloy Composites", Warren, R., in Sintered Metal-Ceramic Composites, Ed. Upadhyaya, Elsevier, Amsterdam, 215-237 pp. (1984)

15. Dewing, E.W. and Iyer, S.P., Met. Trans. 2, 2931-2934 pp. (1971)

16. "Directionally Solidified Composites for Application at High Temperature", Rabinovitch, M., Stohr, J.F., Khan, T. and Bibring, H., in Fabrication of Composites, Eds. Kelly and Mileiko, North-Holland, Amsterdam, 295-372 pp. (1983)

17. "Tensile and Fatigue Behaviour of short Alumina Fibre Reinforced Al Alloys", Harris, S. and Wilks, T.E., in Developments in the Science and Technology of Composite Materials, Eds. Bunsell et al., AEMC, Bordeaux, France, 595-603 pp. (1985)

18. Sakamoto, A., Hasegawa, H. and Minoda, Y., as in ref. 23, 699-707 pp.

19. Dinwoodie, J., Moore, E. Langman, C. and Symes, W.R., as in ref. 23, 671-685 pp.

20. Nishimura, T., Arai, A., Morita, M., Kashiwaya, H., Kiryu, K. and Miyamoto, T., as in ref. 51, 628-633 pp.

21. Nutt, S.R., as in ref. 27, 157-167 pp.

22. Kohyama, A., Igata, N., Imai, Y., Teranishi, H. and Ishikawa, T., as in ref. 23, 609-621 pp.

23. Williams, D.R. and Fine, M.E., as in ref. 23, 637-670 pp.

24. Crowe, C.R., Gray, R.A. and Hasson, D.F., as in ref. 23, 843-866 pp.

25. "SiC Whisker Reinforced Aluminium", Kuylenstierna, C. and Dahlborg, A., (in Swedish) Internal Report R 473/85, Dept. Engineering Metals, Chalmers University, Götenborg, Sweden (1985)

26. Patterson, W.G. and Taya, M., as in ref. 23, 53-66 pp.

27. Donomoto, T., Funatani, K., Miura, N. and Miyake, N., SAE paper 830252 (1983)

28. Krueger, W.H. and Dhingra, A.K., presented at American inst. Chem. Engineers Summer National Meeting, Detroit, August (1981)

29. Hall, I. and Barrailler, V., as in ref. 51, 589-594 pp.

30. as in ref. 27, 3-25 pp.

31. Majidi, A.P., Yang, J.M. and Chou, T.W., as in ref. 27, 27-44 pp.

32. Fulgar, F., Kreuger, W.H. and Gorge, J.G., presented at the
     American Ceramic Society meeting, Cocoa Beach, Florida,
     January (1984)

33. "FRM Aluminium Connecting Rod", ANON., Honda News , June 7
     (1985)

34. Pailler, M., Lahaye and Naslain, R., J. Matsl. Sci., 19,
     2749-2771 pp. (1984)

35. Petrasek, D.W. and Signorelli, R.A., NASA Report TM-82590
     (1981)

36. Winsa, E.A., in Mechanical Behaviour of Metal Matrix
     Composites, Eds. Hack and Amateau, Met. Soc. AIME,
     Warrendale, USA, 283-299 pp. (1983)

37. Petrasek, D.W., Winsa, E.A., Westfall, L.J. and Signorelli,
     R.A., in Advanced Fibres and Composites for Elevated
     Temperatures, Eds. Ahmad & Noton, Met. Soc. AIME,
     Warrendale, 136-155 pp. (1980)

38. Winsa, E.A., Westfall, L.J. and Petrasek, D.W., NASA TM
     73842 (1978)

39. Larsson, L., Proc. V International Symposium on Airbreathing
     Engines, Bangalore 54/1-54/8 pp. (1981)

40. Warren, R., Larsson, L. and Garvare, T., Composites 10,
     126-127 pp. (1979)

41. Warren, R., Larsson, L. and Andersson, C.H., In
     Verbundwerkstoffe, Ed. Ondracek, Deutsche Gesellschaft für
     Metallkunde, Oberursel, 313-324 pp. (1981)

42. "A European Capability in Titanium Metal Matrix Composite
     Technology", Feest, E.A., 1st Brite Technological Days,
     Brussels 14-15 Dec. (1987)

43. "Metal Matrix Composite - Who wants what?", Feest, E.A.,
     Metals and Materials, May (1988)

2.3.2. Ceramic Matrix Composites

R. Warren

Chalmers University of Technology

S-40296 Göteborg

Advanced ceramic materials with high resistance to wear, corrosion and high temperatures, are currently finding application in heat engines, as cutting tool tips and other wear parts, as valve and pump components, as lightweight armour, as well as in a number of aerospace components. They are, however, brittle materials with low fracture toughness; their fracture strength is determined by an inherent flaw population which generally leads to a very wide scatter in individual strength values and a mean strength that falls with component size. They are sensitive to local stress concentration and consequently to contact damage and thermal shock. These factors lead to a poor reliability which is unacceptable for many applications and which can only to a limited extent be offset by improved quality control and advanced computer aided design. Other problems associated with the introduction of ceramics is their poor chemical and thermoelastic compatibilities with metals. Moreover, high temperature strength potential usually cannot be fully realised since sintering aids, added to promote densification also promote premature creep processes.

The promise of fibre reinforcement of ceramics is to alleviate some of these problems, largely through an increase in fracture toughness and, at high temperature, an increase in creep resistance. The proven examples of fibre reinforced cements show that the idea is indeed feasible. However, of the many fibres available for composite reinforcement few are able to withstand the extreme conditions of ceramic processing, suffering not only purely mechanical damage, but also structural degradation or excessive reaction with the matrix at elevated temperatures. Differences in thermal expansion between fibre and matrix may be detrimental or beneficial depending on the sense of the stresses created in the matrix. Some fibres that have been incorporated in ceramic matrix composites with at least partial success are listed in Table 1; some examples of successful fibre/matrix combinations will be discussed below.

Table 1.   Some fibres used in ceramic matrix composites

Fibre	Method of preparation	Commercial name and/ or manufacturer
SiC	(Polymer pyrolysis)	Nicalon/Nippon Carbon
Si-Ti-O-C	(" ")	Tyranno/UBE
SiC	(CVD)	AVCO
Al_2O_3	(sol-gel)	Saffil/ICI
Al_2O_3		FP/Du Pont
$Al_2O_3-SiO_2-B_2O_3$	(sol-gel)	Nextel/3M
$Al_2O_3-SiO_2$		Sumitomo
SiC-whiskers (small)	gas phase reaction	ARCO; Tateho; Tokai and others
SiC-whiskers (large)	vapour-liquid-solid reaction	Los Alamos NL
$\beta-Si_3N_4$		UBE
TiC		GTE
C	(polymer pyrolysis)	various

Fibre reinforced ceramics can be conveniently divided into three groups namely:

I.   Whisker Composites, produced by conventional ceramic processing; examples are SiC whisker reinforced Al_2O_3 (1,2,3), Si_3N_4 (4,5,6,7); Mullite (3), SiO_2 (9), ZrO_2 (10).

II. Long Fibre Preform Composites, produced by infiltration of a preformed multidirectional fibre array. Infiltration can be achieved by chemical vapour deposition, polymer-pyrolysis, sol-gel techniques, reaction bonding. Examples are carbon or SiC fibres in matrices of C (11), SiC (12,13,14), Si_3N_4 (14,15), Al_2O_3 (16), SiO_2 (16).

III. Long Fibre - Slurry/Hot Pressed Composites, produced by
drawing fibre tows through a matrix powder slurry, drying,
stacking and hot pressing. Examples include SiC or C-fibres
in glass (17,18,19), glass-ceramic (17,18,19), Si_3N_4
(20,21).

For a more complete picture of existing ceramic composites,
several recent review papers (7,22-26), appearing in special
issues of journals (27,28,29) can be recommended. Further
details of the methods of their preparation are given in section
3.3.2. In the brief review of properties given below it should be
remembered that the results may be very dependent on details of
processing and not necessarily characteristic of the material
type as a whole.

1.   WHISKER COMPOSITES

Single crystal fibres (whiskers), typically 0.1 - 1 μm in
diameter and 10-100 μm in length, of SiC have been commercially
available for about 5 years (30-32). Their small diameter, their
freedom from grain-boundaries and other defects and their
stoichiometric composition impart extreme strength and high
stiffness. Of equal importance, they have greater stability at
high temperature than most amorphous or polycrystalline fibres.
Since they became commerically available several matrices have
been successfully reinforced with them. Improved mechanical
properties at room temperature as well as at high temperatures
have been reported for matrices of Al_2O_3 (1,2) (Fig. 1), Mullite
(1,2) and ZrO_2 (10),
while the inherently
tougher Si_3N_4 seem to
require either large
diameter whiskers (5)
or a higher volume
fraction (6,8) if
strength and
toughness are to be
improved.
Increased creep
resistance (3), as
well as wear (7) and
erosion (33)
resistance have been
reported. Another
important feature of
SiC whisker
reinforced ceramics

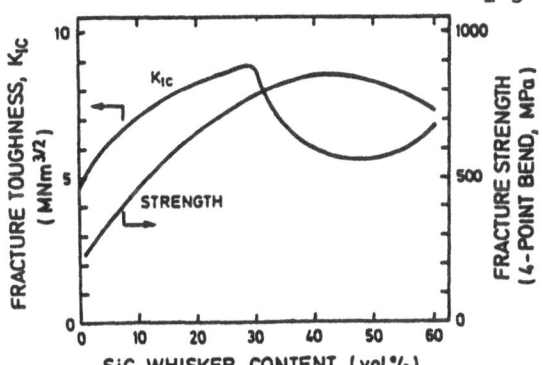

Fig. 1. Showing the effect of SiC whisker
content on fracture toughness and
strength of alumina matrix
composites (based on ref. 2).

is their potential for electric discharge machining, since they
are electrically conducting (8). An interesting question is
whether fibre reinforcement can increase the reliability of

ceramics for example by reducing the scatter of strength. Studies of the scatter for example in terms of the Weibull modulus are as yet sparse and conflicting.

A possible source of weakness in these composites is an increased chemical reactivity under certain conditions. For example, SiC exhibits a greater tendency to react with metals than alumina which leads to a poorer compatiblity of a SiC_w/Al_2O_3 composite relative to unreinforced alumina. Under reducing conditions SiC also exhibits poor oxidation resistance.

SiC-whisker reinforced ceramic composites (mainly Al_2O_3 matrix) have today attained various stages of commercial launching in the USA, West Germany, Japan, Canada and Sweden, mainly as cutting tools, but also for various wear applications. Exhaust manifold liners in SiC_w reinforced SiO_2 have been developed in France (9).

Besides problems of properties and processing (see section 3.3.2.), two factors to consider are toxicity and cost. When less than 1 μm in diameter whiskers must be regarded as an inhalation hazard in handling in view of experience with asbestos (34). The high costs of both whiskers and composite components are predicted to fall as production increases to relatively competitive levels (Fig. 2 (35)).

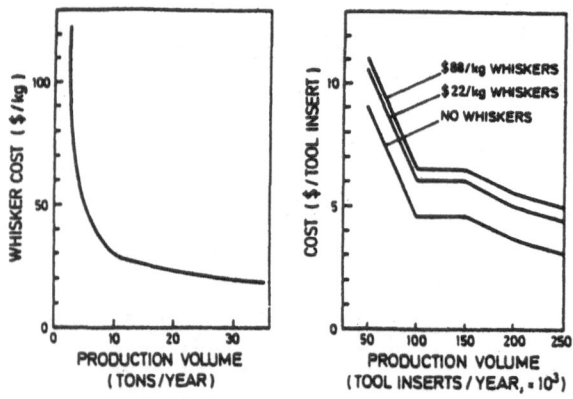

Fig. 2. Cost estimates for SiC whiskers and whisker-reinforced alumina composites (35).

The number of whiskers available today is limited (SiC, α-Si_3N_4, β-Si_3N_4, TiC). Thus, future research will probably be directed towards the synthesis of new types of whiskers as well as the improvement of existing ones. Al_2O_3-whiskers for example would be ideal for reinforcing ZrO_2-ceramics. A recent research proposal from the Battelle, Geneva Institute in fact presents a programme for a multiclient study on synthesis of new whiskers (36).

2.   LONG FIBRE - PREFORM COMPOSITES

The development of continuous fibre reinforced ceramics has

not been as straightforward as for whisker reinforced ceramics.
Although multifilament carbon, SiC and Al_2O_3 fibres have been
available longer than SiC-whiskers, fibre/matrix incompatibility,
fibre degradation at relatively low temperatures, processing
difficulties (see section 3.3.2.) and extremely high cost has
meant that the few composites available today have found
application.

Nevertheless, preform composites such as carbon/carbon (11)
and SiC/SiC (12,13,14) exhibit very tough, pseudo-ductile
fracture behaviour (see fig. 3). They retain a finite strength
to very high temperatures. They have been tested thoroughly (12,37) and it is possible to tailor the mechanical properties by employing various fibre coatings to modify the fibre/matrix bond strength. By controlling the three-dimensional fibre preform structure, desired degrees of anisotropy can be obtained. The high cost of these composites, and the oxidation sensitivity in the case of C/C, has limited their applicability to components as listed in Table 2.

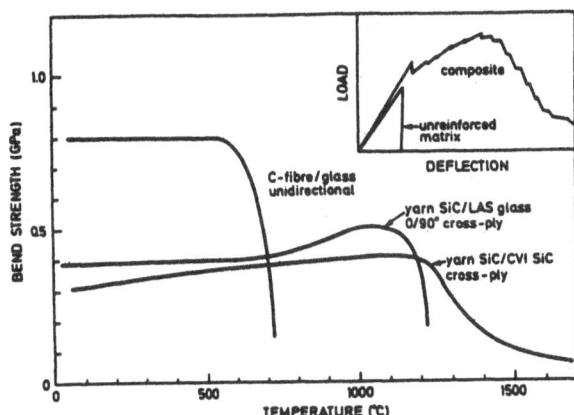

Fig. 3. Effect of temperature on the
bending strength of selected
long fibre reinforced ceramic
composites with around 50 vol.%
fibres. Inset is a schematic
illustration of a load-
deflection curve for a long-
fibre, ceramic matrix composite.
Based on several references.

The future research on preform composites will probably be
concentrated on various aspects of production methods to reduce
costs and production times (see 3.3.2.). Efforts will also be
made to develop multifilament fibres exhibiting structural
stability to higher temperatures.

3. LONG-FIBRE - SLURRY/HOT PRESS COMPOSITES

The use of glass and glass-ceramic matrices may be
considered to be one of the most important, as well as one of the
first (17,18), approaches to composite fabrication (see section
3.3.2.). Fully dense material with high room temperature
strength (ca. 1000 MPa), stiffness (ca. 150 GPa) and toughness
(ca. 15-25 MPam$^{1/2}$) have been obtained with both carbon (17,18)

Table 2.   Applications of Long Fibre Infiltrated Preform
            Composites.

Current Applications

o Brake discs for aircraft, trains, tanks,
   racing cars   C/C
o Missile rocket nozzles   C/C
o US space shuttle nose cone and leading
   edges for thermal protection   C/C
o Surgical implants   C/C

Probable Future Applications

o European space shuttle - thermal protection   SiC/SiC, C/C
o Turbine wheels, hypersonic aeroengines   C/C
o Aeroengines   SiC/SiC
o Components for oxidizing, corrosive
   environments; pumps, crucibles, gaskets   SiC/SiC

---

and SiC (19,38) fibres. Fibre oxidation limits the use of the
C-fibre composites to about 600°C. With multifilament SiC fibres
the mechanical properties are maintained to around 1000°C
(Fig. 4), the limit probably being dictated by degradation of the

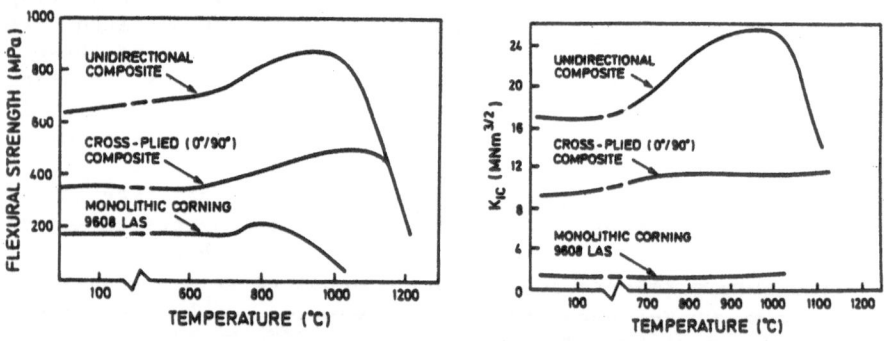

Fig. 4. a) Flexural strength (three-point) for LAS-SiC yarn
           composites.
        b) Fracture toughness, K_{IC}, of LAS-SiC yarn composites
           (from (19,38)).

fibre. Glass-ceramic matrices can be chosen within a wide
composition range making it possible to obtain the desired
melting point of thermal expansion. The fabrication method has a
potential for producing low-cost composites, which has proved to
be an important factor in the success of other types of
composite. Recently the resistance to impact has been

investigated (40) by the US navy indicating a possible interest
in the ceramic composites as armour material. Mention of other
specific applications of SiC-fibre reinforced glass-ceramics are,
however, scarce.

Future effort will probably be directed towards developing
techniques for producing multidirectional laminates and
determining their properties. The development of fibres with
improved high-temperature stability would also improve the
potential of this class of composite.

4. ROOM-TEMPERATURE DENSIFIED MATRICES

It should not be forgotten that ceramics and ceramic composites
have considerable potential for application at moderate
temperatures and not only at very high temperatures. Thus, the
recent reports of room-temperature hardenend "ceramics" such as
Macro Defect Free (MDF) Cement (41), with bend strengths of
150-200 MPa (compared to 5 MPa for ordinary cement paste), as
well as inorganic polymers (42) are very interesting. Such
materials should not be overlooked as matrices for ceramic
composites. Composites manufactured at 70°C, with a
poly-sialate-siloxo- matrix and SiC-fibres for example retain a
bend strength of 180 MPa up to 700°C (42) and offers potential
reduction in cost and energy.

5. SUMMARY AND FUTURE DIRECTIONS

Aside from expected progress in processing, discussed in
section 3.3.2., it is possible to identify a number of other
general thrusts in research and development that are currently in
progress and which are necessary for further successful evolution
of ceramic composites. Firstly, as noted above, there is a need
for alternative and improved whiskers and for continuous fibres
that retain properties to higher temperatures than those
currently available. The main causes of fibre degradation are
oxidation, microstructural changes or impurity effects (24,
section 2.3.1.). Possible developments are the preparation of
long, large diameter whiskers or the development of protective
coatings, e.g. on carbon fibres. At present much effort is being
put into refinement of polymer pyrolysis techniques to produce
continuous fibre with more suitable compositions and structure
e.g. stoichiometric SiC and Si_3N_4 (24).

Further developments can be expected in microstructural
design. This will occur firstly through improved understanding
of the role of the fibre/matrix interface in conjunction with
development of fibre surface modification techniques such as
coating or in-situ reactions. Secondly, understanding of the
role of fibre morphology (volume fraction, length, diameter and

orientation) will improve.

The expected benefits of these improvements are: toughness levels already achieved experimentally but in material produced on a commercial scale; improved reliability expressed as a larger Weibull moduli at higher mean strength levels; improved high temperature capability (above all creep resistance and retained strength).

Thus the potential for improvement of all the composites types discussed above is considerable. Since the four groups described, to a large extent complement each other, a wide range of applications is expected.

A key to the successful application of ceramics will be the use of component-specific design procedures. These normally involve a numerical thermomechanical stress analysis of the component in its service environment coupled with a knowledge of the strength variability of the material. Such procedures will need to be developed taking into account the special properties of composites such as their anisotropy.

6.   REFERENCES

(1) Wei, G.C. and Becher, P.B., Am. Ceram. Soc. Bull. 64, (2), 298-304 pp (1985)

(2) Tiegs, T.N. and Becher, P.F., Proc. Int. Conf. Ceramic Mater. Components for Engines, DKG, Ed. Bunk, W. and Hausner, H, 193-200 pp. (1986)

(3) Porter, J.R., Lange, F.F. and Chokshi, A.H., Am. Ceram. Soc. Bull., 66, 2, 243-347 pp. (1987)

(4) Shalek, P.D., Petrovic, J.J., Hurley, G.F. and Gac, F.D., Am. Ceram. Soc. Bull. 65, (2), 351-56 pp. (1986)

(5) Lundberg, R., Kahlman, L., Pompe, R., Carlsson, R. and Warren, R., Am. Ceram. Soc. Bull., 66, (2), 330-333 pp. (1987)

(6) Buljan, S.T. and Sarin, V.K., Composites 18, 2, 99-106 pp. (1987)

(7) Ishigaki, H., et al., Proc. ICCMVI & ECCM2, Elsevier, Ed. Matthews, F.L., et al., 5.336-5.345 pp. (1987)

(8) Tamari, N., Ogura, T., Kinoshita, M. and Toibana, Y., GIRIO Bulletin, 33, 129-134 pp. (1982)

(9) Sheppart, L.M., Advanced Mater. & Processes Inc., Metal Progress 10, 54-59 pp. (1986)

(10) Claussen, N. and Petzow, G., Journal de Physique Colloque C1 Supple No. 2, 47, 693-702 pp. (1986)

(11) Klein, A.J., Advanced Mater. & Processes inc., Metal Progress 11, 64-68 pp. (1986)

(12) Lamicq, P.J., Bernhart, G.A., Cauchier, M.M. and Mace, J.G., Am. Ceram. Soc. Bull., 65, (2), 336-338 pp. (1986)

(13) Caputa, A.J., et al., Am. Ceram. Bull., 66, 2, 368-372 pp.

(14) Emergent Process Methods for High-Technology Ceramics, Coblenz, W.S., Wiseman, G.H., Davis, P.G. and Rice, R.W., Ed. Davis, R.F., Palmour, H. and Porter, R.L., Plenum, New York (1984)

(15) Crobin, N.D., Rossetti, G.A. and Hartline, S.D., Am. Ceram. Soc. Bull., 64, 12, 1539 pp. (1985)

(16) Fitzer, E. and Gadow, R., conf. on Tailoring Multiphase and Composite Ceramics, Penn. State Univ. July (1985)

(17) Sambell, R.A.J., Bowen, D.H. and Phillips, D.C., J. Mater. Sci., 7, (6), 663-675 pp. (1972)

(18) Bowen, D.H., J. Mater. Sci., 7, (6), 676-681 pp. (1972)

(19) Prewo, K.M., Brennan, J.J. and Layden, G.K., Am. Ceram. Soc. Bull., 65, (2), 305-313, 322 pp. (1986)

(20) Guo, J., Mao, R., Wang, R. and Yund, D., J. Mater. Sci., 17, 3611-3616 pp. (1982)

(21) HIP-ed Carbon Fibre Reinforced Silicon Nitride Composites, Lundberg, R., Pompe, R. and Carlsson, R., to be published

(22) Schioler, L.J. and Stiglich Jr., J.J., Am. Ceram. Soc. Bull., 65, (2), 289-292 pp. (1986)

(23) Cronie, J.A., et al., Am. Ceram. Soc. Bull., 65, (2), 297-304 pp. (1986)

(24) Mah, T-I., et al., Am. Ceram. Soc. Bull., 66, 304-308, 317 pp. (1987)

(25) Marshall, D.B. and Ritter, J.E., Am. Ceram. Soc. Bull., 66, (2), 309-317 pp. (1987)

(26) Davidge, R.W., Composites, 18, (2), 92–98 pp. (1987)

(27) Am. Ceram. Soc. Bull., 65, (2) (1986)

(28) Am. Ceram. Soc. Bull., 66, (2) (1987)

(29) Composites, 18, (2) (1987)

(30) Akiyama et al., German Patent Application No. DE 33 27 101, Tokai Carbon, Tokyo, Japan (1983)

(31) ARCO, United States of America

(32) Tateho, Chemical Industries Co. Ltd., Japan

(33) Sykes, M.T., et al., 18, (2), 163–163 pp. (1987)

(34) Natl. Cancer Inst., Stanton, M.F., 58, 587 pp. (1977)

(35) Karpman, M. and Clark, J., Composites, 18, (2), 121–124 pp. (1987)

(36) "Production of Ceramic Whiskers", Battelle Geneva Research Centres, Experimental Multiclient Study, Research Proposal, January (1987)

(37) Bouquet, M., et al., Proc. ICCM VI & ICCM2, Elsevier, Ed. Matthews, F.L., et al., 2.48–2.59 pp. (1987)

(38) Brennan, J.J. and Prewo, K.M., J. Mater. Sci., 17, 2371–2383 pp. (1982)

(39) Marshall, D.B. and Evans, A.G., J. Amer. Ceram. Soc., 68, (5), 225–231 pp. (1985)

(40) Hassan, D.F. and Fishman, S.G., Proc. ICCM VI & ECCM2, Elsevier, Ed. Matthews, F.L., et al., 2.40–2.47 pp. (1987)

(41) Alford, N.McN. and Birchall, J.D., Mat. Res. Soc. Symp. Proc., Vol. 42, Verhy High Strength Cement-Based Materials, ed. Young, J.F. (1985)

(42) Davidovits, J., et al., Proc. ICCM VI & ECCM2, Elsevier, Ed. Matthews, F.L., et al., 1.462–1.470 pp. (1987)

2.4. Ceramics

2.4.1. Engineering Ceramics and Pyrolytic Materials

D.J. Godfrey

Admiralty Research Establishment

Holton Heath, U.K.

1. <u>INTRODUCTION</u>

The potential of engineering ceramics and pyrolytic materials for high temperature industrial applications (1) has been by no means fully exploited, because it is often difficult to prepare adequately pure void-free materials and deleterious effects from high temperature environments have proved very difficult to combat. Engineering ceramics (2) may be defined as inorganic non-metallic materials fabricated by powder processing techniques, and this terminology is often extended to include pyrolytic materials made by chemical vapor deposition (CVD). It is sensible to consider carbon materials as engineering ceramics, as well as glasses not having metallic components. Since many applications involve air as an environment, often in the presence of H_2O and CO_2, ceramics can be seperated into two classes: the first of materials having reasonable compatibility with air at high temperature, and further separable into oxides and non-oxides having effectively protective surface films; and a second class of non-oxide materials which cannot be used in oxidizing environments.

The ceramic materials potentially useful for engineering applications include oxides, nitrides, and carbides, which differ widely in physical properties and elastic modulus. Strength potential can, in theory, be inferred from E values (in the order of E/20 to E/10) but in practice many factors reduce attainable strengths to a small fraction of this. Thermal stress and thermal shock are serious problems in engineering applications at high temperatures; these are determined primarily by the thermal expansitivity, but are often moderated considerably by the thermal conductivity of the ceramic, expecially when thermal changes are not extremely rapid, but reliable data for this property are usually not readily available for novel materials as the measurement requires great expertise to be done well. Thermal conductivity also is greatly affected by material character and is often difficult to optimize. For instance, it is difficult to obtain SiC ceramics with the very high conductivities which are potentially possible, due to the phonon scattering by impurities and microstructural features. Thermal expansitivity is governed by the atomic bond character and is therefore insentive to processing and degree of densification, unlike thermal conductivity.

2. OXIDES

The high elastic modulus of alumina, its lithospheric abundance, cheapness and availability, and crystal structure stability through a large temperature range have made it one of the most important materials for applications above temperature limits for superalloy use (above 1000°C), although chemical attack by other oxides or elements can cause serious problems (e.g. in steelplant refractories), and its moderately high thermal expansivity often limits its use. The possibility of slip in several crystal directions at elevated temperatures (e.g. by basal slip-generated flaws interacting with grain boundaries above 1050°C) restricts its structural usefulness. The possibility of improving its high temperature properties has not been thoroughly explored; more success might be obtained with fibre reinforcement techniques such as have been developed in SiC/Al_2O_3 cutting tools. Successful strengthening has been demonstrated with magnesioferrite despersions in MgO ceramics, but not at high temperatures. Dispersions of fine ZrO_2 are effective in toughening and strengthening alumina, but the benefit reduces with increasing temperature, and using HfO_2 with a higher monoclinic to tetragonal transition (3) may not be an effective answer. Transformation toughening has been demonstrated to have good potential for toughening beta alumina ceramics ($Na_2O.11Al_2O_3$), which have been intensively studied as a high temperature battery material (4). Improvements in processing technology to yield finer, purer and more sinter-active powers (sol gel and organometallic precursor technology) and smaller and more uniform aggregates (spray and freeze drying) currently provide ceramics with minimal sintering additives and high strength or improved thermal conductivity for applications at moderate temperatures and stresses. In addition to improving properties and reliability, research effort towards reducing processing costs is very important in the competitive international situation. The development of alumina fibre insulation has given major economic benefits for high-temperature industrial technology in furnace and metallurgical applications.

Alumina high strength fibres are in strong competition with SiC fibres for the reinforcement of aluminium metal matrix composites, and research on monofilament and whisker fabrication science, with a strong emphasis on optimising processing economics, is highly desirable.

Zirconia, with a combination of high melting point and high strength potential, but poor thermal stress behaviour as a result of its high thermal expansitivity and low thermal conductivity, has not found many applications at high temperature, although it has been used as a special crucible material for superalloys, and as a glassplant refractory (ZAC). The development of cubic stabilized zirconias containing CaO or MgO (recently both have been used together to improve elevated temperature properties) and more recently Y_2O_3, to prevent the catastrophic cracking which results from the large volume expansion when transforming from the high temperature stable tetragonal phase to the monoclinic phase, has been followed by great activity with cubic materials containing fine dispersions of tetragonal phase. In these crack propagation is hindered when the metastable precipitate transforms, giving significant improvements in toughness and strength. Some toughening can also result from a presence of microcracks in zirconias. A further advance with partially stabilized zirconias (PSZ) has been the development of fine grained materials in which the tetragonal phase persists down to room temperature, known as TZP (tetragonal zirconia polycrystals (5)). However, the use of these materials at even moderately elevated temperature has been complicated by the discovery of thermal degradation of strength in the range 150 - 350°C, due possibly to the effect of moisture on the Y_2O_3 stabilizer, which may be avoidable at sufficiently small grain sizes. Ceria stabilized materials have recently been found to offer either high strength or unusually high toughness, combined with indications from high temperature water exposure testing that they may be more immune to degradation by water. There is clearly still the prospect of further optimising the high temperature strength of these materials by continued research studies. Partially stabilized single crystal zirconia has been shown to be capable of exhibiting high strength up to temperatures as high as 1500°C, with fracture toughness increasing with temperature, and lacks the grain boundaries and their associated phases which allow intergranular failure. Zirconia fibres (6) have been produced with high strengths (1.5 - 2.6 GPa) but applications for these have not yet been found and their interaction with matrices during high temperature processing is obviously a crucial consideration. ZrO_2 whiskers also might be interesting materials, since it has been shown that fine particle size can enhance the stability of tetragonal and cubic phases, and the possibility of fibres giving toughening by a combination of fibre pullout friction and transformation mechanisms is intriguing. The development of near transparent zirconia (Zyttrite (7) and 82 $ZrO_2/8Y_2O_3/10TiO_2$ (8)) ceramics is interesting, since zirconia and titania have a considerable mutual solubility range, which opens up further research possibilities. Such transparency will aid radiative heat transport machanisms and perhaps offset the disadvantage stemming from the poor conductivity of zirconia.

The low conductivity of zirconia has also led to its successful application as a thermal barrier coating for superalloy gas turbine combustion chambers blade platforms etc., and it is now an essential feature of several modern aircraft engines. As coating adherence is a crucial aspect of this application much effort needs to be devoted to ensuring the reliability of thicker and more durable coatings. Thermally sprayed coatings are being studied intensively, and with the recent discovery that a more stable tetragonal phase can be present in zirconia coatings, it is apparent that further improvements in both coatings and monolithic materials may be possible.

Zirconia is an electrolyte for oxygen at elevated temperatures and has attained a very considerable importance as a sensor in both motor car and furnace combustion control systems and in metallurgical processing, and research directed towards optimizing its role in these technologies will continue to be of great importance.

Of the other oxides having significantly higher melting point than Al_2O_3 only magnesia has reached a high level of technological importance, as a refractory brick material for steelplant furnaces. Infilling of porosity with pitch derived carbon to reduce penetrative and disruptive corrosive and improvement of durability by chromium oxide additions have been important technical developments, and although pure theoretically dense magnesias have been made they have not found such a widespread application as the lower density ceramics. MgO can show significant plasticity in some slip directions at room temperature, and this plasticity also restricts its mechanical properties at elevated temperatures; even at room temperature irradiation hardening and chemical polishing were needed to attain a strength of 300 MPa (9), although $MgFe_2O_4$ containing material has given strengths as high as 450 MPa. The rare earth oxides (Y_2O_3 and Sc_2O_3) have high melting points but their lower elastic modulus indicates that their strength will probably be lower than that of Al_2O_3 or ZrO_2, and they have not so far been found to be important in high temperature technology although Ga_2O_3 ceramics are being investigated as reactor neutron absorbers (10). Their combination with other refractory oxides might produce useful materials. Yttria ceramics have been sintered at very high temperature with La_2O_3 additions to give transparent and fully dense materials, but the strength of Y_2O_3 is not high (ca. 50MPa) and it is less creep resistant than Al_2O_3 at 1700°C. Unlike zirconia however, they do not need alloying to prevent destructive phase changes, and improved materials might result from dispersions of a second phase and grain size control.

3. MIXED OXIDES

Although with many oxide combinations the refractoriness of the ceramic is reduced by a lower temperature appearance of a liquid phase, a number of complex oxides are important high temperature materials. Mullite $Al_2O_3.2SiO_2$ and cordierite $Mg_2Al_4Si_5O_{18}$ have lower thermal expansivities than alumina and are therefore more resistant to thermal stress. Considerable effort has gone into the development of cordierite for heat exchanger matrices, and it has been shown to perform far better than spodumene, (lithium aluminosilicate materials), which can lose lithium and deteriorate. Also improved mullites have been made by reacting zircon, $ZrSiO_4$, and alumina to produce a zirconia toughened mullite.

For heating elements operating at high temperatures in air, $LaCrO_3$ materials (11) have been developed (in competition with SiC and especially with $MoSi_2$ elements) following their earlier investigation as hot electrode materials for magnetohydrodynamic electrical power generation (a possible new field of importance for ceramics because insulating and conductive materials resistant to oxidation and corrosion by ash at high temperatures are needed). Vaporisation of CrO_3 is a problem which has been addressed by developing complex formulations e.g. $La_{0.8}Ca_{0.2}Al_{0.25}Cr_{0.75}O_3$. The BaO – ZrO_2 and SrO and ZrO_2 systems offer very refractory materials, but so far practical applications have not emerged and strengths are low; 70 MPa has been reported for $CaZrO_3$.

Fusion reactors are an important potential future field of application for high temperature engineering ceramics. Binary oxide materials containing lithium have been investigated (12) for possible tritium breeder applications e.g. octalithium zirconate (m.p. 1295°C), which has a higher lithium content than the silicates Li_2SiO_3 and Li_4SiO_4 (m.p. 1201 and 1256°C), and the more refractory Li_2TiO_3 and Li_2ZrO_3 (m.p. 1547 and 1600°C).

Aluminium titanate (Al_2TiO_5) (13) has a very low thermal expansivity and despite its rather low strength its potential for diesel engine combustion chambers is the subject of research. Another low expansivity ceramic also being investigated is zirconyl phosphate (14) which appears to be usable up to about 1200°C. Interest in low expansivities seems likely to continue,

as an ideal material with very low expansivity and high strength
has not been discovered.

4. NITRIDES AND CARBIDES

Both silicon nitride (15) and carbide (16) have become
materials of great importance for high temperature engineering
applications. Si_3N_4 has approximately two thirds the thermal
expanssivity and stiffness of SiC, but the thermal stress
advantage that this gives is offset in all but the most rapid
thermal shock situations by the better thermal conductivity of
SiC. Silicon nitride ceramics made by reaction-bonding silicon
powder compacts with nitrogen gas at high temperature have a
valuable and special characteristic, the retention of precise
component dimensions, which is economically advantageous. They
have many applications in welding and furnace technology and in
the handling of molten metals, e.g. pressure die casting of
molten aluminium. Advances in strength and reliability will
result from further research to optimise their fabrication
properties and modify their properties by such techniques as
chromium oxide impregnation, and hot isostatic pressing. Flame
and plasma spraying have been used to produce complex or thin
shapes of porous Si for reaction bonding to RBSN. (17). Further
possibilities exist in the fabrication by reaction bonding of
particulate and fibre-reinforced composite materials, and in the
use of RBSN containing appropriate additives for producing
sintered silicon nitride ceramics by a route which can give much
lower sintering shrinkages than routes using Si_3N_4 powder. Much
research activity with firstly hot pressed and latterly sintered
Si_3N_4 and 'SIALON' (solid solutions of AlO in Si_3N_4) materials
has already established them as novel cutting tool materials of
great value, and for high temperature applications there are
considerable prospects for furnace and metal processing
applications as well as in advanced gas turbines and internal
combustion engines. Recent Japanese work with a sintered Si_3N_4
turbocharger rotor attached by a complex interlayer joining
technique to a steel shaft has shown that such applications can
be made to work effectively, although their economic feasibility
depends on fabrication research reducing their processing costs
and increasing the yield of non-defective parts sufficiently. At
very high temperatures (over 1000 to 1200°C) the strength of
Si_3N_4 ceramics is progressively diminished quite markedly due to
the additives which have been added to make the sintering process
possible. These combine with the SiO_2 on the Si_3N_4 powder
particles to produce a liquid phase which allows sintering by a
solution and recrystallisation process. Al_2O_3 is the best
additive for promoting sintering, but it does not by any means
easily dissolve completely in beta Si_3N_4 to produce a Sialon, and
has a detrimental effect on oxidation resistance as well as
strength. If aluminium nitride containing Si_3N_4 is added to the

powder to neutralise SiO_2, Sialon materials with low silica contents can be formed which have better high temperature properties than high additive level materials (18). Yttria and magnesia have been used as sintering additives to give better high temperature mechanical properties, but care has to be taken to avoid producing materials which can suffer from catastrophic oxidation processes at temperatures below which strength degradation is serious. Another approach is to turn the additives into non-vitreous crystal phases by a heat treatment of the ceramic e.g. to form yttrium aluminium garnet (YAG), $3Y_2O_3.2Al_2O_3$. BeO has been shown to be an effective additive which gives good high temperature strength but the possibility of respiratory poisoning with Be-containing powder has inhibited exploitation of these materials for engineering applications. A very large amount of research on additives to Si_3N_4 has already taken place, with for instance trials with all the practical rare earth oxides and scandia as well as yttria being investigated, but undoubtedly the excellent thermal shock resistance and strength possibilities will encourage much more research on this complex group of materials. Unfortunately pure Si_3N_4 ceramics without additives have proved so far impossible to produce, even by HIP techniques, but could give optimal mechanical properties; the limitation then would be oxidative disruption of the protective film by N_2 evolution at temperatures high enough to decompose Si_3N_4 significantly (e.g. about 1700°C), and contamination of the materials by fuel or operating environment contaminants such as sodium or calcium which in alkaline conditions can severly attack Si_3N_4. Thermal cycling of surface films can produce strength degradation due to cristobalites undergoing a large volume change in the α to β phase transformation, although it has been shown that intentional additions of Ca, Sr or $MgAl_2O_4$ may improve RBSN strength retention after oxidation by producing at the surface silicates instead of cristobalite (19). Another interesting possibility is indicated by the observation that the high temperature strength of Si_3N_4 ceramics can actually be improved after high temperature oxidation because the impurities and additives in the grain boundaries may be reduced by their diffusion into the surface film of SiO_2 formed on the surface.

In addition to these materials, others based on solid solutions in alpha Si_3N_4 and on silicon oxynitride Si_2N_2O have been investigated. Alpha Sialons containing Y_2O_3 may offer useful materials (20). The possibility of replacement of oxygen by nitrogen in high temperature ceramics has led to interest in nitrogen-containing glasses which also may lead to novel stronger and more refractory glass materials with better suitability for high temperature applications.

Silicon carbide is more difficult to sinter to theoretical

density and high strength than Si_3N_4 materials. Reaction bonding of SiC/C mixtures by molten Si is used to produce materials which have good properties for ambient and moderate temperature use.

Effective methods of consolidation which do not leave appreciable quantities of unreacted Si in the microstructure have not been discovered, and this Si restricts their upper use temperature to somewhat below the melting point of Si, 1413°C. The high temperature oxidation resistance of SiC is good, being limited by the SiO_2 film softening or liquefying and by disruption by CO_2 evolution (experiments with single crystals have demonstrated the importance of purity in thes process). The upper use temperature of SiC resistor heating elements appears to be around 1600°C.

SiC has been the subject of extensive research for monofilament fibres and materials are available which meet many of the needs of the rapidly developing technology of aluminium matrix composite materials, although the cost reduction aspect is extremely important and in the future will received considerable research attention. Monofilaments with tungsten or carbon cores prepared by chemical vapour deposition routes, and extruded substrate-less filaments (Nicalon:- organometallic polysilane - polycarbosilane - pyrolysis to SiC route) are currently both receiving much attention; an emphasis will have to be given to improving their high temperature strength if they are to be used as reinforcing fibres in high temperature ceramics. TiC incorporation has recently been used to improve fibre utility and illustrates the innovative approach desirable in this important field. The potential for new discoveries in the technology of silicon carbide materials is illustrated by the discovery that solid solutions of AlN and SiC over a wide compositional range are possible, but uses for these materials have not yet emerged. A most valuable form of silicon carbide, however, would be (as with silicon nitride) a fully dense fine grained pure ceramic densified without an additive (or possibly by a fugitive additive e.g. LiF additions for densifying MgO ceramics). Single crystals of useful size have never been produced in either material, and it would appear that processes of considerable novelty and ingenuity would have to be devised to produce them.

Other carbides and nitride have undergone far less investigation for high temperature applications. Tungsten carbide has insufficient oxidation resistance for high temperature use, despite its excellent mechanical properties. There has been considerable research on TiC coatings for enhancement of wear resistance, but high temperature applications have not so far appeared. Some carbides are very refractory (TaC, NbC, HfC (21) and W/C) but are only suitable for use at high temperatures in non-oxidising atmospheres and effective coatings to prevent oxidation have not been developed.

Aluminium nitride has recently emerged as a substrate for high power semiconductor devices because its thermal conductivity can be five times that of alumina and it can be produced as an insulating ceramic without significant electrical conductivity (although it has been used in semiconducting form as a high frequency transducer material). The strength of presently available material is comparable with all but the strongest fine grained alumina electronic materials. There has been almost no investigation of the high temperature utility of these new AlN ceramics, which are sintered (often with Y_2O_3 additions) in N_2 at about 1900°C. Interesting materials containing added Al_2O_3 have also been developed (22), ('ALON'), which are cubic solid solutions, and Alons can be produced with a large grain size in near transparent form, or with good strength and attractive thermophysical properties as less transparent ceramics. Boron nitride is usable to 1250°C in air, provided the surface film of B_2O_3 is not degraded by contact with other oxides. In inert atmospheres or nitrogen it is a valuable electrical furnace material (e.g. it is the best electrical insulating material for contact with tungsten at 1900°C), but unfortunately it has only a very moderate strength because of its low and anistropic moduli of elasticity, although its flexibility is helpful in resisting thermal stress and shock. It has a quite good thermal conductivity which is much less affected by temperature than many other ceramics. Crucibles of pyrolytic boron nitride for high temperature processing of semiconductor materials are made by chemical vapour deposition. High pressures have been used to prepare cubic and other forms of BN with higher density and strength; they have not so far found high temperature uses, but cubic BN can be used as a special abrasive because of its extreme hardness.

5. OTHER COMPOUNDS

Although considerable research was carried out on the refractory beryllides in the 1950's, they have not been used in high temperature technology, probably because of the reputation of BeO as a respiratory poison (e.g. if the finely divided powder is inhaled). A less-widely apppreciated hazard is the high temperature volatility of $Be(OH)_2$ formed by reaction of BeO and H_2O vapour (e.g. in a very intense fire). Borides of several transition metals are strong and refractory (23), but are not particularly oxidation resistant. However TiB_2 materials are important high temperature electrode materials in aluminium Hall-Herout cell technology, and because of their high elastic modulus are also of some interest for cutting tools and ceramic armour (24). They are difficult to sinter to theoretical density, since oxygen appears to be a detrimental impurity which is difficult to eliminate. Molybenum disilicide has become an important resistor for furnaces operating up to 1700°C in air, and can be used at slightly higher temperatures than SiC furnace elements. Resistance changes during operating life can be less than those with SiC resistors. WSi_2 also has a good oxidation resistance, but has not been used at high temperatures, although it is becoming important for contact applications in semiconductor devices. Its good oxidation resistance has led to many attempts to use $MoSi_2$ coatings for oxidation protection of refractory alloys and carbon fibre/carbon matrix composite materials, but the problem of assuring sufficient reliability in cyclic temperature applications does not yet appear to have been solved, and for the latter application most recent research appears to have concentrated on SiC coatings.

6. PYROLYTIC MATERIALS

Pyrolytic ceramic materials (25) are made by chemical vapour deposition (CVD) techniques, hence the alternative nomenclature of 'CVD Materials'. Most commonly these materials are made as thin coatings attached to a substrate material, but sometimes thick deposits are detached from the substrate to form monolithic shapes. Deposits can have high purity and high density (often reaching the theoretical density of the compound deposited). However sometimes voids are entrapped in the deposit, and sometimes cryptocrystalline or amorphous but void-free deposits are formed with densities substantially lower than the theoretical density of the crystalline compound. Deposits of SiC have been made for high modulus fibres, for reactor fuel particle encapsulation, and for oxidation protection; low pressure CVD (LPCVD) Si_3N_4 is used extensively for surface insulation of silicon semiconductor devices. A complex technology has developed for this application, and many coatings have compositions outside that of stoichiometric Si_3N_4 or have oxygen also present but are not stoichiometric Si_2O_2N, although they be referred to as 'oxynitride' coatings. These specialised deposits have not so far found high temperature applications, but might prove suitable for moderate temperature non-electronic uses (26), and a wealth of research information now exists on their quite complex character. CVD silicon nitride has also been investigated (27), as a pure dense form of this material, which is free of the additives necessary to densify materials made by powder consolidation routes, for high temperature engineering applications.

CVD titanium carbide has recently been developed extensively as a wear-resistant coating for cutting tool applications, but has not been used for high temperature applications, probably because TiO_2 does not protect effectively against oxidation. Production of ZrO_2 coatings by reaction of $ZrCl_4$ and O_2 at 1 bar

(28) is an interesting development. The fact that sintering aids do not have to be used to produce materials with theoretical densities by CVD techniques has led to some interest in their use to produce components for high temperature engineering applications, since the degradation of properties at high temperature by other materials added as sintering aids, which is such a severe problem with materials made by conventional powder consolidation routes, could be avoided. However there are major problems which hinder the effective realisation of this possibility. The microstructure of the CVD deposits is usually mechanical columnar and highly oriented and the mechanical properties of the materials formed are far below those which are theoretically possible. It is also not easy to produce intricately-shaped components or deposits over large areas with a sufficient degree of uniform material character. With some materials considerable problems in controlling stoichiometry are encountered; e.g. fibres of SiC deposited on tungsten wires or carbon monofilaments show serious strength degradation above 1000°C due to the presence of excess of silicon or carbon. Research aimed at solving these problems merits serious attention and could have important consequences for high temperature engineering applications, if successful. Considerable progress has been made in using electrical techniques such as plasma activation and ion-plating to augment thermal energy for assisting the production of the desired deposition species. Competing routes to pure dense ceramic materials do not exist for many materials which sinter poorly or do not melt without decomposition. Methods of growing single crystals of some refractory oxides have been developed (e.g. Al_2O_3 and ZrO_2) but do not offer easy routes to the manufacture or engineering components which are competitive with CVD. Better understanding of depositon chemistry, thermodynamics and aerodynamics will assist the optimisation of chemical vapour deposition technology. The use of CVD to prepare carbon ceramic materials is well established and pyrolytic carbon has been used for fission product barrier coatings in nuclear power technology and for matrices in carbon/carbon composite materials. A recent successful development has been the deposition of the so-called 'diamond-like' carbon coatings (29) by RF plasma activated methods; these can have excellent hardness and adherence, and help protect softer materials against erosion. Considerable interest in using CVD SiC as a matrix material in ceramic fibre composites also exists. An interesting development has been the successful deposition of duplex materials e.g. Si_3N_4 - TiN (up to 28 w/o of TiN). Improvements in deposit microstructure have been affected by using fluidised beds (for SiC coatings), tumbling powder beds (30) and wire brushing during deposition.

7. CRITICAL PROPERTY SHORTFALLS

The fundamental problem with ceramic materials is their lack of toughness. There is some hope that significant improvements in their technological utility will result from the modest but real improvements in toughness, that have resulted from transformable tetragonal zirconia particulate dispersions in zirconia, and other ceramics. So far a really significant improvement in silicon nitride and carbide has not been achieved by this means, but is worth searching for, perhaps using a different transformable dispersant. Toughening is also possible by fibre incorporation; although this impedes densification of the ceramic matrix, and causes anisotropy in properties, so that it becomes very difficult to make materials whose toughness is isotropic. For fibre reinforced ceramics a better choice of silicon carbide, alumina and zirconia fibre materials is needed with good strength and strength retention at and after high temperatures, appropriate fibre diameters to avoid the health problems and precautions that whisker processing poses, and lower cost.

The prospect of transparent ceramics of alumina and zirconia is real and should be pursued.

Rare earth oxides with higher melting points than alumina, could be useful materials if their strength could be improved. The development of aluminium nitride and oxynitride materials offers high conductivity refractories, but their processing needs to be better understood to improve their strength.

In pyrolytic materials, it is not possible to make components, or large area coatings with consistent properties, of pure silicon carbide or nitride. Such a development would offer exiting prospects of improved high temperature capabilities. The development of TiC and diamond-like carbon coatings needs to be better understood and cheapened.

8. FUTURE DEVELOPMENT TRENDS

Transformation and fibre toughening of ceramics will be improved, and costs and processing methods improved.

More refractory and cheaper fibres with high strength and safe processing character will be developed.

Better forms of transparent ceramic will be developed in zirconia and aluminia materials.

Ceramic materials with higher and more reliable strength properties and with less processing and additions will be developed.

Aluminium nitride materials will become more important.

Pyrolytic deposition methods for more uniform coatings over larger areas, and for thicker free-standing components will be developed.

9. NEW MATERIALS

Apart from the transparent zirconia ceramics and rare earth oxide ceramics, the prospect for new materials lies with binary and tenary ceramics, whose potential has not yet been fully explored. However, other tough ceramics are likely to be developed and provided these have good properties and are not too expensive they could be extremely important. Materials processed by novel techniques, such as monodisperse particle, sol-gel processing coprecipitation or thermal spraying may also become important.

10. SIGNIFICANT R + D EFFORTS

In the USA funding for engineering ceramics research and evaluation from the government has grown from $12M in 1971 to the order of $180M by 1986. Currently, the emphasis of US programmes has swung from focussing on engineering applications studies to materials property and reliability optimisation. Currently the best estimate for US government and industry funding on engineering ceramics R and D is $100M / annum.

It has been estimated that Japan is spending between 65 and 130M $ on engineering ceramics and related areas, with the involvement of 500 - 1000 scientists.

The level of European effort is lower than in Japan and the USA and is estimated to lag 3 - 4 years behind them; expenditure probably lies somewhere between 15 and 40M $.

11. FUTURE RESEARCH NEEDS AND PRIORITIES

Research is urgently needed to improve the toughness of ceramics to yield improved resistance to crack propagation in materials which are stronger; have better high temperature capabilities and durability in aggressive environments at lower cost. In particular better fibres, tougher silicon nitride and tougher and stronger silicon carbide are needed. Higher strength alumina and zirconia with enhanced toughness appear possible, but the realisation of this possibility needs to be resourced.

Better processing needs to be developed, not only by

innovatory thinking and new techniques but by painstaking
improvement of existing approaches to give materials with
optimised properties which are defect-free and therefore have
greatly improved reliability. Other research opportunities for
better, lower cost, safely usable fibres for reinforcement of
ceramics (and, importantly, of metal matrix composites),
transparent ceramics, more refractory strong ceramics from rare
earth oxides, stronger silicon carbide ceramics, aluminium
nitride with high thermal conductivity and ALON with transparency
and better coatings and monolithic high purity ceramics made by
pyrolytic deposition of silicon nitride, silicon carbide,
titanium carbide and related materials, and carbon. Quality
assurance of all these materials is very important, and cheap but
effective non-destructive evalutation methods for ceramics and
ceramic powder compacts need to be developed.

REFERENCES

1. Metals and Materials, Godfrey, D.J., 2, (10), 305 (1968)

2. 'Ceramic Science for Materials Technologists', McColm, I.J.,
    Leonard Hill, London (1983)

3. 'Science and Technology of Zirconia' II, Chen, I. and Chiao,
    Y., Ed. Claussen et al, Am. Ceram. Soc., 33 pp. (1984)

4. Idem, Binner, J.G.P., 428 pp.

5. J. Am. Ceram. Soc. 70, Heuer, A.H., (10) 689 (1987)

6. J. Am. Ceram. Soc. 70, Marshall, D.B. et al, (8) C187 (1987)

7. J. Am. Ceram. Soc. 50, Mazdiyasni, K.S. et al, 532 (1967)

8. J. Mat. Sci. 5, Tsukuma, K., 1143 (1986)

9. J. Am. Ceram. Soc. 55, Kruse, E.W. and Fire, M.E., 32 (1972)

10. Powder Met. Internat. 19, El-Houle, S. and El-Sayed Ali, M.,
    (2), 20 (1987)

11. Cer. Bull. Am. Ceram. Soc., Meadowcroft, D.B. and Wimmer,
    J.M., 58, (6), 610 (1979)

12. J. Am. Ceram. Soc., Ortman, M.S. and Larsen, E.M., 66, (8),
    C142 (1983)

13. Yogyi Kyokaishi, Ohya, Y. et al, 91, (6), 289 (1983)

14. J. Am. Ceram. Soc., Yamai, I. and Oota, T., 68, (5), 273 (1985)

15. J. Mat. Sci., Ziegler, G. et al, 22, 3041 (1987)

16. Science of Ceramics 12-Ceramurgia s.r.l., Hausner, H., Faenza, 229 pp. (1983)

17. Special Ceramics 5, Brown, R.L. et al, Ed. Popper, BCRE, Stoke-on-Trent, 345 pp. (1972)

18. J. Am. Ceram. Soc., Bandyopadhyay, S. and Mukerji, J., 70, (10), C273 (1987)

19. Proc. Brit. Ceram. Soc., Godfrey, D.J., 26, 265 (1978)

20. J. Mat. Sci., Mitomo, M., 15, 2661 (1980)

21. Powder Met. Internat., Perry, A.J., 19, (1), 29 (1987)

22. Cer. Bull. Am. Ceram. Soc., Quinn, G.D. et al, 63, (5), 723 (1984)

23. 'Borides, Their Chemistry and Application', Thompson, R., Royal Int. Chem. Lecture Series, No. 5 (1965)

24. J. Am. Ceram. Soc., Baik, S. and Becher, P.F., 70, (8), 527 (1987)

25. J. Mat. Sci., Bryant, W.A., 12, 1285 (1977)

26. J. Mat. Sci. Lett., Moriyama, M. and Kamata, K., 6, 1141 (1987)

27. 'Chemical Vapor Deposition - 5th Internat. Conf. 1975', Gebhardt, J.J. et al, publ. Electrochem. Soc. Inc., 786 pp.

28. J. Mat. Sci. Lett., Yamane, H. and Hirai, T., 6, 1229 (1987)

29. J. Apple. Phys., Vora, H. and Moravec, T.J., 52, (10), 6151 (1981)

30. J. Mat. Sci., Je, J.H. and Lee, J.Y., 20, 839 (1985)

2.4.2. Refractories

Dr. Ir. J.T. Van Konijnenburg

Structural Ceramics - Hoogovens

IJmuiden, NL

1. INTRODUCTION

Over the last decade the demand for refractories has declined significantly in the industrialized world. This trend has also been followed by Western Europe. Production amounted to 7 M tonnes in 1974 in the EEC and is expected to be half of that figure by 1993 (1).

The major reasons for this decline are :

- decline of the production levels of the user industries;
- technological and qualitative improvement of the refractory materials giving longer service life;
- changes in the processes used by the consumer industry using less refractory materials but of higher quality however.

As Table 1 shows, the refractory industry is very much dependent on the iron- and steel industry, which accounts for the major share of the declining demand.

The influence of the technological and qualitative improvements and the changes in the steel processes on the specific refractory consumption are shown in the Figs. 1 and 2.

2. STATE OF THE ART OF REFRACTORIES

The state of the art of refractories is well documented (2, 5, 6, 7, 8).

Table 1 : Refractories sales by end users
(compiled from several sources)

End users	% of sales
- Iron- and steel industry	72
- Non ferrous industry	2
- Glass industry	5
- Cement/lime industry	5
- Chemical/petrochemical industry	2
- Power generation	1
- Ceramics industry	4
- Others	9
Total	100

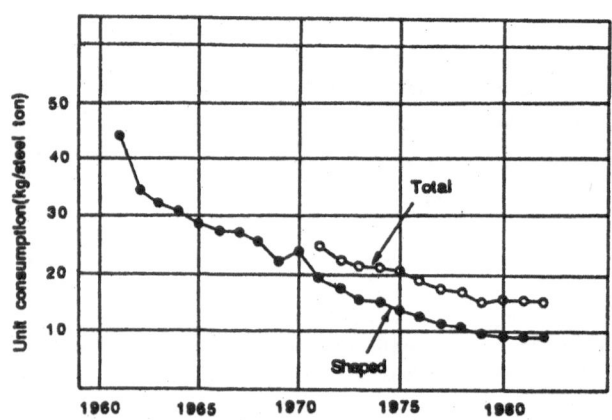

Fig. 1 : Specific refractory consumption (kg/t - crude steel)
in Japan (2).

Over the last decades the relative importance of fireclay
products has declined and the demand for high alumina products
and basic materials has increased.

An other important trend is the change from shaped fired
refractory products to unshaped refractory masses, refractory
castables, ramming mixes and gunning materials, which are formed
and cured in place.

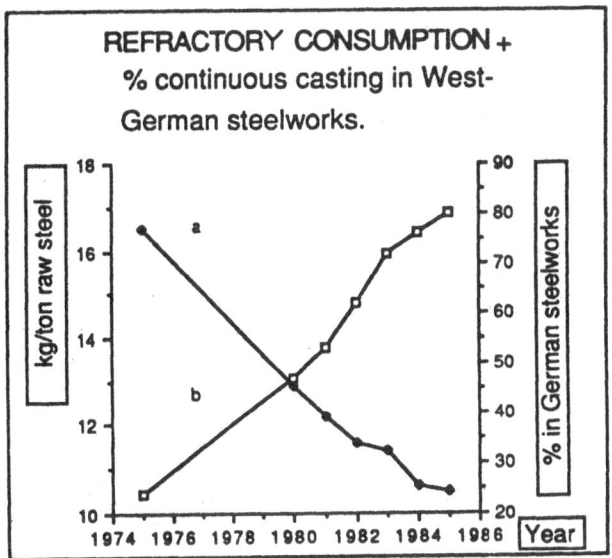

Fig. 2 : Specific refractory consumption (kg/t crude steel) in West Germany, in relation to the increasing use of continuous casting (% of total steel production) (3, 4).

The declining use of refractories and the changes in the types of materials used, has had a negative influence on the general knowledge of refractories. For instance the number of coking plants has declined as has the number of soaking pits in the steel industry, these installations formed the major consumers of silica refractories. Due to this decline many refractory producers have closed their silica production facilities and with that, knowledge of the silica production technology is deteriorating. This problem is detailed painfully in the Japanese round table talk about future trends in refractories (10).

On the other hand new products are being developed to fulfil the specific needs for refractories in new high temperature processes. Particularly for the steel industry new chemically bonded bricks are being developed. These bricks do not need any thermal treatment and have very good corrosion resistance. Their relative influence is shown in Fig. 3.

The need for refractory properties concentrates around the following characteristics :

1. resistance to corrosion;
2. thermo-mechanical behaviour;
3. installation of refractories.

In the subsequent sections these three items will be high-lighted.

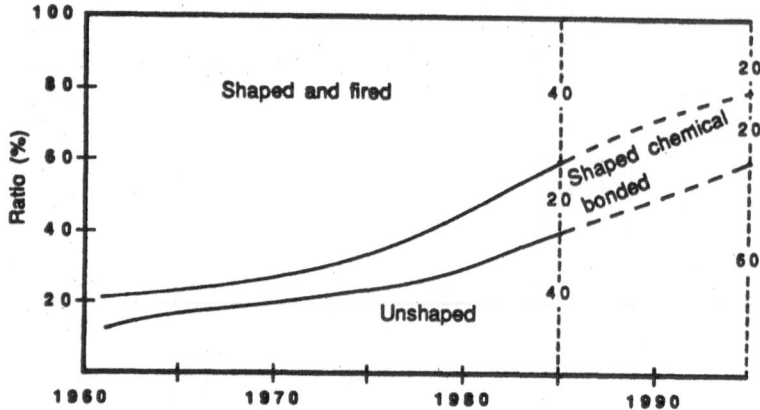

Fig. 3 : The relative change from fired shaped to chemically bonded shaped and unshaped refractories (2).

3. RESISTANCE TO CORROSION

Resistance to slags is of paramount importance for refractories. In this respect the incorporation of carbon into the refractories has been of major importance. Hayashi (2). Fig. 4 shows the situation of the various refractories such as $MgO-C$, Al_2O_3-C, $Al_2O_3-SiC-C$ in relation to their parent compounds.

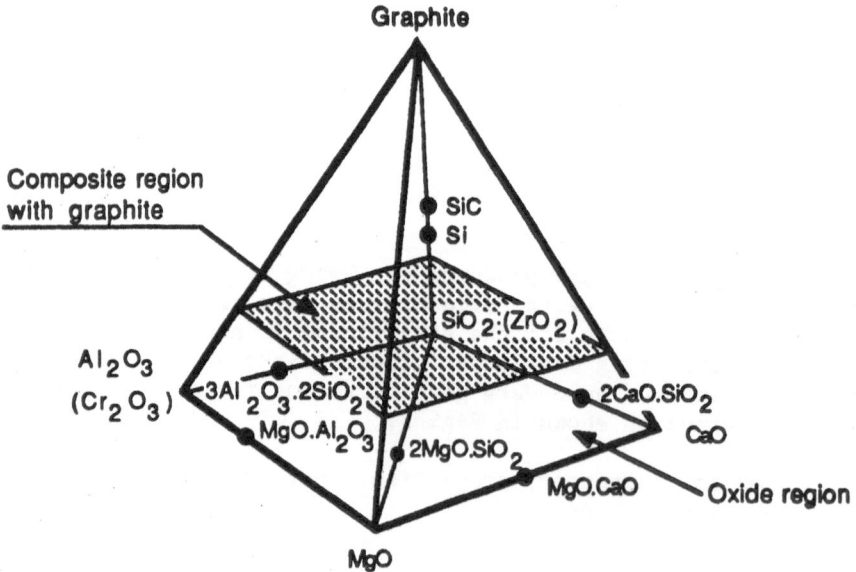

Fig. 4 : Development to non-oxide refractories.

Hayashi remarks here that the graphite is used to separate the grains of oxide materials instead of the ceramic bonding, which is too sensitive to slag attack. This enables the refractory to combine the properties of both material groups, viz. the oxidation resistance of the oxide with the slag resistance of the carbon.

The introduction of carbon in refractories can also be related to the type of behaviour at high temperature. In Fig. 5 the types of refractory A and B are well known. The siliceous material A yields viscous liquids giving rise to deformation and melting phenomena. Type B is called wet-basic or high-alumina because these materials are penetrated by slag or flux, bringing complex damage mechanisms. The liquid phase is hardly involved in the C type refractories, i.e. compositions with carbon.

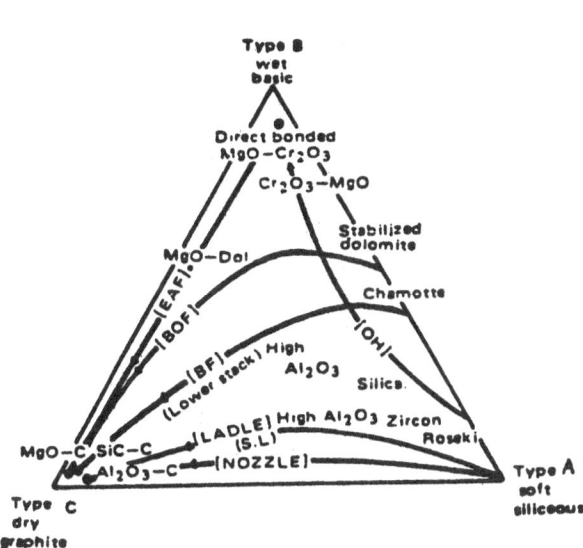

Fig. 5 : Classification by the behaviour of various refractories at high temperatures (2).

The development of MgO-C bricks initially for the BOF steelmaking process (19, 20), and later on also for hot metal pretreatment ladles (21, 22), has been a major step in the expansion of carbon containing refractories.

The introduction of resin-bonded bricks signified a great improvement over the pitch bonded types regarding environmental and refractory properties. Routschka and Majdic (19) showed the overall improved strength during heating up. Corrosion of this type of brick is three-fold : oxidation of the graphite, reaction of the MgO with the slag and volatilization of MgO at high temperatures (26).

The oxidation can be minimized by chosing a high quality graphite and by sealing the pores through reactions with so called antioxidants (for instance Al, Mg or Si) which react to carbides and/or oxides (19, 23).

The slag resistance is improved by using purer types of magnesite. Electro-fused MgO is now being investigated (23, 24, 25). However the addition of antioxidants sometimes decreases the slag resistance (23, 26). The importance of this development needs to be investigated in more detail in the future.

Volatilization of MgO is caused by the reducing atmosphere created by the presence of graphite. The MgO will disintegrate and Mg vapour will oxidize at the brick surface forming a dense surface layer (27). This dense layer forms a protection against further attack. This mechanism needs to be studied further in the near future.

Another development was the incorporation of carbon into alumina refractories, mainly used in the field of continuous casting machines (12, 13, 14) in the steel industry. A further step was the addition of graphite and silicon carbide to high alumina refractories, the so-called ASC-material. This was originally developed for use in torpedo ladles carrying liquid iron (15).

A comparison of slag attack and oxidation behaviour on Al_2O_3-SiC-C (ASC), spinel-C and MgO-C is given by Watanabe, e.a. (16).

The MgO-C specimen (containing 15 % C) showed the highest oxidation resistance. In the slag tests the MgO-C material was the best for both CaO and Na_2O based slags, the ASC materials showed a good resistance to the CaO slag but a rather poor resistance to the Na_2O slag. Spinel-C material showed a good resistance to the Na_2O slag, but was poor in the CaO containing slag. From experiments and trial runs in torpedo ladles Miyagawa, e.a. (17) concluded that the SiO_2 content in the ASC material should be minimized (less than 3 %); further they advised a C-content of 15 % and a SiC-content of 5 % at maximum as the optimal composition. In furnaces using these special bricks it is necessary to use specially developed mortars. For the near future the development of adequate mortars is of the utmost importance as was shown by the investigations of Watanabe (18). It was shown that deep open joints give rise to more severe erosion of the adjacent bricks. The effect of joint width was less marked.

Magnesium-aluminium spinel ($MgAl_2O_4$) refractories are reported to have high alkali resistance (28) as well as high resistance to steel making slags (29). Because of their high resistance to SO_3 and alkali these bricks are recommended for glass-furnace regenerators (32). Phosphate bonded spinel mortars are very resistant to molten aluminium alloys and even liquid magnesium (30).

Submerged nozzles for continuous casting of steel are often made of CaO-stabilized ZrO_2 and graphite. This material shows excellent corrosion resistance. The destabilization of the ZrO_2 causes mechanical wear however. This destabilization is caused by the SiO_2, Al_2O_3 and Fe_2O_3 originating from the ash in the graphite (32, 33).

An analogous problem arises in glass melting furnaces, when only yttrium stabilized zirconia seems to withstand the destabilization by SiO_2 (34).

The above references show that a lot of development work has been carried out in the field of corrosion resistance of refractories over the last few years.

4. THERMAL MECHANICAL BEHAVIOUR

The stability of the refractory construction depends on the thermal stresses in the construction and the bearing loads of the different parts. It is assumed that the process pressure at which an installation is used is not of great importance for the refractory, when it is possible to transfer the pressure to the outer metallic shell of the reaction vessel. When monolithics (castables or ramming mixes) are used, problems may occur since many monolithic materials have closed pores. In that case it is possible that the process pressure is released against the monolithic wall, which may cause cracking.

In modern refractory development two thermal properties are of the utmost importance : thermal conductivity and thermal shock resistance. Naturally other thermo-mechanical properties are also of importance but for this paper the discussion is limited to the properties mentioned.

4.1. Thermal conductivity

In most instances when constructing a refractory lining a choice has to be made between either cooling or insulating of the refractory and the steel structure.

Cooling limits the temperature at the hot face and diminishes slag or metal infiltration. Cooling is often a good solution in cases where slag and/or metal attack occurs and where fast temperature changes are important. In many cases where the furnace conditions are less severe insulation is used in order to save energy.

High thermal conductivity linings may consist of graphite and related materials (35) or silicon carbide (36) or even simply water cooled panels (37). Kennedy reports the influence of cooling under coal gasifier conditions (38). The trend towards cooled structures and accompanying refractories with high thermal conductivity is now well underway and changes the refractory scene significantly as was shown by Hayashi (2). The insulation option has been strongly advanced by the growing use of ceramic fibers of which (39, 40) give a detailed review.

4.2. Thermal Shock

In practice thermal shock often occurs between the service temperature and an intermediate temperature due to instabilities in the process, e.g. slag build-up on the lining of a blast furnace and spontaneous spalling of such slag layers, Winkelmann and Scott (41) and later Hasselmann (42) and Kienow and Fiedler (43) have derived a theoretical equation for the maximum rate of temperature change which a given refractory material can withstand without cracking. Kienow and Fiedler found the following equation :

$$\frac{dT}{dt} = \frac{4(1-\nu)}{3} \cdot \frac{M.\lambda}{\rho.E.C_p} \cdot \frac{1}{F}$$

$\frac{dT}{dt}$ = maximum temperature change allowed,

ν = Poisson's ratio,
M = modulus of rupture,
λ = thermal conductivity,
E = Young's modulus,
C_p = heat capacity,
ρ = density,
F = geometry dependent factor.

The dependence of Young's modulus and modulus of rupture was shown by Kawakemi, e.a. (44), who gave a ranking for Al_2O_3-C materials.

Resistance to thermal shock is enhanced by higher thermal conductivity. Smith e.a. (49) obtained this by incorporating 25 % graphite into a SiC brick.

The introduction of more powerful computers and computational models made possible more complex studies of thermal mechanical behaviour of refractories (45). Typical examples were published by En-Sheng Chen e.a. (46) on an installation for coal gasification and by Yoshimoto e.a. on the hearth wall and bottom of a blast furnace (47).

In this field the use of computer-aided design CAD must be mentioned (48, 49) a development which is especially valuable for very intricate shapes and which yields a direct input for the above mentioned thermo-mechanical calculations. A lot of work remains to be done in this field, especially where complex shapes and structures are used.

5. INSTALLATION OF REFRACTORIES

Furnace design has to be a well balanced arrangement between the expected process behaviour, thermal mechanical behaviour, furnace construction and the most expedient way to line a furnace.

As mentioned previously the total lining life will depend upon the behaviour of the bricks used but also on the joining mortars. The mortar quality is becoming more important nowadays when more sophisticated linings are being introduced and the process circumstances are becoming more severe.

The use of unshaped monolithic refractories is growing steadily. According to (51) in 1983, 30-35 % of the total European refractory consumption was in the form of monolithics, which figure is expected to rise to 40 % in 1993. In Japan that share was already attained in 1983 (52). In Nippon Steel's Yamata Works the figure is already 60 %! Two important books on unshaped refractories have been published recently (53, 54).

A very important development which is not yet complete, is the introduction of low and ultra low cement castables (55), enabling the use of low amounts of water (down to 4-5 %) and reducing the CaO-content considerably (less than 1,5 %) (56). The increasing importance of these castables can be shown with the increasing number of applications. They are now used in the petrochemical industry (57), blast furnace runners (58), continuous casting tundishes (59), etc. Magnesia and magnesia-chrome based, vibratory, thixotropic castables provide the possibility for new and faster lining methods for ladles (60, 61, 62).

The growing importance of the modern castables is producing a decline in the use of ramming materials. For quick repair of furnaces gunning materials are becoming of more importance (63). Special equipment is being developed for hot repair. These methods introduce the need to measure the lining wear. Various methods are currently under development.

6. FINAL REMARKS

As can be seen in the previous section development work in

refractories has been concentrating on corrosion resistance, thermal shock resistance and the increasing use of unshaped materials. For a number of applications synthetic raw materials are under investigation. It will be necessary to identify areas where synthetic raw materials can bring further improvements of a reasonable cost. For the future it is to be expected that in a number of critical areas technical ceramics will be introduced. Examples are rail in reheating furnaces, burner parts, heat exchangers, and breaking rings (boron nitride) for horizontal continuous casters. Many other possible applications will probably be developed in the near future.

An other area of interest is the balancing of bricks and mortars used in constructions. Special attention is required for the development of adequate mortars to go with SiC, Si_3N_4 and resin bonded bricks.

For larger constructions design criteria have to be established in order to avoid thermo-mechanical problems. As mentioned already heavily cooled linings are now under development. Examples are blast furnaces and electric arc furnaces. This trend requires new refractory protection which have to be thermally stable and easy to install in relatively thin layers.

A developmental area which is attracting a lot of attention is the ceramic heat exchanger. This development requires a complete new heat exchanger design with gas tight materials and a reliable joining technique. Such design has to be based upon the specific properties of ceramics. A further innovation is ceramic filters for use in fluidised bed combustion systems.

7. REFERENCES

(1) Industrial Refractories and Ceramics Market in Europe, Frost and Sullivan Ltd., London (1984).

(2) Hayashi, T., "Recent Trends of Refractory Technologies in Japan", Taikabutsu Overseas, 4 (1), pp. 3-19 (1984).

(3) Annual Report 1985, Didierwerke AG.

(4) Klages, G., Klein, A., Rubens, W. and Sperl, H., "Significance of Domestic Dolomite for Steelmaking in the Federal Republic of Germany", MPTCO (2), pp. 26-34 (1987).

(5) Kienow, S., "Refractory Materials", in Ceramics Monographs, A handbook of Ceramics, Part. 2.6., Verlag Schmid GmbH Freiburg (1979-1980).

(6) Harders, F., Kienow, S., Feuerfestkunde, Springer Verlag
     Berlin (1960).

(7) Chesters, J.H., "Refractories - Production and Properties",
     The Iron and Steel Institute, London (1973).

(8) Routschka, G., Jahresübersicht Feuerfeste Werkstoffe,
     Giesserei, 73 (11), pp. 328-337 (1986).

(9) Koltermann, M., Feuerfesttechnik 1980 - Rückblick und
     Zukunftprobleme unter besonderer Beachtung der Bereiche
     Prüfung und Klassification, Energie und Rohstoffe,
     Sprechsaal 113, pp. 663-667 (1980).

(10) A round table talk, Leaders of Steel and Refractory
     Industries Talk about Future Trends, Taikabutsu Overseas,
     4 (3), pp. 53-65 (1984).

(11) Tayoshi, T., Industrial Minerals Supplement, April (1983).

(12) Cooper, C.F., e.a., Trans. J. Br. Ceram. Soc. 84 (2),
     pp. 57-62 (1985).

(13) Ozgen, O.S., e.a., Trans. J. Br. Ceram. Soc. 84 (2),
     pp. 138-142 (1985).

(14) Fujimoto, S., e.a., Taikabutsu Overseas, 5 (3), pp. 3-6
     (1985).

(15) Nishi, M., e.a., Taikabutsu Overseas, 5 (2), pp. 29-34
     (1985).

(16) Watanabe, A., Taikabutsu Overseas, 6 (1), pp. 22-25 (1986).

(17) Miyagawa, S., e.a., Ceram. Eng. Sci. Proc., 7 (1-2),
     pp. 58-74 (1986).

(18) Watanabe, A., Taikabutsu Overseas, 7 (1), pp. 14-20 (1987).

(19) Routschka, G., Majdic, A., Keram Z., 36 (9), pp. 460-470
     (1984).

(20) Refractories for Steel Making, Proc. XXVIIth Int. Colloq. on
     Refr., Aachen 1984, Interceram. Spec. Issue, 34 (1985).

(21) Kyoden, e.a., Taikabutsu Overseas, 5 (2), pp. 13-17 (1985).

(22) Ohishi, J., Taikabutsu Overseas, 5 (3), pp. 21-25 (1985).

(23) Horio, T., e.a., Taikabutsu Overseas, $\underline{6}$ (1), pp. 11-15 (1986).

(24) Miyagawa, Y., e.a., Taikabutsu Overseas, $\underline{6}$ (1), pp. 16-21 (1986).

(25) Naruse, Y., e.a., Ceram. Eng. Sci. Proc., $\underline{7}$ (1-2), pp. 119-130 (1986).

(26) Ameniya, Y., e.a., Taikabutsu Overseas, $\underline{6}$ (1), pp. 26-29 (1986).

(27) Lin, Y.L. and Semlen, C.E., Ceram. Eng. Sci. Proc., $\underline{7}$ (1-2), pp. 27-39 (1986).

(28) Bartha, P., Glastech. Ber., $\underline{58}$ (10) pp. 288-294 (1989).

(29) Carbone, T.J., Interceram Special Issue, $\underline{34}$, pp. 91-94 (1985).

(30) Cisar, A., e.a., in R. Fisher Ed., New Developments in Monolithic Refractories; Advances in Ceramics $\underline{13}$, The Amer. Ceram. Soc., pp. 411-418 (1985)

(31) Kettner, P., e.a., Interceram Special Issue, $\underline{35}$, pp. 14-16 (1986).

(32) Elstner, J., e.a., Fachberichte Hüttenprax. Metall., $\underline{22}$ (5), pp. 450-457 (1984).

(33) Oki, K., e.a., Taikabutsu Overseas, $\underline{4}$ (2), pp. 42-48 (1984).

(34) Aratani, K., Interceram Special Issue, $\underline{35}$, pp. 61-64 (1985).

(35) Van Laar, J., e.a., Ironmaking Proc. $\underline{37}$, pp. 180-192 (1978).

(36) Brown, R., Iron- and Steelmaker, May (1983).

(37) Oberbach, M., e.a., Workd Steel and Metalworking, $\underline{6}$, pp. 188-194 (1984/1985).

(38) Kennedy, C.R., J. Mat. Energy System, $\underline{2}$, pp. 11-20 (1980) $\underline{3}$, pp. 39-47 (1981).

(39) Routschka, G., Majdic, A., Stahl Eisen, $\underline{103}$ (19), pp. 929-934 (1983).

(40) Routschka, G., Majdic, A., Keramische Z., $\underline{33}$ (9), pp. 511-522 (1981).

(41) Winkelmann, W., Scott, D., Ann. Phys., 51, pp. 730 (1984).

(42) Hasselmann, D.P.H., Krohn, D.A., J. Ameri. Ceram. Soc., 55, pp. 208 (1972).

(43) Kienow, S., Fiedler, U., Jeschke, P., Ton Ind. Z., 100, pp. 181 (1976).

(44) Kawakami, T., e.a., Taikabutsu Overseas, 5 (4), pp. 15-20 (1985).

(45) Padgett, G., Refractories, J., 53 (4), pp. 13-20 (1978).

(46) En-Sheng Chen, e.a., Amer. Ceram. Soc. Bull., 64 (7), pp. 982-1000 (1985).

(47) Yoshimoto, World Steel Metalworking, 5, pp. 10-20 (1983).

(48) Lütcke, H., TIZ Fachber., 109 (8), pp. 566-571 (1985).

(49) Burton, H., Fachber. Hüttenprax. Metallw., 27 (4), pp. 537-539 (1986).

(50) Smith, P.L., e.a., SIPRE WB IV meeting April (1987).

(51) Frost and Sullivan Ltd., Industrial Refractories and Ceramics Market in Europe (1984).

(52) Moritama, N., Taikabutsu Overseas, 6 (2), pp. 18-36 (1986).

(53) Nishikawa, A., Technology of Monolithic Refractories, Plibrico Japan, Tokyo (1984).

(54) Fisher, R., Ed., New Developments in Monolithic Refractories, Advances in Ceramics 13, The Amer. Ceram. Soc. (1985).

(55) Lankard, D.R., Lease, D.H., Refractories J., 61 (6), pp. 6-13 (1986).

(56) Seltveit, A., e.a., Ceram. Eng. Sci. Prog., 7 (1-2), pp. 243-260 (1986).

(57) Banerjee, S., Ceram. Eng. Sci. Prog., 7 (1-2), pp. 236-242 (1986).

(58) Howe, R.A., e.a., Ceram. Eng. Sci. Prog., 7 (1-2), pp. 261-266 (1986).

(59) Soejima, T., e.a., Taikabutsu Overseas, $\underline{6}$ (3), pp. 3-8 (1986).

(60) Hösler, M., e.a., Fachber. Hüttenprax. Metallw. $\underline{23}$ (5), pp. 361-366 (1985).

(61) Pitchford, T., e.a., Trans. J. Br. Ceram. Soc. $\underline{84}$ (6), pp. 207-212 (1983).

(62) Klein, W., e.a., Fachber. Hüttenprax. Metallw. $\underline{23}$ (10), pp. 786-794 (1985).

(63) Siegl, W.M., Radex Rundschau 1985 (4), pp. 706-723.

3. MATERIALS PRODUCTION

3.1. Alloys

R. Brunetaud

2 Rue Octave Feuillet

F-75116 Paris

3.1.1. Alloy Production*

1. INTRODUCTION

The production processes used in the preparation of superalloys for the highest temperature components of jet engines have in most respects been ahead of those used for other materials. This section on fabrication is derived largely from the technology of superalloy component manufacture. Much of the technology is transferrable to other materials and it seems likely that over the next decade much R and D in industry will be devoted to doing just that. Not everything is readily adapted. For example titanium is chemically extremely reactive and would pick up impurities from any crucible material. Both titanium and aluminium form impermeable oxide films making sintering of powder a very difficult process. Alloys currently favoured for HT Petrochemical processes are not readily formed either cold or hot. Nevertheless for all these and other HTMs there are some aspects of superalloy production which are being or could be adopted with benefit.

2. INGOT PRODUCTION

2.1. Melting and Casting

After a period of intensive development of new techniques

---

* Conclusions and references for Sections 3.1.1. and 3.1.2. are included after Section 3.1.2.

including ladle refining, continuous melting, vacuum induction melting, vacuum arc remelting and electroslag remelting, the late seventies and the eighties were devoted to detailed improvement of those processes.

Better understanding of the physico-chemical phenomena has been the key to the achievement of greater reliability and more accurate control of micro-structure. In parallel, economic pressure has driven producers to operate bigger units with all their extrapolation problems and to use less raw material by recycling a higher ratio of scrap.

New developments are emerging : increased automation of processes using sophisticated control models, improved instrumentation of furnaces with computer assisted production units, novel techniques such as electron beam and plasma melting or drip casting open new horizons for melting and remelting to give improved cleanliness, soundness and structural control.

Primary Melting

Developments in primary melting are aimed at improving quality by removal of undesired elements, decreasing operating cost and widening the range of materials that are melted (1).

For recycling this implies a reconditioning operation before loading into the melting furnace. Progress has already been made for massive scrap in terms of identification and cleaning. Improvements are needed for light scrap such as machining chips.

Evolution of primary melting techniques addresses both classical routes, i.e. the vacuum induction furnace and the newer techniques such as cold hearth melting and refining with or without an associated induction melting unit (2).

From a thermodynamic point of view these new methods can be directed to :

- reaching the casting temperature and composition faster
- controlling undesirable reactions with the environment (crucible or atmosphere)
- improving removal of unwanted trace elements (3).

These objectives for vacuum induction furnace technology can be achieved by:

- Modification of the shape of the crucible, and choice of frequency to control the stirring of the molten metal to reduce the mechanical-chemical erosion of the crucible (4). More inert refractory linings are needed (bricks of different composition and improved properties obtained by modifying the manufacturing process).

- Improvements of the vacuum system (better pump characteristics, improved filters and better positioning of the ducts) to give the best efficiency for fast reaction but avoiding splashing of the crucible walls to minimize contamination by refractory material.

- A new technique for gas injection through a porous plug at the bottom of the crucible (already tested on several furnaces). This opens new horizons to superalloy melters for modification of the reaction kinetics, the degree of equilibrium control, and the possibility of removal of unwanted trace elements.

Reactions with the crucible lining can be avoided by cold hearth melting, with an electron beam (8), a plasma (5) or a laser. Although these techniques have been used for some time for refractory metals we would expect a more widespread application to superalloy productions in the future. The main problem is the control of vapourisation of elements such as Cr, Mn, Ti. The costs in terms of capital investment, energy and maintenance are still high. Scaling up may also give new problems.

For primary casting, an adaptation of the filtering technique after induction melting is to be expected. Filter materials for casting large tonnages are not yet readily available but this should not discourage manufacturers from designing new tundishes. As long as tundish reactions are unavoidable, the right strategy will be to design the equipment in such a way as to minimise the risk of entrapping dross in the ingot. Even for this apparently simple task it will be necessary to discover and understand all the relevant hydrodynamic and thermomechanical factors.

2.2. Control of the Solidification Process

This is certainly the most complex and diverse aspect of processing. All developments will have the objectives of avoiding contamination, controlling the structure and avoiding major solidification defects especially all types of segregation. Additions to the melt and mould walls to nucleate solidifications have been well developed. The resulting ingots are more readily hot worked.

The production of high quality ingots is by the melting of consumable electrodes (6). Melting under vacuum has now reached maturity and the limitations of the process are known. Improvements are still possible and necessary in terms of regulation of power, heat removal in the crucible as a function of quality of electrode (heterogeneities, porosity pipes, cracks, etc.), and control of heat dissipation during the melting sequences. The process is well established also for production of titanium and zirconium alloys, certain stainless steels, nuclear fuel cladding alloys, and some nickel base alloys. For a very reactive metal such as titanium, consumable arc melting in good vacuum is essential to avoid reaction with refractories or slags.

Electroslag remelting for superalloy ingot production is far from being fully developed. This process potentially gives a better ingot surface, and therefore a better forging yield as well as a better solidification structure less prone to segregation due to the shallower pool of molten metal. On the other hand, it requires a complex strategy of slag management to control chemical reactions between alloy, slag and atmosphere. Some limitations appear also in the control of grain size of large ingots.

Developments should be directed to elimination of defects : segregation, freckles, tree ring effects and white spots (9, 10). This will be achieved through improvements in electrode quality and elimination of mechanical vibration and instabilities of the power generation equipment. Despite any progress that can still be achieved in this field, the consumable electrode remelting techniques VAR and ESR will always be limited in output by the rate at which heat can be removed from the ingot.

In some cases fine as-cast structures are required. Several processes with better control of heat input for melting and heat extraction from the mould, are available or are under development. The VADER process (20) with two horizontal electrodes one of which rotates produces a fine solidification structure.

Another technique using electron beam (7) or plasma refining allows adjustment of the solidification conditions in a slowly rotating mould with permanent "hot topping". Structures are finer but freedom from pores has not yet been achieved. For some products, the drip casting technique might be considered as a partial substitute for the powder route. Despite the definite interest of this approach, there has so far been no report of industrial exploitation.

2.3. Rheocasting/thixocasting and forging,

These two very similar processes cover the range between pouring a molten metal into a mould and hot forging in a die; in

that the mould is filled by gravity or by applying pressure with
the alloy in the partly frozen state. If a static pool of metal
is allowed to freeze partially the solid phase will be in the
form of interlacing dendrites so that the "paste" does not flow
readily and any force applied to the paste is liable to cause it
to crack. If however the melt is vigourously stirred as it
freezes the dendrites break up and spherodise giving a very fluid
structure. Even with a high proportion of solid spheres, the
viscosity is about the same as that of lubricating oil. In rheocasting the partly frozen material is stirred and poured
immediatly into a die or may be pressure die cast. For thixo-
casting the stirred melt is cast into blocks which are
subsequently reheated to cause partial melting before being
formed to shape.

Among the many advantages of the process are:
- no need for heavy forging loads.
- die wear is reduced.
- casting temperature is lower and there is less latent heat
  therefore thermal shock of dies is reduced, and dies operate
  at a lower temperature.

Part of the solidification shrinkage has taken place before
the mould is filled and less remains to occur in the mould,
therefore shrinkage cracks and cavities are less likely to occur.
The type of micro porosity which forms between the dendrites in
conventional casting is eliminated. These processes are very
difficult to achieve reliably in practice. Attempts to use them
for commercial production of copper alloys have been made in the
U.S.A and there is interest in combining them with introduction
of ceramic fibres in France. R & D in this area for producing
improved HTM's should not be overlooked.

3.   POWDER PRODUCTION

3.1. Atomization Processes

The prealloyed powder technique represents and will
certainly remain the main alternative to ingot casting for
wrought superalloys operating in the 500°/800°C temperature
range. The main processes are the following :

- Argon Atomization (currently the major process) : A stream of
  molten metal from a VIM furnace is directed into a jet of
  argon gas and the resulting droplets are blown into a cooling
  tower.

- Vacuum Atomization : The molten alloy, in a crucible under
  argon pressure, is saturated with hydrogen and then forced
  through a ceramic tube into a vacuum chamber. The liquid

metal charged with hydrogen sprays out in droplets as a result of the rapid expansion of hydrogen gas as it enters the vacuum chamber. This process is currently used for gas turbine components. Because titanium would react chemically with any refractor crucible these atomisation processes cannot be used. Alternatives are 1) separation of fines produced as part of the standard metal production process, 2) by converting metal to hydride and then back to metal when a powder is produced, HDH and 3) REP as follows.

- The Rotating Electrode Process (REP) : In an argon filled chamber an electrode turning at great speed (15,000rpm) is melted at the extremity by a plasma or an electron beam arc and the liquid is flung off as droplets, cooling in the cooling chamber. This process is still experimental. An advantage is the uniformity of powder particle size, but the minimum particle size is 150 μm and particle composition varies reflecting the heterogeneity of the electrode.

After forming, the powders are sieved and degasified and put into containers, (all under controlled atmosphere). Cleanliness is essential. In order to minimise the influence of inclusions, an upper limit must be placed on particle size. An acceptable limit from the point of view of the risk of crack nucleation in fatigue at inclusions is 50/80 μm. Atomisation under argon or under vacuum gives a relatively acceptable rejection rate with this size limitation. Larger grain size gives better creep strength needed for example for turbine discs working at over 600°C in the new generation of aero engines. With the expected reduction of size and number of inclusions in future in ingots and reduction in the amount of segregation, etc. interest in the REP will grow only if it can offer improved quality with lower investment and production costs.

Testing methods for estimation of inclusions have been extensively studied (30-36), the most widely used being:

- optical micrography   particle size: 20 - 50 μm.
- elutriation (fractionation
  in liquid or gas)   particle size: 50 - 100 μm.
- scanning electron micro-
  scopy of remelt buttons(35)   particle size: very small
- ultrasonies   particle size: > 90 μm.

3.2. Rapid Solidification

The best known process to obtain rapidly solidified powders (cooling rate 10^6°C/sec, compare with 10^2°C/sec. for classical atomization) is that of spraying a thin jet of liquid metal onto a rotating wheel turning at a very high speed. The jet is dispersed into droplets, centrifuged and cooled by helium (19). Powders so produced of particle size 10 to 100 μm, with very fine micro-structure, begin to melt at temperatures of 45 to 80°C higher than powders with classical structure. Three ODS/ superalloys are currently produced commercially on a several hundred ton p.a. scale, and further developments are expected.

3.3. The Particular Case of ODS Alloys

The introduction of insoluble dispersed oxides to retain mechanical properties to temperatures close to the melting temperature, has for a long time interested metallurgists but the development of this kind of material has been difficult. Any melting which agglomerates oxides must be excluded. The development of the mechanical alloying process in the 70's was the starting point of several ODS alloys (33). The elementary powder or the prealloyed powder is mixed with an oxide, e.g. Y_2O_3 at a concentration of 0,5 to 1,1%. The mixture is put in an attritor (ball mill) arranged either as

- an Archimede's screw containing the balls,
- a drum, rotating at a speed just less than the centrifuging speed of the balls.

Industrially, the second type of machine is now constructed to a size of 2 m diameter, 2,6 m length with 10 tonnes of iron balls for 1 ton powder. The powder particles submitted to this high energy attrition are welded together and successively milled apart and rejoined until they form a relatively coarse-grained powder, in which yttrium oxide spheres of 30 Å diameter are finely and homogeneously dispersed.

4. FOUNDRY CAST COMPONENTS

4.1. Development of Applications and Principal Problems

Centrifugal casting is used for making long heavy gauge tubes of chrome nickel iron base alloys for the chemical and petrochemical industries, etc..
Progress in precision casting, by the lost wax process (11-23) has evolved considerably during the last 30 years, for making turbine blades, nozzle guide vanes and secondary components, complete monoblock turbine rotors, and complex casings.

Major difficulties are as follows:

- Inclusions coming especially from the crucible and from metal handling (improvements have been brought about by the introduction of a ceramic filter);

- Choice of ceramic for the shell mould and the core for good mechanical strength and non-contamination of the cast metal;

- Risk of cracking due to low ductility of the partly frozen or very hot metal and the shrinkage stresses appearing during cooling;

- Macro and microporosity which have deleterious effects on performance in service;

- Heterogeneity of grain size arising from the local conditions of solidification;

- Production of satisfactory welded assemblies or repairs of foundry defects (particularly for large castings).

- The high cost of quality assurance procedures.

4.2. Ceramics for Shell Moulds and Cores

The preferred ceramics are principally zirconia based, supported on silica or alumina, bonded by ethylene silicate gel or colloidal silicates. They are required to have high temperature strength and thermal expansion coefficients relatively close to that of the metal, non reactivity with the metal (difficult to achieve with alloys containing additions such as boron or hafnium) and retain the ability to be shaken out. Further, since wall sections or core may be as thin as 0.4 μm, the ceramics must exhibit reasonable mechanical strength.

The introduction of directional solidification (DS) exacerbates the difficulties. Conventional investment castings solidify in a few minutes; DS however requires higher temperature (approaching 1500°C) over a period of 20 minutes and demands refractory ceramics of high purity with adequate creep strength.

4.3. Microporosity Problems

Elimination of microporosity presents a severe production problem which is increased by the use of recycled material (32). Major studies of the evolution of structure have been undertaken by differential thermal analysis and microscopy of materials quenched during directional solidification (16).

By following the formation of dendrites and carbide precipitation, it is possible to reveal and understand the process of porosity formation. For example, addition of boron decreases the carbide precipitation temperature blocking the interdendritic space and consequently plays a useful rôle.

Conversely only a few ppm of nitrogen can be damaging by forming carbonitrides often connected with oxides. Currently alloy specifications limiting nitrogen to 5 ppm for certain nickel-base alloys are in force. Microporosity can be removed by hot isostatic pressing (HIP) (17, 18) with improvement of mechanical properties. Often, however, the time-temperature cycle for HIP is not compatible with the optimum thermal treatment. Moreover HIP is an expensive technique and is not included in normal production routines, but rather it is used to recover components which would otherwise be rejected owing to excessive porosity.

4.4. Non-Destructive Testing Methods, NDT

Control of quality by the use of NDT is of increasing importance in industrial processing. The principal techniques are :

- radiographic inspection;
- surface crack detection by dye penetrant or, for ferritic materials by magnetic method;
- grain size control by macrography.

The latest developments include:

- X-ray examination with automatic image analysis is used on hollow parts to detect the residues of the core removed by a chemical process. It is also expected to detect smaller defects (e.g. inclusions, porosity) using superconducting X-ray sensors.
- X-ray tomography has been adapted from the medical field to industrial usage to monitor wall thickness or the arrangement of cooling circuits.
- automatic image analysis in surface crack detection by macrographic inspection.

5. FOUNDRY PROCESSES

5.1. Equiaxed Castings

The classical lost wax process will not be described here. Nevertheless the production of shell moulds by building up by repeated immersion of the pattern in a slurry is still a delicate operation which is now automated and performed under controlled conditions of humidity and temperature, to give improved uniformity and reliability.
Grain size can be controlled by the introduction of innoculants, usually cobalt oxide, in the first ceramic layer but remains a problem for thick walled components. Here research is required on rapid cooling techniques, or the development of mechanical methods, e.g. by vibration including ultrasonic vibration of the

melt or shaking of the mould during solidification to break up and disperse the dendrites as they form, so innoculating the rest of the melt (26).

5.2. Directional Solidification

For turbine blades, the presence of grain boundaries has a major influence on high temperature creep and fatigue properties. The development of casting by plane front directional solidification (12-19) (DS) produces elongated grains, parallel with the length of the blade, conferring greater creep strength, with adequate ductility in the direction of maximum tensile stress (37).

For modern industrial production, blades are made by moving the melt slowly from a hot furnace into a progressively cooler zone (at a speed of about 30 cm/h) to ensure a planar growth front of the alloy grains. For particularly small pieces, lateral heat loss is controlled by use of exothermic powder on top of the mould (15). A steep temperature gradient must be maintained in the liquid metal to prevent formation of dendrites ahead of the solidification front.

5.3. Directionally - Oriented Eutectics

DS eutectic alloys are prepared from an original melt of eutectic or near-eutectic composition, by directionally freezing at low speed through a steep temperature gradient. Structures are typically fibres or lamellae of a stiff reinforcing phase, of carbide, oxide or intermetallics, entrained within an alloy, or intermetallic matrix with an orientation orthogonal to the solid/liquid interface during solidification, providing a natural, "in-situ", composite.

Important for alignment is maintenance of a planar solidification front, especially where the ingot cross section changes along its length. For a particular alloy, planarity of the front, and regular defect-free crystallisation conditions are influenced by the ratio of temperature gradient to solidification rate (G/R).

Large temperature gradients give containment problems and moulds with sufficient strength are currently too expensive. These very high cost products were envisaged for applications as gas turbine blades, but despite their early promise 25 years ago, nearly all development work has now been discontinued.

5.4. Single Crystals

Since the mid 1960's, the technique of directional

solidification has been used to manufacture single crystal superalloy components. The alloy melt is directionally solidified on to a seed monocrystal in the mould base, which is either artificially inserted, or is generated by solidification up a helical channel from which all crystal orientations except one are terminated at the channel wall (14, 19, 26, 31). On an industrial scale, normal DS techniques are used, with a maximum temperature gradient across the solidification front, to minimise microporosity and to produce a very fine dendritic structure for improvement of mechanical properties. The current trend is to replace the circular clusters by single blade moulds allowing better control of thermal gradients.

Interdendritic microporosity, which is difficult to detect impairs mechanical properties; it can be reduced by subsequent hot isostatic pressing.

5.5. Rapid Solidification Techniques

Nickel base alloys

The deposition, by spraying molten droplets, either directly from an atomiser or by remelting powders, onto a preform or a cooled mandrel, gives cooling rates of the order of 10^5 to $10^5 °C$ sec^{-1}. Components may be built up by deposition of successive layers, with very fine grain structures and good micro-homogeneity (28). Industry has so far been slow to adopt the technique mainly owing to the incidence of porosity and of high cost of production.

Similarly, the technique of "melt spinning" produces rapidly solidified metal. Here a jet of liquid metal is deposited on a strongly cooled wheel edge. The resulting "tape" of several millimeters width is detached and wound continuously on to a spool. The technique is used to produce nickel-base alloy tapes for brazing. Particular advantages are :

- the production of compositions impossible to form by classical means;
- high ductility, owing to a microcrystalline structure, only obtainable by RST.

6. MODELLING

Design and commissioning the production of a new cast component can be a long and expensive process. This has stimulated the preparation of computer codes using finite element analysis to model the heat flow from a mould filled with liquid metal as freezing takes place. The sequence of temperature patterns can be used to predict the way in which the grains of solid will form and grow as a function of cooling rate and temperature gradients in the various parts of the component. Also the distribution of contraction on cooling can be used to indicate where shrinkage cracks might occur or porosity might develop, (24). In this way the time for development and its cost can be reduced significantly. Studies are in progress to use similar techniques to optimise pouring temperature and rate feeding, hot topping and distribution of chills.

3.1.2. Cold and Hot Working and Finishing

R. Brunetaud

2 Rue Octave Feuillet

F-75116 Paris

1. CLASSICAL METALLURGY

1.1. Billet, bar, sheet metal

Forming processes, not only produce the required shapes but also must be designed to homogenise and refine the structure. The metallurgical processes involved are generally well understood but their application sometimes still runs into difficulties. In particular, the coarser structure of very large ingots can be difficult to break down. Some alloys have a very narrow forging range of temperatures and so require very precise temperature control.

Accurate control of furnace temperature and fast acting presses allow a more uniform hot working sequence by a closer control of heating and heat losses during forging. Better control of working is also ensured by automated process control based on hot working models which take into account thermal and rheological effects of the deformation sequence. This is all the more necessary when the primary working has to be done with the same accuracy as the final deformation to guarantee a structure suitable for subsequent processing and able to meet well defined requirements which are controlled by non-destructive testing for the different types of products from billets to sheet (13).

Extrusion can give in a single operation complete breakdown of the cast structure and a fine, controlled grainsize. A special mention is made of electron beam melted materials refined, drip cast and extruded, and the consolidation-extrusion of powder mixtures.

Austenitic steels lend themselves well to cold forming. In particular, roll forming on a mandrel has been developed to make cylindrical and partly tapered tubular components in one piece.

The limited European market for superalloy sheet makes production in Europe difficult to justify economically.

1.2. Forging

Press forging is often used, with hydraulic presses up to 60.000 t for large pieces and "screw" presses for the smaller sizes. Hot die forging (21) is used for forming over a narrow range of temperature at a normal speed. When superplastic deformation is possible as for titanium alloys or for powder metallurgy nickel base alloys, isothermal forging can be employed at a low speed with a very good metallurgical control giving the possibility to produce "near net shape" parts (21).

For forging large gas turbine blades a preform made by a numerically controlled multihammer gives a precise distribution of metal for final forging. This last is done in such a way that the adiabatic heating compensates for the radiation losses, thus giving in practise almost isothermal forging conditions (34).

An example of the integration of the forging operation with final heat treatment is given by the fabrication of high strength turbine discs, in IN718 DA (Direct Ageing). The final forging sequence at relatively low temperature (900-980°C) gives a fine grain structure with high dislocation density, and is followed directly by ageing treatments. A grain size of ASTM 8/10 with very fine γ' precipitates is obtained. This confers a high elastic limit and resistance to low cycle fatigue up to 550°C.

2. POWDER METALLURGY

2.1. Atomised powders

Powder metallurgy has opened large perspectives (27), but also poses many problems (22). The first fabrication method developed during the 70's for the manufacture of gas turbine discs, was based on extrusion to compact the powder and isothermal forging; it is still the most widely used process. Extrusion requires a high load (> 30,000 tons for bars of 250 mm diam.), to give a fine grain structure. Isothermal forging is easy since the extruded rod is superplastic at the forging temperature, therefore relatively low capacity presses are suitable (only 8000 tons for turboreactor discs). This operation requires however, dies with high mechanical strength up to

1100°C, the molybdenum alloy, TZM, is used in vacuum or argon to prevent oxidation, tools in zirconia have proved to be too fragile, but it is hoped that tools in cast nickel base alloys, heavily alloyed with tungsten or molybdenum will allow forging in air. The nickel base super-alloy IN 100 is already in use for forging titanium alloys.

The fabrication of components to "near-net-shape" by HIP, – hot isostatic pressing – was developed with the hope of considerably reducing costs but has been found to be too sensitive to inclusions and microsegregation for good fatigue resistance. Forging reduces the influence of defects by promoting diffusion. The HIP of superalloys is itself quite expensive since several hours at 1200°C and 100 MPa pressure are required to achieve full density.

The industrial development of near-net-shape sintering has been practically discontinued. Satisfactory results have been obtained by forging preforms made by HIP, but these do not demonstrate superplasticity and need high capacity forging facilities.

The manufacture of preforms by semi-liquid sintering (29) followed by forging has been tried experimentally to replace the expensive operation of HIP, but without great success. Alloys containing boron are suitable for semi-liquid sintering both by HIP or by forging or milling, but have not reached the industrial level of development.

It is possible to produce plate by rolling of very high performance alloys such as Astroloy from semi-finished sheet fabricated by powder metallurgy. The product is fine grained and weldable.

Extrusion of alloy powders with oxide dispersion strengthening (ODS) yields a dense composite with elongated recrystallised grains. Such an oriented structure confers satisfactory ductility in the longitudinal direction (33), by eliminating the transverse grain boundaries which provide a path for fracture.

2.2. Modelling

A limited amount of data can be obtained from tests with model materials such as plasticine or lead-tin alloys, giving empirical data on the distribution of deformation and the stresses on tools. More usually the methods of finite element analysis modelling (38, 39, 40) are called upon to follow deformation, stress and temperature during forming. These methods are complex requiring a system of automatic re-iteration

of finite elements and a good knowledge of the rheological and
thermal characterisitics of the material, as well as the
conditions of friction on the tool. Computer codes have been
developed for tools of axial symmetry, which facilitates the
optimisation of forging conditions to obtain the desired
structures and to assist design. Codes have also been developed
for forging in three dimensions.

3 MACHINING OF SUPERALLOYS

3.1. Standard Machining (41)

The properties of nickel base high temperature materials
that are of importance in machining are:

- poor thermal conductivity giving higher temperatures in the
  cutting zone;
- high strength from room temperature up to the highest
  temperature reached in the cutting zone.
- rapid work hardening.

High speed steels which have sufficient toughness to
withstand local shock loading are conventionally used for
machining. For tools requiring higher performance (tapping),
steels of the type M42 or T15 are used. For drilling and milling
carbide structures obtained by powder metallurgy are showing
20-80 % life improvement and limited deformation during heat
treatment.

Typical examples of cemented carbides are the ISO types K-10
to K-20. They are subject to deterioration by notch wear.
Particles from the cut material stick to the tool surface and
periodically remove carbide fragments. It is necessary to have a
fine dispersion of carbides and ground tool face geometries with
a clean cutting edge.

The application of coatings to tool tips is not advantageous
because they give a rounding of the cutting edge, which causes a
marked work hardening of the material. Cutting speeds of the
order of 20-40 m/min., are judged to be adequate for Ni and Co
alloys, compared with 100 m/min for steels.

The major advances are with ceramic tools which can work at
temperatures approaching 1200°C at the cutting edge, compared
with 700°C for carbides (before oxidation becomes serious). Most
widely used are Al_2O_3 + TiC or Si_3N_4 + Al_2O_3 (sialon) giving
cutting speeds of 130-200 m/min. The brittleness of the
materials, however, restricts their use to very high speed
turning, and surface milling, requiring machines of high rigidity
and power. New developments are in the field of ceramic

composites e.g. Al_2O_3/SiC, (trichytes) (42) which have high
thermal conductivity and lower expansion coefficients, both of
which reduce thermal shock sensitivity; and have some toughness
to resist shock loading.

Surface grinding is used to obtain high surface finish and
precise dimensions, and for alloys with poor machinability gives
a higher productivity than that achieved by conventional turning
or milling; finishing by shot blasting is needed to generate
compressive surface stresses and achieve good fatigue properties.

Contour machining with simple computer control of lathes and
millers in which the tool is positioned by synchronous motors can
also be done. The successful introduction of adaptive control
would be of particular benefit in the machining of high strength
HTM's. For adaptive control instruments sense machine torque,
tool loading, tip temperature and vibration and the measurements
are used via a computer to control spindle speed and feed and
tool position in such a way as to optimise process speed, tool
life and surface finish.

3.2. Chipless Machining

Electro discharge machining (EDM) is simple and flexible,
with electrodes of copper or graphite. Wear of the electrodes is
of the order of 1 % of the machined volume. With a current
density of the order of 100 A/cm^2 the speed of machining is 0.1
to 0.2 mm/min. Cutting using a copper wire electrode gives a
fine accurate cut (thickness of 0.3 to 0.4 mm) at a speed of
ca. 0.5 cm^2/min.. Cutting by a high pressure water jet containing
abrasive particles has similar characteristics. The major
drawback to EDM is the "burning" of the surface layer, which may
therefore contain microcracks so shortening the fatigue life.
The microcracks can be removed by light chemical machining,
followed by shot blasting.

In electrochemical machining (ECM) a stream of electrolyte,
a simple sodium chloride or nitrate solution, passes rapidly
between the tool (as cathode) and the component. The electrolyte
is filtered continuously to remove particles. The tool requires
a suitable design to ensure laminar flow of the electrolyte over
the cutting area. The technique has the advantage or being wear
free. Machining rates are high, 1-2 mm/min. with current
densities of 50 A/cm^2 (large industrial machines deliver
30,000 A). Post machining shot blasting, enhances fatigue life.
The application is suitable for machining of disc faces, bosses
in ring forgings, etc.

The technology is now highly developed for boring complex
arrays of cooling channels in turbine machinery, including

multihole drilling. The automatic positioning of drills and jig borers, with selection of drill size and speed is long established. It can be adapted to all kinds of techniques for producing holes.

4.   FUTURE DEVELOPMENTS

4.1. General Aspects

There must be a continuing improvement in quality and capability of products and/or reduction in costs of production. Strategic aspects must also be considered; with good patents it is possible to control a market with all the economic benefits this implies.

Improvements in process technology by themselves are not enough; it is necessary also to consider :

- modelling methods which permit optimisation of processes;
- automation with registration of the mean parameters which can be checked easily;
- methods of non destructive testing to give more precise indications of defects and capable of adaptation for online inspection.

These problems are not in principle specific to high temperature materials; for instance the basis of modelling systems for castings or for forgings are similar for all metallic alloys. The adaptation and detailed implementation will however present some unique problems, in connection with the complexity of the alloys and of the performances to be achieved.

4.2. Emerging Technologies

In melting a large effort is expected with the object of obtaining cleaner melts and improved control of the ingot structure. Progress in ladle refining allows the use of arc melting in place of VIM for medium performance alloys. In consumable electrode remelting, ESR may replace VAR in order to eliminate oxides or to obtain a finer grain size. Very fine structures are achievable by drip melting as in the Vader process but difficulties in production on an industrial scale are not yet resolved. However to achieve the superclean quality needed to avoid crack nucleating inclusions in many applications of advanced high temperature alloys, more sophisticated (and also probably more expensive) processes will be necessary. Already good results are being obtained by using the electron beam cold hearth remelt technique.

Water atomized powder and cold compaction has already been

used to process highly alloyed stainless steel mainly
for tube production. It is in the area of P.M. nickel superalloys
that further progress is needed. The elimination of inclusions
by filtering alone is not satisfactory and the development of
cleanliness in melting methods is essential. Significant
improvements would have a major impact in that they would allow
manufacturers;

- to use a larger range of powder sizes with a lower
  rejection rate,
- to use rotating electrode processes of interest for high
  temperature applications
- to make it possible to produce components close to the final
  dimensions in a single hot isostatic pressing operation.

The development of chemical vapour treatment of powder in an
halogen atmosphere could facilitate diffusion during the
sintering operation. Studies on sintering in the semi-liquid
state should also be continued.

In spite of the exceptional developments in recent years,
more progress is required in precision casting of small
components as well for large castings of heavy section. Studies
of ceramics for use as mould shells and cores are very
important. New processes including rapid cooling and
rheocasting could in principle, give products with improved
performance; for single crystal solidification, higher thermal
gradients would give better structures.

Forging operations are now considered to be part of the
thermomechanical treatment needed to give high performance alloys.
The major process in progress is isothermal forging of superplastic powder materials; a new die material capable of working
in air to replace TZM would make for easier production and reduce
cost.

For conventional machining the introduction of further
improved high speed cutting materials is expected and will need
large scale testing.

For surface treatment, laser surface glazing and ion
implantation show great promise to improve resistance to
crack initiation under cyclic stresses.

REFERENCES

- From AGARD Congress Materials substitution and recycling 1983

(1) New developments in recycling, Norton, R.C. and Keneham, O.B.

(2) Effets des traitements sous vide sur l'évolution des teneurs en éléments traces dans les superalliages, Wadier, J.P. and Morlet, J.

- From 7th ICVM Tokyo 1982

(3) Production of ultra low nitrogen steels and alloys in vacuum induction, Katayman, H. and Nakamura, Y.

(4) Metallurgical and plan design aspects of vacuum distillation processes, Ellebrecht, C.

(5) Study of physiochemical processes in plasma arc remelting of the surface layer of ingots and billets, Latash, Y.V. et al.

(6) Progress in the vacuum (VIM, VAR) melting of high performance alloys, Sutton, W.H.

(7) New developments in electron-beam melting, Shiller, S. et al.

- From vacuum metallurgy conference 1984

(8) Electron-beam cold hearth refining furnace for the production of nickel and cobalt base superalloys, Hunt, C.D.A. et al.

(9) A mechanisms of "white spot" formation in remelted ingots, Wadice, J.F. et al.

(10) Chemistry and structure control in remelted superalloys ingot, Cordy, J.T. et al.

- From High Temperature Alloys for Gas Turbines, Liege Meeting 1978, D. Coutsouradis et al.
  Applied Science Publishers London 1978.

(11) Quality of casting of superalloys, Bachelet, E. and Lesoult, G.

(12) Progress in advanced directionally solidified and eutectic high temperature alloys, Drapier, J.M.

- From Superalloys 1980 Seven Springs meeting, Tien, J.K. et al.
  American Society for Metal 1980.

(13) Thermomechanical processing of Haynes Alloy 1988 sheet to improve creep strength, Klarston, D.L.

(14) The development of single crystal superalloy turbine blades, Gell, M. et al.

(15) Development of low cost directionally solidified turbine blades, Hoppin, G.S. et al.

(16) Influence of the chemical composition of nickel base superalloys on their solidification behaviour and foundry performance, Ouichou, L. et al.

(17) The metallurgical aspects of hot isostatically pressed superalloy casting, Antony, K.C. adn Radavich, J.F.

(18) Hiping various precision cast engine components in nickel base superalloys. Lamberigts, M. et al.

- From High Temperature Alloys for Gas Turbines, Liege. Meeting 1982, R. Brunetaud et al, D. Reidel Publishing Cy.

(19) Superalloy technology to-day and to-morrow, Versnijder, F.L.

(20) VADER a new melting and casting technology, Boesh, W.J. et al.

(21) The evolution of the forging process on discs, Coyne, J.E. and Couts Jr., W.H.

(22) The relationship between structure, properties and processing in powder metallurgy superalloys, Davidson, J.H. and Aubin, C.

(23) Precision casting of turbine blades and vanes, Drafner, J.M.

(24) Microporosity formation in investment castings of nickel base superalloys: metallurgical effects, thermal modelling and foundry assessment, Ouichou, L. et al.

- From superalloys P4, Seven Springs meeting M. Gell et al. Metallurgical Society of AIME

(25) Development of a conventional fine grain casting process, Would, M. and Benson, H.

(26) Cost effective single crystals, Goulette, M.J. et al.

(27) Superalloy powder processing, properties and turbine disc
applications, Chang, D.R. et al.

(28) Fabricated RSR vane manufacturing technology, Baker, S.H.
et al.

(29) Liquid phase sintering of nickel base superalloys,
Jeandin, H. et al.

(30) Superalloy melting and cleanliness evaluation,
Shamblin, C.F.

   - From High Temperature Alloys for Gas Turbines and other
     Applications, Liege Meeting 1986, W. Betz et al.
     D. Reidel Publishing Cy.

(31) Recent development and potential of single crystal
superalloys for advance turbine blades, Khan, T.

(32) Foundry performance and reverted alloys for turbine blades,
Ford, D.A. et al.

(33) Structure, processing of ODS superalloys, Singer, R and
Artz, E.

(34) Forging of high temperature alloys for gas turbines,
Rydstad, H. et al.

(35) Automated electron beam melting for superalloy cleanliness
evaluation, Jarrett, R.N., Conference on Electron Beam
Melting and Refining, Reno 1984, Bakish Materials
Corporation.

(36) Evaluation de la propreté des materiaux de métallurgie des
poudres pour disques de turbomachines, Raison, G., Materiaux
et techniques, décembre 1987.

(37) Mechanical behaviour and processing of DS and single crystal
superalloys, Khan, T. et al, Journal of Metals, July 1981.

(38) Metal forming and the finite element method, past and
future, Kobayaski, S., Advanced Technology of plasticity
19824, vol 2.

(39) A general purpose KEM code for simulation of non isothermal
forming processes, Wu, W.T. and Oly, S.I., NAMRC XIII
Conference 1985.

(40) Finite element analysis of shaped lead-tin disc forgings, Germain, Y. et al, NUMIFORM Conference 1986.

(41) Usinabilité des alliages refractaires, Vigneau, J., Rapport CETIM 1983.

(42) Influence of the microstructure of the composite ceramics tools on their performance when machining superalloys, Vigneau, J., annals of CIR $\underline{V}$ 1987.

3.1.3. The Role of Computers in Production of

High Temperature Materials

S.F. Pugh

Consultant Metallurgist

Abingdon, Oxon, U.K.

The applications for computers in materials production and processing are legion.

Current applications such as in nuclear power production and aerospace where emphasis is on quality and safety analysis or in components operating at the limits of survival in steam and petrochemical plant, demonstrate the power of computer based techniques. Further development is particularly attractive in view of the rapid increase in power of computers combined with reduction in cost of unit operations.

Among the purely computational applications in HTM production are (1) design of components with respect to performance in service and ease of manufacture. (2) Modelling of fabrication processes such as freezing of castings, forging, and welding with the objective of achieving improved macro and microstructures. (3) Selection of materials and their optimisation from thermodynamic and property data bases. (4) Scheduling of production. Flow control and process planning.

Computers are linked to instruments to achieve on-line real time control of quality by intelligently monitoring process variables such as temperature and taking action when discrepancies appear. Also, linked to NDE instruments, computers will immediately signal the appearance of faulty components and again take some appropriate action to avoid wasting further process time on a faulty component and to prevent more faulty items from appearing on the production line.

The automated production of moulds for casting gas turbine blades has already been introduced to give a greater uniformity of the product. High speed complex forging operations could also benefit from automation to achieve isothermal conditions.

Finally the unmanned factory is inevitably coming and will offer both improved quality and lower cost production; two advantages that have hitherto tended to be mutually exclusive.

3.2. Ceramics and Refractories

3.2.1. Ceramics and Pyrolytic Materials

D.J. Godfrey* and J.F.G. Condé[+]

*Admiralty Research Establishment, Holton Heath, England

[+] Materials Consultant; Broadstone, Dorset, England

1. INTRODUCTION

The term ceramics embraces a wide range of inorganic, non-metallic materials which are processed and consolidated at elevated temperatures to produce dense or porous forms as required. This range of inorganic materials includes two specific specialised groups, the refractories and the engineering ceramics (also known as technical ceramics, advanced, high performance or high technology ceramics and in Japan as fine ceramics.)

The refractories are usually made from the stable oxides of metals but may also include silicon carbide and graphite. They are based on natural minerals including fireclay, kaolin, quartz, olivine, kyanite and graphite. Purified mineral derivates may also be employed including alumina, magnesium oxide (from sea-water), zirconia, silicon carbide, calcium aluminate, mullite and pyrolytic carbon. The engineering ceramics embrace oxides, carbides, nitrides and borides and involve more elaborate chemical and other processing. In general they require greater control over the raw material powders, including purity and quality, and more precise forming techniques and much higher processing temperatures than for most of the traditional refractories. The strict distinction between the two groups has become blurred as a result of advances in manufacturing technology of refractories and in the growing use of the engineering ceramics as refractories for certain applications, particularly in the metallurgical industry, mainly in iron and steel production and processing.

The manufacturing technologies of refractories and technical ceramics have many processing features in common i.e. powder consolidation, sintering or firing, etc. and forming procedures. The earliest use of refractories in pre-historic times was as monolithic fireclay linings and it is in this area where in recent years the two manufacturing technologies have diverged most significantly. Monolithic or formed in-situ refractories now form 40 to 50% of refractories usage (depending on country of use) and prefired bricks or shapes used in conjunction with mortar are declining in importance. The methods for forming monolithic refractories include casting, vibration compaction, ramming and 'flame gunning', i.e. methods significantly different from those adopted for technical ceramics and prefired refractories. Monolithic refractories may also, in some instances, employ chemical bonding as distinct from ceramic bonding.

Refractory fibres for insulation are prepared either by steam fiberisation of molten metallurgical slag and rocks such as granite, limestone or slate (1), or by solution processes, introduced in recent years. (2)

2. COLD CONSOLIDATION

(a)   General

The consolidation of ceramic powders to dense ceramic bodies is the main objective of the majority of ceramic processing technology operations. Theoretical and experimental studies on the packing of spheres demonstrate that near-theoretical packing efficiencies, with solid occupancy fractions of 0.637 (3) can be achieved by vibratory treatments with mono-sized spheres, provided that the container size is at least an order of magnitude greater than the sphere diameter. The void spaces between spheres can, in principle, be infilled by appropriately sized smaller spheres, and the possibility of attaining very high solid occupancy volume fractions has been demonstrated by McGeary (4), when a four size system of metal spheres was vibration packed to a solid fraction of 0.951.

Packing of irregularly-shaped particles, such as those typical of ceramics made by the customary crushing and fracture comminution routes, is generally very far from the optimum packing possible, and it is difficult to attain green densities of more than 50-60% of theoretical density with many ceramic powders.

b)   Cold Uniaxial and Isostatic Pressing

Compaction by pressure is a very important technique used

for powder compaction, although consolidation by liquid removal
from powder suspensions (slip casting) is extensively used with
clay products, and with alumina and other oxide ceramics.

Uniaxial pressing employs a rigid die and powder is
compacted by pressure applied through a piston or plunger. The
powder may be dry or wet with 10 to 15% moisture. Isostatic or
hydrostatic pressing employs powder enclosed in an impermeable
bag or membrane made of rubber or other flexible material,
pressure being applied through a pressurized fluid contained in a
pressure vessel. The isostatic pressure enables uniform
consolidation of larger volumes of powder than die pressing and
ameliorates problems related to powder and die wall friction.
The process is suitable for forms of larger length/ diameter
ratio than can be compacted successfully by die pressing. In
pressure compaction operations, the use of free flowing powders,
such as those resulting from evaporative spray drying or
spray-freeze drying is desirable, and vibration is helpful for
producing a preliminary compaction of the powder in the die.
Pressing additives known as binders are frequently used to aid
the pressing operation and optimise the green density obtained.
Binders are usually added to the powder during its preparation
for pressing, since it is important to disperse them effectively
within the powder, and avoid local binder rich spots or
laminations which could produce serious defects in the ceramic
part after sintering. Binders have three main functions: first,
to lubricate the movement of particles relative to each other in
the compact under pressure, secondly, to lubricate the outside of
the compact relative to die wall and minimise forces which
otherwise might crack the quite fragile pressed compact during
the pressure reduction part of the pressing cycle, and thirdly
they increase the strength of the pressed compact and help
prevent damage from die wall friction effects during pressure
release and in subsequent handling of the compact. Binders are
usually organic chemicals or natural products, often of a
polymeric nature, although water and simple organic liquids such
as ethanol or propan-2-ol are sometimes used and can be quite
effective. In die-pressing, flaws in cylindrical geometries and
conical in form, originating from the upper and lower cylinder
rims and often called 'capping' cracks, are frequently
troublesome unless an effective binder is used. Such flaws,
which arise from die wall friction during pressure release (5)
can be avoided by tri-axial compression of the powder in a
flexible envelope in a fluid medium, by using isostatic pressing
technology, but is this usually less convenient and more
expensive. The mechanics of powder pressing have been discussed
in detail by Thompson (6), by Messing et al using a pressure
density curve approach (7), and by Matsumoto (8) (Powder
Compaction Response Diagrams). The uniformity of powder
compaction has been discussed by Strijbos (9) and the behaviour

of alumina by Frey, DiMilia and Reed (10,11,12). Van Groenon and Lissenburg have compared experimental data for density uniformity with finite element calculations (13).

Binder formulation and performance have been discussed extensively in the literature (14 to 22). Particular aspects which have been studied have included inhibition of binder migration during drying and improvement of compact shape and fabrication by extension or pressing, avoidance of defects by close control of binder character. Studies have also covered dispersants for tape casting, binders for spray drying and dry pressing to enhance green density. Strengthening of green compacts (23) to permit machining of complex shapes (possibly by numerically controlled machines) as an alternative to injection moulding may be useful for prototype quantities of high quality engineering components for evaluation and optimisation. Injection moulding may require considerable optimisation to produce complex mouldings free from significant defects.

Work on producing high quality ceramics for mechanically demanding applications in turbines and diesel engines in the recent USA and German government sponsored research programmes has shown that it is very difficult to produce defect-free high quality components quickly, reliably and economically. The emphasis in recent years has shifted from in-engine testing of ceramic components to research on the processing of ceramics to give high quality strong and reliably complex shaped components. Its solution lies in researching the understanding of how powder may be compacted, with the aid of appropriate lubricants and compact-strengthening agents, but also in how to produce powders which are easy to compact to high densities.

Recent research on ceramic fabrication has highlighted the importance of aggregation phenomena (24 to 30). Adhesion between individual ceramic particles causes aggregates, which lead to inhomogeneity in particle packing in the green compact, and result in a non- uniformity of sintering in the fired compact, with consequent inhomogeneity in void content and grain-size and the creation of localized internal stresses, as described by Evans (24) and further discussed by Lange (25) using a pore coordination - number distribution concept. Soft aggregates can be dispersed by surfactants but mechanical processes are required to disperse hard aggregates. In the sintering of alumina, aggregates may decrease isothermal shrinkage rates by a factor of up to ten (26).

c)   Slip Casting

Slip casting is an alternative to pressure consolidation methods and is an effective and long established technique for fabrication of articles from clay. As in injection moulding, a liquid is used to consolidate powder particles. Aqueous suspensions with surface active additives to optimise the zeta-potential double ionic layer charges on particles have been developed successfully and can yield very high density compacts. Porous polymeric liquid absorbants have been used as effective replacements for traditional plaster of Paris. Pressurisation of the slip may help to optimise properties. Liquids other than water may assist dispersion and a capacity for hydrogen bonding is an important property of such liquids (31). Ion exchange may also be helpful in optimising slurries (32). Development of silicon powder casting technology to high densities has been shown (33,34) to make effective nitridation to RBSN difficult although addition of Si_3N_4 'grog' helps nitridation (35).

The injection moulding technique (29) mentioned previously is one of a group of plastic forming techniques (36) which also includes transfer moulding, compression moulding, extrusion and tape forming. All of these plastic forming processes employ powder compounded with additives to form a homogeneous mixture which is plastically deformable. The compounding is usually achieved by use of a high-shear rate mixer (36). The addition of water to clay base materials normally provides adequate plasticity. Other powders, particularly the non-oxides may require about 30% of an organic polymer, usual a thermoplastic which can be formed to shape by heat and pressure. Injection moulding of silicon powder to attain high green density has shown that the melt flow index used to characterise plastic formulations was not successful for ceramic powder mixes and pointed-up more useful treatments.

Transfer moulding is similar in essence to injection moulding and employs a preformed 'bullet'; which is heated before injection into the mould. It has similar shape-forming characteristics to injection moulding but the equipment is generally more simple. It is not a high volume production process and has similar binder removal problems to injection moulding. Compression moulding is similar to die pressing. The 'bullet' is placed in the die; heated and then deformed under pressure to the form of the mould. The shaping potential is limited, it is slow compared with injection moulding but requires less binder and this leads to reduced shrinkage.

Lengths of thin ceramic sheet or strip can be formed by various tape forming processes. The ceramic powder may be combined with a polymer and rolled or extruded or a low ash paper may be coated with a slurry and the paper burned off prior to firing. Both these forms can be roller corrugated and used for heat exchange matrices. A further variant is to coat an impervious polymer film with a slurry. This form can be cut or punched to shape and then fired.

The slurries employed in the tape forming routes are somewhat complicated requiring careful control of ceramic powder particle size distribution and binder chemistry to achieve consistent results.

3. HOT COMPACTION (DENSIFICATION)

(a)   Sintering

The consolidation of compacted powders at high temperature is dependant upon two fundamental processes: solid state diffusion, and liquid solid interactions. Theoretical studies show that sintering rates should be proportional to the inverse of the third power of the sintering particle size, and it is well established that fine particles sinter better than coarse particles. Another approach to optimising packing and sintering has been to form uniformly-sized fine powder e.g. TiO_2 (44) and Ta_2O_5 (45). It is often found that sintering is activated by admixture of another substance at a minor concentration, and it is difficult to establish whether a liquid is formed locally (liquid sometimes occurs in apparent binaries below the solidus temperature in the relevant phase diagram, due to the presence of impurities, which are often silica containing). In the presence of even a small liquid fraction sintering behaviour is often profoundly modified; usually sintering consolidation is favoured. Consolidation is achieved because of the thermodynamic benefit of reduced surface area, and it is believed that the thermodynamic 'pressure' effect causes grains to dissolve in liquid where they contact each other, and this is followed by recrystallization of the dissolved material into a solid form again; with the surface area of the compact being reduced, and also its void content reduced. Recent studies on sintering have focussed on the role of aggregation in the powder compact, and in a series of papers Lange and collaborators have discussed its relevance to the formation of defects (39 to 43).

Sol-gel techniques providing a uniform distribution of small particles have also been shown to yield materials which sinter well at reduced temperatures; techniques developed for silica (46) have been applied successfully to other ceramics.

Solid state sintering is sensitive to the surface behaviour of the powder which governs the strength of the aggregates which may impede uniform sintering. The diffusion behaviour is also affected by the chemical modification of the powder surface, and especially by sinter active additives, which do not always function by a liquid phase mechanism (e.g. the beneficial effect of MgO additions on the sintering of alumina).

Liquid phase sintering has been the subject of research (47,48). Liquid phase sintered products have less likelihood of possessing optimum high temperature strength and durability, because of the possible effects of traces of liquid since the properties of two phase systems are determined by the weaker phase. Recent ceramics, such as the high purity fine-grained high strength alumina Vitox developed for in-body protheses, is a well sintered product which shows that pure material can densify well. High strength alumina has also been obtained by using optimised powder and high shear processing techniques with specially selected organic binders.

The extensive use of low heat capacity fibre insulation in modern furnaces has permitted the use of shorter firing times ('fast firing') and this has yielded useful cost reductions in ceramic material processing, and also necessitated research to determine the minimum times necessary for effective sintering. Other methods of direct heating of the ceramic are becoming available; in addition to the long-available high frequency electrical induction method, in which a susceptor is often used, although some ceramics like zirconia may suscept when hot, laser and focused infrared radiation from power lights and recently microwave heating (49) have received some attention and show promise for reducing heat cycle time and cost. Another advantage of modern sintering furnace environments is the reduced sodium content of fibre insulation as opposed to refractory bricks, and as with the pore novel heating techniques just described, may lead to less formation of beta alumina $(Na_2O)_{12} Al_2O_3$ lower melting phase in alumina ceramics.

b   Reaction Bonding

The most important materials covered by this term are reaction bonded silicon nitride (RBSN) and carbide (RBSC). RBSN is made by compacting silicon powder and then consolidating this to a ceramic by reaction with nitrogen at high temperatures; negligible dimensional change takes place, unlike the shrinkage experienced with sintered ceramics. Research on RBSN has recently been concentrated on optimising processing details such as powder size, improving compact density to yield ceramics with as little as 12% porosity ($2.8g/cm^3$), and optimising reaction schedules. Fracture strengths of up to 350 MPa with Weibull moduli of around 20 have been achieved (50).

A major problem has been to make reliable components by injection moulding, without serious defects arising from into-mould flow problems or from voids arising when the considerable plastics content of the moulded items are removed by heating. There has also been considerable interest in post-sintering of RBSNs containing oxide sintering additives (Al_2O_3 or

Y_2O_3, sometimes with other oxides as well), because the overall reduction in sintering shrinkage possible helps with maintenance of dimensional tolerances. Densities as high as 98.5% of theoretical have been achieved, with shrinkage of only 5-10%. The cost advantage of reduced machining needs is however offset by the increased expense of the two heating stages, although it might prove possible to combine these. High densities have also been obtained by the slip casting route, although defects originating from air and hydrogen bubbles in the viscous slips have been a problem. Thermal spraying has been used successfully to make high density porous shapes suitable for reaction bonding (51) and perhaps the highest compact densities of all are attainable with this method, since by choice of spraying parameters deposit densities varying up to virtually full density are in principle possible. The method has indeed been proposed for densification of oxide and other compound ceramics, but suffers from the need for a freely flowing powder to be available if the usual powder spraying techniques are employed. The wire type of spraying pistol can be used with powder compacted into flexible cord form, but again the cost of powder treatment has prevented its wider use. A novel application of the reaction-bonding principle has been used (52) to produce low density silicon nitride/zirconia ceramics by reaction bonding mixtures of silicon and zirconia powder. Up to 80% ZrO_2 can be incorporated before the negligible dimensional change feature of normal RBSN begins to be affected, and materials can have thermal diffusivites similar to those of zirconia ceramics, whilst strengths are at least twice those of aluminium titanate ceramics. As the zirconia content is increased, the oxidative degradation of strength (1000°C oxidation), which with normal RBSN can be about 40%, is progressively reduced. This oxidative degradation is a serious problem which has been given insufficient attention, although surface treatment with oxides (e.g. CaO, SrO) to prevent cristobalite formation (which is the reason for the strength degradation) can be helpful.

The reaction bonding of silicon carbide is usually accomplished by the infiltration of liquid silicon into a powder compact mixture of silicon carbide and carbon. Improvements in properties and processing compatibilities in the last decade have been quite modest, but the hardness and corrosion resistance of the product have led to its use in may technological fields (e.g. sewage treatment machinery). The appreciable unreacted silicon content is only a disadvantage at temperatures above or near the melting point of silicon; since silicon is a quite hard and oxidation resistant material in its own right. Its potential for further strength improvement however seems limited, although the rival sintered silicon carbides are proving difficult to develop to strengths comparable with those readily attained with sintered silicon nitride.

Reaction bonding is of importance in refractories where silicon carbide refractories may utilize an oxide bond between the SiC grains. Silicon carbide is fired with clay at 1500°C and a silicate glass forms at the grain boundaries. Higher strengths can be achieved by use of SiC and silicon where an oxynitride (Si_2ON_2) bond or nitride (Si_3N_4) can be developed during firing depending on O_2/N_2 ratio in the firing atmosphere (54). Cement-free refractory concretes employing a chemical bond are a further example of reaction bonding (55).

c)   Hot Pressing

Development in the ability to sinter 'difficult' materials to high density in the last decade has rather diminished the importance of this technique in attaining theoretical density in ceramic materials; nowadays near theoretical densities in alumina and silicon nitride and carbide ceramics can be achieved by hot pressing. The most effective hot pressing conditions for alumina were recently described (56).

By the use of hot pressing, effective densification of silicon nitride can be achieved with minimal addition of MgO. The product may be used at temperatures of 1000-1100°C without a decrease in strength. Orientation effects on strength and thermal conductivity of up to 20% are introduced. Although progress has been made in hot pressing to near net shape of simple-shaped parts, the limited life of carbon dies, and the high cost of machining the dense ceramic have limited its attractiveness as a processing technique in recent years. Some years ago, interest in the USA developed in 'forging' oxide ceramic shapes to form hollow or curved components products by hot pressing in moulds by a process probably involving creep mechanisms. The possibility of 'superplastic' - deformation, as with metallic alloys if realisable technically, could well become important for ceramic materials, particularly for large shapes.

Undoubtedly, the most important development in the hot pressing field has been the application of hot gas isostatic pressing (HIP) to ceramic materials.

The HIP technique has been applied to the processing of ceramics with considerable success including materials such as alumina (57), zirconia and beryllia, titanium nitride and boron nitride, silicon nitride and carbide (58).

Pressures up to 300 MPa have been used, up to ten times greater than in hot pressing, and temperatures in excess of 1700°C are available in modern equipment. With silicon nitride, HIP has been applied to Si_3N_4 powder compacts, to RBSN, and to sintered silicon nitride and sialon materials. With porous powder compacts and RBSN, a gas barrier is essential, and special materials such as glasses have been developed for this, and fully densified materials have been produced using minimal quantities of additives to optimize high temperature properties (59).

Encapsulation is not required with sintered material, but the additive level has to be sufficient for open porosity to be eliminated in the pre-sintering. Dimensional tolerance retention can be quite good, but the great advantage is the potential to reduce the size of, or render unimportant moderate defects in materials, thus improving their reliability. Most HIP work is carried out in argon atmospheres, although other atmospheres are possible. Oxidizing environments present special problems with furnace materials in the pressure vessel, and are currently the subject of development e.g. for non-structural ceramics such as the superconductor $YBa_2Cu_3O_7$, in which oxygen levels are critical and difficult to achieve internally in thick components.

The first cost of HIP equipment is large, but its use has proved economic for gas turbine parts because of the increased reliability its defect minimisation confers. It is undoubtedly a valuable technique which will be applied to some materials where high reliability is a problem, e.g. in gas turbines, although it remains conjectural whether it will be used with automotive engine parts, dental ceramics or load-bearing surgical ceramic joint implants, where close control of atmospheric sintering processes and proof-testing to detect defects at room temperature may prove to be sufficient.

Significant R & D Efforts

Significant R & D efforts on ceramics consolidation and processing exist in the USA, Japan, Sweden and the EEC countries in Europe. Problems encountered in production of high quality components for demonstrator projects have switched emphasis to ceramics processing. Many heat engine and energy related projects have appreciable elements relating to ceramics production but it is difficult to make quantitative budget estimates. However it seems likely that the Japanese R & D effort in this field is still the largest and amounts most probably to several £10M's per annum.

Porous reaction bonded nitride ceramics were evolved before the dense forms and hence comparatively more is known about them. The good refractory properties of RBSN are related to the absence of additives which are required for sintering processes. Porosity is a limiting factor in that it leads to degradation of strength properties by oxidative degradation and hence research aims at achieving higher density and optimising processing. Programmes are less significant in volume than in other areas of ceramics processing. In the wider context of reaction bonding in the refractories field, proprietary manufacturers in the USA, Europe and Japan all have active interest in low cement monolithics where the bond develops by reaction between the constituents in service.

There are significant although perhaps limited programmes on hot processing in the USA and Sweden. In the USA (60) the sinter HIP process development is considered very important and emphasises the importance of scaling-up to commercially viable size.

Future Research Needs and Priorities

Ceramics with well controlled processing are needed having fine grain size and high and reliable strength properties. Where high temperature is envisaged processing-related additives should be minimized or eliminated. These considerations apply particularly to silicon carbide. HIP processing may be crucial in this.

For powder consolidation by pressing or moulding, better organic additives are required, which lubricate the compaction process and equipment, which strengthen the compact to resist pressure-release damage, and which are easily removed thermally without causing defects.

Specific areas relating to consolidation of powders include improved powders of high purity and control of particle and agglomerate size distribution (36). For slipcasting, slip development together with better moulds and mould coatings are required and for injection mouldings better blending of consituents, alternative binders and well designed burn-out schedules are necessary. R & D is required on shrinkage during sintering in order to attain high reproducibility and minimise machining. Higher throughput to enable lower unit costs is also important and further work on 'fast firing' is required in this context as well as to reduce processing energy costs. Alternative methods of heating must also be investigated particularly those which together with modern insulation technology can provide reduced thermal capacity. The adaption of high grade refractory fibrous insulation is also essential to improve the purity of the furnace environment and further research on alternative refractory fibres is desirable in this context.

In the field of porous technical nitride ceramics there is need to develop surface treatments and/or coating processes (e.g. impregnation or pyrolytic treatments) to inhibit chemical (oxidative) or slag attack. There is also a continuing but low

priority need for improved materials and greater reproducibility
as well as processing optimisation to achieve enhanced properties
combined with reduced processing costs.

The potential for production of more complex technical
ceramics such as the silicon nitride/zirconia material is an area
of reaction bonding which merits further research. There is also
scope for innovative use of reaction bonding in the refractories
field to produce both pre-fired shapes and monolithics.

The high capital cost of HIP equipment and its suitability
for mainly specialised componenets suggests that future research
needs should be assessed critically in relation to competing
research needs in other areas of processing which may ultimately
prove adequate and more cost effective.

REFERENCES:

1.   Annals of Occupational Hygiene, Okberg, I., 31 (4B),
     529-545 pp. (1987)

2.   Handbook of Composites Vol. 1, Birchall, J.D. et al., Ed.
     Watt, W. and Perov, B.V., North Holland, 115-154 pp. (1985)

3.   Communications Amer. Ceram. Soc., Prost, H.J. and Ray, R.,
     C19 pp., Febr. (1982)

4.   J. Amer. Ceram. Soc. 44, McGeary, R.K., (10), 513 pp.
     (1961)

5.   Special Ceramics, 1962, Long, W.M., Academic Press, London,
     327 pp. (1963)

6.   Bull. Am. Ceram. Soc. 60, Thompson, R.A., (2), 237 pp.
     (1981)

7.   Bull. Am. Ceram. Soc. 61, Messing, C.L., et al, (8), 857
     pp. (1982)

8.   J. Am. Ceram. Soc. 69, Matsumoto R.L.K., (10), C216 pp.
     (1987)

9.   J. Am. Ceram. Soc. 62, Strijbos, S., et al, (1), 57 pp.
     (1979)

10. J. Am. Ceram. Soc. 64, DiMilia, R.A., (4), 667 pp. (1983)

11. J. Am. Ceram. Soc. 67, Frey, R.G. and Halloran, J.W., (3),
     199 pp. (1984)

12. J. Am. Ceram. Soc. 66, DiMilia, R.A. and Reed, J.S., (9), 667 pp. (1983)

13. Van Groenen and Lissenburg

14. Bull. Am. Soc. 62, Sarkar, M. and Gremiger, G.K., (11), 1280 pp. (1983)

15. Bull. Am. Soc. 58, Salamone, A.L. and Reed, J.S., (6), 618 pp. (1971)

16. Powder Tech. 34, Krycer, I., et al., 39 pp. (1983)

17. Advances in Ceramics – Forming of Ceramics, Mikeska, K. and Cannon, W.R., Ed. Mangels, J.A., Am. Ceram. Soc., Columbus, Ohio, 164 pp. (1984)

18. Advance in Ceramics, Nies, C.W. and Messing, G.L., 58 pp.

19. Powder Tech., 44, Miller, T.A. and York, P., 219 pp. (1985)

20. Bull. Enging. Exp. Stn., Jenike A.W., Utah 123 (1964)

21. Bull. Am. Ceram. Soc. 65, Schuetz, J.E., (12), 1556 pp. (1986)

22. Bull. Am. Ceram. Soc. 66, Gurak, M.R., et al., (10), 1495 pp. (1987)

23. Ceram. Age, 82, Teter, A.R., (8), 30 pp. (1966)

24. Am. Ceram. Soc. 65, Evans, A.G., (10), 497 pp. (1982)

25. Am. Ceram. Soc. 67, Lange, F., (2), 83 pp. (1984)

26. Am. Ceram. Soc., Dynys, W. and Halloran, J.W., (4), 596 pp. (1984)

27. Bull. Am. Ceram. Soc. 65, Ciftcioghu, M., et al., (12), 1591 pp. (1986)

28. Am. Ceram. Soc. 70, Ciftcioghu, M., et al., (11), C329 pp. (1987)

29. Bull. Am. Ceram. Soc. 62, Mangels, J.A. and Williams, R.M., (5), 601 pp. (1983)

30. Bull. Am. Ceram. Soc. 61, Mizutas, U.C., et al., (8), 872 pp. (1982)

31. J. Matl. Sci. 20, Parish, M.V., Garcia, R.R. and Bow, H.K., 996 pp. (1985)

32. J. Am. Ceram. Soc. 68, Crosbie, G.M., (3), C83 pp. (1985)

33. Bull. Am. Ceram. Soc. 62, Williams, R.M. and Ezis A., (5), 667 pp. (1982)

34. Bull. Am. Ceram. Soc. 63, Sacks, M.D., (12), 1510 pp. (1984)

35. J. Am. Ceram. Soc. 67, Williams, R.M., Ezis, A. and Coverly, J.C., (4), C64 pp. (1984)

36. Silicon Nitride Based Ceramics: Fabrication, Processing and Properties; Research and Development Priorities, Davidge, R.W., Riley, F.L., Evans, D.C. and Wordsworth, R.A., AERE R 12276, Commercial, CEC, JRC, Petten (May 1986)

37. Bull. Am. Ceram. Soc. 62, Mangels, J.A. and Williams, R.M., (5), 601 pp. (1983)

38. J. Matl. Sci. 12, Edirisinghe, M.J. and Evans, J.R.C., 269 pp. (1987)

39. J. Am. Ceram. Soc. 66, Lange, F.F., (6), 396 pp. (1983)

40. J. Am. Ceram. Soc. 66, Lange, F.F., (6), 402 pp. (1983)

41. J. Am. Ceram. Soc. 67, Lange, F.F., (2), 83 pp. (1984)

42. J. Am. Ceram. Soc. 65, Evans, A.G., (10), 497 pp. (1982)

43. J. Am. Ceram. Soc. 67, Dynys, F.W. and Halloran, J.W., (9), 59 pp. (1984)

44. J. Am. Ceram. Soc. 67, Fegley, B., (6), C113 pp. (1984)

45. J. Matl. Sci., 21, Ogihara, T., 2771 pp. (1986)

46. Bull. Am. Ceram. Soc. 64, Johnson, D.W., (12), 1597 pp. (1985)

47. J. Am. Ceram. Soc. 65, Lange, F.F., (2), C23, 1982 and 66 pp., (2) C33 pp. (1983)

48. J. Am. Ceram. Soc. 64, Singh, V.K., (10), C133 pp. (1981)

49. J. Matl. Sci. Letters 5, Meek, T.T. and Blake, R.D., 270-274 pp. (1986)

50. J. Mat. Sci. 22, Ziegler, G., et al., 3041-3086 pp. (1987)

51. Special Ceramics 5., Brown, R.L., et al., Ed. Popper, P., BCRA, Stoke on Trent, 345-359 pp. (1972)

52. British Patent, Godfrey, D.J.

53. Proc. Brit. Ceramic Soc. 26, Godfrey, D.J., 265-278 pp. (1978)

54. Refractory Uses - Practicality of High Technology Ceramics, Fisher, G., Bull. Am. Ceram. Soc. 66, (7) (1987)

55. Recent Progress in the Use of Monolithic Refractories in Europe., Kuonert, W., Advances in Ceramics, 13, New Developments in Monolithic Refractories, Ed. Fisher, R.E., The Am. Ceram. Soc., Columbus, Ohio (1985)

56. Bull. Am. Ceram. Soc., 64, Leshkivich, C.J. and Crayton, P.M., 684-86 pp. (1985)

57. Bull. Am. Ceram. Soc. 64, McCoy, J.K., (9), 1240-49 pp. (1985)

58. Werkstofftech., 14, Ziegler, G., 189 pp. (1983)

59. Sci. Ceram. 11, Heinrich, J.J. and Böhmer, M., 439 pp. (1981)

60. US Congress Office of Technology Assessment, OTA-TM-E-32, September (1986)

3.2.2. Powder Production for

Structural Ceramics

F. Cambier and A. Leriche

Centre de Recherche de l'Industrie Belge de la Céramique

Mons - Belgium

1. INTRODUCTION

The need for advanced ceramics has increased in recent years leading to development of new raw materials (1). It is now well established that the ultimate strength of sintered bodies is strongly linked to microstructure and theoretical density. Such a target implies the use of thermally active powders showing special characteristics :

- Extremely fine particle size of the primary crystals (2).

- Sintering at the lowest possible temperature to reduce grain growth (3).

- Adequate morphology (4). Spherical particles give fluidity and suppression of grain growth phenomena while small equiaxed particles avoid large aspect ratio. A few particles larger than the others in a monodisperse powder can be the origin of very long distance defect and monodisperse particle packing is limited to a lower green density (maximum 67 % of theoretical density) than heterodisperse particles. Agglomerates leading to defects at the green and the fired stage (5) must also be avoided.

- High purity level. Impurities lead to amorphous or crystalline grain boundaries which reduce high temperature strength and corrosion resistance. The nature and location, of impurities is strongly related to the synthesis route.

The aim of this section is to review the various synthesis

routes currently used industrially to produce raw materials for the structural ceramics silicon nitride, silicon carbide, zirconia and alumina.

2. SYNTHESIS ROUTE FOR STRUCTURAL CERAMIC POWDERS

2.1. Silicon nitride

Silicon nitride exists in two crystallographic phases α and β, the latter being the high temperature phase. The dissociation of silicon nitride (*) reactive powders at high temperature requires that additives are necessary to produce densified materials.

Four different routes are currently used to produce commercial silicon nitride powders.

i)   Silicon nitridation : $3Si + 2N_2 \longrightarrow Si_3N_4$

ii) Carbothermal
     reduction of silica : $3SiO_2 + 6C + 2N_2 \longrightarrow Si_3N_4 + 6CO$

iii) Vapour phase reaction
     synthesis : $3SiCl_4 + 16NH_3 \longrightarrow Si_3N_4 + 12NH_4Cl$

IV) Thermal decomposition
     of diimides : $SiCl_4 + 6NH_3 \longrightarrow Si(NH)_2 + 4NH_4Cl$
                              $3Si(NH)_2 \longrightarrow Si_3N_4 + 2NH_3$

Table 1 presents the chemical and physical characteristics of powders synthesized by the four chemical routes which show that the residual impurities are related to the synthesis route.

Direct nitridation of free, unreacted, silicon involves the retention of other metals such as aluminium, calcium and iron (catalyst). During the reaction, silicon and silicon nitride sinter forming strongly bonded aggregates. Vigorous milling is necessary to produce irregularly shaped and sized particles, Fig. 1, and an increase in oxygen content. Further washing procedures with acids to remove metallic impurities can introduce contamination by, for example, fluorine (6).

The only commercial silicon nitride powder produced by silica carbo-reduction contains high carbon (0.9 %) and oxygen (2.65 %) levels and very few β grains (1 %). It is characterised by a regular crystal shape (Fig. 2) and a large agglomerate size distribution. However, the tap density and therefore the green density obtained after pressing is very low.

---

*   1875°C under 0.1 MPa nitrogen pressure.

Table 1 : Properties of typical silicon nitride powders from the
4 synthesis routes

Powder	STARCK LC10	TOSHIBA A	GTE Sylvania	TOYOSODA TS 7
Raw material	Si	SiO_2	$SiCl_4$ (vapour phase)	$SiCl_4$ (from diimides)
Purity (%)	96.5	96.2	-	99.9
O_2	1.34	2.65	1.00	1.04
Si (free)	<0.1	-	<0.2	-
C	0.17	0.9	-	0.1
Fe	0.015	0.01	<0.005	0.005
Cl	-	-	<0.05	0.1
Al	0.05	0.2	<0.01	<0.001
Ca	0.006	0.01	0.001	<0.001
Agglomerate size (µm)	0.4	1.3	1.6	0.4
Surface area (mg^2/g)	11.0	8.7	4.9	9.9
Tap density	0.67	0.5	0.4	0.7
(%)	6	1	12	15
Crystal size (µm)	0.5	0.3	-	0.5
Shape of agglomerates	clusters 5	chains 100 balls 20	whiskers	clusters 3
Frequency of whiskers	none	none	many	a few

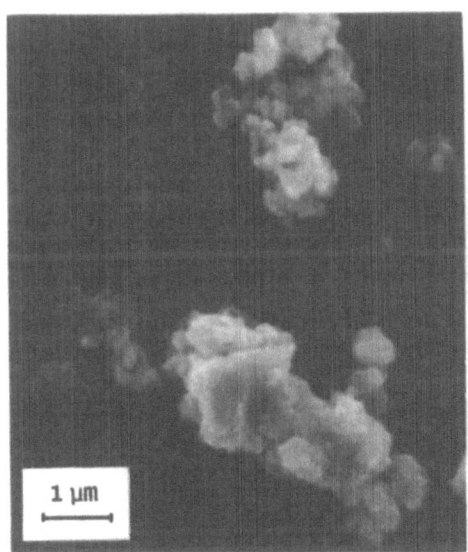
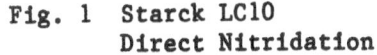

Fig. 1 Starck LC10
        Direct Nitridation

Fig. 2 Toshiba A
        Silica Carboreduction

The two last synthesis methods from $SiCl_4$ lead to the

presence of residual chlorine. These powders present the highest
degree of purity (up to 99.9 %), and small crystal size. The
powder prepared by vapour phase reaction consists mainly of
whiskers (Fig. 3) whereas the powder synthesized by thermal
decomposition of diimides is characterised mainly by globular
crystals (Fig. 4).

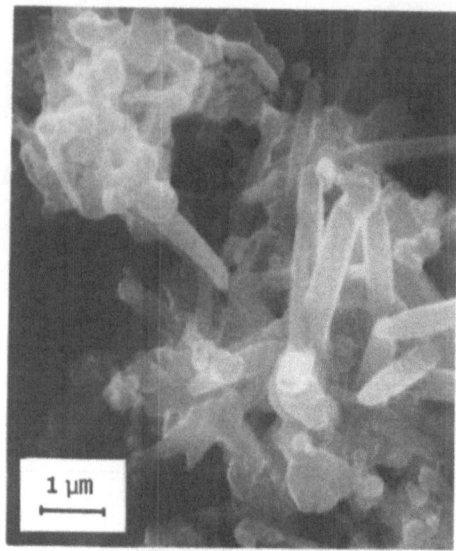

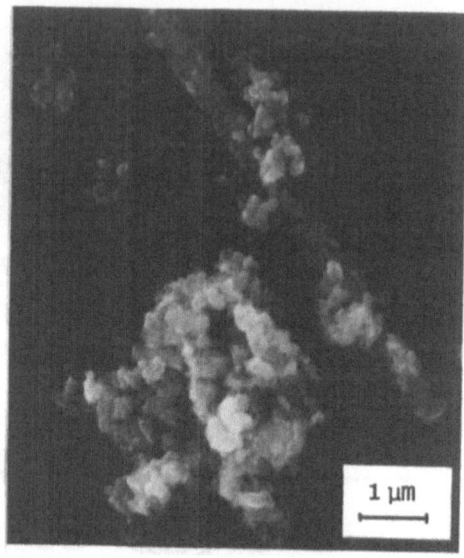

Fig. 3 GTE Sylvania SNE10
   Vapour Phase Reaction

Fig. 4 Toyosoda TS-7
   Diimide Decomposition

2.2. Silicon carbide

Silicon carbides exists as two main allotropic phases. At
room temperature, the stable form is the cubic β transforming to
hexagonal α at approximately 2100°C. At normal pressure SiC
sublimes and dissociates at temperatures higher than 2500°C.

Silicon carbide powders are usually produced by four
methods :

i) Solid phase reaction between silicon and carbon,

$$Si + C \longrightarrow \quad -SiC \ (1408°C)$$

ii) Carboreduction of silica - (Acheson's process)

$$SiO_2 + 3C \xrightarrow{Fe} \quad -SiC + 2CO \ (2000-2300°C)$$

After HF-acid washing, the Acheson-SiC powder purity is

about 98 to 99 %, the main impurity being iron.

iii) Reaction between silicon chloride and methane

$$SiCl_4 + CH_4 \longrightarrow \quad -SiC + 4\ HCl \text{ (vapour phase}$$
$$\text{plasma reaction)}$$

This process produces β-powder of high purity and relatively uniform particle size.

iv) Thermal decomposition of organosilicon compounds

$$CH_3SiH_3 \longrightarrow \quad -SiC + 3H_2$$

This vapour phase reaction is carried out at 1600°C using argon as the carrier gas.

Silicon carbide powders can also be synthetized by sol-gel processes (e.g. alkoxide process). The resulting ultrafine and amorphous oxides are reacted with ammonia or methane at 600-1350°C to obtain very fine amorphous carbides. On heating the SiC powder to 1450°C in H_2, the powder crystallises as β-SiC.

In general, silicon carbide powders prepared by vapor phase processes and the sol-gel process have very high purity (up to 99.99), submicron size and are in the β-form in contrast to Acheson-type powders which are less pure (96 to 99 %), have to be milled and are composed of α-phase particles.

Table 2 presents the chemical and physical properties of some silicon carbide powders synthesized by different routes.

Both β-SiC and milled α-SiC sinter at temperature higher than 2000°C in the presence of small amounts of carbon, boron, aluminium, beryllium...

The β-SiC grains are usually small and equiaxed whereas α-SiC are plate-like grains. Hence, when β-SiC is used as starting powder at temperature higher than 2000°C, special attention has to be paid to microstructure development because large platelike crystals of α-SiC can occur in a matrix of fine-grained β-SiC due to the α → β transformation.

2.3. Alumina

Of the many allotropic phases of Al_2O_3, the α-phase is the stable one at high temperature (> 1000°C).[3] Alumina is the most widespread technical ceramic. Products related to this range of raw materials are bioceramics, cutting tools, translucent ceramics, wear parts, etc. There are six different commercial

Table 2 : Properties of silicon carbide powder from different synthesis routes

Powder	IBIDEN Standard Betarundum	LONZA UF-10	SUPERIOR GRAHPITE HSC 100 GL
Synthesis route	Vapor phase process	Acheson process	Direct reaction
Purity (%)	99.2	97.0	97.4
Si (Free)	-	< 0.2	< 0.3
C (Free)	0.4	< 0.4	< 2.0
Fe	0.05	< 0.1	-
O_2	0.1	< 0.7	< 0.3
Agglomerate size (μm)	0.32	1.3	1.5
Surface area (m^2/g)	11.5	10.0	10.0
Crystallographic phase			

methods for producing high purity α-alumina (4, 9) :

1. The Bayer and the modified Bayer process; (Fig. 5) by alkaline digestion of bauxite followed by hydrolysis to form aluminium hydroxide. The purity level depends strongly on this precipitation step, the maximum purity by classical methods being about 99.8 % and the main impurities being sodium, silicon and iron. By multiple precipitation-dissolution procedures (modified Bayer process), the purity increases and can reach 99.98 %. Aluminium hydroxide has to be calcined slowly to avoid grain growth. However during heat treatment, strong aggregates appear which have to be ground, with as a consequence slight contamination of the product. Typical primary crystals are disk shaped as shown in Fig. 6.

2. The thermal cracking of ammonium alum; (Fig. 5) uses the aluminium hydroxide of Bayer process as ammonium alum precursor. Almost all of the impurities are removed by a recrystallisation process and the ammonium alum is thermally cracked to obtain very fine and vermicular shaped α-Al$_2$O$_3$ particles (Fig. 7). Since the volume of the alum molecule is about ten times that of alumina it permits very small crystal size to be obtained but all of the impurities present in the alum remain in the alumina.

259

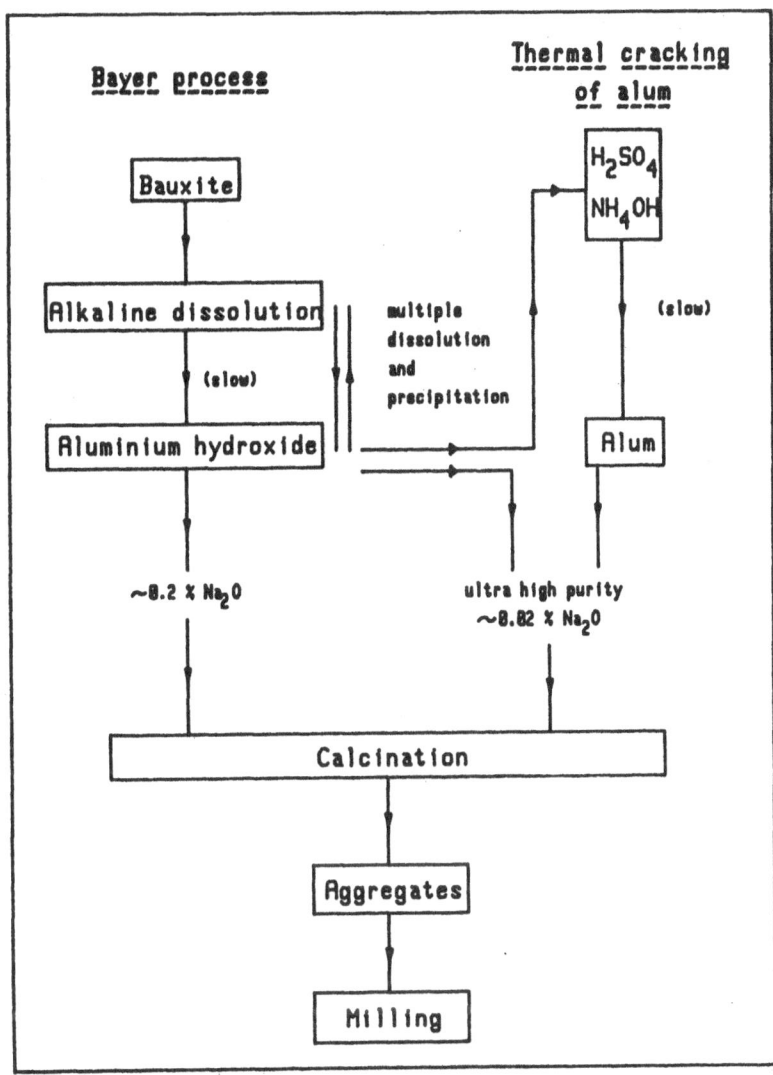

Fig. 5   Scheme describing the alumina powder synthesis by Bayer
         process and by cracking of ammonium alum.

3. The hydrolysis of organometallic compounds; such as AlR_3 or
$Al(OR)_3$ (R = alkyl) which can be easily purified by distillation.
The hydrolysis method gives very fine and pure particles with
narrow distribution and a high powder bulk density which induces
less shrinkage during sintering than classical powders.

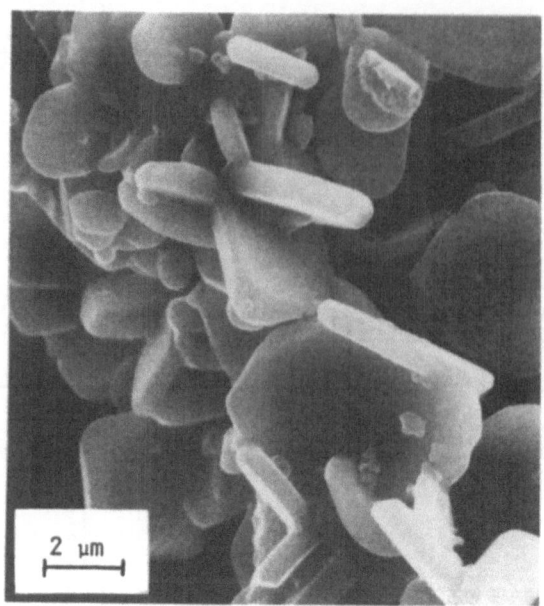

Fig. 6   Typical Bayer primary crystals obtained
after rapid calcination.

Fig. 7   Vermicelli structure of ex-alum
aggregate

Table 3 : Chemical and physical properties of some alumina powder prepared by different routes

Production process	Modified Bayer	Thermal cracking of alum		Hydrolysis of organometallic compound	Ethylene chlorodrin	Spark discharge
Powder	Sumitomo Chem Co	Criceram	Showa	Sumitomo Chem Co	Sin-Nihon Chem Co	Iwatani Chem
	A-HPT	A6Z	Ua-5105	AKP-30	L30	RA-40
Crystal phase	α	α	α	α	α	α
Purity (%)	>99.9	99.98		>99.99	99.99	99.99
Si (ppm)	8	40	6	<40	<10	50
Na	15	48	10	<10	10	<5
			5			
Mg	4	<3	<1	<10		
Ca	2	<5	<2			
Fe	7	10	4	<20	10	30
Cu				<10		5
Ga			<2		<10	
Cr			<2			
Agglomerate size (μm)	0.4	0.3	0.25	0.4	0.4	0.08
Surface area (m^2/g)	5–6	6	10	7.5	5–40	20
Tap density	1.2	0.8	10	1.3	0.6	1.0
Bulk density	0.5	0.6	0.6	0.85	0.3	

4. The ethylene chlorohydrin process; an aqueous solution of
sodium aluminate is neutralized with ethylene chlorohydrin to
form aluminium hydrate which is aged to give solid boehmite and
calcined to produce γ or α alumina. This process can be
considered as a variation of the modified Bayer process allowing
easier separation of sodium, iron and silicon species.

5. Spark discharge process; the aluminium hydrate to be calcined
is formed by electric discharge in aqueous solution with metal
(Al) electrodes. The powder obtained shows irregular primary
particle sizes and a wide size distribution. Purity is strongly
dependent on the quality of the aluminium raw material.

6. Thermal cracking of ammonium aluminium carbonate; to yield
high purity alumina with a particles size of 0.1-0.5 μm and good
sinterability. An advantage of this method in contrast to the
ammonium alum process is better control of the particle size
because the thermal cracking reaction does not involve melting of
the material.

Table 3 gives the chemical and physical characteristics of
some alumina powders synthesized by different methods. It can be
seen that of all the methods give high purity and very fine
powders.

2.4. Zirconia

Zirconia presents three polymorphs : the monoclinic,
tetragonal and cubic phases. The tetragonal and cubic phases
(stable at high temperatures) can be partially or fully
stabilized at room temperature by the addition of oxides such as
magnesia, yttria, calcia and ceria. More than 125 zirconia
powders were investigated for this review. Large grain size
zirconia for refractory use will not be considered here.
Zirconia powders for structural ceramics are produced from the
impure free oxide "baddeleyite" and the silicate compound
"zircon",

The purification procedures used are :

1. From baddeleyite mineral; the main impurities to be removed
are silica, titania, ferric oxide, and phosphoric anhydride. The
mineral is washed by acids, filtered, dried and the remaining
metallic impurities are eliminated by magnetic separation. High
purity baddeleyite when available, can be simply electrofused and
milled to obtain zirconia powder.

2. From zircon mineral; the chlorination and thermal
decomposition methods are described in Fig. 8. The various
methods aim to obtain $ZrCl_4$ which is hydrolysed by different

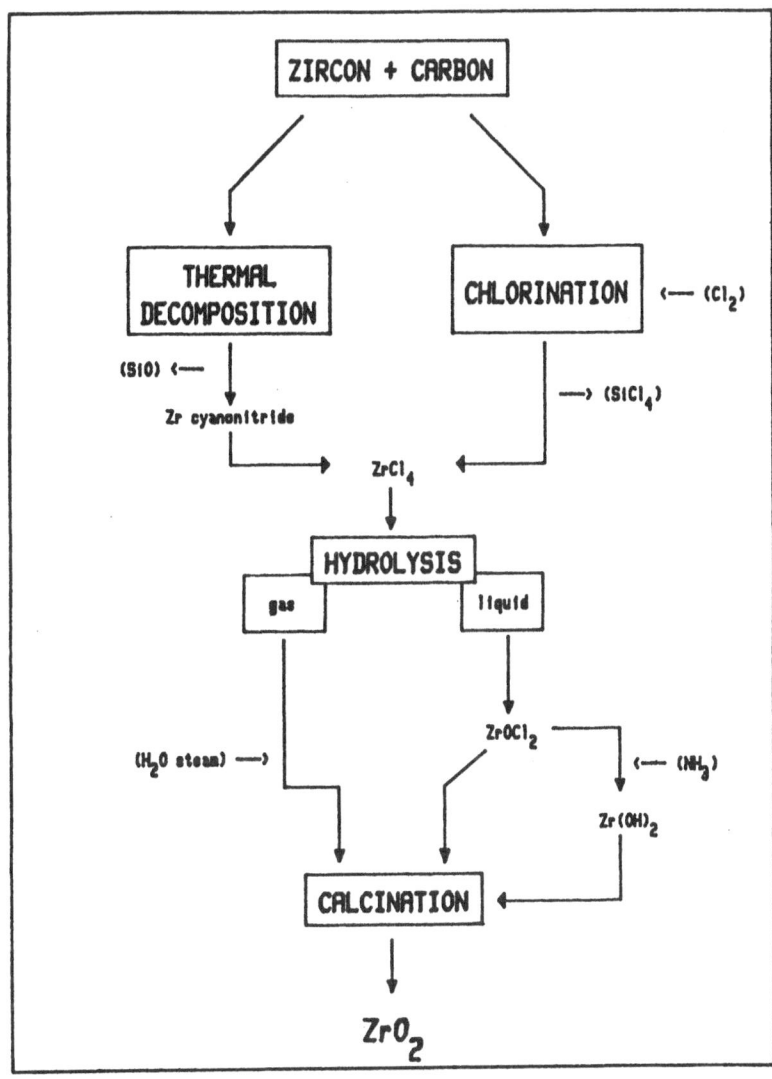

Fig. 8 Zirconia powder synthesis by thermal decomposition
or chlorination of zircon.

means, the product being calcined to obtain ZrO_2. The alkali
oxide decomposition is presented in Fig. 9.

Zircon can react with sodium hydroxide or carbonate at
temperatures higher than 600°C to form sodium zirconate or sodium
zirconate silicate which give zirconia after calcination. To
avoid impure oxide, purification is carried out either by

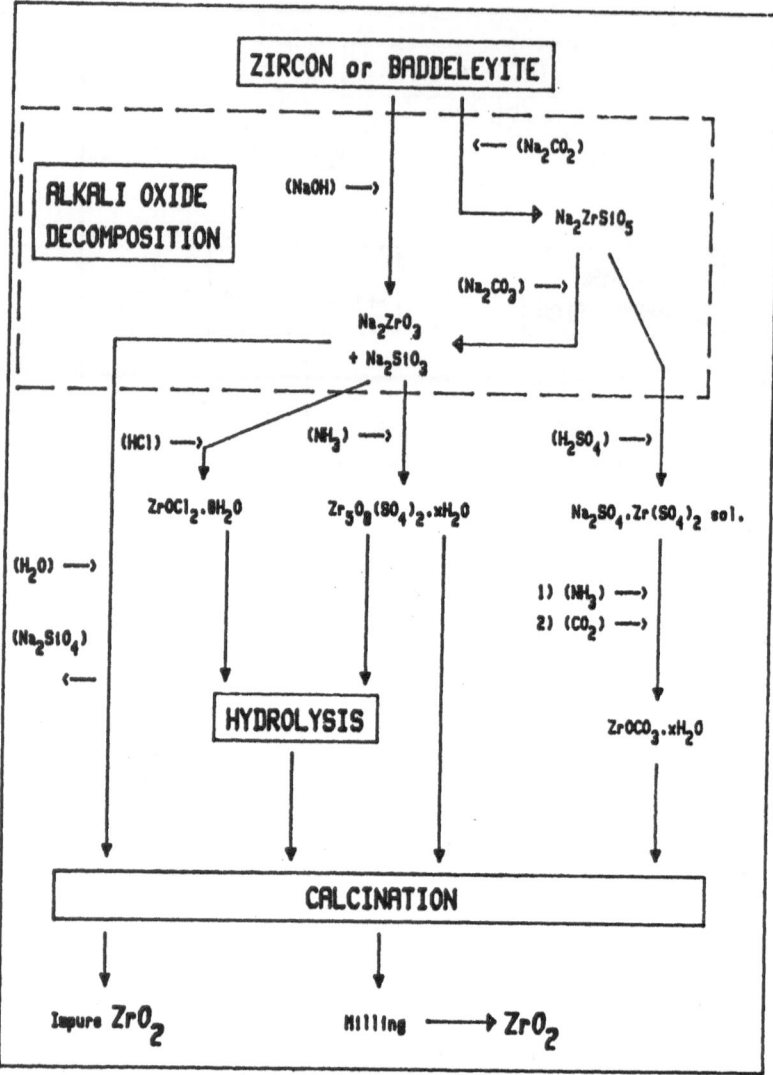

Fig. 9   Zirconia powder synthesis by alkali oxide
         decomposition.

sulfuric or hydrochloric acid leaching.

Lime (or doloma) can also be used for fusion. The following
reaction is the preferred one, because it permits hydrochloric
acid leaching of calcium silicate :

$$2CaO + ZrSiO_4 \xrightarrow[1600°C]{} ZrO_2 + Ca_2SiO_4$$

Table 4 : Chemical and physical properties of some zirconia powders

Supplier	Grade	phases	purity	SiO_2	TiO_2	Fe_2O_3	Na_2O_3	CaO	particle size (µm)	surface area (m²/g)	Synthesis route
Daaichi Kigenso	DK5	mono	99.2	0.3	0.18	0.04	–	0.03	4	–	impure baddeleyite
Unitec	PSZ-Y6*	mono/tet/cubic	99.5	0.15		0.10	0.10		0.9	7	electrofusion of pure baddeleyite
Pechiney	UPH	mono/tet	99.9	0.003	0.0015	0.004	0.001	0.003	0.05	–	chloride
Dynamit Nobel	Dyna-zirkon x F-5Y	mono/tet		0.01	0.01	0.05	0.03	0.05	0.35	6	alkali oxide route (sulfonation)
Toyosoda	TZ-3Y °	mono/tet	99.9	0.006	–	0.003	0.002	–	0.7	15	alkali oxide route (chlorination)
Z-Tech	SM994	mono	99.5	0.25	0.06	0.03	–	–	0.7	–	plasma

* : 6 w/o Y_2O_3

x : 5.6 to 4.9 w/o Y_2O_3

° : 5.2 w/o Y_2O_3

A further method is the plasma route. Zirconia particles are injected into an arc where they melt and dissociate into zirconia and silica. The cooling process allows primary solidification of zirconia dendrites surrounded by vitreous silica. This last compound is leached with boiling caustic soda and washed to remove sodium metasilicate. As a consequence of the special morphology of zirconia, easy reduction can be carried out to very fine crystals (0.1 µm). Further improvements of this technique have been achieved recently; by optimum control of dissociation temperature (2200°C), liquid silica can be separated directly from solid zirconia, silica traces are then leached out with sulfuric acid. This route leads to a different zirconia powder shape : compact clusters of very small particles (0.01 to 0.1 µm).

Generally, the stabilizers are added during the production process either by coprecipitation, on by electrofusion in order to obtain a homogeneous mixed powder.

Examples of zirconia powders obtained by the various processes are presented in Table 4.

3. FURTHER DEVELOPMENT IN CERAMIC POWDER PRODUCTION

One of the most important parameters responsible for the high cost of structural ceramic components is the price of raw materials.

Established production routes such as the Bayer process for alumina, Acheson for silicon carbide, arc melting for zirconia or direct nitridation for silicon nitride lead to powders of lower price because of the production volume and the precursors used. Such powders are less expensive than the raw materials developed recently. Silicon nitride is perhaps the raw material for which the non-conventional methods will become most rapidly competitive : indeed silicon chloride is a quite cheap precursor. Although, today, prices of $SiCl_4$-based silicon nitride are still at the same level as very fine milled Si-based powders, it seems that prices will fall as production volumes increase. It must also be noted that a high proportion of the price of very fine classical powders can be attributed to the milling operation (7). However, structural applications require not only fine powders but also close control of shape, size and purity of the particles in order to decrease the number and size of the defects and therefore to improve the quality of the ceramics and the fabrication yields.

Classical routes do not allow such control : shapes of particles are irregular, size reduction is costly, purity decreases as size decreases. Consequently development of non conventional processes including vapour phase reaction, solution-precipitation or sol-gel processes must be encouraged and intensified.

Some examples of recent developments have been (8) :

- very fine (< 0.1 μm) silicon carbide and silicon nitride powders produced by laser reaction of silane with ethylene or with ammonia respectively

- fine (0.3-0.6 μm) oxide powders precipitated from metal alkoxides reacting with an alcohol water solution. The powder is sinterable and gives sintered bodies with more than 99 percent of theoretical density

- the sol-gel process also starts from metal alkoxide precursors in alcohol which are hydrolysed to produce a gel. After drying, very fine powders (0.003 to 0.1 μm) can be obtained.

The cost of these new synthesis methods remains very high, mainly due to the use of expensive raw materials but this should decrease in the future to become accessible for industrial purposes. R&D has to be devoted not only to ceramic powders but also to the synthesis of lower price precursors. Several large chemical industries in Europe are working toward this.

Another advantage of non conventional routes is that they often allow the synthesis of mixed powders. Indeed ceramic materials are rarely constituted by a pure chemical. Additives such as sintering aids, inhibitors of crystal growth, phase stabilizers are often used.

Moreover composite materials (with particles, whiskers) showing improved mechanical properties are being investigated widely. In both cases, but especially for composites, perfect homogenization is needed. The up-to-date tendency of some advanced ceramic powder suppliers is to synthesize mixed powders such as zirconia-alumina, zirconia-mullite... These powder are prepared by sol-gel or coprecipitation methods which guarantee good homogeneity. The availability of such mixed powders would allow the users to avoid the most delicate processing step during the composite material fabrication : the powder mixing homogenization. Research in this area should also be encouraged.

REFERENCES

1. D.W. Johnson, Jr., "Non-conventional powder preparation techniques", Am. Ceram. Soc. Bull., 60, (2), pp. 221-224, 243 (1981).

2. A. Kato, "Study on powder preparation in Japan", Am. Ceram. Soc. Bull. <u>66</u> (4), pp. 647-649 (1987).

3. M.P. Harmer and R.J. Brook, "Fast Firing – Microstructural Benefits", J. Brit. Ceram. Soc. <u>80</u> (5), pp. 147-148 (1981).

4. F. Cambier, "Raw materials available for conventional processing of engineering ceramics : alumina and silicon nitride powders", Proc. Advanced Ceramics pp. 20-36, Ed. J.S. Moya and S. de Aza, Soc. Esp. Ceram. Vidr. Arganda del Rey – Madrid, Spain (1986).

5. F.F. Lange, "Sinterability of Agglomerated Powders", J. Am. Ceram. Soc. <u>67</u> (2), pp. 83-89, 1984.

6. A. Leriche, V. Vandeneede, D. Libert, F. Cambier, "Powder Characterization and Optimization of Fabrication and Processing of Silicon-based Engineering Ceramics", Final Report Contract SUT 117-B (RS). Commission of the European Communities. Substitution and Materials Technologies and Ceramics. July 1986.

7. E. Rothman, J. Stitt and H.K. Bowen, "A Look at Ceramic Powder Production Processes – Old and New", Ceramic Industry, pp. 24-29, May 1985.

8. "Synthesis of Ceramics by New Techniques" Special Issue of Yogyo-Kyokai-Shi, Vol. 95, N° 1, 1987.

9. "Production Process and Characteristics of High Purity Alumina", S. Horikiri, pp. 23-31, F.C. Annual Report 1986, edited by Japan Fine Ceramics Association.

3.3.1. Metal Matrix Composites

R. Warren

Chalmers University of Technology

S-40296 Göteborg

General Principles

The technical and patent literature contains countless descriptions of methods for preparing metal matrix composites. This reflects the difficulties involved; in general the preparation of MMC is more involved and expensive than that for a plastic matrix composite with similar form and reinforcement morphology. This is largely due to the high temperature necessary for the manipulation of the metal matrix.

To be successful, a composite preparation process must achieve penetration of the fibres by the matrix whilst maintaining the desired reinforcement morphology and at the same time achieve consolidation of the matrix without excessive damage to the fibres. The second of these conditions necessitates the use of high homologous temperatures and consequently the likelihood of reinforcement matrix interactions, as discussed below. The preparation process must usually also have the potential for low- cost, large-scale production.

Possible reinforcement/matrix interactions that can lead to deterioration of fibre and/or matrix properties include: reinforcement dissolution (and reprecipitation), chemical reaction, activated crystallization or recrystallization of fibres, embrittlement of fibre (or matrix) by atoms from the matrix (or reinforcement phase), unfavourable matrix precipitation at the reinforcement/matrix interface, thermal expansion mismatch. In general, these effects will increase with increasing temperature and time at temperature. Interactions in specific reinforcement/matrix combinations are described in more detail in section 2.3.1.

Most preparation processes can be assigned to one of three types:

1) Initially separate, reinforcement and matrix are combined; at some stage during the process the matrix is in the <u>molten state</u>.
2) Initially separate, reinforcement and matrix are combined; the matrix is in the <u>solid state</u> throughout the process.
3) <u>In-situ methods</u> in which the composite microstructure is developed during the process itself (for example directional eutectic solidifaction or co-extrusion of powders).

Molten matrix methods are attractive since they offer a rapid penetration of the reinforcement by the matrix and effective densification of the composite. However, they are normally restricted to non-reactive reinforcement/matrix combinations having low intersolubility, since any interaction occurs extremely rapidly when the molten state is involved. For non-reactive systems, on the other hand, it is sometimes necessary to develop means to improve the wetting and bonding between reinforcement and matrix by e.g. surface treatment of the reinforcement or active additions to the matrix. An important example is Al and its alloys which exhibit poor wetting (contact angle > 90°) on ceramics up to a transition temperature around 1000°C (1). This is probably due to the presence up to this temperature of an oxide film on the melt rather than poor chemical affinity between alloy and ceramic. An important consequence of poor wetting is that spontaneous infiltration of a reinforcement preform cannot be achieved. The problem has been overcome e.g. by surface treatment of the reinforcement (C fibres in Al (2)), by active addition to the matrix (alumina fibres in Al-Li alloys (3)) or by pressure assisted infiltration (see below).

Examples of composite preparation methods are given in more detail below for specific composite systems.

Particulate Composites

The currently most important particulate composites are Al- and Mg-base alloys reinforced with SiC in the form of either angular abrasive particles or flakes. Typical volume fractions are 10 to 30 vol.% with particle sizes around 20 μm. Methods of preparation include:

1) Mixing particles into the melt followed by conventional casting.
2) Pressure-assisted infiltration of a reinforcement preform to give a billet suitable for extrusion and/or other working processes.

3) Mixing of alloy and reinforcement powder followed by hot
   pressing of the mixture to give a billet suitable for
   extrusion and/or other working processes.

An advantage of particulate composites is that they are
amenable to most conventional working processes. Billet sizes of
around 100 kg produced. In the immediate future it is to be
expected that the above methods will be scaled up and also
optimised to improve microstructural uniformity of the
particulate phase. The cost of preparing large billets is
predicted to be below $ 20/kg.

Recently, a spray-forming method has been successfully
applied to the preparation of SiC-Al composites (4). The
reinforcement particles are fed into the atomised spray of the
matrix alloy melt. The mixture is sprayed onto a suitable
substrate (e.g. a rotating cylinder to form tubing) giving a
composite with practically full density and good reinforcement
uniformity. A drawback of the method is that spraying of Al must
be carried out under inert conditions, necessitating relatively
elaborate equipment.

Short-fibre reinforced Metals

The most common short-fibre reinforcements are SiC whiskers,
δ-alumina, and mixed Al_2O_3-SiO_2 fibres (see section 2.3.1.).
Whisker reinforced Al-alloys are produced either by a powder
process, similar to that described above for particulate
composites, or by pressure assisted infiltration of fibre
preforms. Where appropriate, the composites produced in both
methods are extruded and worked to improve tensile properties
(section 2.3.1.) which leads to a reduction of the whisker length
to between 1 and 10 µm (5). The δ-alumina and mixed-oxide fibre
composites are normally prepared by pressure-assisted
infiltration methods the most common at present being so-called
squeeze casting. This process as applied to the preparation of a
locally reinforced, engine-piston is illustrated schematically in
Fig. 1. A fibre preform is fixed in a preheated die cavity, a
measured quantity of alloy melt is poured into the die and a ram
applies pressure during solidification. Pressure of less than
10 MPa are sufficient to achieve infiltration of most alloys but
final pressures of around 70 MPa are usually applied since these
have been found to give high quality castings when the method is
used for unreinforced alloys. The die casting and preform are
designed to give as near net-shape as possible.

The popularity of the squeeze casting method for composite
preparation is that it is a technique already developed for
unreinforced Al-alloys and was moreover easily modified to
include fibres. For Al-alloys, the casting can be carried out in

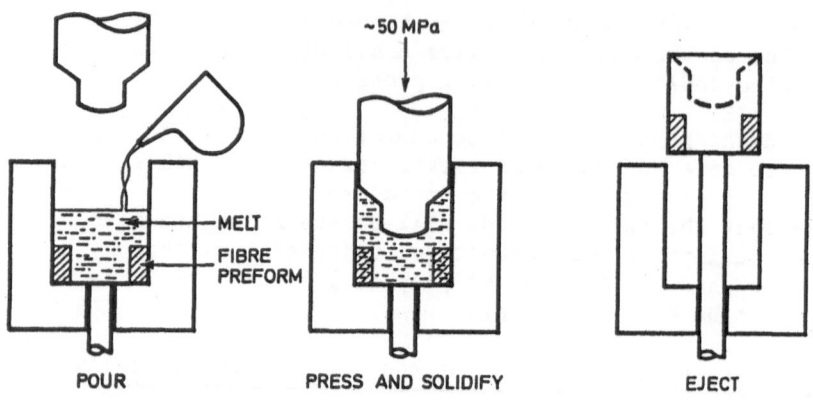

Fig. 1. (sections 3.3.1.) Schematic illustration of squeeze
        casting.

air. However, other casting techniques both existing and new are
being investigated for the preparation of this type of composite.
A recently reported example is a vacuum/gas-pressure infiltration
method suitable for small-series component production (6). A
development to be anticipated in the future is the infiltration
of continuous preforms to produce e.g. bar and plate. Alumina
and mixed oxide fibre composites are normally used in the as-cast
form. However, preliminary studies of their extrusion indicates
that by careful choice of extrusion parameters, increased
orientation of the fibres can be achieved without excessive fibre
breakage (7).

Most immediate developments can be expected through
improvements in preform preparation techniques. Short-fibre
preforms are normally prepared by letting the fibres settle out
from a water-based suspension in a suitable die and then drying.
The volume fraction can be adjusted between about 5 and 60 vol.%
without excessive fibre crushing by applying suitable pressure
during the settling stage. The fibres tend to settle
horizontally and so take up an orientation that is roughly two-
dimensionally random (8).

Both organic and silica binders are usually added to the
preform at the suspension stage to subsequently impart strength
to the preform during handling, during any necessary machining
operations and during infiltration. Prior to casting the
preforms are fired to fix the silica binder. Improvements in
this method or the development of new techniques can be expected
to lead to a reduction in harmful non-fibrous inclusions,
improvements in the uniformity of the spatial distribution of
fibres, greater control of fibre orientation and the ability to
predetermine local variations of volume fraction.

Although significant commerical production of short fibre composites has been restricted to Al-alloy matrix materials, preform/casting methods are generally applicable and could with suitable modification well be extended to e.g. Mg- and Cu-base composites.

Multifilament fibre composites

Examples of methods developed to prepare long, multifilament fibre composites include:

1. Melt-drawn wire preform method. A continuous multifilament strand of reinforced composite is drawn from the melt and is subsequently stacked or woven to give a larger configuration that can be hot-pressed to give the required component. The method was developed above all for C-fibre reinforced Al-alloys (2). For these it was necessary to surface treat the strand with a Ti/B mixture using a vapour process to promote wetting of the fibre.

2. Slurry methods. The fibre strand is impregnated with matrix metal powder by drawing it through a slurry of the powder. The impregnated strand can be stacked and hot-pressed in a similar fashion to the melt-drawn strand. The method is used for reactive fibre/matrix combinations that would not tolerate molten metal methods (see also section 3.3.1.).

3. Vacuum infiltration. Fibre strands or a suitably prepared preform (see below) are placed in a mould and infiltrated with a molten matrix alloy containing an active addition to promote wetting. Vacuum casting is usually necessary since active additions are usually sensitive to oxidation. The method was developed primarily for the preparation of Al reinforced with alumina fibres with 2-3% Li as the active addition (see section 2.3.1.).

4. Pressure-assisted infiltration. A suitably prepared preform is placed in a casting mould and infiltrated with molten matrix by the application of pressure. As in the case of short-fibre composites squeeze-casting is a commonly used method.

With infiltration methods it is desirable that the fibres are prepared as a preform having to configuration required in the cast component. Thus an important area of development for these composites is the preparation of long-fibre preforms. Techniques available include traditional methods of weaving. To produce three-dimensional, multidirectional preforms special three-dimensional weaving processes as well as methods based on braiding are being developed. A problem in all such structures

is that of fibre uniformity both at the level of individual
fibres (fibre contact and clumping) and at the inter-strand
level. Methods must be developed to ensure fibre separation;
theoretical and experimental studies must be made to understand
the effects of strand-to-strand non-uniformity on properties and
methods developed to optimise the structures accordingly.

Monofilament fibres

The most important monofilament fibres are CVD forms of
boron and SiC and tungsten wire (see section 2.3.1.). Boron
(usually coated with SiC) is used mainly with Al-alloy matrices.
SiC with Al and Ti-alloy matrices and tungsten with superalloys.
Since after preparation the composites are extremely difficult to
work it is necessary to prepare the composite directly in the
required component shape. The underlying principle of most
methods of preparation is to hot press ('diffusion bond') a
suitably cross-plied laminate structure built up of monotapes of
single layers of parallel filaments. The matrix alloy can be
incorporated in the monotape (see below) or in the form of foil,
alternatingly stacked between the fibre plies before pressing.
Diffusion bonding is carried out in the solid state close to the
melting point of the matrix. To achieve full densification
a pressure of 10 to 100 MPa held for up to about an hour is
suitable. In non-reactive systems pressing can be carried out
above the melting point in a closed die.

The nature of the process means that the composites can only
be prepared in simple laminate geometries such as plates and
beams. As an example, a method to produce a turbine blade is
shown schematically in Fig. 2.

Many methods have been suggested for producing monofilament
monotapes. Examples are: a) monofilaments attached to a backing
foil or plastic or matrix alloy using an organic adhesive to be
burnt off after stacking; b) monofilaments held together by woven
cross-fibres; c) monofilaments bound by a plasma-sprayed layer of
matrix alloy; d) monofilaments together with a layer of matrix
powder all bound with an organic binder to be burned of after
stacking; e) monofilaments impregnated with matrix by melt
drawing (non-reactive combinations). Methods involving the
burning-off of organic binders usually leave a residue of carbon
which can prove harmful to tungsten fibres and to certain
matrices.

An alternative method for producing monofilament composites
that avoids the restriction of laminate methods is to clad the
fibres individually; the clad fibres can be collected in suitable
containers and hot isostatically pressed (HIP). An advangage of
the HIP technique is that relatively complex shapes can be

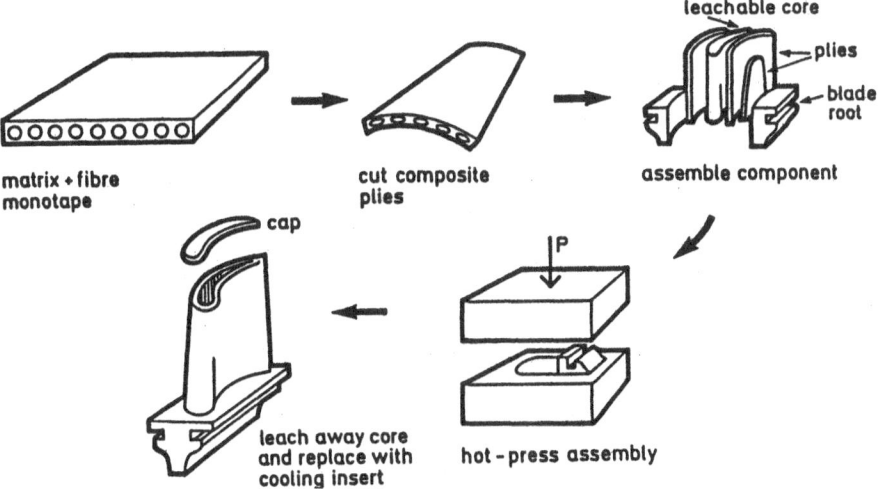

Fig. 2. (section 3.3.1.). Schematic illustration of
fabrication of WRS turbine blade by a composite
tape/hot pressing method (based on ref. 9).

formed. Methods of fibre cladding include coating methods (e.g.
electrolytic coating), and insertion of the fibres into fine
tubes (10).

Because of the high cost of monofilaments and difficulties
involved in the fabrication of these composites, they are
expected to remain expensive and restricted in use to exclusive,
albeit strategic, applications. Thus most technical development
will be related to the solution of design problems for specific
applications, in particular the integration of the material into
larger structures. Taking WRS turbine blades as an example,
problems of leading edge design and blade attachment will have to
be solved.

Future Development

SiC-particle reinforced Al-alloys have already proved to be
a class of competitive engineering materials; current production
capacity is estimated to be around 500 tons per annum. An
important development in this composite type is expected to be
optimisation of the particle morphology i.e. shape, size
distribution and uniformity of spatial distribution, leading to
improved ductility and toughness. Extension of particle
reinforcement to Mg-base alloys can be expected.

A promising area of application for short-fibre composites
appears at present to be as local reinforcement in Al-alloy
castings to impart wear resistance and dimensional stability.

However, if they are to compete more generally with the particulate composites, reductions in price and/or improvements in mechanical properties of the composite must be achieved. Property improvement can be expected through the development of new fibre compatible Al-alloy matrices. Sufficient cost reduction may be achieved by the development of cheaper fibres and the development of processing techniques, such as continuous casting and liquid phase forging (see section 2.3.1.).

Long fibre composites will probably always suffer from a higher price and restriction in methods of fabrication. However, improvements in fibres and processes will make them competitive for special applications. Developments can be expected in the preparation of long fibre preforms e.g. multi-directional woven structures and laminates. A rewarding avenue of study could be the development of hybrids consisting of more than one fibre or composite type.

In the development of high-temperature composites, the search for structurally stabler fibres and more compatible matrices will continue. Matrices of interest include Cu, which is inert to most fibres, and intermetallic compounds such as Al Ni and Ti Al in which the reactive metals can be expected to exhibit lower activity.

References

(1) Köhler, W., Aluminium 51, 244-250 pp. (1975)

(2) Renton, W.J., (editor), American Institute of Aeronautics and Astronautics, New York

(3) Riewald, P.G., Kreuger, W.H. and Dhingra, A.K., US Patent 4,012,204 (1977)

(4) Willis, T.C., White, J., Jordan, R.M. and Hughes, I.R., presented at Conf. on Solidification Processing held in Sheffield (1987)

(5) Nair, S.V., Tien, J.K. and Bates, R.C., International Metals Reviews 30, 275-290 pp. (1985)

(6) Mykura, N., presented at Institute of Metals Conference, Metal Matrix Composites: Structure & Property Assessment, London, November (1987)

(7) Clyne, T.W., in Proc. Sixth International Conference on Composite Materials, Eds. Matthews et al, Elsevier Applied Science, London, vol. 2 (2)275-(2)286 pp. (1987)

(8) Van Hille, D., Bengtsson, S. and Warren, R., submigged for publication in Composites Science and Technology

(9) Winsa, E.A., Eds. Hack adn Amateau, Met. Soc. AIME., Warrendale, USA, 283-299 pp. (1983)

(10) Warren, R., Larsson, L.O. and Garvare, T., Composites 10, 121-125 pp. (1979)

3.3.2. Ceramic Matrix Composites

R. Warren

Chalmers University of Technology

S-40296 Göteborg

Fibre-reinforced ceramics are a very promising class of structural materials (see section 2.3.2.). Compared with monolithic ceramics they exhibit higher toughness, fracture energy absorption and reliability at equal or higher strength levels (see sect. 2.3.2., refs. (2; 7, 12, 25, 33, 37, 39, 40)). The development of the various types of ceramic composites has, however, been impeded by severe processing problems such that progress has been possible only in limited directions. Even for these successful materials it has become clear that processing techniques that can achieve better fibre/matrix distributions, more optimum interfacial properties and a wider choice of constituent phases are necessary for further improvements in the properties and applications of fibre reinforced ceramics.

There are several general reviews on ceramic matrix composites (see section 2.3.2. refs. (22-26)), and an extensive review on production methods (section 2.3.2., ref.(23)). The following discussion will refer mainly to three distinct classes of ceramic composites, viz: i) whisker composites, ii) long fibre preform composites, iii) long fibre - slurry/hot press composites.

1. WHISKER COMPOSITES

The superior properties of SiC whiskers has led to a rapid development of various whisker reinforced ceramics which exhibit improved mechanical properties (sect. 2.3.2., refs. (1, 2, 5, 6, 8, 10)). The rapid success of whisker reinforced ceramics has been possible only because these composites may be produced by conventional ceramic processing techniques. Most currently-used processing "flow sheets" includes the following stages (see sect. 2.3.2. refs. 2,5,8,10):

1. Whisker "cleaning" i.e. Wet dispersion followed by sedimentation (1) or sieving (sect. 2.3.2., ref. (8)) to remove whisker bundles or other impurities.
2. Wet mixing of matrix powder and whiskers, typically by ball milling (2) (sect. 2.3.2., ref. (2, 10)), ultrasonic homogenizing (sect. 2.3.2., ref. (2)) or by high speed homogenizing (sect. 2.3.2., ref. (5)).
3. Hot pressing

Since whiskers inhibit the sintering of the matrix (3) it is very difficult to densify whisker composites without using high pressures. Uniaxial hot pressing is a relatively straightforward method but it is restricted to simple geometries necessitating expensive machining operations after pressing. Production rates are low. Nevertheless, hot pressing has been used in many of the reported investigations (2) (sect. 2.3.2., refs. (1, 4, 7, 8, 10)). More interest is now being focussed on other methods to produce more "near-net-shape" parts. This has led to a need for studying green forming methods such as injection moulding and slip casting.

Combinations of forming and densification methods currently being studied include:
- cold isostatic pressing (CIP) + sintering + post-sinter, hot isostatic pressing (HIP) (unencapsulated) (4)
- CIP, slip casting + HIP (glass-encapsulated) (5)
- slip casting - reaction bonding (6)
- injection moulding (7)
- slip casting (8)
- sintering + post-HIP-ing (9)

Pressureless sintering, to a state of closed porosity, followed by HIP-ing has so far only been possible with low whisker contents, for example, 10 v/o in an Al_2O_3 matrix (2.3.2., ref. (2)) and 20 v/o in a Si_3N_4 matrix (9). This and the fact that higher temperatures have to be used leading to increased whisker/matrix chemical interaction, limit the toughening effect achievable, as compared to hot pressed 25-30 v/o whisker composites. Hitherto, reports of mechanical properties of whisker reinforced composites have been variable and in some cases conflicting. In most instances this can be attributed to variations in whisker quality and/or variation in processing leading for example to variations in defect content or the spatial homogeneity of the whiskers.

Future research will have to be directed towards: synthesis of new improved and lower cost whiskers; mixing techniques to permit better control over fibre morphology; shaping techniques such as slip casting and injection moulding followed by pressureless sintering or HIP-ing; understanding and describing the complex fracture mechanics (10, 11) of whisker reinforced ceramics. More attention must be given to the sciences underlying the technology viz. inorganic chemistry, colloid science, rheology and solid state chemistry. To produce successful composites, an integrated approach to processing technology is essential.

2. LONG FIBRE-PREFORM COMPOSITES

This class of composites includes materials such as carbon/carbon and SiC/SiC composites but is not limited to these fibres or matrices. Thus it has been possible to produce composites on the basis of multidirectional woven preforms of multifilament carbon, SiC, Al_2O_3 or $Al_2O_3-SiO_2-B_2O_3$ fibres (see Table 1, sect. 2.3.2. for description of fibres). These preforms are subsequently infiltrated by a choice of ceramic matrices using various deposition techniques, all of which are conducted at temperatures significantly lower than normal ceramic sintering temperatures but which unfortunately require substantially longer processing times.

Chemical vapour deposition (CVD) or infiltration (CVI) (12, 13) sect. 2.3.2., refs. (12, 13)) can be used for several matrices (e.g. SiC and TiC). They are formed in situ by a gas phase reaction and so can be deposited in the pores of a fibre preform penetrated by the gas mixture. A problem is to control the deposition in such a way as to avoid closing the porosity at the preform surface before a sufficiently high density is attained in the bulk. Typical deposition temperatures reported are SiC - 1200°C (2.3.2., ref. (13)) and TiC - 950°C (13). The process for production of a component is illustrated schematically in Fig. 1. The infiltration can be carried out as a batch process in which many components are loaded into a large reaction chamber. The infiltration time, with intermediate surface removal, is typically in the range of weeks. Small samples have, however, been produced in a continuous set-up in less than 24 hours (2.3.2., ref. (13)). Densities in the range 80-90% of theoretical density with <10% open porosity are normally attained.

Alternative, lower-temperature methods of depositing the matrix involve liquid impregnation (+ pyrolysis) or liquid sol impregnation (+ sintering). Polymer pyrolysis of polycarbosilanes or polycarbosilazanes have been successfully utilized to form matrices of SiC and Si_3N_4 respectively (14, 15) (2.3.2., ref. (14)). A pyrolysis method has been used extensively in the production of carbon/carbon composites (2.3.2., ref. (11)). The

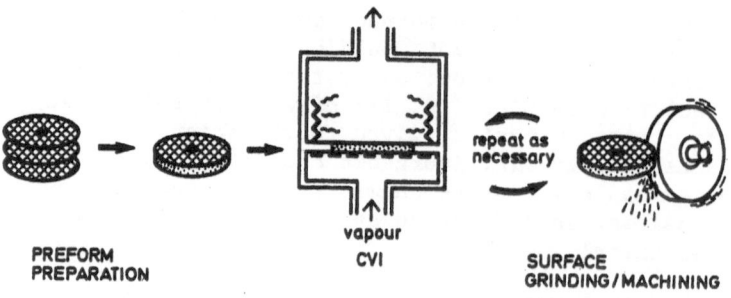

Fig. 1. Schematic illustration of CVI preparation of
fibre reinforced ceramic component.

advantage of pyrolysis methods is the low temperature (compared
to ceramic sintering temperatures). Moreover, it is easier to
infiltrate a fibre preform with a liquid than with a powder
slurry. A drawback is, again, long process times; pyrolysis
yields are low and repeated impregnations/pyrolyses are needed to
achieve adequate densities. Another drawback is the high cost of
the polymer precursors. Carbosilanes or silazanes are today
produced only in small cost-ineffective quantities. Precursors
for carbon (phenolic resins, pitch) have become less costly
however.

High cost and repeated impregnations/sinterings are also
necessary in sol-gel processes. Sol-gel techniques are today
mainly used for oxide ceramics such as Al_2O_3, ZrO_2, SiO_2 or
mullite (16) (2.3.2., ref. (16)). A sol may be described as a
colloidal suspension of entities having a size between large
molecules and particles (in this case inorganic molecules in a
liquid). The sol can be polymerized (gelled) either chemically
or by raising the temperature. A low-density gel is formed,
which after drying can be sintered at considerably lower
temperatures than ceramic powder compacts.

Both polymer pyrolysis and sol-gel techniques can be
combined with CVI or powder methods. For example, the first
infiltration may be with a polymer or sol filled with ceramic
powder, subsequent infiltrations with either pure polymer or sol,
and the final infiltration filling the smalles pores with CVI.

The preform may also be infiltrated with a solid or liquid
which after deposition is reacted with a gas or liquid to form
the desired ceramic matrix (reaction bonding). Preforms have,
for example, been infiltrated with silicon powder slurries and,
after drying, nitrided to form Si_3N_4. Composites of monofilament
(CVD) SiC fibres in a matrix of reaction bonded Si_3N_4 have been
reported with very good mechanical properties (2.3.2.,
ref. (15)). Multifilament SiC fibres such as Nicalon, having

excess surface carbon, are more reactive towards the silicon at nitridation temperatures (1300–1400°C) and degrade seriously (17). This might be avoided if the nitridation time is reduced significantly as in the method developed by Pompe et al. (18). An important advantage of reaction bonding is that the reactions usually involve volume increases (e.g. $3Si + 2N_2(g) \rightarrow Si_3N_4$) and thus more effectively fill the preform pore structure than the pyrolysis or sol-gel techniques. Hitherto, however, fully dense composites have not been obtained with any of the preform infiltration methods.

Important future research must include:
- Preform development to obtain higher fibre volume fractions and multidimensional preforms with a narrower pore size distribution (i.e. smaller variations in fibre spacing).
- Development of fibres with higher temperature capability, lower cost, and in other ceramics.
- Improved understanding of the sciences underlying: polymer synthesis and pyrolysis; sol-gel chemistry; gas-phase reactions; and rheology of molten polymers, sols or suspensions (i.e. organic chemistry, inorganic chemistry and fluid mechanics).
- Development of low-cost composites with appropriate fibres for lower temperature, non-structural applications, for example, with matrices densified at room temperature such as macro-defect-free (MDF) cement (2.3.2., ref. (41)) or inorganic polymers (2.3.2., Ref. (42)).

3. LONG FIBRE – SLURRY/HOT PRESS COMPOSITES

A class of ceramic matrix composites has been developed utilizing the fact that a mixture of glass or ceramic powder and fibres may be densified to almost full density, if the powder melts and a pressure is applied during sintering (2.3.2., refs. (17-21, 38-40)). Composite laminate structures, for example, can be produced as illustrated in Fig. 2 (2.3.2., ref. (19)). A fibre yarn is passes through a slurry, containing matrix powder and various additives (binder, dispersant), and wound on a drying drum. Unidirectional "pre-pregs" can then be cut, stacked and hot-pressed after binder burnout. Hot-pressing is carried out at temperatures wher viscous flow is the dominant densification mechanism. Lack of low cost, high temperature and oxidation resistant fibres has led to relatively low melting point materials being the most successful matrix materials. Glasses and more recently glass-ceramics have been used. The latter can be melted and formed as a glass but crystallize on heat-treatment to yield a polycrystalline ceramic material (2.3.2., refs. (17-19)). A wide range of melting points and thermal expansions can be obtained by adjustment of the composition of the matrix. The most commonly studied matrices are taken from

the systems Lithium-Aluminium-Silicate (LAS) or
Magnesium-Aluminium-Silicate (MAS) (2.3.2., ref. (19)).

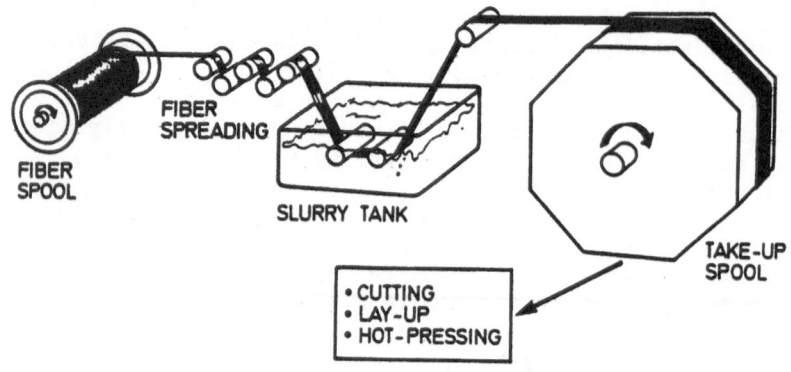

Fig. 2. Schematic illustration of slurry method for preparing long fibre reinforced glasses and ceramics.

It is also possible to use the slurry/hot press method with
matrices that densify by liquid-phase-sintering. For example,
C-fibre reinforced Si_3N_4 containing sintering additives has been
hot-pressed (2.3.2., ref. (20)) or HIP-ed (2.3.2., ref. (21)) to
full density. The slurry/hot press technique can be applied to
configurations other than laminates. For example, filament
winding of the slurry-impregnated yarn is possible.

A new technique now being pursued is the melt infiltration
of three-dimensionally woven preforms (in a process similar to
the squeeze-casting of metal matrix composites (2.3.2.,
ref. (19)). This is a very promising way to produce ceramic
composites since high density, flaw-free composites can be formed
in a single processing step, with small dimensional changes from
preform to final product and with virtually any fibre
reinforcement geometry. However, success with this approach is
limited by fibre-damaging chemical reactions at the higher
temperatures required and by the very high melt viscosities of
ceramics and glasses (order of magnitude larger than for metals)
which demand very long infiltration times.

Techniques involving densification by viscous flow are
probably used with best advantage in the production of moderate
temperature (<1000°C) materials, for example in applications
where low thermal expansion, high thermal insulation, chemical
stability and some high temperature capability coupled with the
greater reliability offered by fibre reinforcement, is needed.

The future research needs for this type of ceramic composite
include:

- Development of slurry infiltration techniques for
  3-dimensional fibre preforms, perhaps followed by HIP-ing
  instead of uniaxial hot pressing.
- Investigations of the feasibility of pressureless sintering of
  long fibre composites.
- Expanded exploration of melt infiltration of fibre preforms.
- Development of low-cost fibres for moderate temperature
  (<1000°C) applications.
- Development of ultra high performance fibres for hot pressing
  of non-oxide ceramic composites.
- Development of combined processes, for example, slurry
  infiltration, reaction bonding, polymer impregnation,
  sol-impregnation, CVI, hot press or HIP.

FUTURE NEEDS - SUMMARY

The greatest needs in the processing of ceramic composites
can perhaps be summarized with three words: DENSITY, SHAPE and
ECONOMY. It is desirable to develop processing techniques that
yield dense, near net shape components preferably in a rapid,
single step process without the need for expensive raw materials.
Much remains to be done, however, before ceramic matrix
composites reach this stage. Processing experience obtained in
the field of polymer and metal matrix composites will be valuable
but, a large part of the development will remain very specific
for ceramics.

The need for an integrated approach encompassing basic
sciences such as inorganic chemistry, colloid science, organic
chemistry, solid-state chemistry and fluid mechanics, for
example, must be stressed. Applied sciences such as textile
engineering (for preforms), chemical engineering and mechanical
engineering must also contribute to the development of new
processing methods for ceramic matrix composites.

In spite of improvements brought about by whisker and fibre
reinforcement, ceramic composites do not possess the built-in
reliability of metals. One approach to reducing the scatter in
mechanical behaviour of these materials is the use of very
stringent quality assurance introduced at each stage of
processing. Important parameters to monitor are the size of
internal pores and other defects remaining after consolidation,
any fibre degradation that might occur during handling and during
high temperature consolidation, the uniformity of whisker and
fibre distribution and finally the state of the surface of the
final product. In these respects, the need for the development
of more sensitive techniques of non-destructive evaluation for
monolithic ceramics applics equally to the composites.

286

4. REFERENCES

1. Lundberg, R., Nyberg, B., Willander, K., Persson, M. and Carlsson, R., Composites, 18 (2) 125-127 pp. (1987)

2. Inoue, S., Niihara, K., Uchiyama, T. and Hirai, T., Proc. Int. Conf. Ceramic Mater. Components for Engines, DKG, ed. Bunk., W. and Hausner, H., 609-617 (1986)

3. Clegg, W.J., Alford, N.McN. and Birchall, J.D., Proc. Int. Conf. Engineering with Ceramic, London, in press (1986)

4. Tiegs, T.N. and Becher, P.F., Am. Ceram. Soc. Bull., 66 (2), 339-342 pp. (1987)

5. Lundberg, R., Nyberg, B., Williander, K., Persson, M. and Carlsson, R., Proc. First Int. Conf. on HIP, Luleå, Sweden, June in press (1987)

6. Starr, T.L. and Harris, J.N., as ref. (2), 217-224 pp.

7. Pujai, V.K., Willkesn, C.A. and Corbin, N.D., presented at Am. Ceram. Soc. 89th Ann. Meet., abstr. 37-C-87 April (1987)

8. Nagel, A., Hoffman, J., Greil, P. and Petzow, G., as ref. (7), abstr. 38-C-87

9. Hoffmann, M.J., Greil, P. and Metzow, G., Proc. Int. Conf. Science of Ceramics 14, Canterbury, England, in press (1987)

10. Becher, D.F., Tiegs, T.N., Ogle, J.C. and Warwick, W.H., 4th Int. Symp. Fract. Mech. of Ceramics, Blacksburg, USA (1985)

11. Kageyama, K. and Chou, T.W., Proc. ICCM VI & ECCM 2, Elsevier, Ed. Matthews, F.L., et al., 2.60-2.69 pp. (1987)

12. Cristin, F., Naslain, R. and Bernard, C., Proc. 7th Int. Conf. CVD, Ed. Sedwick, T.O. & Lydin, H., Electrochem. Soc., Princeton, USA, 499 pp. (1979)

13. Rossignol, J.Y., Quenisset, J.M. and Naslain, R., Composites, 18 (2) 135-144 pp. (1987)

14. Walker, Jr., B.E., et al., Am. Ceram. Soc. Bull., 62, (8), 916-923 pp. (1983)

15. Fizer, E. and Gadow, R., Am. Ceram. Soc. Bull., 65, (2), 326-335 pp. (1986)

16. Pierre, A.C., Uhlmann, D.R. and Hordonneau, A., Rev. Int. Hautes Temp. Refract., <u>23</u>, (1), 29-35 pp. (1986)

17. Fischbach, D.B. and MacLaren, D., NASA-Report, DOE ET 13389-T1 (1982)

18. Lundberg, R., Pompe, R. and Carlsson, R., As ref. 11, 2.33-2.39 pp. (1987)

3.4. Joining

3.4.1. Joining of Metallic Materials

T.G. Gooch

The Welding Institute

Abington Hall, Abington, Cambridge CB1 6AL, UK

1. PREAMBLE

The service performance of high temperature equipment depends on the particular characteristics of the materials of construction, on the joining procedures employed in fabrication, on the joint property data available and on the design criteria involved. As far as possible, these individual aspects are considered in each section, but it must be borne in mind that they are closely interrelated.

2. PRESENT STATE OF ART

At the present time, joining of materials for high temperature service is most commonly accomplished by fusion welding. Solid state joining is used to a lesser extent, as are brazing and mechanical fastening. The dominance of fusion welding stems from a number of considerations, especially :

i) it offers flexibility to design and economy in manufacture,
ii) in many cases, tolerance on machining requirements can be relaxed, and appreciable component misfit can be permitted, and
iii) it enables a continuous load bearing path to be achieved in the structure.

In addition, for many high temperature alloys, the use of a fusion joining process obviates problems associated with surface refractory oxides, etc., that restrict alternative methods in which the component parts remain solid.

The vast majority of high temperature plant is currently made from a heat resisting steel, whether a ferritic material alloyed with chromium and molybdenum or an austenitic grade. Certainly, this is the case for the petrochemical (1) and power generation (2) industries, although for gas turbines other alloys are employed, notably nickel and titanium based systems (3, 4). The fusion welding characteristics of such established materials are well understood in principle in terms of metallurgical changes that take place and the various forms of defect that can occur (5, 6). To a large extent, measures necessary during welding to avoid the formation of defects, whether operator/-process induced (slag, porosity, lack of fusion, etc.) or cracking at high or low temperature, have been defined and can be incorporated into fabrication procedures.

A main thrust of operating plant is to increase service temperatures and this is leading to study of newer materials such as directionally solidified and monocrystalline components, and oxide dispersion strengthened (ODS) and intermetallic alloys, quite apart from advanced composite materials. A number of approaches have been examined for welding such systems (3, 7, 8); none has achieved commonplace production status, although fusion welding has been employed for low stress applications (7).

Plant for the power generation and petrochemical industries is normally welded using an arc as the heat source, with shielding of the joint area from the atmosphere being accomplished by a flux (e.g. metal inert gas (MIG), tungsten inert gas (TIG) welding) (Fig. 1). The TIG process finds application also for gas turbine construction, together with resistance, laser and electron beam fusion welding (Fig. 2) and solid state friction welding, depending on the component concerned. Brazing and mechanical fastening are utilised, to a greater extent relative to arc welding than in power plant and similar equipment (3, 9).

Over the years, a diversity of further welding processes has been developed (10), of actual or potential application to high temperature materials. In the former category, flash butt welding has been used for boiler tubing, etc., while diffusion bonding entirely in the solid state or with a transient liquid phase is receiving consideration for situations where conventional fusion welding is not suitable.

The salient features of the different joining methods are known, submerged arc welding offering high joint completion rates and productivity in heavy section material, and TIG welding being employed for fusion joining operations requiring greater precision, for example (Fig. 1) (10). Viewed overall, and certainly for the more common methods, there is a good practical

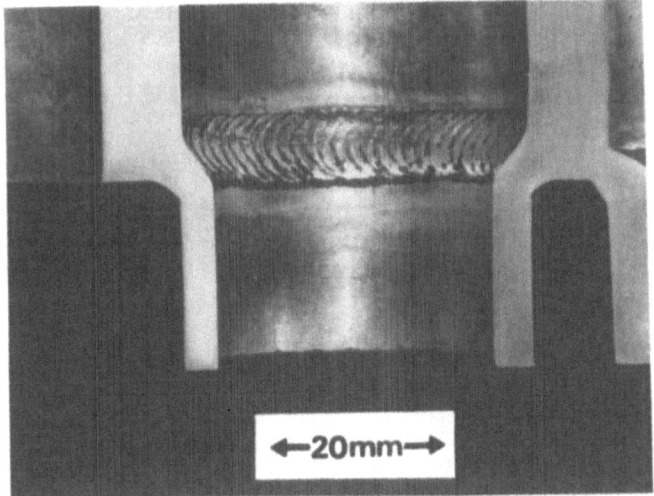

Fig 1   Section through backface tube/tube sheet weld in austenitic stainless steel made by pulse TIG procedure. Neg No 42626

Fig 2   Electron beam welds in nickel alloy gas turbine blades. Mag x20. Neg No 34964.

understanding of the effects and relative significance of the various process parameters requiring control to achieve sound welds, and a wide range of welding equipments is commercially available. Increasingly, fusion welding is being mechanised and/or automated to reduce dependence on operator skill and to improve productivity, and this has led to the development of ancillary equipment for overall control of the welding operation (Figs. 3 and 4) (11). There are, moreover, well defined procedure qualification codes in use by industry to ensure production of sound joints, and a number of non-destructive testing (NDT) techniques can be employed.

Substantial data have been generated on the service properties obtaining from welds in high temperature materials that can be employed for design purposes, and there has been considerable effort (2) devoted to quantifying the mechanical impact of welds, with regard to geometric (12) and residual stress (13) effects.

3. CURRENT LIMITATIONS

Fusion welding presents two particular drawbacks (5, 6). Firstly, the inherent thermal cycle with exposure to temperatures up to the solidus will induce metallurgical change in most metallic materials, while a chill cast weld metal is produced : the joint properties may therefore differ significantly from those of the unaffected parent material. Secondly, depending on the material/process combination involved, a number of defect types may be induced.

Considering the latter of these, the propensity for fusion welding to cause defects dominates the derivation and specification of welding procedures : the prime aim is almost invariably the achievement of joints meeting specified requirements on soundness. Even though property optimisation may be desirable, some reduction in joint characteristics due to welding is normally accepted rather than having defective welds placed in service. The various mechanisms by which defects form have been identified, although not always adequately quantified (14), and fusion welds can normally be produced reliably in most established high temperature materials. However, the situation is much less satisfactory for newer alloys developed to enable operating temperatures to be increased, since to a large extent their inherent characteristics are inimical to the production of sound welds of high properties. For example, materials based on precipitation hardening practice, such as nickel or cobalt based superalloys, have a high level of alloying elements and thus show a general tendency to the formation of low melting point phases which induce cracking in the weld area at elevated temperature, the problem being exacerbated by the high matrix stiffness (6).

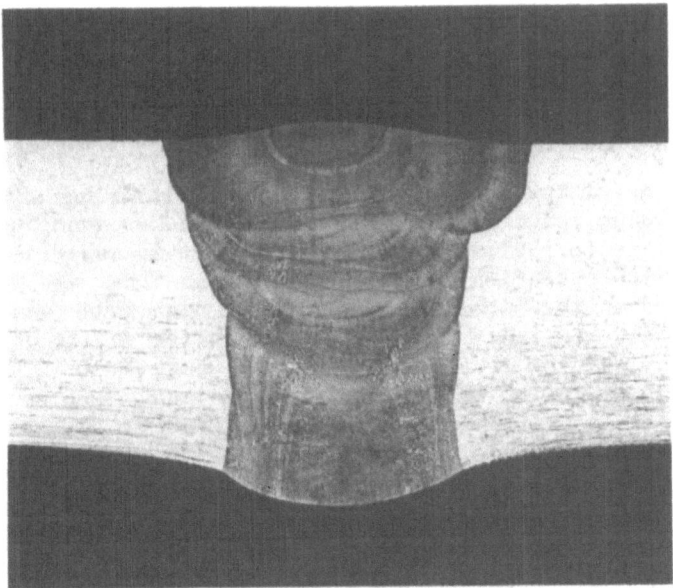

Fig 3   Section through austenitic stainless steel TIG weld made
         using fully automatic multipass procedure.   Max x10.
         Neg No OY3978

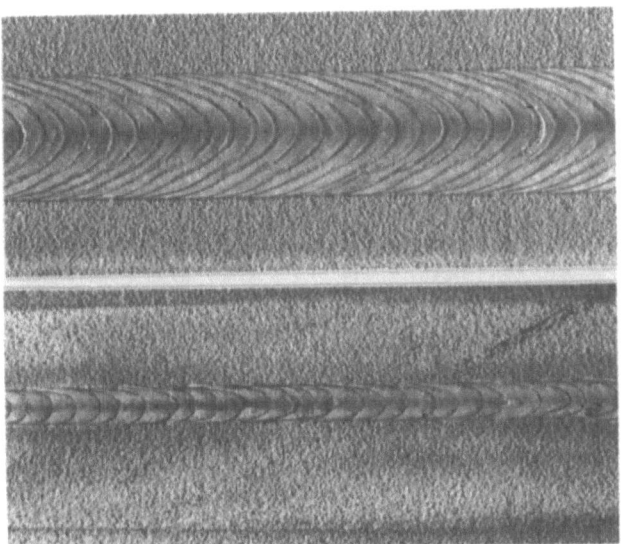

Fig 4   Pulsed plasma welds in austenitic stainless steel showing
         excellent bead profile:
         a)   Top surface. Mag x5. Neg No 38464
         b)   Underbead. Mag x5. Neg No 38465

Comparable problems may exist with more recent metallic materials, both ODS (15) and intermetallic (16) systems. Thus, with many high temperature alloys, it is questionable whether conventional fusion welding will offer sufficient assurance of joint integrity.

Because of the economic attraction of retaining existing fusion welding practice for fabrication, studies continue on the welding characteristics of new high temperature materials (7, 8, 15, 16), with attention to optimising conventional arc welding methods and to the potential advantage of high power density electron beam and laser processes in which the volume of material affected by welding is minimised (3, 7). Nonetheless, the limitations of fusion welding are clearly recognised by industry and increasingly consideration is being given to alternative joining methods (17-19). Some scope is afforded by brazing (9), provided that the temperature limitations of brazed joints are recognised and can be allowed for in design, but solid state joining would seem to be of greater potential application. Two methods in particular can be identified, friction welding and diffusion bonding. The former is already widely employed in general engineering (10), and is known to produce sound joints in most high temperature materials, but at present is limited in its areas of usage by geometric constraints while there is a lack of information on the effects of the inevitable plastic metal flow on joint properties. Similar comment pertains to diffusion bonding, with the further drawback that achievement of bonding over the whole joint area may be problematic in materials designed for intrinsic high temperature strength and scaling resistance (17) : bonding may be limited by the resistance to local metal flow so that full interfacial contact is not obtained, and by refractory surface oxides. Moreover, as yet no satisfactory procedure has been derived for structure sensitive materials which gives complete continuity of structure across the bond. Hence, whilst joints in the solid state or with a liquid phase can certainly be made in new (and established) materials of construction, process development is required before the approach becomes a production reality.

There is a general requirement for improved reliability and sensitivity of NDT. However, particular problems arise with solid state joints in that there is no fully acceptable technique for locating defects at the bond line.

Given that sound joints can be produced, it remains a fact that even for common materials the data base on weldment properties available to the designer is very much less than for the parent materials. This is the case for both mechanical characteristics and environmental behaviour such as scaling resistance, etc. Some loss of properties at welds must be expected, although this is not always explicitly recognised by design codes, and appropriate data are required across the entire gamut of high temperature equipments to obtain more exact and reliable plant design. Further, current design philosophy tends to assume that welded plant functions as an integral unit. Bearing in mind the disparate microstructural changes at welded joints, this is a major oversimplification and effort is needed to develop a comprehensive methodology for welds (2), allowing for local metallurgical and hence property variation and for differences in weld geometry associated with different welding procedures.

4. FUTURE DEVELOPMENT PROSPECTS

Little significant change is seen in the welding behaviour of alloys for conventional petrochemical and power plant. Some progressive change is likely in the materials employed to obtain increased operating temperatures (e.g. usage of 9Cr/1Mo steels, as opposed to lower alloy variants), but this should not introduce any "new" weldability problems. At the same time, appropriate quantification of behaviour will be required so that economic, reliable welding procedures can be defined. This is achievable with present technology, and because of the widespread usage of "conventional" alloys its importance should not be underestimated.

To some extent, it will be possible to join newer materials by existing techniques : already, the successful application of TIG welding to intermetallic alloys has been demonstrated (8), although substantially more work is required. Greater problems will attend those materials relying for their properties either on a directional structure or on the presence of a second phase, whether ODS or metal matrix composites. Both strengthening mechanisms are completely disrupted by conventional fusion welding, and while welds and brazed joints will almost certainly be possible, some local reduction in properties must be expected. For this reason, the newer materials are more likely to be joined by more novel techniques.

Advances in conventional welding processes will follow the trends evident in other fabrication areas, with moves to enhancing productivity and reliability such as adoption of narrow gap procedures (Fig. 5) and increased automation (Fig. 3). High power density electron beam and laser welding (Fig. 6) will be employed, in view of the high joint completion rates offered. In addition, greater control of the welding operation can be foreseen. Recent years have seen substantial effort devoted to mathematical modelling of the welding operation (20, 21), and to some degree it is now possible to control welding conditions and local thermal cycle so as to promote specific metallurgical

Fig 5   Narrow gap submerged arc welding of heavy
section steel showing small weld metal volume
required. Neg No 57735.

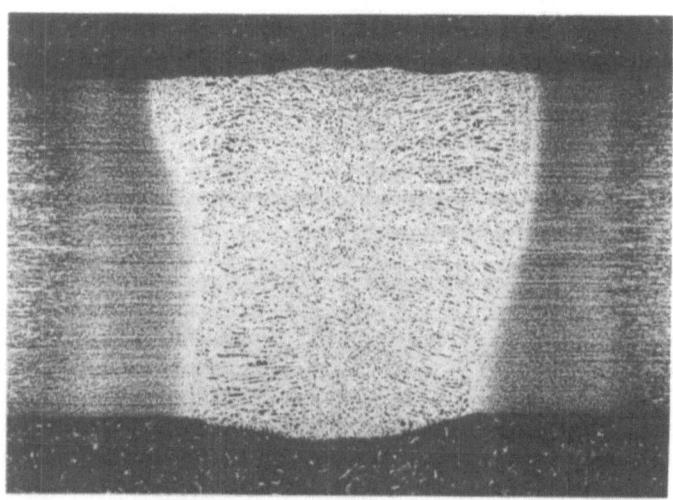

Fig 6   Laser weld in 2mm martensitic stainless steel produced at
1250 mm/min. Mag x32. Neg No OJ1782

transformation in the weld area (22). With improved understanding, especially of the factors controlling the thermal cycle at temperatures close to the material melting point, this approach will be more widely employed to tailor the weld area microstructure to give required service properties.

A number of process development areas are being explored for new structure sensitive alloy systems. Solid state joining remains attractive, especially for precision joining of components which are in near final form. Friction welding may well become more frequently used, with the inception of forms of motion other than rotation so that a wider range of component geometry can be handled (Fig. 7), although there must be close evaluation of the effect of the inherent plastic flow parallel to the faying surfaces. Diffusion bonding clearly offers promise, provided engineering requirements on flatness to achieve contact over adequate joint areas can be fulfilled. A number of process variants can be envisaged, involving perhaps a preplaced filler (21) or modification of the faying surfaces by ion beam bombardment (24) or sputter coating (25, 26) to enhance subsequent joining, with or without a transient liquid phase (27, 28). In principle, such an approach can be adapted to most new materials, but two caveats much be noted. Firstly, material-specific study will be necessary to ensure that high temperature bonding can be achieved in a production environment without unacceptable degradation of properties. Secondly, although the general principles of diffusion bonding with or without some form of surface activation are understood, the factors controlling the joining operation have not been fully quantified (29) : more detailed attention to modelling the physical changes involved is necessary for improved prediction of joining behaviour.

Alternative high energy input rate joining systems are being examined, such as explosive (7) or capacitor discharge (30) welding. It is difficult to predict how far these approaches will be viable in a production environment, but the latter method has shown promise in joining metal matrix composites with minimal disruption to the material.

A range of different treatments can be employed after joining to obtain enhanced properties, such as hot isostatic pressing to remove internal flaws (3, 27). Further, substantial scope exists for the application of controlled postweld heat treatment to obtain continuity of microstructure across the joint line of welds produced in the solid state or with only a local liquid phase.

Improved prediction of service behaviour and hence plant reliability requires generation of good property data, and this is being tackled on a fairly broad front for both creep (2) and

Fig 7   Linear friction weld in 10 x 25mm titanium
alloy. Neg No 56390.

environmental (1) characteristics. However, in large part such
data are produced by the parent material suppliers; substantially
more effort is needed on welds, although it is difficult for
material producers to allow for the wide range of welding
processes that may be applied by industry. What is wanted is
derivation of standardised specific short term tests so that most
promising material/weld combinations can be identified for
further longer term study (14). This is a vexed question, since
any test must essentially reproduce service behaviour implying
that the service behaviour is understood and can be adequately
modelled. If service involves varying load/temperature
conditions, with say creep/fatigue interaction or a significant
influence of the environment on failure, definition of a short
term test procedure becomes problematic.

Insofar as experience has consistently indicated welds to be
preferential sites for failure of operating plant, the particular
features of joints must be well understood and appropriate
allowance made. Most predictive systems and constitutive
equations developed to date tend to assume material homogeneity.
This is clearly not the case for welds (2). Nonetheless,
advances have been made, and more precise behavioural prediction
in the future can be expected, although this will require both
comparison of short and long term material characteristics and
also confirmation of the reliability of predictive models by
undertaking suitable realistic, probably close to full scale,
tests.

5. NEW TECHNOLOGY

In the context of welds in conventional heavy section
materials as employed for power and petrochemical plant, the
major new technology of interest is likely to be the achievement
of fully automated welding systems, using current arc processes,
with attendant improvement in reliability of welding (11). The
utilisation of newer joining methods will require clear
demonstration either that they present significant cost reduction
in fabrication with no loss of weldment properties, or that they
offer major technical advantage. Nonetheless, it can be argued
that the full potential of electron beam welding (and, by
analogy, laser welding) has not been defined and further study is
warranted because of the attraction of single shot welding under
well controlled conditions.

Greater change in joining technology is seen with new
materials. Solid state methods are needed, and in principle
diffusion bonding in one guise or another should be developed.
This will necessitate study of surface behaviour, and technology
can be identified for modifying the surface physically and
chemically (18, 19, 25), including deposition of suitable

Fig 8   Sputter coating of ODS specimens prior to bonding. Neg No 03338.

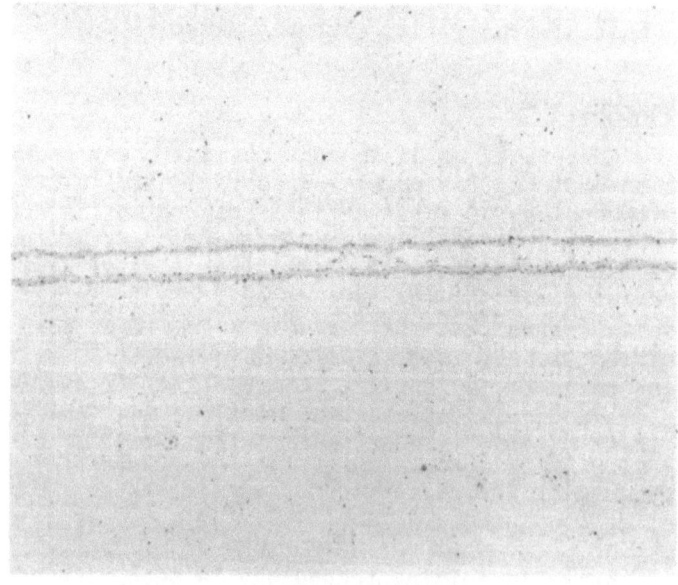

Fig 9   Brazed joint in nickel-base ODS material, produced using a modified BNi5 filler deposited by sputtering. Note interfacial second phase and oxide agglomeration. Mag x400. Neg No OX3672.

coatings, to facilitate joining (Figs. 8 & 9). However, development is required before such an approach attains production status.

6. INTERNATIONAL RESEARCH AND DEVELOPMENT

A range of individual programmes is being undertaken internationally to advance joining of materials in all sectors of high temperature plant. Most are fairly restricted in aim, for example studies to improve weld metal creep behaviour (31) undertaken by welding consumable manufacturers (32). Within the UK, there is a wide ranging commitment by the Central Electricity Generating Board to improving all facets of reliability of power generating plant : in the main this follows existing joining technology, but with significant attention to better prediction of service performance of welds (33, 34). The work recognises the particular problem of dissimilar metal welds (35), and is paralleled by study in the USA and elsewhere (35, 36).

Work on joining new materials in the USA has been reported, including intermetallic (8) and ODS (7, 15) alloys, and metal matrix composites (30). As far as is known, attention is currently being paid to the first and last of these in particular. Similar effort elsewhere has been limited by material availability, but within Europe the COST 501 project has sought to join ODS materials by a diffusion bonding or brazing approach in which controlled composition interlayers are being deposited by sputtering (25, 26). Continuing work is in progress by turbine manufacturers, evaluating diverse joining methods for materials of present and future application (3).

Effort on advanced materials can be identified in Japan (27, 28), including directionally solidified metals and ceramic/metal joining, and there is also considerable attention to development of arc welding systems. Ceramic/metal joining is being further pursued within the USA (37) and Europe (38).

7. RESEARCH NEEDS

It is considered that equal priority should be given to joining both established and new materials. In the former category, effort should be largely, but not exclusively, on ferritic and austenitic steels. The problem areas with conventional welding processes are well defined, and there should be attention given to improving reliability of welding and of service performance. Work on new materials should aim more at derivation of alternative joining techniques (which may well be applicable also to current alloys). The specific work undertaken will depend on the view taken regarding which material types will become of future practical importance, but in the short term at least atten-

302

tion should be paid to ODS, intermetallic alloys and metal matrix composites. Various forms of diffusion bonding/brazing are envisaged : consideration should be given also to the application of such techniques for joining dissimilar materials, most specifically metal/ceramic combinations. Again, attention to process reliability is needed, especially in view of the uncertainty regarding NDT.

On this basis, the following needs can be identified, in decreasing order of priority for each material class.

Established materials

1. a. Greater automation of welding
   b. Usage of high power density electron beam and laser welding
   c. Improved prediction of service behaviour of joints, involving modelling of degradation processes and correlation with large scale test data.
2. Expansion of available data base on weldments properties
3. a. Establishment of short term test methods for welds
   b. Improvement modelling of the fusion welding operation for enhanced microstructure and property control.

New materials

1. Development of solid state joining processes, especially
   a. Diffusion bonding and allied joining procedures
   b. Friction welding, including linear and other motion forms
2. Development of NDT methods
3. Improved modelling of diffusion bonding and allied joining procedures
4. Usage of electron beam and laser welding
5. Study of alternative high energy input rate joining systems.

8. ACKNOWLEDGEMENTS

The view expressed above are those of the author. In formulating them, he received considerable assistance from a number of people, both at the Welding Institute and elsewhere. Particular thanks are given to Dr. I.A. Bucklow, Dr. J.A. Williams and Mr. R.H. Jeal.

9. REFERENCES

(1) Krikke, R.H., Proc. Conf. "Behaviour of Joints in High Temperature Materials", CEC, JRC Petten Establishment, Pub. Applied Science Publishers, UK, 1982, pp. 49-57?

(2) Williams, J.A., idem, pp. 187-212.

(3) Grünling, H.W. and Schneider, K., idem, pp. 5-43.

(4) Jeal, R.H. and Gupta, S., Proc. Conf. "International Gas Turbine Congress", Tokyo, Japan, Oct. 1987, pp. III-279-285.

(5) "Metallurgy of Welding", Lancaster, J.F., Allen & Unwin, London, 3rd Edition, (1980).

(6) Davin, A. et al., vide Ref. 1, pp. 87-110.

(7) Shoemaker, L.E., Proc. Conf. "Advances in Welding Science & Technology", ASM, Gatlinburg, USA, May 1986, pp. 371-377.

(8) David, S.A. et al., Weld. J., 65, 4, April 1986, pp. 93S-98S.

(9) Christensen, J. and Sheward, G.E., vide Ref. 1, pp. 117-161.

(10) Houlcroft, P.T., "Welding Processes", Pub. Cambridge University Press, UK, (1967).

(11) Cary, H. and Barhorst, S., vide Ref. 7, pp. 783-794.

(12) Fidler, R., CEGB Report TPRD/M/1583/R86, Aug. 1986.

(13) Ibid, CEGB Report TPRD/M/1558/R86, March 1986.

(14) Gooch, R.G., vide Ref. 1, pp. 167-180.

(15) Kelly, T.J., Proc. Conf. "Trends in Welding Research in the United States", ASM, New Orleans, USA, Nov. 1981, pp. 471-485.

(16) Santella, M. and David, S.A., Weld. J., 65, 5, May 1986, pp. 129S-137S.

(17) Moore, T.J. and Glasgow, T.K., Weld J., 64, 8, Aug. 1985, pp. 219S-226S.

(18) Spinat, R. and Honnorat, Y., Proc. Conf. "High Temperature Alloys for Gas Turbines and Other Applications", Liège, Belgium, Oct. 1986, D. Reidel Publising Co, Holland, pp. 151.

(19) Haufler, G. et al., idem, pp. 8001.

(20) Szekely, J., vide Ref. 7, pp. 3-14.

(21) Goldak, J.A. et al., Met. Trans. B., 15B, 2, June 1984, pp. 299-305.

(22) Alberry, P.J. et al., Met. Tech., 10, 1, Jan. 1983, pp. 28-38.

(23) Duvall, D.S. et al., Weld. J., 53, 4, April 1974, pp. 203S-214S.

(24) Funamoto, T. et al., Quart. J. Jap. Weld. Soc., 3, 4, Nov. 1985, pp. 881-886.

(25) Jahnke, B. and Dannhäuser, K., vide Ref. 18, pp. 175.

(26) Bucklow, I.A., Annual Report No. 4, (1986), COST 501, Project UK5.

(27) Anon, "Welding in Japan '86", Ed. Baba, A., Sampo Publications Inc., Japan, pp. 80-86.

(28) Ibid, idem, pp. 22-28.

(29) Derby, B. and Wallach, E.R., Met. Sci., 16, 1, Jan. 1982, pp. 49-56.

(30) Devletian, J.H., Weld. J., 66, 6, June 1987, pp. 33-39.

(31) Farrer, R.A., Proc. Conf. "Stainless Steels '84", Gotherburg, Sweden, (1984), Pub. Inst. of Metals, London, UK, pp. 336-342.

(32) Lefebvre, J. et al., idem, pp. 330-335.

(33) Gooch, D.J. and Kimmins, S.T., Proc. Third Int. Conf. "Creep and Fracture of Engineering Materials and Structures", Swansea, UK, April 1987, Inst. of Metals, pp. 689-703.

(34) Williams, J.A., idem, pp. 721-740.

(35) Nicholson, R.D. et al., Proc. Conf. "Dissimilar Welds in Fossil-Fired Boilers", New Orleans, USA, (1985), EPRI, CSD-3623.

(36) Various, Proc. Conf. "Joining Dissimilar Metals", EPRI, Pittsburg, USA, Aug. 1982.

(37) Mizuhara, H. and Huebel, E., Weld. J., <u>65</u>, Oct. 1986, pp. 43–51.

(38) Nicholas, M.G., Brazing & Soldering, No. 10, Spring 1986, pp. 10.

3.4.2. Joining of Advanced Ceramics for

High Temperature Applications

K.M. Ostyn[+], S.D. Peteves[°], A.G. Vinckier[+*]

[+] Research Center of the Belgian Welding Institute, Ghent,
Belgium
[*] State University Ghent, Belgium
[°] Joint Research Centre Petten Establishment, Petten, The
Netherlands

1. INTRODUCTION

For many structural applications, where ceramics could be
extremely suitable for critical components because of their
specific characteristics, the success of their application hinges
to a large extent on the possiblity of forming ceramic to metal
or ceramic to ceramic joints of adequate quality. Conventional
fusion welding of ceramics poses enormous problems because of
their high melting points, a tendency to decompose rather than to
melt and the effects of the severe thermal shocks associated with
welding. The realization of reliable joints is a complex task
especially in the case of metal-to-ceramic joining. Specific
problems are the thermal expansion mismatch, the chemical
non-compatibility and the large difference in elastic-plastic
behaviour between the materials to be joined. Specific
properties of some ceramics and metals are shown for comparison
in Table 1.

High thermal stresses followed by cracking, set up by
thermal expansion and/or shrinkage, either during jointing or
"in-service" thermal cycling, could be reduced or even eliminated
using interlayers with graded thermal expansion coefficients.

TABLE 1

Material	Melting Point (°C)	Therm.Expansion Coeff. ($10^{-6}K^{-1}$)	Thermal Conductivity $Wm^{-1}K^{-1}$	R.T. fracture strength MPa *	E-modulus at R.T. GPa
Al_2O_3	2050	9	8	350–450	360–405
α-SiO_2	1710	22.2	1.5		
ZrO_2(part. stab.)	2800	11	2	600	215–205
TiC	3140	7.7	32	~1000	
HP Si_3N_4	1900 (subl)	3	25	650–800	310–320
RB Si_3N_4	-	3	5–12	200	200
α-SiC	2650 (subl)	4	50–200	460–500	420–405
Sialon	-	3	20–25	800	300
Al	660	23.5	238	45	68–78
Cu	1083	17.0	397	209	117–124
Fe	1536	12.1	78		
Ni	1455	13.3	88	317	
Mo	2615	5.1	137	435	365
Nb	2467	7.2	54	240	
W	3387	4.5	174	550	400
mild steel	1400	15	63	460	210

(*) Bend strength is quoted for ceramics and tensile strength for metals.

Easily deformable interlayers, with low moduli of elasticity could also be applied in order to minimise the effect of lack of ductility exhibited by industrial ceramics which leads to high stresses where contacting surfaces do not conform.
Possible joining techniques and some examples of innovative ideas will be reviewed in this article. Firstly, some basic principles of interfacial phenomena will be discussed.

2. INTERFACIAL PHENOMENA

The phenomena that take place at the interface of two dissimilar materials, such as a ceramic and a metal, are of great importance during joining processes. The understanding of these phenomena and a knowledge of the structure of the resultant interface are essential for controlling the bonding process and hence for achieving bonds that meet technological requirements. The chemical and physical aspects of these interfacial interactions will be discussed in some detail below.

2.1. Physical interactions:

Upon close contact of a ceramic and a metal surface, characterized by the free surface energies σ_c and σ_m, an interface is formed that requires an energy σ_{cm} to be stable. The driving force for the physical interactions resulting in the interface formation is the work of adhesion, W_{ad}, defined as (1)

$$W_{ad} = \sigma_c + \sigma_m - \sigma_{cm} \qquad (1)$$

Since an experimental determination of W_{ad} is not possible, its evaluation requires the measurement of the surface free energies and the interfacial energy σ_{cm}. However, the determination of σ_{cm} is also inherently difficult as revealed by the only valid experimental technique (2). In practice W_{ad} is derived from the classical wetting or sessile drop technique. By measuring the contact angle θ, W_{ad} can be calculated from the following relation (3):

$$W_{ad} = \sigma_m (1 + \cos\theta) \qquad (2)$$

It should be noted that W_{ad} can also be estimated from the solid state equivalent of the sessile drop on a ceramic substrate (4). Nevertheless, the quantity W_{ad} for the liquid metal/ceramic does not correspond to W_{ad} for the solid metal/ceramic interface, and experiments have shown that any correlation between the two is invalid. Furthermore, the concept of adhesion as presented earlier is valid only for true equilibrium conditions and in the absence of any diffusion or reaction at the interface. Under these conditions W_{ad} can be accepted as a reasonable measure of the strength of the joint as discussed elsewhere (5). Several

investigations have studied the wettability of metals to
ceramics, in particular to oxides such as Al_2O_3 and SiO_2 (6).
There are rather few studies on adhesion of metals to Si_3N_4 or
SiC (7-10). The reported contact angles of the latter materials
with several metals are given in Table 2. Collected values for
adhesion on oxides can be found in several review papers on this
subject (11).

Improved wetting conditions result when the interfacial
energy is decreased, which for a given system implies that the
chemistry of the interface must be changed. The contact angle of
adhesion can however also be greatly influenced by decreasing the
roughness of the ceramic surface. Studies have clearly
demonstrated that roughening a substrate decreases its
wettability, although this effect is somewhat lessened at higher
temperatures (12). This is in contrast with metals where wetting
or capillarity is enhanced by a "satin" or lightly ground surface
finish.

Other factors which affect the wetting behaviour are the
temperature and the atmosphere. By increasing the temperature
for example, the contact angle decreases in accordance with the
negative temperature coefficient of the surface and interfacial
energies. However, the sharp transition from non-wetting to
wetting conditions found in many metal-ceramic systems at high
temperatures cannot be attributed to this temperature effect
(13). This transition may better be ascribed to either the onset
of interfacial reactions or the decomposition and/or evaporation
of the surface metal oxide that prevents the physical contact
with the ceramic. Adsorption of oxygen on the surface of many
liquid metals causes a sharp decrease in their surface energy
values, while on the contrary, it may promote adhesion in several
metal-ceramic oxide systems. For these systems, particularly for
SiO_2 and Al_2O_3, it can be deduced from a large amount of
experimental data that wetting and adhesion increase as the
affinity of the liquid metal for O_2 increases. These
observations led to successful modelling of the metal-oxide
wetting behaviour (14). However for nitrides or SiC the
situation is not sufficiently clear to allow for adequate
modelling of their adhesion to metals. It is hoped that the
availability of pure single crystals (via CVD processes) of these
ceramics will simplify experimental work on their wetting
behaviour.

2.2. Chemical reactions:

Chemical reactions occuring at the interface of a ceramic/
metal system, once a physical contact is established, further
decrease the free energy of the system. The decrease in the
interfacial energy will result in enhanced adhesion and bonding.

TABLE 2 : Contact angles (Θ) of metals with SiC, Si₃N₄

Metal	Si$_3$N$_4$	SiC	T (K)	Ref.
Cu	145		1373	(9)
Cu-20%Ti	15		"	(9)
Cu-40%Ti	8		"	(9)
Cu-30%Ti	60		"	(9)
Cu-50%Ti	5		"	(9)
Al	120		1200	(10)
Al	46		1400	(10)
Al	25		1500	(10)
Cu	125		1400	(10)
Ag	135		1350	(10)
Al		106	973	(7)
Al		75	1003	(7)
Al		58	1043	(7)
Al		80*	1043	(7)

(*) indicates value for oxidized SiC.

The formation of the reaction film usually promotes the wetting behaviour but depending on its thickness it also alters the interfacial stresses. The change in stresses will depend on the volume change associated with the chemical reactions, on the difference in thermal expansion coefficient and on the crystallographic structure of the phases. Furthermore, each reaction product has its own characteristics insofar as strength, brittleness and thermal or oxidation resistance is concerned.

Whether the joining technique is active metal brazing or solid state bonding, the reaction which will proceed can be predicted with the aid of thermodynamic and kinetic models. Although the actual bonding process is a non-equilibrium one the chemical reactions can be considered as equilibrium reactions, since the initial interest lies in whether or not the reaction will take place, and partly in predicting it from thermodynamic data. Suppose the reaction between silicon carbide or nitride and a metal Me takes place as follows:

$$SiC + Me = Me\text{-}Carbide + Si$$

$$Si_3N_4 + Me = Me\text{-}Silicide + 2N_2$$

By determining the ΔG (Gibbs free energy change) for the reactions, it can be predicted whether such a reaction is thermodynamically favoured. Table 3 lists the heats of silicide formation (at room temperature $\Delta H_f = \Delta G$) of selected transition metals (15). The stability of some of the silicides should also be taken into account, for example, it has been pointed out that the formation of complex silicides such as Me_5Si_3 is energetically less favourable than that of $MeSi$ or $MeSi_2$. Thus, the silicide formed initially will tend to pick up, if kinetically permitted, increasing amounts of Si until a Si-rich silicide is formed. It depends on the specific joint application and careful consideration of the properties of the intermetallic component, whether this chould be allowed to take place by increasing the joining temperature and/or time. Obviously the same principle holds for the choice of metal Me. For example the temperature of the phase transitions and chemical reactions in the Nb-Si system are considerably higher (>1850°C) than those of Ni and Fe, which lie in the range of 1000-1200°C. Based solely on these criteria it appears that joining of Si_3N_4 to Nb should be recommended for high temperature applications (16).

For the silicon nitride-metal system a metal nitride could also form via the reaction:

$$Si_3N_4 + Me = Me\text{-}Nitride + Si$$

The product free Si, can possibly diffuse through the

TABLE 3 : HEAT OF FORMATION ($-\Delta H_f$) OF TRANSITION METAL "SILICIDES" (Kcal/mole M)

Metal	M_3Si	M_2Si_3	M_6Si_3	MSi	MSi_2
Cr	11.0		15.6	19.0	29.4
Ti			27.8	31.0	32.0
Zr		25.0	27.6	37.0	38.0
Mo	8.0		13.4		26.0
Mn	10.9		11.0	23.2	
Fe	7.5		11.7	17.6	19.4
Co		13.8		24.0	24.6
Ni	11.2	17.7		20.5	20.9
Nb			23.2		33.0

nitride layer and react with the excess metal to form silicides.
The nitrides of Zr, Ti, Al, Ta can, theoretically, form at all
temperatures since they are more stable than Si_3N_4.
Nevertheless, nitride formation is certainly not expected to
happen at low temperatures. Moreover, this depends on the
ambient partial pressure of O_2 and N_2. If the ceramic surface is
oxidized, i.e. if a film of SiO_2 covers the Si_3N_4, and the metal
is a strong oxide former (e.g. Ti), then a Me-rich oxide will
form rather than a nitride (17).

Thus, it can readily be appreciated that the chemical and
physical reactions at the ceramic/metal interface are quite
complex and many factors remain unknown. Only by gaining further
understanding of the surface and bulk thermodynamics or kinetics
of the reaction, will it be possible to set up useful guidelines
for the selection of materials and techniques for a specific
joining application.

3. JOINING TECHNIQUES

3.1. Liquid Phase Joining Techniques:

Liquid phase joining techniques, in particular brazing, are
currently the most highly developed processes for joining
dissimilar materials. Conventional vacuum brazing technology,
carried out at moderate temperatures and under low pressure, can
be applied.
Brazing of metals is relatively uncomplicated and fast. However,
brazing of ceramics is considerably more difficult. The two main
reasons are:

- ceramic surfaces are relatively 'inert' and are not easily
  wetted by conventional liquid metal alloys.
- there is usually a large difference in the thermal expansion
  coefficients of the ceramic and the metal or metallic braze.

Several techniques have been developed to obviate these problems.

The wettability problem can be overcome by first applying a
metallic coating to the ceramic substrate. Many electronic
applications are based on such a technique using e.g. a moly-
manganese (Mo-Mn) metallisation on Al_2O_3 prior to brazing.

Direct metallisation of Si_3N_4 has been developed by Toshiba
(Japan). Heating Si_3N_4 in a vacuum at a low nitrogen partial
pressure brings the decomposition temperature of Si_3N_4 into
metallic Si and nitrogen-gas down to about 1200°C. The
metallisation process is accelerated by introducing Ni and Cr
powder onto the surface prior to the heat treatment. Complete
metallisation of the surface with eutectic Ni-Cr-Si phases

happens after a few minutes.   A conventional Ag-Cu braze alloy can subsequently be used to achieve a joint between the metallized Si_3N_4 and Mo.
Although satisfactory results have been obtained with metallized Si_3N_4-Mo joints, problems still exist with the thermal expansion mismatch between the Ni-Cr-Si phase and the ceramic.

In another technique use is made of "activated" brazes which contain active-metal alloying additions (e.g. Ti, Zr, Nb) and wet some ceramic materials directly without the need for prior metallisation. Ti is a very common "activator" and is used with Cu-Ag braze alloys. Both oxide and non-oxide ceramics can be joined to metals in this way providing some precaution is taken to accommodate thermal mismatch. Maximum joint strength is obtained with an optimum Ti-concentration (approx. 5%) in order to avoid the formation of brittle intermetallic silicide compounds and brazing at 880°C for 30 min. Stress-relieving interlayers are sometimes used to reduce the build up of residual stresses.
Successful Si_3N_4 - metal joints have been obtained with a thin Al-layer sandwiched between two very thin (40 μm) layers of Al-10% Si alloy (18). Similarly SiC-Cu or stainless steel joints can be realized with a brazing alloy of Cu+35% Mn and a Cu/C composite compliant interlayer (19). By varying the amount of carbon fibre in the composite it is possible to tailor the thermal expansion coefficient and the Young's modulus. These two last examples are usable up to 300°C.

An important drawback of the above mentioned techniques is the temperature-limit imposed by the braze material which is much lower than the maximum temperature capability of the ceramic or metal partner.

3.2. Solid State Diffusion Bonding:

Solid state bonding is achieved by atomic diffusion, by solid state reactions at the interface or by forcing surfaces together under high pressure. This method seems to be the most promising, certainly for high-temperature applications. The bonding strength of such a joint is expected to be higher when compared with brazing or mechanical joining. The basic restraint on the success of solid state bonding is perhaps the need for sophisticated and expensive installations for providing high temperatures and/or high pressures in high vacuum. A summary of solid state bonding techniques, some of which are already being applied commercially, while others show commercial potential but, as yet, have not been exploited, is listed in Table 4.

Surprisingly very little work has been done on diffusion bonding of ceramics to ceramics. Here, problems related to

TABLE 4 : SOLID STATE JOINING TECHNIQUES

Material	Interlayer	Remarks
Al_2O_3) Si_3N_4) to Cu ZrO_2) SiC)	aluminium alloy and transition metals	
Al_2O_3 to steel	Nb/Mo	
Al_2O_3) AlN, TiN) to steel TaC, TiC)		simultaneous sintering of ceramic and bonding to metal using hipping
Si_3N_4 to steel	Al/Invar Ti/Al/Fe	
Si_3N_4 to Si_3N_4	Al_2O_3	hot pressing technique

thermal expansion mismatch are virtually not present. Some good results have been obtained with reaction-bonded-siliconized SiC containing a considerable amount of free Si (20).

Examples of ceramic-to-metal solid state bonding are more numerous. In most cases the reaction mechanism is very complex and not well understood. Results are often based on "trial-and-error" experiments. Reaction-bonded-siliconized SiC (RBSiC) has also been successfully joined to high nickel alloys such as IN617 (54Ni - 22Cr - 12Co - 9Mo) and Alloy 800H (48Fe - 32Ni - 20Cr). An example of such an interface obtained between RBSiC and IN617 after a heat treatment at 1100°C for 1 hour under vacuum, is shown in Fig. 1. The reaction zone between RBSiC and 800H seen in Fig. 2 is much more complex.
Si_3N_4 can be joined to itself with or without Al_2O_3 powder between the two parts. The addition of Al_2O_3 results in a considerable strength increase up to 800°C (21).

Much effort is going into the design of joints and methods for overcoming residual stresses by the use of interlayers (22). Calculations using finite-element-analysis programs are useful to gain an insight into residual stress-distributions (23-24). A particularly good example is the bonding of SiC or Si_3N_4 to Nimonic 80A (25). A sound joint between these materials with low residual stresses and free of cracks can be obtained if bonding is performed using Ni as the insert metal for bonding and a low thermal expansion metal (W or Kovar) and a soft metal (Cu) as insert metals to reduce the thermal stresses. Good tensile strengths are obtained with this composite.
A special technique is friction welding, which has been successfully applied to join alumina bars and pipes (28).

3.3. Solid State Bonding using Hot Isostatic Pressing (HIP):

In this technique, probably the most expensive of all, a joint is realised under isostatic pressure rather than under uniaxial pressure. The pressures applied are usually about 10 times larger than the uniaxial pressures used for solid state reaction joining. The temperatures are very high (0,9 Tm) in order to cause plastic deformation of at least one of the components and obtain a very intimate contact at the interface. Such a HIP treatment can be used to form the joint or, after pressureless reaction joining, to densify the composite. Using the optimum parameters a substantial increase in joint strength can often be gained. Satisfactory results have been obtained with RBSiC, with Si_3N_4 and Al_2O_3.

The need to exert pressure during diffusion bonding causes most joints to be of the face seal type which is not well suited

318

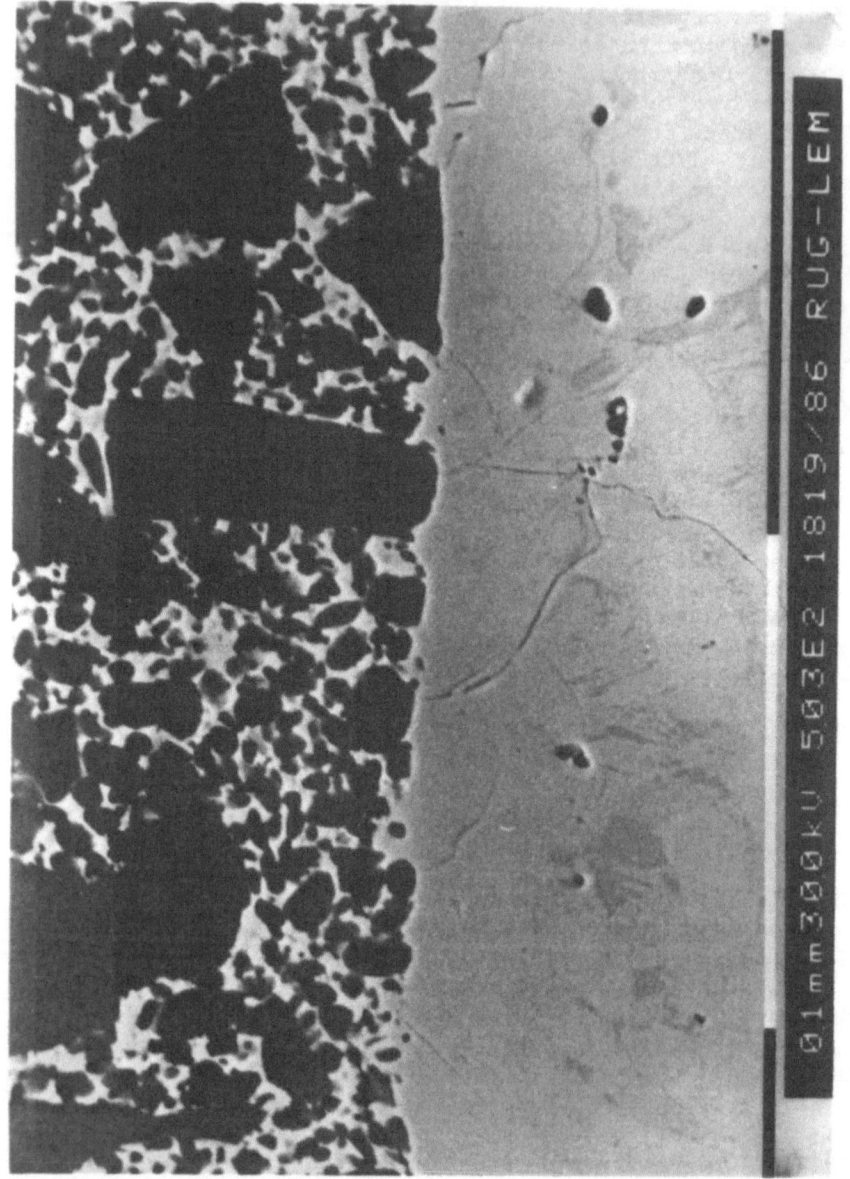

Figure 1 : Interface between reaction bonded siliconized SiC (top) and IN 617 (bottom) (1819/86)

319

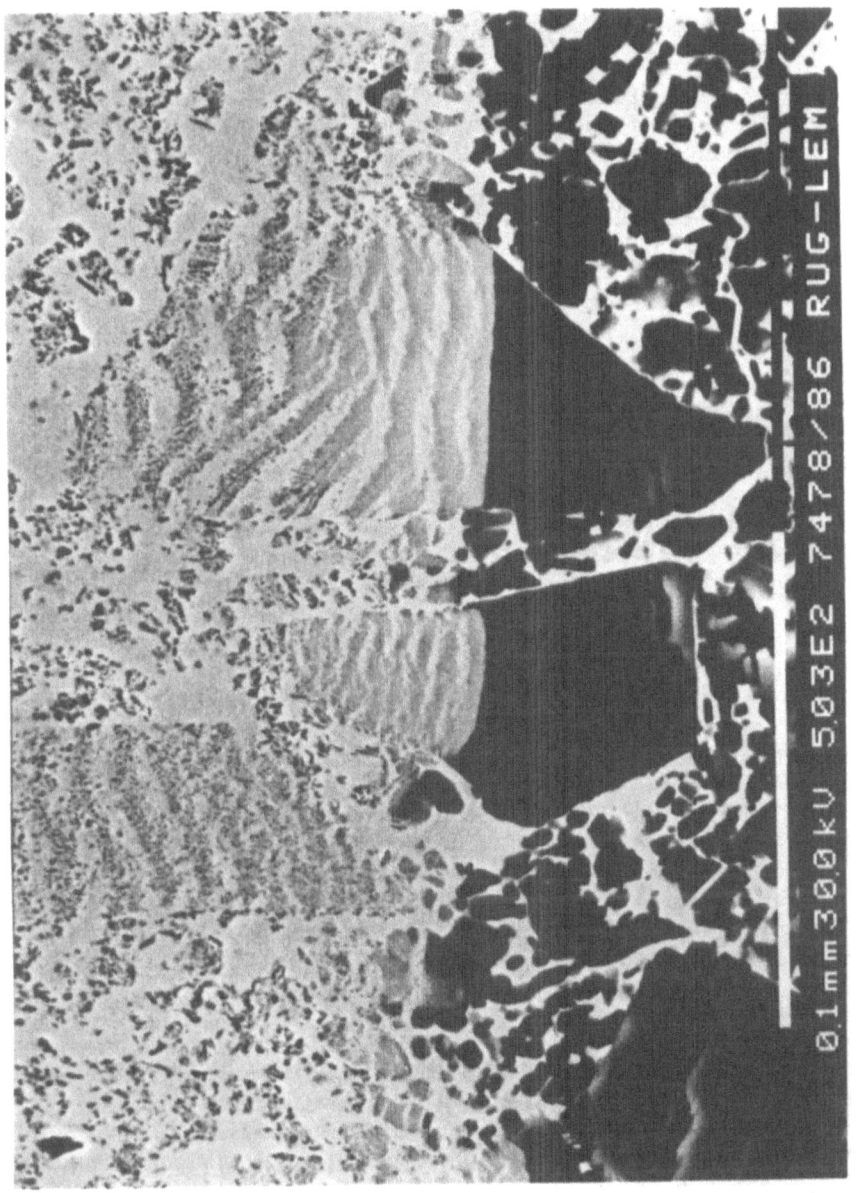

Figure 2 : Interface between reaction bonded siliconized SiC (bottom)
and Alloy 800H
(7478/86)

for accommodating mismatches in thermal expansion. Therefore, face seal type joints are mostly used for small components, thus mitigating the mismatch stresses.

4. HIGH TEMPERATURE APPLICATIONS

The two main fields which are very attractive for the use of ceramics are the automotive industry and the general energy-conversion industries.

4.1. Automotive industry:

Metal to ceramic joints have been achieved for the following applications:

- turbocharger rotor
- piston cap
- brake pad heat insulation
- rocker arm tip
- diesel glow-plug and inlet port (hot plug)

For these applications commercial production already exists, albeit mostly on a limited scale. Of those applications, the piston cap will most probably be commercialized within 5 years.

The geometry of the turbocharger rotor is a relatively simple butt joint between an Si_3N_4 rotor and a steel shaft (26), for which an activated Ag-Cu braze material is used. There are however a number of interlayers between the ceramic rotor shaft face and the steel shaft face. The interlayers consist of metals with different properties, one which provides a low thermal expansion coefficient and the other which provides a low Young's modulus and is therefore easily deformable. It should be noted that the location of the joint is at a point where the maximum service temperature is well below the capability of the joint and that it maximises the use of metal in the critical shaft area which is exposed to bending stresses.

4.2. General Energy-Conversion Systems:

The main applications will be found in advanced gas-turbine applications, heat exchangers and tiles for magneto-hydrodynamic channels or for the first wall of the next European Torus fusion reactor. For the first application mainly Si_3N_4 is considered while for the two other applications SiC is the prime candidate. Solid state bonding is envisaged as the only way to proceed. Most of the systems are still subject of further work to improve the quality and the reproducibility of such joints and to fully understand the process.

Reaction-bonded silicon carbide infiltrated with silicon (RBSiC) has been used for several components (mainly tubes) of the GAST-project (Gas Cooled Solar Tower Power Plant) (27). Successful RBSiC-to-RBSiC joints have been made in two different ways: by solid state diffusion in an inert atmosphere and by brazing using silicon with additives. Both joining techniques resulted in gastight joints with a strength between 30% and 100% of the bulk material.

5. CONCLUSIONS AND SUGGESTIONS FOR FURTHER DEVELOPMENT

A brief review of interfacial phenomena and possible joining techniques has been given which are applicable to ceramic-ceramic or metal-ceramic joints in the field of engineering ceramics for high temperature applications.

Brazing is a method widely employed particularly for joining oxide ceramics. However, the temperature capability of the brazed joint is limited. Solid state joining, which involves the application of pressure at elevated temperatures, is the most promising joining technique for applications at very high temperatures. With uniaxial hot-pressing the joining is limited to components with relatively simple shapes. Hot isostatic pressing is more suitable for complex shapes. The principal problems however are the stresses caused by differing thermal expansion coefficients and brittle complex structural transitions.
Actual research is focused on the transition region between metal and ceramic studying extensively the specially developed graded layers and powder mixtures. Such a transition joint resulting in a metal-ceramic joint with a minimum of residual stresses is usually a very complex system. For each specific combination an optimum configuration of intermediate layers of powder mixtures needs to be determined. The results of ongoing laboratory projects on this subject and its ultimate success will have a far reaching influence on the expansion of the ceramics market.

Standardization in measuring engineering data and quality control (non-destructive evaluation) are areas where further research and development are necessary. There is still a lack of sufficient data, e.g. concerning static, dynamic and thermal fatigue properties of metal-ceramic joints. Such data are necessary to check and improve existing simulation models and computing programs for determining three-dimensional temperature and stress distributions.

The reliable use of ceramics as structural components requires effective flaw-detection capabilities. The small critical flaw size (of the order of 1 μm) of ceramics precludes the use of standard non-destructive evaluation (NDE) techniques that have been developed for detection and characterization of critical flaws in metals. The techniques employed currently for ceramic NDE are microfocus X-radiography and fluorescent dye penetrant testing. For the evaluation of ceramic joints, ultrasonic techniques are becoming popular, as well as acoustic microscopy which allows the detection of defects with dimensions of 50 μm. For rough surfaces or irregular geometries, acoustic techniques may not be effective. In those instances holographic interferometry is a viable alternative especially when only a limited area of a component, e.g. a joint, needs to be examined.

For short term industrially exploitable results, it is necessary to bridge the gap existing between the academic approach from materials scientists and the industrial needs of development engineers.
In the U.S.A. a coordinated action is taken within the frame of the Heat Engine Programmes, while in Japan the emphasis is put onto car-engine applications. In Europe, coordinated actions, such as BRITE and EURAM stimulate collaboration between university centres and industry.

REFERENCES

1. Interfacial Phenomena in Metals and Alloys, Murr, L.E., Addison Wesley Publ., Reading MA (1975)

2. J. Mat. Sci., McLean, M. and Hondros, E.D., 6, 19 pp. (1971)

3. Surfaces and Interfaces in Ceramic and Ceramic-Metals Systems, Pask, J.A. and Tomsia, A.P., ed. Pask, H.A. and Evans, A., Plenum, NY, 411 pp. (1981)

4. Phil. Mag., Pilliair, R.M. and Nutting, J., 16, 181 (1967)

5. Z. Metalkd., Elssner, G. and Krohn, U., 70, 71 (1979)

6. Oxides and Oxide Films, Beruto, D., Barco, L. and Passerone, A., ed. Vijh, A.K., Marcel Dekker, NY, Vol. 6, 1 pp. (1981)

7. J. Mat. Sci., Laurent, V., Chatain, D. and Eustathopoulos, N., 22, 244 pp. (1987)

8. Silikatetechnik, Muller, K. and Rebsch, H., 17, 279 pp. (1966)

9. Trans. JWRI, Naka, M. and Okamoto, I., 14, (1), 29

10. J. Mat. Sci. Lett., Naka, M., Kubo, M. and Okamoto, I., 6, 965 pp. (1987)

11. See for example: Mater. Chem. and Phys., Passerone, A, 15, 263 pp. (1986), Pat. Res. Soc. Proc., Klomp, J.T., Vol. 40, 381 pp. (1985)

12. J. Mat. Sci., Hitchcock, S.J., Caroll, N.T. and Nicholas, M.G., 16, 714 pp. (1981)

13. J. Amer. Soc., Brennan, J.J. and Pask, J.A., 51, 569 pp. (1968)

14. J. de Chimie Phys., Chatain, D., Rivollet, I. and Eustathopoulos, N., 83, 561 pp. (1986)

15. Mat. Lett., Purarka, S.P., 1, 26 pp. (1982)

16. J. Mat. Sci. Lett., Naka, M., Saito, T. and Okamoto, I., 6, 875 pp. (1987)

17. J. Appl. Phys., Taubenblatt, M.A. and Helms, C.R., 53, 6308 pp. (1982)

18. "New Bonding Technique of Ceramics to Metals", Okade, S., Hioko, S., Kohno, A. and Yamada, T., - Hitachi Research Laboratories.

19. Quart. J. of the J.W.S., Okamura, H., Miyazaki, K., Marsuzake, T., Shida, T., Ura, M., and Okuo, T., 4, No. 2, 198-205 pp. (1986)

20. Proc. Int. Conf. on Diffusion Bonding, Ostyn, K.M., Cranfield (UK), 183-193 pp. (1987)

21. Yogyo-Kyokai-Shi 91, Kanzaki, S. and Tabata, H., No. 11, 520-522 pp. (1983)

22. Brazing and Soldering 10, Nicholas, M.G., 11-13 pp. (1986)

23. J. Am. Cer. Soc., Suganuma, K., et al., Comm. C-265 pp. (1984)

24. Quart. J. of the J.W.S., Hamada, K., et al., 3, 73 pp. (1985)

25. Nippon Kokan Technical Report Overseas No. 48, Yamada, T., et al., 67-74 pp. (1987)

26. The turbocharger rotor has been developed by Nissan Motor Co. Ltd. in collaboration with NGK Spark Plug Co. and NKG Insulators Ltd.

27. "Ceramics in Advanced Energy Technology", Röttenbacher, R., et al., European Colloquium, Petten, The Netherlands, 231 pp. (1982)

28. Trans. of Jap. Weld. Soc., Kanayama, J., et al., Vol. 16, no. 1, 95 pp. (1985)

4. Materials Constraints in the High Temperature

Industrial Technologies

4.1. Energy Production and Conversion - Fossil Energy

4.1.1. The Combustion Technologies

Dr. B. Meadowcroft

CERL Materials Division

U.K.

This topic is concerned solely with those aspects of combustion technology where materials are exposed directly to the combustion gas. It will be considered in two parts; first, materials requirements within the combustion zone itself, and second, requirements for materials exposed to downstream combustion gases. It excludes the detailed requirements of gas turbines which are used to expand pressurised combustion gases such as in oil and coal fired combined cycle applications, and in direct coal-fired gas turbines, and also excludes gasification systems, and M.H.D.. The requirements of materials relevant to their exposure to steam, such as in power generation systems, are dealt with in section 4.1.2.

1.   MATERIALS REQUIREMENTS IN THE COMBUSTION ZONE

In general, combustion chambers are lined either with refractory or with economiser/evaporator tubes. Refractories are used more in smaller units where heat loss would otherwise be significant, and particularly in circulating fluidised beds were their erosion resistance is of importance. Improved refractories able to give increased lifetimes economically between overhauls would be of great value.

Economiser/evaporator tubes are not strictly in the remit of this volume as their metal temperatures range from 200–500°C. However, as the corrosion and erosion processes are generically similar to those which apply at higher temperatures, and the rates of degradation are such as to be technology limiting, their requirements will be mentioned briefly. Corrosion is most severe when incompletely combusted gases hit the furnace tubes, and is primarily a solid-fuel-combustion problem. Stoichiometric combustion is more easily controlled with liquid fuels. While this is basically a design problem in conventional boilers (although often "cured" by the selection of a more corrosion resistant material) staged combustion for NOx control will increase the occurrence of such conditions. As also discussed in the gasification section (the atmospheres are similar), and from UK pulverised coal boiler experience, highly alloyed materials are necessary to gain any reduction in corrosion rates, and more effective materials solutions are still sought.(1). Coextruded tubes, current practise often uses AISI 310 as outer layer, or coated tubes, generally have to be utilised to combine the corrosion resistance of the improved alloy with the required waterside stress corrosion resistance.

Only in bubbling fluidised beds is it usual to have heat transfer surface within the combustion zone. Both evaporator and superheater tubes are required because the heat capacity of the downstream gas (limited to ~900°C to avoid ash sintering in the bed) is insufficient to provide all the heat energy for superheating/reheating. In-bed air-heater tubes have also sometimes been proposed.

Most significant problems again occur in the sub 600°C range, indeed sub 300-400°C when erosion of the tubes can be particularly severe. Recent experience (in the pilot plant for the T.V.A. 170MW plant) showed an increase in erosion rates of evaporator tubes by a factor of ten on changing the tube bank and type of coal, and stressed the importance of this problem and the possible need for materials solutions.(2). It is now known that if the surface temperature of tubes is increased sufficiently (depending on the erosiveness of the bed) oxide can form which effectively eliminates erosion.(3). 9% and 12% Cr steels are particularly effective, but why, and what the optimum materials solutions are, is not known and is an urgent requirement for this technology.

At superheater and air heater tube temperatures and for uncooled components - e.g. hangers - deposits form on the tubes giving further protection against erosion. However, the deposits can encourage sulphidation particularly when limestone is added to the bed to reduce SOx emissions. Whilst sub-critical steam production at 540°C is probably viable now, higher steam temperatures and supercritical steam conditions (meaning thicker tubes and higher surface temperatures) need further validation. Improved materials could be required, possibly in the form of coextrustions or coatings of corrosion resistant outer layers over mechanically strong inners.

Air heater materials are a seperate consideration as the

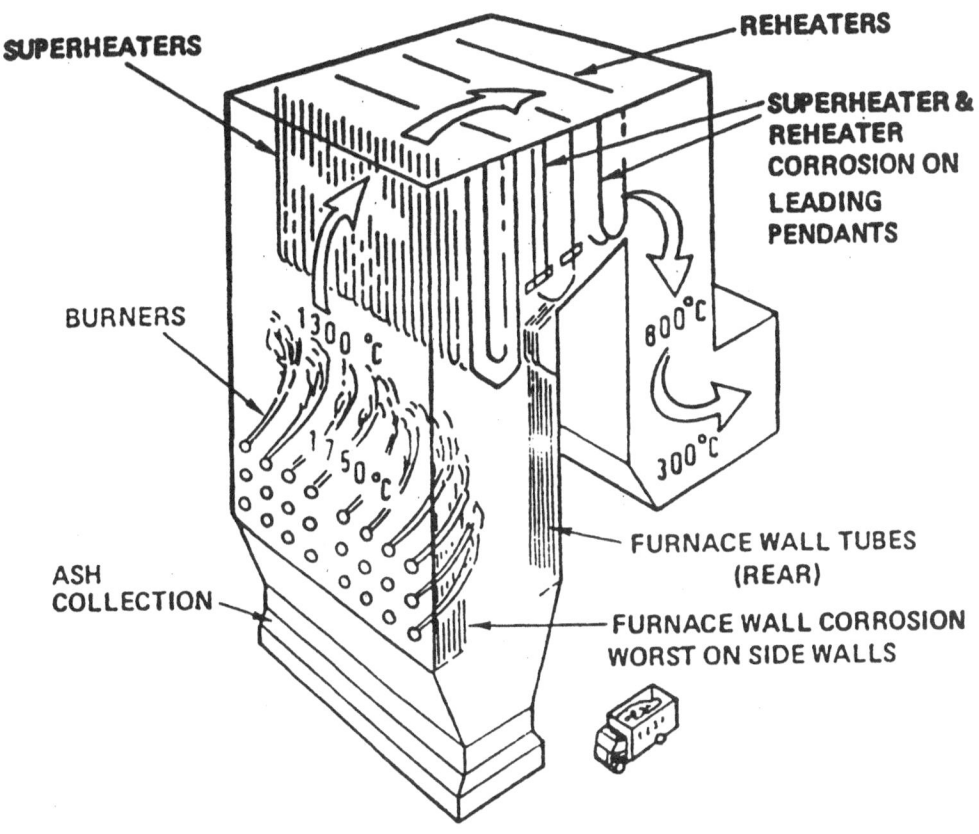

Fig. 1.   An impression of a typical boiler

internal pressures are much lower, comparable with the combustion
gas pressure. However, such systems seem currently out of favour
compared with out-of-bed air heaters.

2.   MATERIALS REQUIREMENTS IN THE DOWNSTREAM COMBUSTION GASES

As indicated earlier, in conventional pulverised coal plant
all of the superheater and reheater tube surface is downstream of
the combustion zone.(Fig. 1). This means that the combustion
gases are fully combusted and oxidizing, but the metal surface
temperatures can be such that the volatile species in the gas can
form molten deposits. These enable aggressive conditions to be
established beneath the deposits which can lead to rapid
corrosion of conventional superheater alloys. The main
contaminants which deposit are alkali metal sulphates, which form
molten deposits by reacting with the otherwise protective oxide.
The melting point is affected by the particular alkali metals
present, typically sodium and potassium, and their ratio in the
deposits, which is dependent on the amounts released from the
coal feed during combustion. Corrosion with crude and residual
oils is generally more severe than with coal, the presence of
vanadium and the much lower ash levels in oil being major
factors. Oil causes corrosion of austenitic steels more severely
than for ferritics and can limit steam temperatures to 540°C, up
to 30°C below the current coal fired values. In U.K. coals the
chlorine content in the coal is high (0.25% average) and a linear
correlation has been derived between plant corrosion rates and
coal chlorine content. In addition the importance of both metal
and gas temperatures on corrosion rates has been demonstrated.
(4).

The mechanical properties of the currently available alloys
limit the required temperatures of operation. Two points must be
realised when considering the working temperature of heat
exchanger tubes. First, because of the heat flux through the
tube and the steam boundary layer, the mid wall metal temperature
is appreciably higher than that of the steam; typically 30-70°C,
superheater tubes often being hotter than reheaters. Second, it
is not possible to design a boiler in which all the tubes have
exactly the same outlet temperature and therefore the designer
must assume a higher outlet temperature than the nominal value at
the turbine stop valve (TSV) - typically 20-30°C higher.

Thus, for turbine steam temperatures of 540°C, currently
typical in much of Europe, mid-wall metal temperatures are up to
a maximum of about 600°C, and ferritic steels such as 2.25% Cr or
12% Cr can be used. For steam temperatures of 570°C, such as
have been standard in the U.K. for 20 years, creep resistant
austenitic steels such as 316, 321, 347, or Esshete 1250 are
appropriate. Even so, for the aggressive coals in the U.K.

corrosion rates can be such as to cause premature failures and the extensive use of coextruded tube has resulted. This is a sophisticated cladding technique to provide a thick, corrosion resistant, but mechanically weak, outer layer metallurgically bonded to a mechanically strong inner layer.(6). Typical current outer layers are AISI 310 (Fe25%Cr20%Ni) and 50%Cr50%Ni, whilst inner layers commonly used include Esshete 1250 and Alloy 800H.

As will be discussed further in section 4.2.2. and is summarised there in Table 1, the potential efficiency improvements to be obtained by increasing steam pressures (particularly supercritical) and temperatures are leading to significant activities throughout the world. Major efforts are being directed to high temperature materials developments. Fig. 2. shows the way permissible design stresses fall off with temperature for all established boiler tube materials, the strongest alloy 12R72 has welding difficulties and is not currently commercially available. All such alloys have comparable corrosion resistance, and for coals which are not aggressive chemically can give design lifetimes ($\sim 1.5 \times 10^5$ h) up to steam temperatures (TSV) of about 600°C and pressures of about 30 MPa with tubes of reasonably economic dimensions.(6).

For more advanced steam conditions or more aggressive fuels the achievable lifetimes with these alloys decrease and improved materials will be necessary (some increase in conditions can be obtained if regular replacement of the final superheater/reheater tubes is allowed for). Two improvements need to be obtained – first to obtain materials with improved high temperature creep strength, and second to improve the corrosion resistance, probably by increasing the chromium content of the alloy. The two improvements can be approached conjointly or separately. The improved corrosion resistant alloy must be suitable for application by an effective coating technique (e.g. coextrusion), to an improved creep resistant inner layer. Whilst coextrusion has proved an effective coating technique able to provide 100% protection of the substrate, it requires the coating alloy to be suitable for hot working. Alternative processes able to achieve equivalent integrity would be valuable. These requirements are already identified in Japanese and U.S. programmes to develop more advanced plant and will require extensive efforts in Europe over the next decade, in order to maintain a competitive position.

In fluidised bed combustion the restricted combustion temperature limits the release of alkali salts to far lower levels than occur in conventional plant - indeed to levels where in pressurised systems it is possible to consider the operation of gas turbines on coal combustion gases simply cleaned of particulates. The off-gas temperature of 900°C means that turbines with uncooled blades can be used but the aggressiveness of the gas can still be such that sophisticated coatings are necessary to limit corrosion and surface degradation rates of aerofoil surfaces (generally Type II hot corrosion - (7). The requirements of metallic duct surfaces and uncooled components in the same atmosphere appear less severe. Not only is the acceptable degradation greater but the effective insulation behind the duct lining reduces the driving force for deposition of the alkali. At steam-cooled superheater tube temperatures corrosion should also not be significant. In circulating bed systems there may be localised erosion problems which could be difficult to solve with materials, particularly in the case of recent trends in design of air heater to provide clean gas for a gas turbine. Pressurised air heaters can also impose stringent mechanical property specifications and lifetime requirements which are difficult to satisfy at present.

In pressurised fluidised bed plant, systems currently being developed need to be considered separately from the "second generation" systems currently being evaluated in design studies. At present whilst cyclonic clean up of the gases (the final stage and subsequent ductwork being metal lined to avoid refractory debris entering the turbine) may be adequate for turbine protection it is not so for environmental particulate emissions. Therefore positive filter clean up systems are being developed which would avoid the need for both high and low temperature gas clean up and to cover the possibility that such systems are necessary for turbine protection. The intimate contact between filter material and gas requires it to be very resistant to attack by the aggressive contaminants in the gas. Ceramic based materials are receiving detailed consideration, but their structural reliability, chemical inertness and blockage susceptibility with time leave room for major improvement. Cross flow filtration may be necessary to prevent the last problem, but improved ceramic composistions and fabrication methods are likely to be necessary to provide technically and economically viable positive filter media.

In some second generation conceptual designs currently under consideration, methods for removing the gas turbine temperature restrictions are being sought, for example, by heating the cleaned combustion gases using a clean fuel (perhaps from a coal gasifier) burned in the gases. As the resultant gas temperature will be above the softening temperature of the ash, in this case positive filtration of the gas before reheat will be essential. Thus these second generation systems increase the importance of work on positive filtration systems, and improved refractory filter materials.

Finally it is possible to consider heat exchangers suitable

for indirect firing of gas turbines. Such systems have been
built in the past, and have used various fuels, but materials
limitations of the metallic heat exchangers limited the turbine
inlet temperature to 750°C. Oxide dispersion strengthened
materials (e.g. MA 956) now offer much higher metal temperature
capability with good corrosion resistance. The development of a
heat exchanger based on these materials is in progress within the
European collaboration scheme COST 501.

3. SUMMARY

A summary of the requirements of materials developments
identified in this section is given in Table 1. Many of these
areas are currently being addressed in European collaborative
research programmes.

TABLE 1

Requirements for R&D in Structural Materials for Fossil Fuel
Combustion Plant

Component	R&D Requirements
Combustion chamber lining	Refractories of longer lifetime
Economiser, evaporator tubes	Tubes of improved corrosion resistance in incompletely combusted gases
Tubes in fluidised beds	Improved erosion resistance
Superheater and air heater tubes	Coextruded tubes or equivalent Coatings for oxidation resistance
Superheater tubes for higher steam temperature and pressure	More creep resistant ferritic/ martensitic steels, austenitic steels, coextruded tubes or equivalent
Superheat tubes for plant burning aggressive coals	Coextruded or coated tubes
Possitive full flow filters	Refractory filter materials
Very high temperature (>1100°C) heat exchanger	Development of ODS ferritic alloy tubes

4. <u>REFERENCES</u>

(1) Meadowcroft, D.B., Mats. Sci. and Eng., <u>88</u>, 313-320 pp. (1987)

(2) Vincent, R.Q., Poston, J.M. and Smith, B.F., Proc. Int. Conf. on Fluidised Bed Combustion, ed. Mustonen, J.P., ASME, 672 pp. (1987)

(3) Rademakers, P.L.F., Bos, L., van Wortel, J.C. and Kolster, B.H., Proc. Conf. Fluidised Bed Combustion - Is it achieving its Promise? Institute of Energy, London (1984)

(4) "The Control of Fireside Corrosion in Power Station Boilers (Third Edition)", Laxton, J.W., Meadowcroft, D.B., Clarke, F., Flatley, T., King, C.W. and Morris, C.W., Central Electricity Generating Board, Newgate St., London (1988)

(5) Flatley, T. and Thursfield, T., J. Mats. Energy Systems, <u>8</u>, 92-105 pp. (1986)

(6) Asbury, F.E. and Brooks, S., Proc. Conf. Advances in Materials Technology for Fossil Power Plants, American Society for Metals, Chicago (1987)

(7) Meadowcroft, D.B. and Stringer, J., Mats. Sci. and Tech., <u>3</u>, 562-570 pp. (1987)

(8) Electric Power Research Institute; Technical Brief to Contract RP 2387-3 "Indirect-Fired Gas Turbines" (Oct. 1987)

4.1.2. Steam Cycle Power Plant

Dr. B. Meadowcroft

CERL Materials Division

U.K.

The purpose of this Section is to describe the high temperature materials factors which limit the future development of advanced steam cycles for electric power generation. The requirements of the materials exposed to the combustion gases were discussed in section 4.1.1. here the concern is all the other materials exposed to the steam, including headers, main steam pipes, and all parts of the turbine.

The current steam conditions (540-570°C, 16-24 MPa) common throughout the world have stabilised for about two decades because of materials limitations, identified at a few stations built with advanced super critical steam conditions. Over the last few years it has become clear that efficiency gains (Table 1) which would be obtained from increased steam conditions (pressure, temperature), and double reheat are now likely to be achievable from concerted materials development programmes.(2). As a consequence major development programmes are in progress in Japan and the U.S.A., and a significant effort is underway in Europe. This Section will consider explicitly only requirements for advances in conventional plant. Because over 70% of the power in a pressurised fluidised bed combustor (PFBC) combined cycle is generated from the steam cycle, advances in efficiency by increasing steam conditions will benefit that cycle about as much as for conventional plant. On the other hand, for gasification combined cycle systems the gas turbine provides the majority of the power and its outlet temperature is generally <600°C, so advances in steam conditions for that cycle are neither feasible nor required.

The increases in efficiency identified in Table 1,

Table 1.　Predicted improvements in efficiency of conventional
steam plant form increases in steam conditions (1)

Steam Conditions TSV Inlet (Pressure/superheat/reheats)	Final Feed Temperature	Percent Efficiency Improvement
160bar/565°C/565°C	253°C	Reference
240bar/565°C/565°C	290°C	2.3%
240bar/565°C/565°C/565°C	288°C	4.1%
310bar/565°C/565°C/565°C	288°C	4.7%
310bar/593°C/593°C/593°C	325°C	7.2%
345bar/650°C/593°C/593°C	344°C	9.3%

* Percent efficiency improvement is improvement in efficiency
compared with reference divided by reference efficiency and
multiplied by 100.

particularly the inclusion of double reheat, may cause
availability and reliability problems with some potential
materials of construction if mid-merit (i.e. operating for only a
restricted number of hours each day) operation rather than base
load is required. The designer also has to envisage the whole
life scenario for proposed plant, which is unlikely to be always
base load. Thus, materials sensitive to thermal fatigue damage
may be unacceptable for an economic design.

The following paragraphs discuss the various components in
turn. For boiler tubes failures considered to be acceptable,
since replacement is not prohibitively expensive and failure does
not give rise to safety problems, the escaping fluid being
contained within the boiler. In contrast, the safety
implications of pipework or turbine failure are of major
significance and plant design must take this into account. All
materials are working in the creep regime and extensive long term
creep testing of any novel materials is essential to ensure that
plant lifetimes (150,000 to 250,000 h frequently) can be
predicted with confidence.

For pipework and steam headers the current materials used
are alloy steels, such as 1/2CrMoV, but subcritical conditions
and steam temperatures up to about 570°C are their limit,
particularly for pipework, without economically and technically
unacceptable tube thicknesses. Advances in conditions beyond
these values will require stronger materials. Higher chromium
ferritics, particularly the 9Cr alloys currently under
development, are likely to offer acceptable properties up to

about 600°C steam temperature. Further advances in conditions will demand austenitic materials, but problems have already been reported due to thermal ratchetting and creep fatigue interactions at the early advanced supercritical plant, and methods of avoiding these problems will have to be identified. It should be noted that as with boiler tubes only a small fraction of the turbine will be subject to the advanced steam pressure and temperature, whereas all headers and pipework will be subject to those conditions.

For valves and valve chests the considerations above apply except that the alloys are forged or cast. In addition, because of their thick and complex sections, cyclic operation will accentuate any thermal fatigue limitations of a candidate material.

Again for the first stage turbine casings and nozzle boxes the limitations of current materials and the future developments required are as above.

Turbine rotors are major components which may be cooled or uncooled. Currently low alloy steels, for example 1Cr1Mo 1/4V, are typically used uncooled, and their operating range could be extended somewhat by cooling. At higher temperatures more highly alloyed materials will be required and 12Cr steels are receiving detailed attention. Because of the size of these components fabrication difficulties are significant. Alternatively austenitic alloys could be considered, but limitations would be, casting ingots of sufficient size and avoiding thermal fatigue problems associated with load following.

Advances in steam conditions will mean turbine blades of nickel alloys. However there is extensive experience of the use of such alloys in blading applications and no developments in blade materials for forseeable advances in steam conditions are envisaged.

Finally, bolting developments will be required. The current ferritic alloys will have to be replaced by nickel based alloys (e.g. Nimonic 80A) for ferritic casings, but for austenitic casings, suitable austenitic bolting will have to be developed.

In summary, advances in steam conditions could give significantly increased efficiency and are dependent on material developments which are already underway, worldwide. These developments must however, take into consideration the operational requirements of the plants for which they are intended.

REFERENCES

1. Private Communication, Davidson, B.J., (C.E.G.B., U.K.)

2. 1987, Conf. on Advances in Materials Technology for Fossil Power Plants, Townsend, R.D., American Society for Metals, Chicago, October 1987

4.1.3. Coal Gasification Materials for

Plant Construction

D. Lloyd

British Coal Corporation

Stoke Orchard, Gloucester GL52 4RZ, UK

1. INTRODUCTION

Over the last decade, or so, significant world-wide research and development has been directed towards establishing novel and more efficient ways of utilising coal, in order to :

 i) maintain the diminishing and strategic supplies of natural oil and gas by the production of synthetic substitutes,

 ii) increase power generating efficiencies, and

iii) decrease environmental pollution.

Research and development into advanced coal gasification processes represents one area where attention is being focussed. Rather than relying on the inefficient practices employed in the past, when coal gas was manufactured by its thermal decomposition in a closed retort, efforts are now being directed towards 'total gasification' methods where virtually complete consumption of coal takes place. Because of the advanced nature of these new processes, the materials used to construct such plant often have to withstand aggressive environments that frequently involve high temperatures and pressures. Materials degradation due to corrosion and wear and their conjoint interaction represent common problems. For this reason it is important that appropriate materials and proficient engineering design practices are employed during plant construction to ensure safe and reliable operation.

In the following sections the principles of coal gasification, the various designs of gasifier and their industrial application will be described. Special emphasis will then be placed on critical plant components that are exposed to high temperatures and the degradation mechanisms that they are expected to resist. The availability of materials suitable for such component manufacture will be reviewed and areas requiring further research in the area of high temperature materials will be identified.

2. GASIFICATION REACTIONS

The gasification of coal is brought about by its high temperature endothermic reaction with steam. The two most important reactions are :

$$C + H_2O \longrightarrow CO + H_2, \text{ and}$$

$$C + CO_2 \longrightarrow 2CO$$

The heat required to promote these reactions is usually produced by the in situ reaction of oxygen, or air, with some of the coal;

$$C + 1/2 \ O_2 \longrightarrow CO$$

$$CO + 1/2 \ O_2 \longrightarrow CO_2 \text{ (relevant in the cooler gas of the gasifier).}$$

A possible, less common, alternative is to supply the heat indirectly from an external source, such as a combustor or nuclear reactor (1,2). When air is used to support combustion a low calorific value gas (approximately 3 to 4 MJm^{-3}) is produced which is heavily diluted with nitrogen. On the other hand, if oxygen, or an external heat source, is employed a high calorific value gas (approximately 10 MJm^{-3}) is formed.

3. GASIFICATION SYSTEMS

3.1. Introduction

To ensure efficient coal gasification it is important that good contact is achieved between the solid carbon and the reacting gases. Clearly, this will be affected by the size distribution of the coal particles in the system. Moreover, bituminous coals soften when heated to temperatures between 400 and 500°C and particles is this condition may coalesce to form an agglomerated mass. Efficient gas-solid contact may, therefore, require careful size preparation of feedstocks, the selection of non-caking coals, or the pre-treatment of coal in order to destroy its caking capacity.

The coal will also contain mineral matter which can be withdrawn from the gasifier as a dry unfused ash, as a clinker or as a molten slag depending on the mineral composition of the coal (which can vary considerably) and the temperature of the process. It is also desirable that this ash should be removed and disposed of with the minimum carbon content in order to maximise the conversion efficiency.

Because of these important considerations the various gasification systems that have been developed fall into three basic categories that are defined in terms of the way in which the solids contact the reaction gases, i.e. in a fixed bed, in an entrained gas/solids stream or in a fluidised bed.

3.2. The Fixed Bed Gasifier

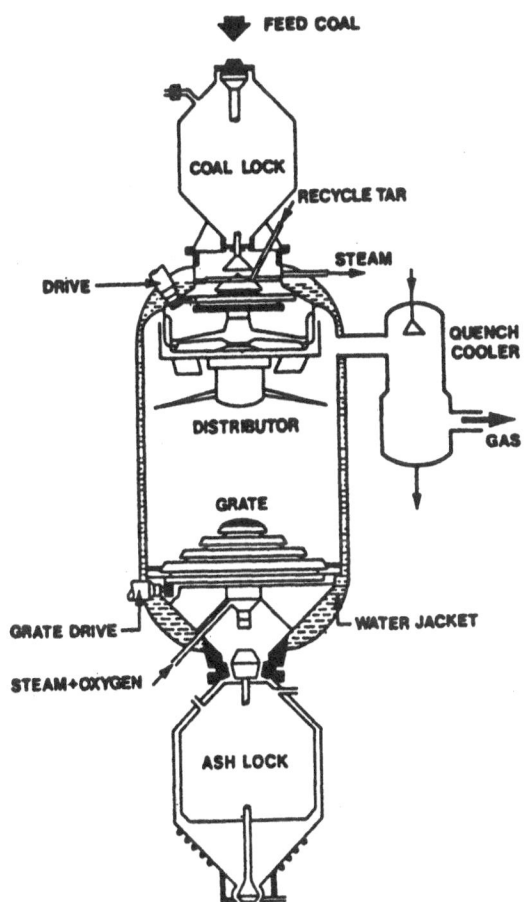

A typical illustration of a fixed bed gasifier is given in Fig. 1. The gasifier operates under pressure with the bed of coal slowly descending against an upward flowing stream of oxygen and steam. Hot gases rising from the combustion that takes place in the lower part of the gasifier heat the incoming coal. Some tar is produced as a result of coal volatilisation in the upper part of the gasifier.

The Lurgi system has the advantages of good heat economy, operation under pressure (thereby increasing throughput, reaction kinetics and cycle efficiency while decreasing capital building cost) and little dust carry-over. The main disadvantages are the need for non-caking lump coal, the

Fig. 1 A schematic illustration of a Lurgi gasifier

production of liquid tars and liquor effluents, high steam consumption and a low coal throughput.

The original Lurgi gasifiers operated at a sufficiently low temperature such that a dry ash was produced. However, it was recognised that appreciable improvements in efficiency and economics could be achieved if the gasifier could operate in the slagging mode with the ash being withdrawn as a liquid slag rather than a dry ash. This way of operating is particularly attractive for relatively unreactive bituminous coals. This objective has successfully realised in a demonstrator facility.

3.3. The Entrained Flow Gasifier

For an entrained flow gasifier (Fig. 2) pulverised coal

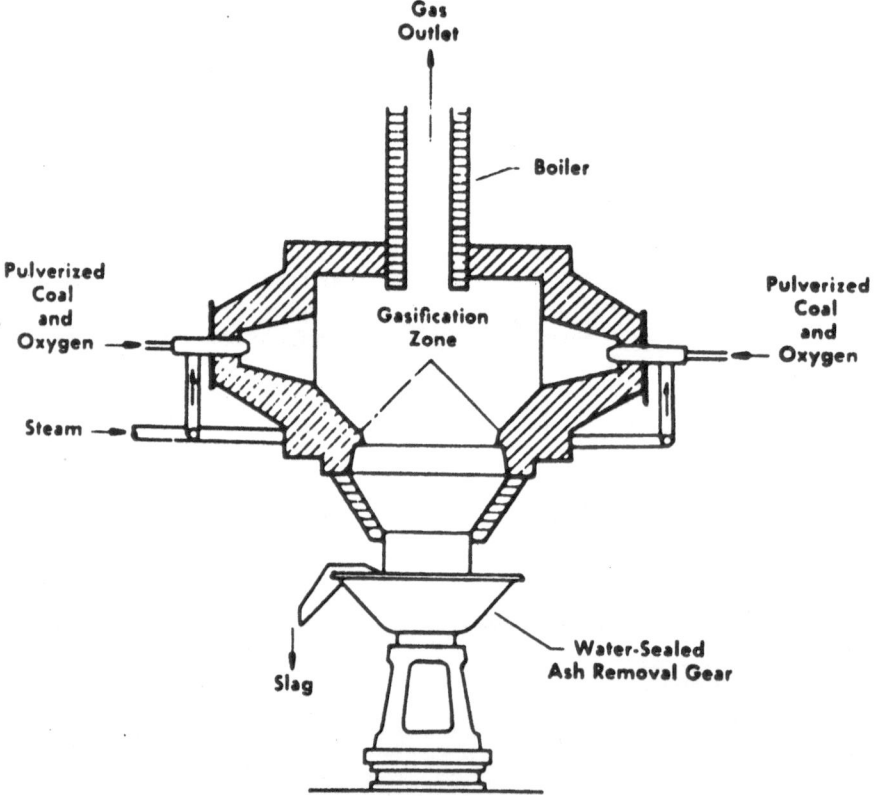

Fig. 2 A schematic illustration of an entrained flow gasifier.

which may be fed dry, or as a coal/water slurry, is injected along with oxygen and steam into the reactor at a high velocity. The gasification reactions take place rapidly and the carbon is almost totally consumed with little tar production. Because of

the high reaction temperatures; 1500°C or more, the gas is
relatively rich in carbon monoxide, while the ash that is
generated runs down the walls of the gasifier as a slag.

The feasibility of this approach is demonstrated by slagging
entrained flow gasifier currently in use as a gas producer at a
combined cycle power generation demonstrator plant.

3.4. The Fluidised Bed Gasifier

A fluidised bed gasifier (Fig. 3) consists of a 'dry'
coal/char bed which is agitated by an upward stream of steam
and oxygen or air introduced through a base distributor plate. The
process is reliable and easy to start-up and turn-down.
However, significant particulate carry-over in the product gas can
occur. Also this method of gasification is best suited for
lignites and the more reactive non-caking coals that produce an
ash with a high

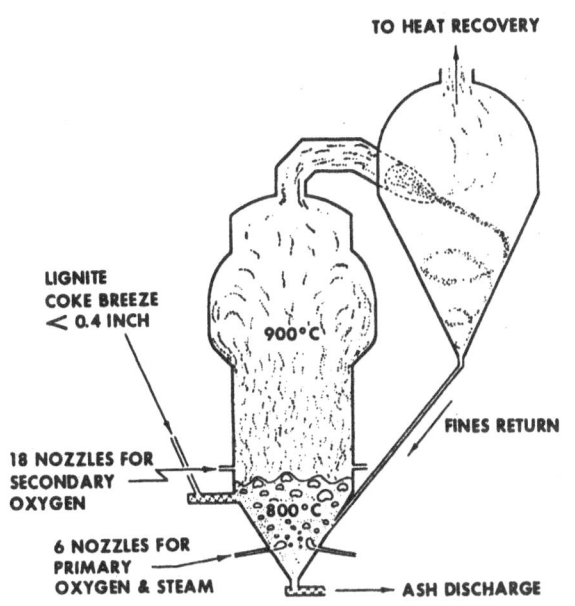

Fig. 3 A schematic diagram of a fluidised
bed gasifier

softening temperature. For bituminous coals improved utilisation
of carbon can be achieved by feeding the gasifier char to a
fluidised bed combustor (1).

4. COAL GAS USAGE

4.1. Industrial Applications

Coal gas used for most industrial applications has a low
calorific value and if produced by the reaction of air and coal
will contain appreciable quantities of nitrogen, thereby making
it unsuitable for distribution over long distances. It is,
therefore, utilised close to the source of its production.

Typically, the gas, which may need to be free of solids and gaseous contaminants, is used for steam-raising, space heating and for firing kilns used in the manufacture of bricks, cement and pottery, etc.

4.2. Synthesis Gas Production

Synthesis gas because of its high calorific value is suitable for long-range distribution. It is, therefore, an ideal substitute for natural gas. However, before it is suitable for this purpose, it needs to be free of solid and gaseous impurities and its composition has to be adjusted by a shift reaction of the form :

$$CO + H_2 \quad \text{-->} \quad CO_2 + H_2$$

and then subjected to a methanation reaction;

$$CO + 3H_2 \quad \text{-->} \quad CH_4 + H_2O.$$

These fragments produce a gas that is suitable for substitution into the natural gas distribution network.

Synthesis gas is also used extensively for the manufacture of many chemical feedstocks such as hydrogen, methanol and ammonia that can then be employed in the production of other chemicals such as petrol and fertilisers. It is also envisaged that synthesis gas, and for that matter industrial fuel gas, will be utilised to fire a gas turbine in a combined cycle power generation plant.

4.3. Combined Cycle Power Generation using a Coal Gasifier

Substantially better thermal efficiencies than those of conventional steam raising power stations are potentially attainable in combined cycle power plant comprising both a gas and a steam turbine. Fig. 4 is a simple flow diagram demonstrating how a pressurised coal gasifier can be employed in such a cycle. Additionally, sulphur emissions, mainly as H_2S, can be reduced to relatively low levels by low temperature scrubbing with certain organic chemicals or, in the case of fluidised bed gasification, by the addition of limestone, or dolomite, to the reactor thereby gettering the sulphur as calcium sulphide.

If the gas, that is to be combusted in the gas turbine, is produced by an entrained flow, or a fluidised bed gasifier, it is important that the appreciable sensible heat of the off-gas is utilised by raising steam (Fig. 4); the temperature of the off-gas from a Lurgi gasifier (approximately 500°C) is too low to make heat recovery worthwhile. Cycle efficiency calculations

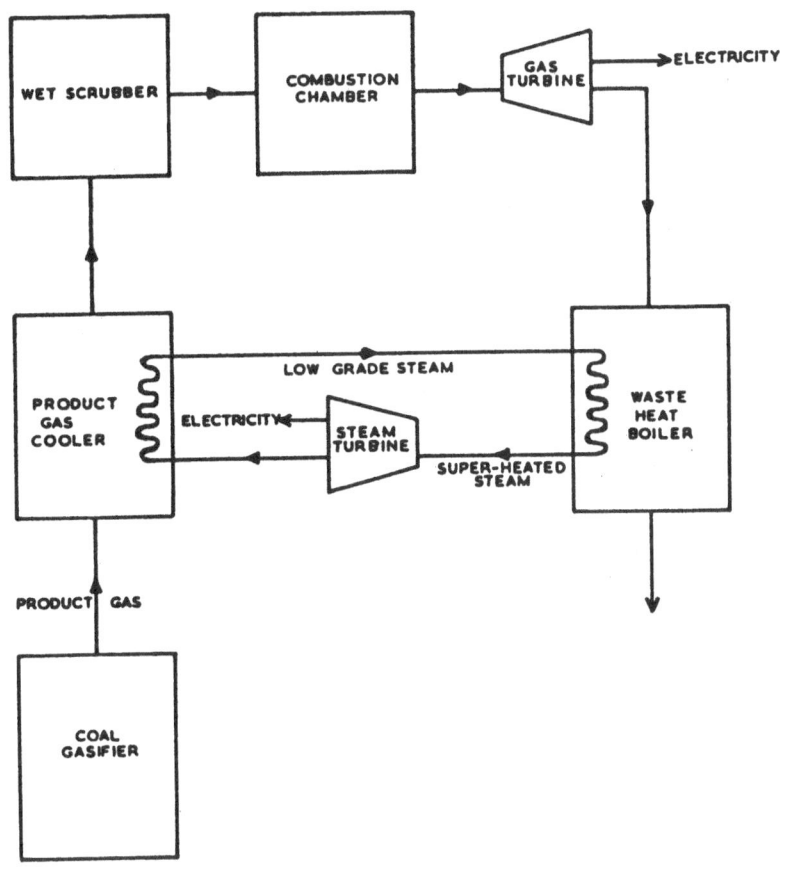

Fig. 4 A flow diagram of a combined cycle power lay-out
        incorporating a gasifier

have demonstrated that ideally superheated, rather than low
grade, steam should be produced at this stage. However, this
means that a raw gas heat exchanger would have to operate in a
reducing atmosphere at metal temperatures close to 650°C. As
illustrated in Fig. 4, steam is also produced in the oxidising
exhaust of the gas turbine.

Finally, economic/cycle efficiency calculations suggest that
it is more profitable to employ oxygen rather than air to provide
the heat necessary to gasify the coal even though the cost of the
associated oxygen plant is not insignificant.

5. MATERIALS OF CONSTRUCTION FOR GASIFICATION SYSTEMS

5.1. Introduction

When designing and constructing a coal gasification plant it is necessary to consider the requirements of the complete system and not just those associated with the gasifier, alone. Clearly, the greatest demands will be placed on those materials that are in direct contact with the gasifying reactants, the hot reducing gas and waste products; the magnitude of these demands will, however, vary and will be a function of the type of gasifier and system employed.

Before discussing the reported behaviour of materials exposed in commercial and pilot plant gasifiers and laboratory rigs it is necessary to consider briefly the design of gasification systems and the types of constructional materials required.

5.2. Gasification Systems - Design and Materials Requirements

Much of the technology and equipment involved in the production of the gasifier's feedstocks such as the coal, coal/water slurries and oxygen already exist. However, the transportation of coal and its slurry into the gasifier, especially when the reactor is working at pressure, will place heavy demands on pumps and lock hopper valve seals (5). However, it is anticipated that these problems can be accommodated by the experience gained in other areas of the petro-chemical and power generating industries.

The requirements placed on materials used to construct the gasifier and other pieces of hardware, such as cyclones and hot solid/gas transport lines, are demanding because of the high temperatures and pressures and the corrosive and abrasive materials environments encountered. However, by careful design and appropriate materials selection it should be possible to build safe, reliable and cost effective systems.

It is recommended (6) that the gasifier vessel should be manufactured from a low alloy ferritic steel, typically 2.25 wt % Cr - 1 wt % Mo steel. To protect this steel from the aggressive gasification environment, it should be refractory-lined and possibly water-cooled. However, to avoid aqueous corrosion of the shell's inner surface, the steel should be maintained at a temperature (approximately 300°C) in excess of the dew-point of the coal gas that may permeate though the refractory lining. If extensive refractory cracking occurs hot spots may develop on uncooled parts of the steel shell which may result in hydrogen attack at elevated temperatures and hydrogen embrittlement if the reactor cools too rapidly during shut-down. It is essential that the refractories employed to line the gasifier should possess good mechanical strength, corrosion and wear resistance. Because of the highly corrosive nature of the gasification environment, few alloy components are used within the gasifier and those that are needed, e.g. stirrers and tuyere nozzles must be effectively cooled.

For most applications (see section 4) it is necessary to produce a gas that is free of solid, volatile and gaseous contaminants. As explained in section 3.4. the gas produced by a fluidised bed reactor is heavily laden with solids and these have to be removed using a series of cyclones constructed in a similar fashion to the methods adopted for the gasifier, i.e. refractory lining and water cooling. It should also be recognised that current work on second generation particulate removal systems is considering the use of ceramic candle filters (7). To remove the volatile compounds produced by the Lurgi system and the gaseous impurities common to all gasification systems gas quenching and wet scrubbing techniques using chemicals and water are available. In these areas of plant the major materials problems will be ones of aqueous corrosion that have already been addressed and catered for in other related industries.

Because the raw product gas that leaves fluidised bed and entrained flow gasifiers possesses significant amounts of sensible heat it is necessary to make use of this heat prior to gas cleaning. This is achieved by raising steam in a heat exchanger system (Fig. 4). To minimise the risk of high temperature corrosion these heat exchangers are arranged such that the coolest part, i.e. the evaporator section, is in contact with the hottest, inlet gas. Even so, the low alloy steels used used for this purpose are still likely to require coating with a more corrosion resistant material, based possibly on Al_2O_3. Current practice recommends that the production of superheated steam takes place in the exhaust of the gas turbine where the atmosphere is oxidising and far less corrosive. However, for cycle efficiency improvements it is desirable to superheat in the hot raw gas product. As will be discussed in section 5.3., the identification of alloys, or possibly structural ceramics, suitable for the construction of a durable heat exchanger capable of producing superheated steam at 565°C and 160 bar is a major challenge facing the materials engineer in this field of work.

Once the product gas is clean it is then suitable for utilisation. If it is to be used as a synthesis gas, or converted into a substitute for natural gas, it will need further processing. However, the numerous process routes have already been established e.g. shift and methanation, and present few materials problems. If the gas is to be used in a combined cycle power generation cycle, it will be burned and expanded through a gas turbine (Fig. 4). Since the overall cycle efficiency for power generation will be largely influenced by the gas turbine entry temperature it is desirable to employ a turbine that can accept very hot gases. Currently, the most advanced gas turbine for industrial applications has an entry temperature of ca. 1250°C. However, further increases in inlet temperature would be possible if strong, more temperature resistant, ceramics could be developed.

5.3. Behaviour of High Temperature Materials in Gasification

5.3.1. Introduction

To identify areas requiring further high temperature materials research and development, thereby promoting the refinement of advanced coal gasification systems, it is necessary to review the relevant studies that have already taken place. In the following discussions the behaviour and degradation mechanisms of refractories, ceramics and alloys that have been exposed in commercial and pilot plant gasifiers as well as the simulated environments produced in the laboratory will be reported. It will be apparent to the reader that little experience of materials behaviour in commercial systems is presented because of the commercial value that is placed on this information.

5.3.2. Refractories and Ceramics

The selection and performance of refractories and ceramics in coal gasification systems has already been considered in depth and reviewed by several workers (8-13). Although little information of their behaviour in demonstrator plant is reported it is commonly believed that materials are available that can be utilised successfully.

For dry ash applications the demands placed on refractories and ceramics are not so great as those of slagging gasifiers where the molten ash accentuates corrosion and erosive attack. For this reason refractory and ceramic selection and performance under dry ash and slagging conditions will be considered separately.

5.3.2.1. Dry Ash Applications

The main functions of refractories used in dry ash conditions are to keep the outer steel containment, e.g. gasifier and cyclone shells and transport lines cool (approximately 300°C) and to offer protection against erosion. Dial (8) states that these requirements are best met by employing a multi-layered refractory lining. He suggests that a suitable material for the layer next to the steel could be an insulating, low iron, high silica castable. The hotter layer, facing the gasification atmosphere (at approximately 1000°C), might consist of a low iron, low silica, dense alumina castable refractory; if particularly severe erosive conditions prevail it is recommended that this layer should be fired (14). To improve the mechanical integrity of the refractory linings, it is good practice to weld corrosion resistant V-shaped pegs to the containment walls with their legs penetrating into, and thereby keying, the refractory (14). Because of the large differences between the thermal expansion coefficients of the refractory linings and the steel containment cracking of the former during curing or in-service operation is always a real threat. Practical studies by Anderson et al. (15) have considered this problem in terms of optimising the initial drying-out and heating-up procedures for monolithic refractories. Their recommendations include the use of a dense 50 % Al_2O_3, hot face refractory having a low cement content, a low shrinkage coefficient and 310 SS fibre reinforcement, (however, Stringer (14) does not advocate the use of metal fibres in gasification atmospheres because of the risk of refractory cracking being induced by fibre corrosion). It is also suggested (15) that the stainless steel pegs used to key the refractory to the outer shell should be pre-coated with asphalt that will burn away during heating thereby leaving room for expansion. Slow rates of heating, typically 20°C/h are also advised. Analytical work (16), calculating the mechanical stresses that develop in refractory linings during heating has also been performed in order to optimise their design and modes of gasifier operation. In general these results (16) support the findings of the practical studies (15).

As well as straightforward mechanical failure, refractories used in dry ash environments may also be susceptible to the following chemical, or physio-chemical, mechanisms of degradation.

(a) Alkali Attack

This form of degradation occurs due to the pick-up of alkali species by the refractory and the subsequent formation of molten phases or, as is more usually the case, the growth of voluminous solid reaction products that promote refractory cracking and spallation. While this problem is common in iron-making blast furnaces few failures have been reported in coal gasifiers. However, it is reasonable to assume that if coals containing high alkali metal contents are utilised, e.g. lignites, alkali attack may well be troublesome. Kennedy and Schlett (9) recommend the use of high density, high alumina refractories, which will limit the penetration of the alkali species, or high silica (60 %) refractories, that react with the alkalies to form a glaze that seals the hot surfaces of such materials. However, they recognise that the choice of these refractories will either reduce the linings' thermal shock resistance or the maximum operating temperature to less than 1000°C. For higher temperature applications (1000 to 1300°C) dense magnesia-chrome, or magnesium-aluminate spinel, refractories are proposed.

(b) Carbon Bursting

When some types of refractory are exposed to atmospheres containing carbon monoxide deposition of carbon can occur due to the decomposition reaction;

$$2CO \longrightarrow C + CO_2$$

Carbon deposition takes place over the temperature range 400 to 700°C and, therefore, commonly within the bulk of refractories where such temperatures prevail. The carbon depositon is catalysed by the presence of iron, or iron compounds, within the refractories (9) and, therefore, the incorporation of steel fibre reinforcement of refractories should be used cautiously. Carbon deposition also appears to be accelerated by alkalies but retarded by H_2S (9).

Few instances of carbon bursting have been reported in the literature. Even so, Kennedy and Schlett (9) claim that for low temperature applications dense alumina castables with low iron contents or high-fired fireclay or alumina refractories should offer good resistance to this form of attack.

(c) Silica Volatilisation

It is well known that sufficiently high temperature, in excess of 1200°C, the formation of volatile SiO (due to the reduction of SiO_2 by H_2) and $Si_2O(OH)_6$ (due to the reaction of steam with SiO_2) can lead to refractory degradation. However, for dry ash applications process temperatures are less than 1200°C and no volatilisation problems are envisaged. Bakker (17) has reported that at relatively low temperatures (500°C and less) changes in strength occur when refractories are exposed to steam and hydrogen. For example, at about 500°C, reductions in the strength of some silica-free, alumina concretes, due to the leaching of lime from the cement phase, was observed. He (17) also noted strength increases for some high alumina, dense concretes when exposed at approximately 300°C. From these observations Bakker (17) concluded that 50-60 % alumina concretes, with high or even intermediate purity calcium aluminate cements, are the preferred linings for dry ash gasifiers, transport lines and cyclones.

5.3.2.2. Slagging Applications

In addition to the degradation mechanisms cited for refractories and ceramics used in dry ash environments, the materials used in slagging gasifiers also run the risk of attack from molten coal slag, iron bursting and the combined effects of corrosion/erosion. The attack by molten coal slag is seen to be the most aggressive form of degradation and the most difficult to resist.

(a) Molten Coal Slag Corrosion

The slags that are formed from coal are usually acidic, i.e. the CaO + MgO + FeO : SiO_2 < 1 (10) and can induce liquid/solid reactions (commonly referred to as corrosion). Such attack can occur via a simple dissolution mechanism or it can be more complex involving interactions with certain impurities found in the ceramic (refractory) or slag. Experience (9) has shown that alumina and alumina-chromia refractory bricks are attacked rapidly when exposed to such slags. However, laboratory tests carried out at the Argonne National Laboratory (9) have shown that dense, high chromia refractories, such as the chrome-spinel $MgCr_2O_4$, do exhibit good resistance to slag attack over the temperature range 1200-1600°C. When the slag is in a turbulent state the demands placed on refractories and ceramics are even greater due to the combined effects of erosion and corrosion. Therefore, whenever possible the gasifier should be designed, and operated, in such a way as to avoid such turbulence. Evaluation of refractories and ceramics in molten coal slag (9) has also shown that low porosity is another important factor in reducing slag penetration and corrosion. The need to limit slag penetration has been discussed by Bakker and Stringer (10) who observed the failure of a 90 % Al_2O_3 - 10 % Cr_2O_3 refractory due to the corrosion of the cement binding phase which in turn led to accelerated deformation a phenomenon as pyro-elastic creep. In addition to employing dense refractories, Bakker and Stringer (10) suggest that slag penetration can be reduced by imposing a steep temperature gradient across the material, which encourages slag freezing. They also state that the presence of corundum (Al_2O_3) and periclase (MgO) in the grains of the refractory should be avoided because they are preferentially attacked by coal slag. Unfortunately, the dense chrome-based refractories that are seen to offer good resistance to slag attack have poor thermal shock resistance and can be susceptible to iron-bursting (9) (see the following discussion). Therefore, it is important to ensure that when they are employed, account is taken of these weaknesses by avoiding rapid cooling and heating and endeavouring to keep the iron content of the slag low.

For slagging zones that are in excess of 1600°C large

amounts of refractory cooling are essential if a reasonable
service life is to be expected, i.e. 3 years or more. This is
achieved by securing the refractory lining to a water wall. The
design is such that the temperature of the hot refractory face is
cool enough to freeze a layer of coal slag on to the surface
thereby affording good corrosion resistance. A further approach
is to lower the operating temperature of the gasification process
yet still maintaining a molten slag; this can be achieved by the
use of suitable fluxes or coals that possess low melting point
ashes (9).

(b) Iron Bursting

Failure of refractories containing spinel phases, such as
$MgCr_2O_4$, can occur due to their reaction with the iron oxide
present in coal slag. The reaction products that form tend to be
voluminous and promote cracking and spalling of the refractories.

To avoid this form of attack it is recommended that dense
refractory bricks are employed since these materials resist slag
penetration (9).

5.3.3. Alloys and Coatings

The behaviour of alloys and coatings in gasification
environments has been the subject of research for over a decade.
Early studies (18) considered the performance of alloys in pilot
plant and laboratory furnaces at relatively high temperatures,
typically 800°C. This work demonstrated that few alloys were
capable of even short life-times under such conditions because of
catastrophic sulphidation, oxidation and carburisation. In many
of these studies (19) the use of thermodynamic phase stability
diagrams were employed to account for the behaviour observed.
Additionally, some account of corrosion kinetics has been made
and dynamic boundaries, delineating the differences between
protective and non-protective behaviour, have been superimposed
on to the thermodynamic stability diagrams (20).

More recently efforts in this field have been concentrated
on evaluating the performances of low alloy and stainless steel
and coatings, such as aluminised and chromised, in real and
simulated gasification environments at low temperatures
representative of those typical of a raw product gas heat
exchanger, i.e. 300-600°C. The findings are encouraging.
Lewis (21) reports that at 400°C, in a simulated syngas
environments, aluminised (> 18 wt % Al) and chromised
(> 25 wt % Cr) coatings and stainless steel cladding on low alloy
steel performed reasonably well. However, intergranular
corrosion did occur along coatings cracks and in chrome denuded
zones, such as those produced by welding. Similarly, Perkins et

al. (22) have shown that up to 500°C, in a simulated medium calorific value gas, high chromium (> 23 wt Cr %) ferritic and austenitic stainless steels, FeCrAl, and titanium, and some of its alloys, exhibited good corrosion resistance. Grabke et al. (23), reporting the findings of a European COST research programme have described the corrosion behaviour of Fe - 32 wt % Ni - 20 wt % Cr and Fe - 25 wt % Cr - 20 wt % Ni alloys under varying oxygen and sulphur partial pressures within the temperature range 700 - 800°C. They concluded that the initial stages of oxidation/sulphidation are important and if sufficiently high oxygen and low sulphur, activities can be achieved during these early stages of exposure, the growth of sulphide scale are significantly suppressed. In practice, however, it would be essential that any pre-oxidation treatment is carried out under controlled conditions. Such a treatment will not eliminate corrosion but it will retard it for some time.

Clearly, the identification of alloys and coatings suitable for the construction of a raw product gas heat exchanger is a major materials issue. Since the majority of tests evaluating alloys and coatings have been performed under simulated laboratory conditions it is essential that this information is linked with the few reported performances observed in real systems. In particular, it is necessary to establish what effects the presence of carbonaceous deposits and coal slag have on corrosion rate. Additionally, the added influence of HCl, formed during the gasification of some coals, on corrosion needs to be considered further. For example, the interesting observation of Saunders and Schlierer (24), showing that under certain conditions the presence of carbonaceous deposits can alleviate the corrosive influence of HCl, needs to be investigated in more detail.

Finally, it appears that high temperature alloy corrosion of heat exchanger materials in coal gasifiers can be exacerbated due to aqueous corrosion that occurs during periods of shut-down. When these systems are off-line condensation can occur liberating aqueous, corrosive compounds, such as sulphuric and polythionic acid (25). These acids then act to disrupt the protective scales that are necessary for high temperature corrosion resistance. In a commercial plant attention would have to be paid to this problem and during periods of shut-down measures to avoid condensation, e.g. nitrogen purging, may be required.

6. RECOMMENDATIONS FOR FURTHER RESEARCH AND DEVELOPMENT INTO MATERIALS FOR GASIFICATION SYSTEMS

The preceding discussions have demonstrated that by using current technology and available materials it is possible to build advanced gasification systems. However, with further materials research and development more efficient and reliable systems will emerge needing less attention and maintenance.

The following recommendations for further work are identified :

(i)   There is need to define the safe operating limits for heat exchangers that are used to recoup the sensible heat contained in the raw product gas of entrained flow and fluidised bed gasifiers prior to gas cleaning. It is highly likely that the performance of conventional engineering alloys and coatings in such environments is already known. However, because of the commercial value that is place on such information, it is not publically available.

(ii) It is desirable to develop new alloys, coating systems or structural ceramics that will facilitate the manufacture of a durable raw product gas heat exchanger capable of raising superheated steam at approximately 565°C and 160 bar. A suitable alloy would need to possess good resistance to oxidation/sulphidation, and possible carburisation and adequate high temperature strength. A coating system (probably based on Al_2O_3) in addition to having corrosion resistance would also have to be defect free and maintain good contact with its alloy substrate.

It is possible that SiC could serve as a suitable structural ceramic, possessing adequate corrosion resistance, for the manufacture of a raw product gas heat exchanger. However, the mechanical properties of ceramics are notoriously unreliable because of their flawed structure. Therefore, it is in the areas of ceramic processing and quality control where more work is required.

(iii) Work evaluating the performances of refractories and ceramics in molten cool ash slags should continue with emphasis being placed on the use at high density, high chromia containing refractories. Additionally, the influence of coal ash composition on slag attack and the possible use of fluxes on lowering slag melting points should be investigated for European coals.

(iv) Finally, to ensure safe and reliable plant operation the further development of on-line, non-destructive monitoring techniques is called for. These techniques would be required to detect the onset of cracking and catastrophic corrosion in metallic, refractory and ceramic materials.

REFERENCES

(1) "Gas from Coal", A National Coal Board Report, (1983).

(2) Schroter, H.J., Schendler, W. and Weber, H., High
Temperature Materials Corrosion in Coal Gasification
Atmospheres, Edt. Norton, J.R., Elsevier Applied Science
Publishers, (1984).

(3) The British Gas/Lurgi Slagging Gasifier : Status,
Applications and Economics, Evans, R. and Hiller, W.,
A British Gas International Consultancy Service
Publications, (1986-1987).

(4) "The Texaco Coal Gasification Process - Synthesis Gas for
Chemical Feedstocks", Crouch, W.B., International Coal
Conversion Conference, (1982), Pretoria, South Africa.

(5) "Materials of Construction for Advanced Coal Conversion
Systems", Nangia, V.K., ESCOE, Noyes Data Corporation,
(1982).

(6) "Design Properties of Steels for Coal Conversion Vessels",
Gabe, D.E., 4th Annual Conf. on "Materials for Coal
Conversion and Utilisation", U.S. DOE, Gaithersburg,
Maryland, USA, (1979).

(7) "The Behaviour of High Temperature Filter Materials in Hot
Gasifier and Combustor Atmospheres', Oakey, J.E. and
Reed, G.P., I. Chem. E. Symposium Series 99, Gas Cleaning at
High Temperatures, Pergamon Press, (1986).

(8) Dial, R.H., J. of the Canadian Ceramic Soc., Vol. 43,
pp. 65, (1973).

(9) "Refractories for Coal Gasification - The State of the Art
in the U.S." Kennedy, C.R. and Schlett, P.E., Ceramics in
Advanced Energy Technologies, Proc. of the European
Colloquium, J.R.C., Petten, NH, (1982), D. Reidel Publishing
Co.

(10) "Materials Requirements for Coal Gasification Combined Cycle
Power Plants", Proc. of Int. Gas Research Conf., Los
Angeles, pp. 454, (1981).

(11) "The Corrosion of Refractories in Coal Gasifiers at Elevated
Temperature", Yurek, G.J., Conf. on Corrosion/Erosion of
Coal Conversion System Materials, NACE Pub., (1979).

(12) "Design of Refractories for Coal Gasification and Combustion Systems", Vojnovich, T., EPRI Rept. No., AF1151, (1979).

(13) "Design of Refractories for Resistance to High Temperature Erosion-Corrosion", Vaux, W.G., EPRI Rept. No., AP-1955, (1981).

(14) "Materials of Construction II - Refractory and Ceramic", Stringer, J., High Temperature Materials Corrosion in Coal Gasification Atmospheres, Edt. Norton, J.R., Elsevier Applied Science Publishers, (1984).

(15) "Improvement of the Mechanical Reliability of Monolithic Refractory Linings for Coal Gasification Linings", Anderson, E.M., Glasser, R.P., Schroedl, M.A., 5th Annual Conf. on Materials for Coal Conversion and Utilisation, U.S. DOE, Gaithersburg, Maryland, U.S.A., (1980).

(16) Bray, D.J., Smyth, J.R. and McGee, T.D., J. American Ceramic Soc., Vol. 59, No. 7, pp. 706, (1980).

(17) "Refractory Applications in Coal Gasifiers", Bakker, W.T., 3rd Annual Conf. on "Materials for Coal Gonversion and Utilisation', U.S. DOE, Gaithersburg, Maryland, U.S.A., (1978).

(18) "The Properties and Performance of Materials in the Coal Gasification Environment", Ed. Hill, V.L., Am. Soc. of Metals, (1981).

(19) "Thermodynamic Phase Stability Diagrams for the Analysis of Corrosion Reactions in Coal Gasification/Combustion Atmospheres", Hemmings, P.L. and Perkins, R.A., EPRI Rept. No., FP539, (1977).

(20) "High Temperature Alloy Corrosion in Coal Conversion Environments", Natesan, K., High Temperature Corrosion, NACE Pub. 6, (1981).

(21) "Evaluation of Coated and Clad Heat Exchangers for Syngas Coolers", Lewis, E.C., EPRI Rept. No., AP4406, (1986).

(22) "Evaluation of Alloys for Fuel Cell Heat Exchangers", Perkins, R.A. and Vonk, S.J., EPRI Rept. No., EM-1815, (1981).

(23) "Materials Behaviour in Coal Gasification Environments", Grabke, H.J., Norton, J.F. and Casteel, F.G., High Temperature Alloys for Gas Turbine and Other Applications, (1986), Edt. Betz, W. et al., D. Reidel Publishing Co.

(24) "Sulphidation of Coal Gasifier Heat Exchanger Alloys",
Saunders, S.R.J. and Schlierer, S., High Temperature
Corrosion in Energy Systems, Edt. Rothman, M.F., A.I.M.E.,
(1985).

(25) "High Temperature Alloy Requirements for Coal Fired Combined
Cycles", Davidson, B.J., Meadowcroft, D.B., Stringer, J.,
High Temperature Alloys for Gas Turbine and Other
Applications, (1986), Edt. Betz, W. et al., D. Reidel
Publishing Co.

4.1.4. Fuel Cells

K. Joon, S.B. v.d. Molen, E.H.P. Cordfunke

Netherlands Energy Centre

Petten, The Netherlands

1. INTRODUCTION

Fuel cells are devices in which, on the basis of conventional electrochemical principles, the chemical energy of a fuel is converted directly, i.e. without combustion, into electrochemical energy. This is brought about by catalytic oxidation of the fuel in a porous anode structure, transfer of the liberated electrons through an outer circuit to a porous cathode in which they reduce oxygen, also by catalytic reaction, and appropriate charge and mass transfer through an electrolyte separating both electrodes.

For a fuel such as hydrogen, for example, the overall chemical reaction is:

$$H_2 + {}^1/_2\ O_2 \rightarrow H_2O + energy$$

Electricity production in this way is continuous so long as fuel and oxidant are available to the cell. To increase power production the cell area can be enlarged, individual cells can be connected in series to form modules or stacks, which can be further connected in series or in parallel. In an ideal stack individual cells are stacked and separated by a bipolar metal plate serving the functions of electron conduction, gas separation and distribution, edge sealing and structural support.

Fuel cells offer numerous advantages compared to other, conventional and advanced, power production technologies. From energy conservation considerations, the efficiency is the most important parameter. Because of the direct conversion of fuel to power without an intermediate Carnot cycle, electrical efficiencies of 80% are possible theoretically. Additional conservation opportunities derive from the facts that this efficiency is maintained over a very wide range of loads and that fuel consumption at idle is extremely low (only 1% of full power consumption).

These operational characteristics together with the modular construction of fuel cell stacks and the variability of the power to heat output, make this technology suited for a wide range of applications: base load, part load and load following operation for central or dispersed generation in grid-connected or stand-alone stations for power or combined heat and power production in any desired unit size that, moreover, may be expanded according to need. The DC-output can easily be converted to any AC specification.

Fuel cells also have considerable environmental advantages. There is no emission of particles and emission levels for SO_x and NO_x are (far) below 10% of current US-standards. In addition, fuel cells are extremely quiet; the cell stack itself makes no noise. The only sound generated is from small blowers for air supply to and cooling of the cells. Because of their compact construction, also use of land is limited. In addition, the modular construction allows prefabrication with as advantages lower manufacturing costs, short on-site construction times, reduction of installed overcapacity and optimal quality control and -assurance.

To the advantages mentioned above, others may be added, such as fast response, high reliability, low maintenance, long life, and fuel flexibility. When fully developed, these will of course be reflected in a positive influence on power production economics. However, at present a number of these items remain to be demonstrated technically.

2. TYPES OF FUEL CELLS

Fuel cells are identified by and named after the electrolyte on which they are based. The best known are the alkaline cell (AFC) using potassium hydroxide as electrolyte and operating at about 80°C, the phosphoric acid fuel cell (PAFC) operating at about 200°C, the molten carbonate cell (MCFC) which operates at about 650°C and the solid oxide fuel cell (SOFC) at 1000°C. Only the last two are genuine high temperature cells and will be dealt with further in this chapter.

2.1. MCFC

The molten carbonate fuel cell uses a mixture of alkali

metal carbonates as the electrolyte, for which at present the
62-38 mol % eutectic of lithium and potassium carbonate is used.
The electrolyte is contained in a highly porous, electrically
insulating matrix whose pores are completely filled, while it
only wets the surfaces of the larger pores in the electrodes.
The most important characteristics of state-of-the-art porous
components are as follows:

	Anode	Matrix	Cathode
Material	Ni	$LiAlO_2$	NiO
Porosity, %	65	50	65
Pore size, μm	3-5	< 1	5-7
Thickness, mm	0,8-1,5	0,7	0,4

In fact, only the anode has some strength. These
components, therefore, have a flat plate design. They can easily
be stacked to form cells and to form stacks with separators
between individual cells. If the separator is a bipolar metal
sheet, power can be obtained from the end plates of the stack and
individual external connections omitted.

Molten carbonate fuel cells operate at about 650°C and
pressures up to 10 bar. Power generation efficiencies may reach
65% at the design point of 0,8V/cell at 160-200 mA/cm². If
wetting of the appropriate catalyst by carbonate can be
prevented, internal reforming of hydrocarbons etc. is possible.
The cells accept both hydrogen and carbon monoxide as fuel as
shown by the main reactions:

anode: $H_2 + CO_3^= \rightarrow H_2O + CO_2 + 2e$

$CO + H_2O \rightarrow CO_2 + H_2$ (water-gas shift)

$CO + CO_3^= \rightarrow 2CO_2 + 2e$

cathode: $CO_2 + \frac{1}{2} O_2 + 2e \rightarrow CO_3^=$

These reactions also show that the carbonate ions
participate in the power generation, but the system can be made
self-supporting in this respect.

MCFC's are generally insensitive to contaminants in the
gases with the exception of sulfur: whereas H_2S poisoning of the

fuel electrode is reversible, SO_x poisoning at the cathode is not.

The gases flow laterally along the surface of the electrodes and gradually deplete towards their exits. Thus, their relative velocities and distributions (cross-, counter- or co-flow) are important as these determine the local power and heat production and thus the temperature distribution. Even in the most favourable case of co-flow with 80% fuel utilisation, lateral temperature differences of 100°C or more can be expected.

2.2. SOFC

The SOFC's have much in common with the MCFC: they accept H_2 and CO as fuels, internal reforming of hydrocarbons and alcohol is possible, they tolerate to a certain extent contaminants in the gases except sulphur, they operate in the same efficiency (60%) and pressure range and in most applications they may compete with the MCFC.

The solid oxide fuel cell operates at 1000°C at which oxygen is reduced at the cathode. The oxygen ions formed are conducted through the electrolyte to the anode where they combine with the fuel and liberate the electrons again. This high temperature would enable spontaneous reforming of hydrocarbons in the cell and provides high quality waste heat for cogeneration purposes or bottoming cycles. In addition, at 1000°C the basic cell reactions are fast, allowing thin component layers to be used and high volumetric power densities to be reached.

At present, the porous cathode is made of strontium-doped lanthanum manganite, the dense electrolyte is yttria-stabilized zirconia and the porous anode is a cermet of nickel with stabilized zirconia. The fragility of these components requires a thick support structure of (porous) calcia- stabilized zirconia, while for interconnection of cells magnesium-doped lanthanum chromite is used.

The structural design of SOFC's differs from all other fuel cells: instead of a flat plate, at present, tubular and monolithic or honeycomb designs are being pursued. In their manufacture this means that the individual components are not fabricated separately and then assembled, but that they are deposited successively by thin layer techniques, one on top of the other and in order of decreasing sintering temperature.

Materials requirements for SOFC's are very stringent. In addition to chemical stability in fuel and oxidant environments, chemical compatibility with the other component materials and electronic/ionic conductivity, compatability of thermal expansion is of the utmost importance to avoid separation of the ceramic layers. On the other hand, as the electrolyte is a solid, problems regarding liquid leakage and liquid electrolyte induced corrosion do not occur.

3.   MATERIALS RESEARCH NEEDS FOR MOLTEN CARBONATE FUEL CELLS

The research and development results of the last 10 years show that several subjects need to be further investigated in order to increase the long term performance of MCFC's. The most important materials problems are:
- cathode dissolution
- mechanical stability of porous materials
- high temperature corrosion
- electrolyte management
- degradation of catalyst/support structures for internal reforming.

3.1. Cathode dissolution

The operation time of a MCFC with a lithiated NiO cathode is limited because of gradual dissolution of NiO and the migration of Ni-ions and precipitation of metallic Ni in the electrolyte between anode and cathode. This process gets more serious with increasing operating pressure. Too high concentration of Ni in the matrix will cause cell shortage.

To prevent this dissolution and precipitation problem new cathode materials with characteristics superior to NiO need to be developed. The new materials should be chemically and mechanically stable in carbonate and should have the required conductivity of $> 0,1 \ \Omega^{-1}.cm^{-1}$ as porous material. Such a combination of properties in one material is difficult to find. For instance, lanthanum perovskites have high conductivities and are stable in the carbonate melt but they react with the $LiAlO_2$ matrix. Development of conducting ceramics should get a high priority. The most promising known materials, Mg-doped Li_2MnO_3 and Mn-doped $LiFeO_2$, have not yet been tested in long term in-cell tests.

Alternatives to new materials could be reduction of the NiO dissolution by either additives to the NiO itself or changes in the electrolyte composition.

3.2. Mechanical stability of porous materials

In a fuel cell stack the dimensional stability of the active porous components is very important. The holding force in a stack is distributed between the seal and the active area. The strength and toughness and thus the shrinkage and creep of the

porous components under operating conditions translate into
changes in the pore structure and, therefore, electrolyte
distribution which influence directly the cell and stack
performance. In addition, design options are needed for
accomodating the shrinkage to prevent separation of components
and avoid gas leaks.

Pure Ni anodes are relatively weak. Currently, therefore,
chromium is added to improve the anode strength and creep
behaviour. Recently, very good results were reported with oxide
dispersion strengthened material of Ni-Al-O. To ultimately solve
the problem of anode creep, deformation and shrinkage, new
materials will have to be developed either in combination with Ni
or as its replacement, such as alternative metals or conducting
ceramics.

The state of the art lithiated-NiO cathode material is very
weak. Strengthening of the structure can be obtained by
developing carbonate compatible mixed oxides with a sufficiently
high conductivity or conducting ceramics with the dual purpose of
also solving the NiO dissolution problem.

3.3. High temperature corrosion

Corrosion of the separator plate is one of the life limiting
factors for MCFC's. It is caused by parasitic consumption of the
electrolyte and causes degradation of the plate's conductivity
and integrity, all of which reduce cell performance. Bilayer and
one-side plated metals such as Ni-coated stainless steels (SS310
and SS316) allow the use of a single separator in the two
fundamentally different atmospheres. Up to now no single
candidate material has been identified for the different
environments. Materials research should be concentrated on the
development of a single, cost-effective, tailored alloy which is
stable with respect to carbonate in a reducing as well as an
oxidizing atmosphere.

3.4. Electrolyte management

The electrolyte distribution in a cell and in a cell stack
will be determined first of all by the wetting properties, pore
size distribution and porosity of each active porous component.
Parasitic consumption of electrolyte by other components results
in a change of electrolyte inventory and distribution during
operation and will influence the long term performance of the
cells. Thus, corrosion and wetting of oxide layers on separator
plate and gas distribution channels outside the active area are
the most important processes. The loss of electolyte can be
improved by developing non-wettable ceramic materials, conductive
as well as non-conductive, for construction or as coatings on

separator plate materials.

3.5. Degradation of catalyst/support structures for internal reforming

The degradation of the catalyst/support structures for internal reforming of CH_4 is mainly due to the wetting of the Ni-catalyst with carbonate. The dissolution of CH_4 in carbonate is very limited and so this wetting reduces the activity of the catalyst significantly. To improve the performance of the internal reformer the development of a conductive ceramic support material with Ni and non-wetting properties is necessary.

3.6. Priorities for R&D

The technical aims of the different MCFC programmes in the world are the demonstration of commercial IMW class units around 1996 preceded by units that are an order of magnitude smaller by 1992. These are developed from even smaller experimental stacks down to about 1kW. However, a simple upscaling of state of the art technology along such lines will not yield the highly competitive commercial product needed. The long term performance will have to be improved. One way would be to further develop the state of the art of materials, design concepts and/or fabrication techniques but it may well be that entirely new thinking in these areas will be necessary. A rather impressive R&D effort will therefore be necessary on subject areas related to the problem areas described in the previous paragraphs. To indicate priorities right now would be inappropriate and premature because all still need development.

4.   MATERIALS RESEARCH NEEDS FOR SOLID OXIDE FUEL CELLS

The SOFC is regarded as the third generation fuel cell. As such it is much less developed than the MCFC, as may be illustrated e.g. by the life times and conversion efficiencies (~ 45%) reached to date. This means that materials and materials properties of individual components as well as compatibility of materials of different components have not yet been optimized. In addition, structural and mechanical problems are aggravated by the high fabrication and operating temperatures. Thus, apart from applied R and D, a lot of fundamental research on materials properties and fabrication technology is needed in the categories:
- durability
- efficiency
- environment
- fabrication

For most specific subjects these are strongly inter-related.

4.1. Durability

In this category two main topics can be distinguished, i.e. gradual degradation of electrochemical performance and mechanical strength. As far as gradual degradation is concerned the stability of pore and grain morphologies of especially the porous components has to be improved. In addition, the chemical compatibility and inter-diffusion of materials between components are important research subjects. This requires modification of existing or development of new materials and fabrication processes for the individual components, taking into account cell operation parameters such as temperature, pressure, gas composition and contaminants.

Mechanical strength is of the utmost importance because of the large temperature differences experienced during temperature cycles, either for sintering during fabrication, during cell operation or during power excursions. Stress and fracture behaviour of single components but especially of laminated, multilayer structures should, therefore, be investigated with high priority to develop modified or new materials, design configurations and fabrication processes resulting in strong and thermomechanically compatible structures.

Another way of reducing the problems stemming from high temperature cycles would be to develop electrolyte materials with enhanced oxygen ion conductivity at lower temperatures. Candidates for modification of the stabilized zirconia in this respect are the 3d transition metals which at the same time could also positively affect the sintering characteristics.

4.2. Efficiency

The R and D efforts of the previous paragraph should not only result in improved endurance and long term performance but should preferably also include a superior initial electrochemical efficiency. On the other hand, the opposite way could also be followed: develop new materials with better efficiency and superior other characteristics.

An important parameter that determines efficiency is the active area available for the catalytic reactions, i.e. the three phase boundary between electrode, electrolyte and gas. This area is at present per definition very limited and might be enlarged to the entire gas-electrode interface if electrode materials are developed with mixed electronic and ion-conducting properties. Particularly the anode reactions could profit substantially from such a solution.

In addition, there is in general room for improving

polarisations, internal resistances and contact resistances in which certainly the interconnector should not be overlooked.

4.3. Environment

In relation to operating conditions considerable materials R and D remains to be done. Temperature related studies have already been mentioned and it should be realized that any change in this parameter has ramifications for all cell materials and components. Even more efforts may be expected with respect to pressure: extended operation at increased pressure up to 10 bar has not yet been demonstrated although higher pressures are known to increase cell efficiency.

In so far as the gases are concerned, hydrogen and oxygen diffusion through the interconnection as well as their influence on the interconnector's conductivity will have to be studied prior to further materials development. The same is true with respect to contaminants and reaction products in relation to practically all cell components. Already there is a clear incentive for development of sulphur-resistant electrode materials, especially at the fuel side.

4.4. Fabrication

There is an urgent need for fundamental ceramics research to elucidate the influence of fabrication process parameters on material properties. From such basic understanding it should be possible to select the correct materials, to match materials for composite structures, to select proper fabrication routes and to reproducibly manufacture materials/components/cells having the desired characteristics.

4.5. Priorities for R&D

As the future potential and economic impact of solid oxide fuel cells in power generation is strongly dependent upon its cost effectiveness, which needs considerable improvement, materials research should primarily be directed toward higher efficiency, longer lifetimes and reduced cost. Priority should be given to the realisation of simple thin multilayer structures of improved state of art products or of entirely new materials production techniques and/or cell design concepts. In fact, a massive effort will be needed regarding both electrode materials and oxygen ion conductors, preferably for reduced operating temperatures.

5. <u>CONCLUSIONS</u>

The renewed interest in fuel cells is due to their potentially interesting characteristics and to recent developments in high temperature materials technology. The latter allow or promise the economic production of ceramics with strong components and a high active surface to bulk volume ratio.

The success of high temperature fuel cells will depend strongly on the further development of suitable materials in which high conductivity and chemical and mechanical stability are combined.

To date, practically none of the available materials includes all of the desired properties and characteristics, which, in combination with the market prospects of fuel cells, is sufficient reason for a considerable effort in basic and applied materials research and in development of fabrication technology.

6. <u>REFERENCES</u>

For general as well as more detailed information reference is made to:

1. Assessment of Research Needs for Advanced Fuel Cells, DOE-report DOE/ER/30060-T1, (Nov. 1985)

2. Fuel Cells, Technology Status Reports for 1985 and 1986, report DOE/METC-86/0241 and DOE/METC-87/0257, Morgantown Energy Technology Center, (1986 and 1987)

3. Various fuel cell conferences, seminars and workshops:
   10-14 June 1985, Ravello (Italy), (Unesco);
   7-8 October 1985, Noordwijkerhout (Holland), (CEC/PEO);
   21-22 January 1986, Tokyo, (Institute of Applied Energy);
   26-29 October 1986, Tucson (Arizona), (National Fuel Cell Coordinating Group);
   4-5 June 1987, Taormina (Italy), (CEC/ENEA/CNR);
   26-29 October 1987, The Hague (Holland), (PEO).

4.1.5. Magneto-hydronamic Energy Conversion

Prof. F. Negrini

University of Bologna

Italy

1. INTRODUCTION[*]

Magnetohydrodynamic (MHD) energy conversion is a process that generates electrical power by passing an electrically-conductive high temperature gas stream, usually obtained from the combustion of fossil fuels, through a strong magnetic field at high speed. The MHD generator has no moving mechanical parts. Coal, gas or oil are burned at temperatures in the range 2700 to 3000 K with preheated air and/or oxygen enrichment; small amounts of an easily ionizable compounds (called "seed", such as alkali, typically potassium, salts) are added to the combustion gases to increase ionization and therefore the electrical conductivity. Such gases are then expanded at high speed through a channel with a strong transverse magnetic field. An electrical voltage is thus obtained between two opposite conducting walls of the channel (electrodes) and d.c. electrical power can be extracted. Hot gases leaving the channel are then used to generate steam, from which additional electrical power is obtained by a conventional steam turbine plant. This combined cycle MHD-steam is known as an open cycle MHD power plant. Magnetohydrodynamics can provide, therefore, direct fossil fuel fired power generation with higher efficiency, lower environmental pollution and lower costs than existing conventional energy conversion plants.

---

[*] by Francesco Negrini (Electrotechnical Inst.) and Franco Sandrolini (Applied Chemistry and Materials Science Dept.), Engineering Faculty, University of Bologna (Italy).

2. ENERGY PRODUCTION, CONVERSION AND UTILISATION

The direct conversion of thermal energy by MHD method and devices into electrical energy is wholly dependent on high temperature materials performances and endurance in the operating conditions. The MHD energy conversion methods are indeed the more effective the higher is the working temperature of the ionized gas. Every effort is therefore made to increase the hot gas ionization by adding ionizing chemicals (or seed, such as potassium salts) and increasing the process temperature. Chemical stresses add therefore to thermal, electrical and mechanical stresses on materials.

Plant components

The schematics of an open cycle MHD plant is shown in Fig. 1. Electrical power is produced in both the MHD part of the process, called the topping cycle, and in the boiler system/steam turbine generator, called the bottoming cycle. The process requires some critical components, such as the combustor, the nozzle, the channel and the strong magnet, usually of the superconducting type. Other somewhat less critical components are the heat and seed recovery plants, where additional electrical power is obtained through the steam turbine generator. Heat is recovered through oxidant preheating and seed is recovered by condensation and electrostatic precipitation. Additional components are designed in order to decrease the So_x and NO_x content of exhaust gases and purify the seed.

As it can be seen, the most critical components for MHD technology for materials performances are undoubtedly the combustor and the channel, particularly in coal-fired MHD plants, but also preheaters of air, with or without oxygen enrichment, require high performance materials, due to high temperatures needed for the high efficiency MHD process. Other more conventional components of MHD power plants, such as steam boilers, metal-tube heat exchangers, pipes, etc., may exhibit serious materials problems, due to the temperatures involved in the presence of ionizing chemicals, combustion gases, vapours and liquids (such as molten slags) and by-products in the forms of vapours and fogs formed by reaction with each other and with component materials. General constraints involved in the various MHD components will be briefly examined.

- Combustors

To obtain high efficiencies in MHD power plants, combustors must supply gas flows at very high temperatures (2750-3000 K) and moderate pressures (5-7 bar) with high electrical conductivity (10 S/m).

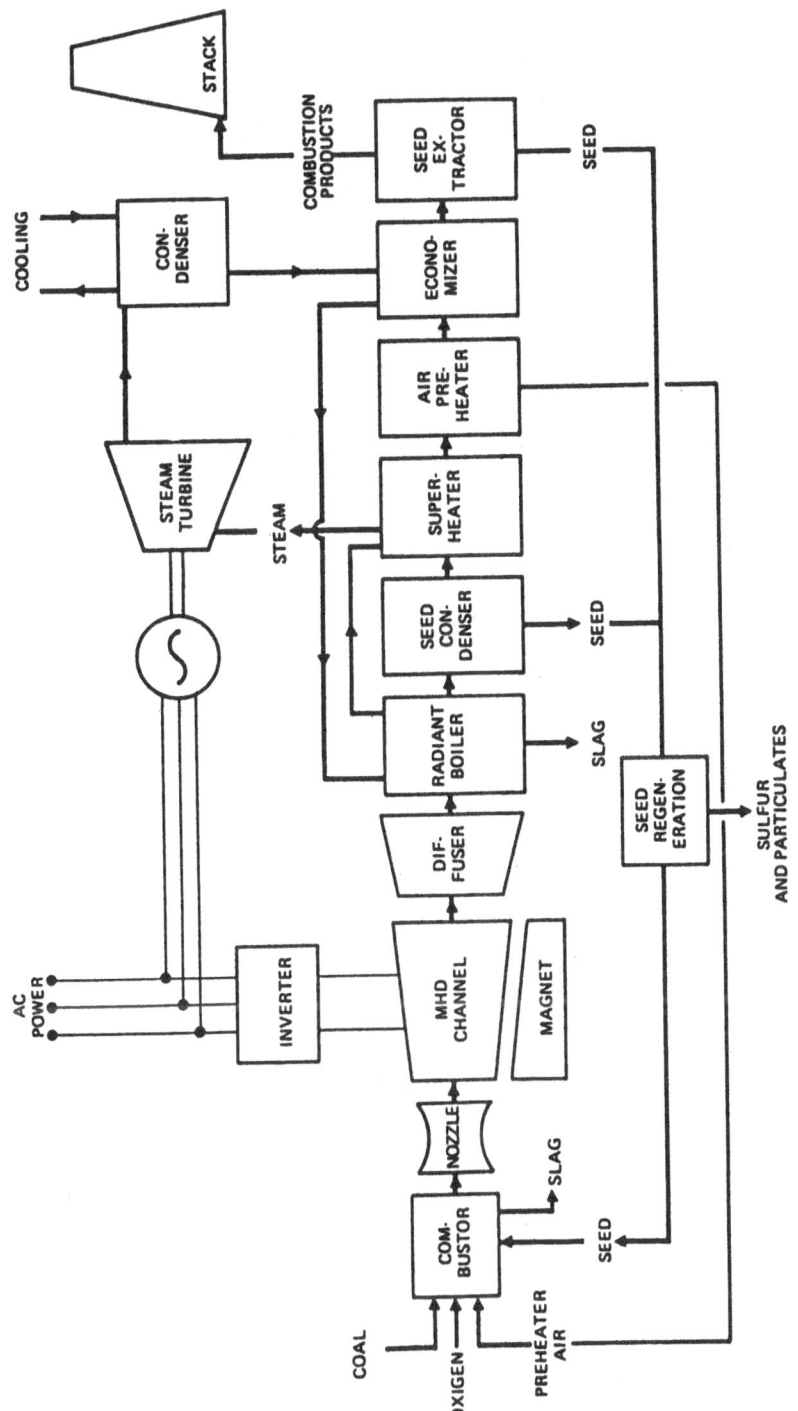

Fig. 1: MHD plant (schematic)

Combustor materials are therefore subjected to high temperature creep, erosion/corrosion effects caused by ionizing chemicals, combustion gases, vapours and by-products, high electrical fields due to floating of the combustor at the full value of the Hall voltages arising in the electrical generator, which may be altered by the action of molten slags, ionized gases, combustion oxygen, etc.. Moreover, starting and pausing transients can create thermal shock problems.

- Channels

MHD generator channels must basically perform the function of draining the electrical current generated by the highly ionized gas stream, which must flow, in addition, at the highest possible rate.
Channel materials must therefore provide the functions of both electrical insulation and conduction in the presence of very high temperatures and strong electrical and magnetic fields (the latter up to 6 T). Further, they are continuously in contact with a high speed mass flow of a strongly ionized combustion gas. Moreover, in the case of coal-fired MHD power plants also the effect of molten slags must be taken in account. Erosion is therefore the most dangerous effect of the gas flow, which can be further enhanced by flow asymmetries, velocity peaks, etc., caused by the electromagnetic forces. Molten slags increase erosion effects and bring electro-chemical problems, due to their electrical conductivity, which can attain 100 S/m depending on coal and seed composition, seed amount, etc.. Slag deposition on electrodes can result in a decrease of insulation between electrodes, and hence a decrease of process efficiency due to cathode wall resegmentation, and in slag/electrode interactions leading to electrochemical attacks and complex oxidation-reduction equilibria and subsequent dangerous erosion and electrical arcing. Finally, mechanical stresses are of course, always present.

- Other MHD power plants components

MHD combustors require proper air preheating to ensure the high temperatures needed by the channel. Preheating temperatures usually range from 800 to 1900 K, depending on the oxygen enrichment of the combustion air. Materials must therefore withstand average working temperatures higher than in other industrial technologies, in the presence, however, of some other heavy parameters, such as erosion/corrosion effects, reducing conditions, etc..

Performances and materials

The most critical performances are required in the MHD

channel, where two kinds of high temperature materials must be used: electrical conductors (electrodes) and insulators. Electrodes must exhibit the following main performances:

- <u>Current density</u> up to 3-4 A/cm^2, without remarkable Joule heating.
- <u>Electrical conductivity</u> higher than 10 S/m.
- <u>Thermionic emission</u> high enough to promote diffuse electrical current transfer to electrode surfaces, to prevent electrical arcing.
- <u>Good resistance to corrosion/erosion</u> of the high temperature gas flow, which can, moreover, contain vapours and fogs due to ionizing seed and molten slags.
- <u>High thermal conductivity and mechanical strength</u> to withstand thermal stresses arising from the thermal gradient (up to 300 K/mm) existing between the surface exposed to high temperature gas and channel cooling system and from operation transients.
- <u>Chemical and mechanical compatibility</u> with electrical current lead-out devices and joining interface materials.

Electrodes which have actually been applied in commercial MHD power plants or tested in experimental MHD facilities are usually classified according to the technical solution given to problems of channel cooling as hot, continously working at temperatures higher than 1900 K, semi-hot, working at 1400 to 1900 K, and cold, working below 1400 K.
Materials for hot and semi-hot electrodes are usually ceramics or ceramics containing composites (e.g., cermets), while materials for cold electrodes are usually high melting metals and alloys.
Ceramic materials based on Y and Ca-stabilized zirconia, alkaline and rare earth zirconates and rare earth oxides have been widely and succesfully experimented in hot electrodes.
Rare earth chromites, mixed chromites of Y, Mg, Al, Fe, La, etc., Mg-Fe spinels with Fe_3O_4 additions to increase electrical conductivity, pure and mixed carbides, borides and silicides (such as ZrB_2, SiC, $MoSi_2$) have been usefully applied in MHD power plants for semi-hot electrodes. Various kinds of composite materials containing ceramics (such as SiC/Ti, ZrO_2/Cr/W, etc.) have also been used for semi-hot electrodes.
Copper, noble metals such as Pt and Pd, special stainless and nickel steels, tungsten alloys with Cr, Cu, Ag, Cr and Ni/Cr based alloys have been actually used for cold electrodes.
Table 1 summarizes the main conducting materials successfully used in MHD power plants, with some remark on their real behaviour in the working conditions.

Channel insulators must exhibit the same thermal and mechanical characteristics of electrodes. In addition, however, they must fulfil the following main performances:

Table 1:   Electrode materials

Type		Cycle and remarks
HOT ELECTRODES (>1900 K):		
Ceramics:	based on Y and Ca-stabilized Zirconia, alkaline and rare earth zirconates, etc.	Gas and coal cycle; external collectors in Pt, W, Cu, refractory alloys, etc. Zirconia is a good ionic conductor, with good resistance to corrosion and erosion
	based on Zirconia and and Ceria mixtures	Gas and coal cycle; Ceria supplies electronic conductivity. Possible composite electrodes with ceramic electrical collectors
SEMI-HOT ELECTRODES (1400-1900 K):		
Ceramics:	based on La and Sr chromites and mixed chromites of Y and Mg, Y, Mg and Al, Al and Fe, etc.	Gas and coal cycle; high electrical conductivity (10^3 S/m). Decomposition may occur in hygroscopic $CrO_x + La_2O_3$. Low resistance to erosion
	based on spinels: $MgAl_2O_4$, $FeAl_2O_4$ mixed with Fe_3O_4	Gas and coal cycle, slag wetted. Compatible with Al_2O_3 electrical insulators
	Carbides, borides and silicides: $ZrB_2 + SiC$; SiC; $SiC + MgO$; $MoSi_2$	Coal cycle; (easily oxidized in gas cycle)
Cermets:	$SiC + Ti$; $ZrO_2 + Cr + W$; $LaCrO_3 + Cr$; $MgO + NiCr$; $ThO_2 + Cr$	Coal cycle; (metals are easily oxidized in gas cycle). Possible composite electrodes with ceramic insulators
COLD ELECTRODES (<1400 K):		
Pure metals:	Pt, Pd, Cu (T<775 K)	Gas and coal cycle
Alloys:	W-Ag and W-Cr alloys, stainless steels, Haynes alloy, Inconel, Hasteloy, etc.)	Coal cycle; (easily oxidized in gas cycle)

- **Electrical resistivity** higher than 10^4 $\Omega \cdot m$ in the working conditions, to provide proper electrical segmentation of the channel.
- **Dielectric strength** higher than 10^4 KV/m, to withstand the electrical fields due to Faraday and Hall voltages.

Insulating materials used in power MHD plants are all ceramic: MgO, Al_2O_3, Mg spinel, the so-called SIALONS, ceramics based on Si, Al N and O in various ratios, and oxygen-free ceramics, such as BN (highly resistant to thermal shocks), Si_3N_4, etc..

Table 2:   Insulating materials for MHD channels.

Type	Cycle and remarks
MgO	Gas and coal cycle; variable density, high resitivity ($10-10^5$ $\Omega \cdot m$ at 1975 K), high chemical stability, high mechanical strength
$MgAl_2O_4$	Gas and coal cycle; variable density
Al_2O_3	Gas and coal cycle; sedd interactions
Si_3N_4	Coal cycle; decomposes at high temperatures
BN	Coal cycle; high thermal shock resistance
SIALON ($Si_{6-z}Al_zN_{8-z}O_2$)	Coal cycle

Combustor materials must fulfil the same requirements of channel materials, as quoted above. Ceramic materials such as ZrO_2, MgO, Al_2O_3 are therefore used in these components.

In air pre-heaters, steam boilers, heat exchangers and other less critical MHD power plant components, various kinds of metals and ceramics have been successfully used. In the presence of corrosion/erosion problems due to molten slags and seed linings and coating technologies with ceramic materials may be applied.

As to manufacturing technologies, ceramics and composites are usually processed by means of advanced powder technologies involving milling, forming and sintering, or a combination of the latter as in hot pressing, in proper sintering atmospheres. In some cases, deposition methods on metal electrodes have been successfully applied, such as hot spraying, vapour deposition, etc.. Metals are usually processed by conventional metal working processes or, particularly for heat resistant alloys, powder metallurgy.

3. MHD TECHNOLOGIES DEVELOPMENTS

Channel materials deserve most efforts for future development of MHD technologies, due to the process temperature limitations they exert. But other problems exhibit an about equal importance in the future development of MHD technologies, such as slag rejection and seed recovery, which can affect the overall efficiency of MHD process, and the NO_x and SO_x environmental pollution control.

Moreover, full achievement of MHD retrofit energy conversion process can lead to an economic impact on European power generation which can be estimated as it follows:
- an installed power increase by more than 50 %, regardless of fired fuel;
- an overall energy conversion efficiency increase;
- an environment impact decrease, by a decrease of exhaust gases by 25 %, NO_x emissions by 80 %, SO_x emissions by 95 %, hot water emission by 25 %.

Finally, the MHD plants can lead to an economic re-use also of conventional electrical power plants by applying topping units, so that the conventional power plants could act as the bottoming unit. The overall efficiency of such a process, briefly named "retrofit", can thus be increased to 40-45 %, in comparison with the average efficiency of an existing electrical power plant of 32-35 % (as it is the case for fossil fuel fired plants which are operating since 15 years and more).

An immediate field of MHD development may be recognized in the construction of the first demonstrative "retrofit" MHD plant in Italy (about 250 MW_{th}). By this plan, the European industry will be able to grow up at a world level in this as well as in related innovative technological fields, such as high temperature materials and technologies, superconducting and cryogenic technologies, power electronics, industrial diagnostics, plasma physics and technologies. An export activity from Europe of MHD components may thus be reasonably envisaged. A marketing investigation identified at least 10 conventional energy conversion plants in Italy to be profitably retrofitted. In Europe, likewise, the number of such plants may be evaluated as about 50.

Industrial priorities may be therefore recognised chiefly in very high temperature materials (VHTM) development:
- controlled sintering of powdered materials (to control material microstructure and withstand mechanical stresses in

the channel and combustor operation conditions)
- controlled doping of electrode materials (to control
  electrical high temperature conduction)
- brazing and soldering techniques between different materials
  (to fasten electrodes and insulating elements to channel
  mainframe)
- coating technologies (chemical deposition, cathode sputtering,
  plasma sputtering, flame spraying, etc.) (to manufacture and
  repair in service channel elements)
- multilayer metal-ceramics technologies (to control thermal and
  electrical efficiency of electrode as well as fastening of
  electrodes to channel walls)
- testing facilities for service life testing and evaluation.

4.2. Materials Constraints in Petrochemical Plant

F.J. Vaes

Dow Chemicals B.V.

4530 AA Terneuzen, The Netherlands

INTRODUCTION

The range of interests in petrochemical processing is for pressure retaining components operating in the temperature range 900°C to 1200°C. High strength to weight ratio, so important for aircraft is here not essential. A material is required with adequate strength, good corrosion-oxidation resistance, a stable structure, easily machinable and weldable even after years of service, readily available in all shapes and sizes (plates, tubes, castfittings) and not too expensive.

The components

-Cracking coils
In this section of the plant, steam and naphtha are heated in tubes fabricated by centrifugal casting. Tubes about 3m long are welded together to make the required length, and connected by static cast bends or formed centricast tubes. Tube diameter is about 100mm and the thickness 10mm. A typical coil consists of 7 lengths of 10m and about four coils are built into a furnace. The castings are made of HK 40 or of better materials such as HP50 Nb (W). The skin temperature, in the radiation section of the furnace may reach 1100°C and above.

-Vessels or reactors
The shell is a cylinder typically 8 m high and 5 m in diameter with hemispherical top and bottom. Inside is the tube/tube-sheet heat exchanger, an array of tubes each containing a catalyst. Around the tubes is a heat-conducting fluid to control the temperature. Wrought alloys such as 304 H, 800 H, IN 617 are used, and the temperature may go up to 900°C.

-Heat exchangers

The energy recuperators are generally of the once through, tube/tube-sheet type. The shell outer diameter is 1m, the tube length 6m containing typically a thousand tubes. Generally the temperature conditions of the exchanger are lower than those of the reactors, but the temperature difference between the shell side and tube side may be higher. The tube/tube-sheet joints for high temperature application is no longer of the well known expanded tube type, but is made by welding, preferably of the internal bore type.

-Transfer lines

For smaller diameters, extruded tubes of austenitic materials are much used. For the larger diameters, tubes fabricated from austenitic sheets, or centrifugally cast tubes of HP 15 Nb are used, with static cast bends or formed centrifugally cast tubes of the same material. Diameters of about 0,5 m and lengths of above 10 m are not unusual.

Of course, in a plant, equipment with moving parts will be necessary such as pumps and valves. Instrumentation recording the processes will also be needed. All must operate at high temperatures, thus creating problems. The materials of plant construction must remain machinable and weldable after years of service not only for maintenance purposes. It is very common in chemical plant to make major modifications with the hope of achieving improved operation.

Mechanical properties requirements

A calculation is needed to determine a minimum thickness for a component. The simplest way is to use the design pressure and the creep strength for 10^5 hr life. at the design temperature of the material used. The shortcoming of this approach is that thermal stresses are ignored. Those superimposed stresses may cause rapid failure. The finite element method can include thermal stresses, but a complete calculation is expensive, and time consuming. Start up and shutdowns or partial breakdowns will also create high thermal stress transients that should be considered in the design. These calculations require values for the E-moduli and the instantaneous thermal expansion coefficient as a function of temperature of the materials. (Data which up to now seems to be missing for the well known alloys.)

Apart from calculation there are some common sense design rules that can be followed to avoid peak stresses. For example, welds are generally not considered in design calculations and their physical and mechanical properties are therefore of no direct interest except their creep strength. The well known alloy 800H with potential for high temperature application is such an example. Due to the hot cracking sensitivity of the alloy, the recommendation is to use the high nickel alloy 600 as filler metal for welding and for the higher temperature region the alloy 625 which has higher creep strength values. The difference in expansion coefficient is sufficient to explain why so many failures occurred with those weldments in the past.

The need for filler material of nearly equal expansion coefficient and with a creep strength equal or better than that of the base material, is a must for welding for high temperature operation. For dissimilar welds between two different alloys with different properties, the choice of intermediate thermal and creeps properties seems to be the best choice for the filler metal.

Accelerated creep tests

100,000hr (11.4 year) creep tests are not a practical proposition but the accelerated conventional creep testing by increasing the temperature or the load to reduce the time for rupture are generally not reliable, for chemical plant materials. The reason is that the austenitic materials commonly used depend for their strength on a range of carbides whose solubilities are very steeply temperature dependent in and just above the service temperature. A morphological study of the carbides of HK 40 from 950 to 1050°C in steps of 25°C shows how sensitive the precipitate is to temperature. The conclusion reached was that the testing temperature has to be as close as possible to the operation temperature. Certainly for remaining life evaluation.

The second parameter that can be increased is the load. But to achieve short life times at operation temperature relatively high loads have to be used. The creep failure mechanism may then be completely different. There remains one parameter that could be changed; the strain rate. This changes with temperature or load variation, but the idea now is to impose a strain rate. The strain rate occurring in the 100,000h creep strength is of the order of 10^{-9} to 10^{-10} sec^{-1}, while the minimum strain rate that can be imposed and measured with the present sophisticated tensile testing equipment is 10^{-8} sec^{-1}.
If now a test piece at temp. T_1 is pulled at a velocity of 10^{-8} sec^{-1}, the registered load will go up to a kind of saturation stress σ_1. At the same temperature T_1, different strain rates in the range 10^{-8} up to 10^{-5} are imposed, a log-log graph of the corresponding σ_1 to the $\dot{\varepsilon}$ values at T_1 can be plotted giving a straight line. This line for material A can now be compared to the measured line for material B at the same temperature, and a qualitative evaluation of both metals can be done.

This simple testing procedure, has proved to be of great value for checking different batches of an alloy; for finding the best matching electrode with respect to strength and to evaluate the strength of material after exposure in plant. The constant strain rate testing method as it is called nowadays, is still under development (1). This new typical high temperature mechanical property has been of great help in short time decision requirements, which are common in the industrial sector.

The casting industry is working hard to improve their products by use of proprietary additions to the melt before casting. That titanium is used as an deoxidant is admitted, but what they do more is not known. The creep values in the high range of the temperature scale are nearly doubled. Proving those treatments by chemical analysis cannot be done, therefore the constant strain rate method is used here with success, to ensure that the purchased material had the high creep values claimed. Strong material needs strong welds again with the CSR-method the weldments of a specially made electrode were shown to have the required high creep strength.

(1) University of Ghent, Laboratorium Soete

Future developments – Higher Temperatures.

Alloys in present use are already operating at their upper temperature limits of strength and temperature. Plant operations would, however welcome stronger materials with higher temperature capabilities, provided the new alloys were not much more expensive. The refractory metals such as Mo and W are stronger at higher temperatures but are not readily worked, are expensive and have poor corrosion resistance. Ceramics may become interesting, SiC has a good thermal conductivity, a vital property where heat transfer occurs. But the reliability of ceramic components has not yet reached an acceptable standard. Refractory insulation is already used to protect furnace walls and can be used in some reactors. Thermal barriers coatings of e.g. ZrO_2 may be plasma sprayed with success on parts of rotative symmetry which have to sustain higher temperatures than the bare metal could support. ZrO_2 is used also for its anti-sticking properties at high temperatures e.g. in those parts of a valve, which are not continually moving. Protection against corrosion or oxidation by coating may be more cost effective than choosing an alternative more expensive alloy for example a high cobalt alloy for protection against sulphur. Coatings, plasma sprayed or by chemical vapour deposit (or by slurry) can be used. Coatings are generally substrate dependent due to outward diffusion of alloy constituents and may also embrittle the substrate due to inward diffusion.

Vapour deposition is economically more suitable for the chemical industry than plasma spraying to give a protective layer. In chemical plant the catalytic effects of surfaces on the process can be important. Build up of coke layers in ethylene furnace coils is a big problem since it has to be burned out at intervals to keep the process going. The mechanism of this build up is not yet understood, surface catalysis does however play a part. Coating of the substrate may be a solution.

Research needs and priorities

The creep values used in codes are generally conservative. A better material quality assurance could raise those values. The constant strain rate method may be of great use.

Centrifugally cast materials are used more and more. A better understanding of the effect of the structure should be critically analysed. (equiaxed versus columnar). Also the cast to cast variation in the E-moduli values as function of the temperature is still not entirely explained.

Coatings may be helpful. Self healing properties of a coating are desirable.

Weldments are still not completely under control. The matching philosophy appeared very successful, but low cycle fatigue testing of weldments, particularly for dissimilar metal combinations are still necessary to prove the fitness of the choices.

4.3. High Temperature Materials Problems in Fusion

and Fission Power Generation

S.F. Pugh

Consultant Metallurgist

Abingdon, Oxon 14 1EG, UK

1. FISSION TECHNOLOGY

Even in established thermal reactors with pressurised water coolant temperatures of about 300°C, the need to demonstrate safe shut down during accident conditions, has involved studying materials behaviour at high temperatures. For example in a postulated loss of coolant accident, zircaloy clad behaviour up to 1000°C has to be known. In all reactor systems fuel behaviour during power ramps and coolant loss is an important factor in fuel and core design. Mechanisms of fission product movement in ceramic fuels up to well over 2000°C and escape of volatiles are relevant to normal operation and accident conditions. The vitrification of high level wastes can also involve high temperature processes.

In fast fission reactors, projected sodium outlet temperatures have been trimmed back steadily over the past twenty years so that containment material developed for other power plant applications can operate under standard conditions. Currently the 9-12 % chromium martensitic steels are being developed for application as in-core fuel element wrappers. By suitable control of composition and heat treatments they remain ductile after irradiation at temperatures above 350°C and retain strength up to 550°C. The inherent resistance to void formation is a big attraction. With such an efficient coolant as liquid sodium thermal shock can still be a problem, and the high degree of reliability required in the sodium to water heat exchangers calls for good design and careful control of fabrication, but again using well established materials. Experience here will be gained from operation of the large French fast reactor power station at Creys Malville. The European Electricity Authorities are to sponsor a two year design and safety study of a fast breeder demonstration power reactor followed by a three year detailed design project. The study commencing in 1989 will be undertaken jointly by European companies and if successful will lead to the building of 3 1.500 MWe commercial power plants. The need to replace 4 GWe per annum of PWRs by FBRs starting in 2010 is foreseen. Clearly the joint European R&D programme based on the 1984 agreement must continue to support the design effort. The big new nuclear power developments which seem likely to be limited severely by lack of suitable high temperature materials are described in the following sections.

Intermediate and High Temperature Gas Cooled Nuclear Fission Reactors

The A.G.R.s

In the British advanced gas cooled reactors, (AGR) only the fuel elements operate above 600°C i.e. at "high temperatures". The highest local cladding hot spot temperature is a little over 800°C, but the average is much lower e.g. about 650°C in the gas outlet region. The thin fully austenitic 20Cr25Ni steel suffers plastic strain cycling during reactor start up and shut down, hence considerable R&D has been necessary to develop high ductility and resistance to irradiation embrittlement. More recent studies have aimed at reduction in plastic strain amplitude, by hardening the steel with nitrogen to reduce collapse of the cladding under the coolant pressure. Other developments have sought to reduce carbon deposition on the outer surface of the fuel elements by controlled oxidation and by coating with materials which do not catalyse the carbon deposition reaction. Thick carbon layers impede heat transfer from the fuel element thus increasing the clad and fuel temperatures unduly.

Since AGRs are now in operation, the main future effort will involve large scale trials of improved fuel cladding with careful post-irradiation examination of spent fuel to assess progress. A gas cooled loop is being installed in one of the Harwell MTRs as part of the programme of fuel testing and coolant chemistry development. The intention is to test under extreme power cycling and upset conditions, in experiments where clad rupture is a distinct possibility.

High Temperature Gas Cooled Reactors, HTRs

Introduction

European involvement in the development of helium cooled

reactors has included the building of a reactor demonstration
unit at Winfrith Heath in the UK, funded and manned as a joint
OECD project, together with a shared programme of materials
testing. The large amount of data generated is still being
sifted and interpreted.

The first experimental high temperature reactor in the FRG –
the AVR at Jülich has operated safisfactorily for more than 20
years. The follow up 300 MWe THTR at Hamm–Uentrop was handed
over to the owners and operators on 1st June 1987, having
operated on trial since November 1985 during which more than one
billion kWH electricity was generated. The next stage planned is
for a 550 MW HTR offering the generation of electricity and/or
process steam at 530°C. The reactor is expected to compete
economically with the much larger PWRs and is suitable for siting
near industrial areas. The HTR is also the only system currently
extant which has any chance of supplying process heat without the
intermediate step of electricity generation.

High Temperature Reactors (HTR) Steam Cycle

There is no metal cladding in an HTR core, the fuel
particles or lumps are clad in graphite and are more uniformly
distributed through the graphite moderator than in AGRs. Since
graphite can survive at temperatures up to 2000°C, the core
temperature is limited only by corrosion of the graphite and loss
of fission products from the fuel. To allow the graphite core to
run at high temperatures while keeping corrosion to an acceptable
level, the chemically inert gas helium is used as coolant at a
high pressure with a core outlet gas temperature of 750°C. The
pressure vessel containing the core and the main pressure circuit
ducts can be thermally insulated from the hot gas and externally
cooled, they therefore do not present a high temperature
materials problem. The heat exchangers must however be exposed
to the coolant on one side and to high pressure water/steam on
the other. The initial aim has been to achieve conventional
steam parameters as in a modern fossil fuelled power station
namely superheated steam at > 500°C.

Compared with the environment of the fire box of coal fired
plant the requirement may seem very modest since in the HTR the
gas temperature is more uniform and the coolant composition well
established and constant, with absence of aggressive ash slagging
components and hot spots from flame impingement. There are
however some features of the nuclear boilers that can provide
novel problems that are difficult to solve.

1. Access to heat exchangers becomes difficult and expensive when
   they have become contaminated with radioactive materials. The
   heat exchanger becomes radio active from dust and corrosion

products carried through the reactor from the primary containment structure and the heat exchangers.

2. The coolant has a very low oxygen partial pressure such that protective oxide films do not form on materials exposed to it. Internal oxidation and carburisation can then occur.

3. Very high integrity of the heat exchangers is required to prevent water in leakage to the reactor since that would cause corrosion of the core material.

Ideally the plant should operate for the full life of the reactor installation without need for repair. To achieve that goal considerable development of components and prototype power producing systems must be done. The turbines pumping the primary coolant gas can be located in the cooler part of the circuit reducing creep rates and chemical attack.

The Helium Gas Turbine

The HTR is deemed capable in principle of raising the temperature of helium at high pressure to a level which would drive gas turbines to produce electricity efficiently. A big advantage is the elimination of the need to raise high pressure steam, and hence inleakage and maintenance problems of high temperature heat exchangers can be avoided. The need for efficient and reliable intercoolers would remain, but those materials problems are less difficult to solve. Turbine inlet temperatures of 900-950°C would allow the use of conventional gas turbine blades and discs but some development of coatings to prevent carbon and oxygen pick up from the non oxidising helium coolant would be necessary, since protective oxide films do not form.

Nuclear Process Heat from HTRs

A consensus has been reached that the direct use of the heat output from a nuclear reactor of optimum size is not viable for use in the chemical or steel industries because either the heat output is too large; or the temperature attainable is too low; or the need for reactor outages would be unacceptable or present severe economic penalties. Production of a secondary fuel might however be economic. Three such processes have been studied, (1) the reaction of water and methane to give hydrogen, (2) steam gasification of coal and (3) thermal decomposition of water. These are all endothermic reactions absorbing and storing chemically large amounts of energy.

The major technological problem is in developing suitable heat exchanger materials and possibly also to evolve novel heat

exchanger designs to use brittle refractory materials.
Transferring heat from one gas to another involves a large
temperature drop unless the area of heat exchanger surface is
very large and hence the rate of heat flow per unit area is low.
Large heat exchangers are however costly especially as high
quality expensive materials must be used. Some designs of plant
include an intermediate circuit and so impose two drops in gas
temperature together with consumption of more power to circulate
the coolant gas. The minimum temperature for steam reforming is
800°C but coal gasification needs significantly higher
temperatures. It is doubtful whether with reasonably sized heat
exchangers the temperature drop would be less than 100°C giving a
reactor gas outlet temperature of well over 1000°C with a first
wall heat exchanger at about 1000°C in some regions.

HTR High Temperature Materials Problems

Material behaviour in HTR systems will be a strong function
of the working environment, which may be divided into three
regions :

i)   the helium primary coolant;
ii) the helium intermediate coolant;
iii) the process gas.

In essence i) and ii) are very-high-purity helium and
although its nuclear and "working fluid" properties are
unaffected by the traces of impurities present, they dominate and
control its chemical properties.

In addition to residual air and other contaminants on
reactor start-up, these impurities may originate from in-leakage
of air and water from the atmosphere, cooling circuits or process
streams; from out-gassing of insulating materials and graphite;
and from reduction of metal oxides. Hydrogen may diffuse in from
waterside corrosion of boiler tubes, or from process streams.
Oil vapour may arise from lubricated bearings or diaphragms and
will pyrolyse in the core to give hydrogen and methane.
Maintenance and refuelling operations will introduce further
impurities to the circuits. In the primary circuit these
impurities react with the hot graphite to produce an atmosphere
in which hydrogen and carbon monoxide are the predominant
impurities, but with a significant level of methane. The
operational efficiency of the continuous purification plant must
be considered for each chemical species, in order to estimate the
equilibrium levels of impurities.

Since the primary circuit gas in a process heat system must
be well above 800°C and be compatible with the hot graphite of
the reactor core, it would carburise and possibly internally

oxidise all conventional heat exchanger materials. A protective
oxide film is not formed. A surface protective coating is
therefore required. Molybdenum which forms MoO_2 in reactor gas
might be suitable; otherwise refractories such as silicon,
carbide, graphite or boron carbide could be used. On the
secondary side the helium composition could be controlled to form
a normal protective film of oxide. Careful pressure balancing
and control of start-up temperature gradients, together with good
design, might eliminate the need for the primary heat exchanger
material to withstand high stresses at high temperatures thus
avoiding the need for creep strength of a high level.

Providing the intermediate heat exchanger gas is
continuously cleaned, the secondary heat exchanger should not
become contaminated with levels of radioactivity which cause
problems in maintenance, or contaminate the final product. The
secondary heat exchanger would however need to give more reliable
long term service than is common at present in conventional
plant. Even if both heat exchangers remain free from leaks there
remains the possibility that hydrogen could diffuse into the
reactor primary circuit and re emerge as the radioactive tritium
isotope. Clean up of the intermediate circuit gas would however
prevent the chemical products emerging from the plant from being
contaminated with tritium.

As with conventional heat exchangers fouling of the surface
could cause a deterioration in heat transfer and require
occasional clean-up.

The materials and engineering problems of providing nuclear
process heat at temperatures of 800°C and above are quite
formidable. The alternative of producing nuclear electricity for
supplying process heat or providing hydrogen by the electrolysis
of water is quite easy by comparison. It is not therefore
obvious that the pursuit of nuclear process heat at 800°C and
above will receive much international support over the next ten
years.

2. NUCLEAR FUSION TECHNOLOGY

The heat energy of the sun and many stars is obtained mainly
by fusion of nuclei of the light atoms to form heavier stable
nuclei. The main reaction is fusion of four atoms of hydrogen to
form helium. At very high temperatures collisions between ions
occasionally have sufficient energy for fusion to occur. The
rate of such fusion reactions increases with temperature, (giving
higher mean ionic energy), and with the density of the gas, since
smaller ionic distances will allow more collisions in unit time.
Achieving the right high temperature conditions for fusion in the
laboratory has proved very difficult owing to very high heat

losses and rapid decompression during the reaction. The very hot gas is highly ionised – it forms a plasma – so that the escape of ions can be retarded by a suitable magnetic field. For over 25 years the quest for controlled release of fusion energy has sought to achieve high enough density and temperature of a plasma for sufficient time for fusion of an adequate fraction of the nuclei. This has involved building bigger and yet bigger machines. The present generation machines including the Joint European Torus, JET at Culham, UK, should achieve the easiest nuclear fusion reaction namely the fusion of the two hydrogen isotopes tritium (T) and deuterium (D).

$$D + T \rightarrow He + n \quad \text{or,} \quad {}^{2}_{1}H + {}^{3}_{1}H \rightarrow {}^{4}_{2}He + {}^{1}_{0}n$$

For the post-JET era the hope is that Europe will collaborate with the US, Japan and Russia to design INTOR or will design and build a next European torus, NET, machine with the aim to produce a self sustaining fusion reaction, in which the alpha particle energy reabsorbed in the plasma can maintain a high enough temperature to allow further thermonuclear reactions to take place. The plasma will need to be refuelled with D and T while generated helium is exhausted during the burn.

If the INTOR or NET projects reach their objective a demonstration power reactor is planned to follow. Its objectives apart from generating fusion energy in long "burns" will include

1. a demonstration that more tritium can be formed and extracted from the blanket than is consumed in the burn, and
2. That electricity can be generated efficiently.

It will be necessary to continue to place emphasis on study of plasma physics since, unless a "burn" can be achieved there will be nothing for technological exploitation. It may seem premature to discuss a materials and engineering programme for fusion power generation. Experience with fission systems has, however, indicated that large efforts and long time scales are involved in such work, particularly with the high degrees of safety and reliability that must be demonstrated to licence for operation, plant containing large amounts of radio-active material.

The First Wall Problem

The most advanced system for approaching the controlled generation of fusion energy is the Tokamak, in which a hollow segmented ring shaped "vacuum" vessel containing hydrogen forms the secondary of a transformer. When a large transient electric current is passed through the primary windings the gas in the torus is ionised and passes a large current which raises the plasma temperature to approaching 10^5 K. Magnetic interactions also cause the plasma to be compressed into a ring of small cross section in the centre of the vessel. The plasma rapidly cools and decompresses after each input pulse of energy.

Other systems for producing high plasma temperatures and pressures have also been studied, of these only the reverse field pinch - which is also toroidal and the ICF (Inertial Confinement Fusion) systems are still considered. Other systems, e.g. mirror machines, have almost been abandoned world wide.

The primary materials difficulty, interaction of the plasma with the "first wall" i.e. the inner surface of the toroidal vessel presents a special problem. Protons, electrons, alpha particles and electromagnetic radiation, escaping from the plasma give up their energy entirely in the first few μm of the incident surface. Thus a thin layer of material suffers thermal shock, is eroded by sputtering and is implanted with hydrogen and helium. In addition the whole of the first wall is irradiated by 14.1 Mev. neutrons which give up some energy and cause nuclear transmutations and atomic displacements. Very considerable international effort is allocated to study of this problem.

Since the plasma containment vessel must maintain its structural integrity it is likely that in more advanced devices it will be protected from direct interaction with the plasma by coating or cladding. There are a few other components suffering even more severe exposure to the plasma. Fixed and movable limiters which are located closer to the plasma may have to withstand heat loads as high as several tens of MW/m² such that highly refractory materials of good thermal conductivity will be required. If additional heating by neutral beam injection is used then a local first wall armour may be required to give protection from neutral beam 'shine-through'. Mo and W alloys are currently favoured for these applications since tests have indicated that they are likely to survive longer (suffer lower rates of erosion).

In achieving a "burn" considerable advantage can be obtained from preheating the hydrogen isotopes in the torus and/or post heating to maintain temperature by injection of large amounts of RF energy through a window, either to induce ion-cyclotron resonance (20-100 MHz) or electron cyclotron resonance (100 Giga Hz). Keeping the window temperature below 400°C is difficult and may require liquid nitrogen cooling. Above 400°C radiation causes a large increase in RF energy absorbtion. Even below 400°C a twofold increase is likely and exacerbates the cooling problem. There is as yet an unexplained variation in susceptibility to radiation damage from one specimen of alumina to another. The use of fine grain material has extended the dose before cracking to $> 10^{26}$ n/m².

First wall shields

To survive high surface temperatures needs materials having high melting point and low vapour pressure such as those found in the first half of the first row of the periodic table, namely Be, B, C and compounds of B and C with strongly electropositive metals such as aluminium and titanium. The covalent compound of four valent silicon with carbon or nitrogen are also a well recognised refractories. The light elements are favoured also because they cause less cooling of the plasma if sputtered into it, than do the heavy elements, and are not strongly radioactivated.

Providing a satisfactory first wall structure is so difficult that the present trend is towards operation at "low" temperatures for next generation machines, i.e. 200°C or less for the bulk temperature. It is then possible to use an austenitic stainless steel such as 316L or copper tubes without undue problems from radiation swelling, creep, fatigue or loss of ductility. Most of the fusion energy would pass through to the blanket as fast neutrons where sufficiently high temperatures to drive an efficient heat engine would still be generated. The criteria for choice of first wall cladding would include : ease of application and repair, good thermal contact with the heat sink, low evaporation rate and absence of spallation when subject to flashes of energy and ions during a burn, low hydrogen isotope and helium retention and high re-emission rates, low rate of long-term build-up of damage and an acceptable degree of mechanical interaction with the substrate. Fine-grain graphites and thin, multilayer ceramics are being developed and tested.

The Breeder Blanket and Heat Engine

Most of the energy of D + T fusion reactions appears as 14.1 MeV neutrons, a very penetrating form of radiation which passes through the first wall and its shield with very little loss of energy. Therefore in a power reactor it will be necessary to provide behind the first wall about 1 metre of blanket to absorb the neutrons' energy and to carry it to a heat engine. The neutron itself will need to react with lithium to form tritium which can be used to refuel the reactor. To prevent reaction of tritium with further neutrons and avoid loss by decay it will be necessary to extract the tritium rapidly, efficiently and continuously from the blanket. The nuclear physics of tritium production is already established from its production in thermal reactors. The technological problems for a fusion system are however immense.

During the many collisions that the 14 MeV neutrons have with atoms of the blanket, coolant and pipe work not only will their energy be dissipated but they will also cause atomic displacements, and generate helium by spallation reactions, two well known forms of radiation damage.

For high thermodynamic efficiency of the power cycle the fusion heat should be extracted from the first wall/blanket at the highest permissible temperature. In practice the maximum source temperature will be governed by the durability of the blanket structural materials as a function of the operating temperature and also by the need to achieve a given service lifetime.

From the point of view of simplifying demands on materials, containing the lithium as a solid refractory compound in a can cooled externally by helium is attractive. The hot helium gas could then be circulated through external heat exchangers to raise steam. The difficulty of removing solid breeder capsules and replacing them in the blanket has, however, directed considerable effort to development and testing of microspheres and porous compacts to release tritium rapidly into a helium gas stream to be subsequently extracted outside the fusion machine.

In an alternative engineering design the lithium is contained in a liquid metal such as lead + 17 at % lithium MP = 235°C, or a low melting point salt mixture such as lithium and beryllium fluorides (FLIBE). The lithium bearing liquid can be circulated slowly and continuously through the neutron flux and then through a tritium extraction plant. The coolant circuit is quite separate; the coolant could be high pressure water, liquid sodium or high pressure helium gas circulated through the blanket region then through a heat exchanger.

In view of the long timescale required for design, for developing materials and for testing prototype breeder blanket and coolant circuits, even without the added complication of radiation damage, it is necessary to narrow down the bewildering range of breeder materials, coolants and structural materials.

However, even when a well established material has been chosen there will remain three major differences in exposure in the nuclear fusion system although, happily, not necessarily all occurring simultaneously in all components.

1. Presence of a high energy radiation field.
2. Exposure to an unusual chemical environment.
3. Occurrence of an excessive number of stress transients due to temperature and pressure variation of the coolants and including thermal shock.

There are also the important cross terms, i.e.

4. The behaviour of irradiated material at elevated temperatures under stress and during stress transients.
5. The influence of chemical environment on fracture mechanisms.

Items 2, 3 and 5 can be tackled using less expensive facilities than those required for handling irradiation or irradiated materials.

Tritium Containment and Efficient Recycling

Tritium has two very unfortunate properties. It is radioactive and it is an isotope of hydrogen – and hydrogen is well known to dissolve and be very mobile in very many materials, and the more so the more elevated the temperature. Tritium is therefore likely to leak out of any fusion system and constitute a radiological hazard if suitable and adequate steps are not taken to contain it. This "containment problem" is of considerable magnitude because of the large quantities of tritium in proposed reactor systems.

Tritium decays with a half-life of 12.3 yrs to give an 18.5 kev (max) β and inactive ^{3}He, equivalent to an activity of 10^7 Curies per Kgm of T. In typical reactor designs there will be a total T-inventory of 5-10 Kgms (5 to 10.10^7 Curies), a daily consumption of 0.3 Kgms, a production and extraction rate from the blanket at least equal to this, of course, and a daily plasma throughput of 6 kgms. (These figures are for a 1000 MW power plant with a 5% burn-up per charge). Radiological safety requirements for a fusion reactor have not yet been specified very firmly in terms of emission, but is is unlikely this will be allowed to exceed 10 Curies per day, which is about 10^{-6} to 10^{-7} of the inventory.

Expressed simply, this means that there will be shifted each day many Kgms of tritium around a very complex system, many parts of which are at high temperature and under intense particle irradiation, with the requirement to contain the tritium within the total system to better than 1 part in a million. It will be essential to demonstrate that this can be done before the construction and operation of a T-reactor can be licensed.

To achieve adequate containment requires choosing processes and materials and designing components to minimise the inventory and leakage rates. To estimate these quantities, we need quantitative data on tritium solubility constants (S), diffusion coefficients (D) and permeation constants (P) in presence of neutron irradiation or at least in irradiated material for all materials likely to be employed in a reactor system, with their dependence in many cases on temperature up to ~ 800°C, say, and on pressure down perhaps to 10^{-10}–10^{-12} Torr. Also temperature gradients may quite markedly increase or decrease rates of transport depending on the size and magnitude of Q, the "heat of transport". Thus Q is another quantity of interest, at least for materials likely to be in regions of steep temperature gradient.

Similar transport data are needed in considering the development of specific items of a reactor system - for example for estimating tritium extraction rates from tritium breeding materials and storage systems and in connection with such devices as membrane systems for purification of plasma exhaust and flushing box gases. In these cases we seek materials with high permeation rates are needed.

There is a considerable body of data on the solubility, diffusion and permeation rates of hydrogen in many materials, mostly metallic, but similar data for tritium itself are unfortunately very sparse indeed. However, from what is known, and taken with the rather more plentiful measurements that have been made with deuterium, it is clear that, at least at elevated temperatures, isotope effects are small. The heavy isotopes move more slowly but only by factors of, very roughly, $1/\sqrt{2}$ and $1/\sqrt{3}$ of that of hydrogen. Experiments have indicated that some oxide films on structural metals can prove effective barriers to hydrogen permeation. The possibility of leakage through joints and seals must not however be overlooked.

Nuclear Fusion-Irradiation by Fast Neutrons

All the components of a fusion reactor are irradiated by fast neutrons from the first wall shield to the external magnetic field coils, but here attention is given only to those materials which will be required to operate at a high temperature, i.e. ca. > 600°C. At that temperature in steels radiation creep gives way to conventional thermal creep, and displacement damage such as that which causes room temperature embrittlement of b.c.c. metals would be removed by annealing. The remaining damage for austenitic steels includes 1) dimensional changes due to void and bubble formation, and 2) loss of creep ductility due to helium generation by the $^{10}B(n,\alpha)$ reaction or high energy neutron-induced spallation reactions which also form helium. Void and bubble formation under high levels of fast neutron irradiation have been studied intensively on a world wide basis for over 20 years because of its relevance to fuel cladding and in-core component behaviour of fast fission reactors, while loss of creep ductility has been studied for 30 years but on a somewhat smaller scale being of most relevance to the cladding behaviour of AGR fuel. The higher neutron energies, 14.1 MeV, produced by fusion reactors gives a greatly increased helium yield compared with that of fission neutrons which have energies of only 2 MeV. In addition the fuel and core components of a fast reactor are small and can be replaced every year or so while the breeder sections of a commercial fusion reactor are very large, expensive and difficult to replace. They might be required to survive the full life of the reactor for economic power generation. That would involve irradiation exposures well beyond those achievable in the next several years on large specimens.

Coolants and Breeder Materials

Experience with PWR, BWR, Candu and SGHW fission power reactor systems provides much useful information relevant to use of pressurised light water or heavy water, or boiling water as a coolant and working fluid. For both ferritic and austenitic steels there should be adequate information on carry over of radioactivated corrosion products to the heat exchangers or generating plant. Presumeably with such coolants provision of nuclear superheat would not be economic. Gaseous coolants such as helium or carbon dioxide are economic only if pressurised and operating at above 500°C. The main problem in their use would therefore be in demonstrating the integrity of the structural pipe work operating at high stress, high temperature and in an intense field of fast neutrons. The same comments would apply to the use of liquid sodium as coolant but here the high stresses would arise from thermal shock, highly probable with a material of such good thermal conductivity, and also from high magnetohydrodynamic stresses when rapid changes in magnetic field occur.

The best solid breeder Li_2O will corrode steels severely in the presence of water vapour but not when dry. Whether dryness could be guaranteed at all times depends on success in preventing air in-leakage or leaks in the heat exchangers or tritium extraction plant. The lithium silicate, Li_2SiO_2, is more tolerant of water vapour. Of the liquid breeders, lithium and and lithium lead eutectic are the leading candidates. The emphasis of current compatibility work is on martensitic and austenitic steels with an increase in the number of studies of their "low activation" variants and vanadium base alloys to reduce the radwaste disposal problem.

The generation of large magnetohydrodynamic pressures when uncontrolled magnetic field transients occur may also be relevant to the choice of metallic breeder materials. The high pumping pressures in steady operation are not of such concern if the rate of flow is low. For metallic as well as other liquid breeders, materials information is required on mass transfer effects in thermal gradients, tritium solubility, and tritium escape characteristics into and through the walls of the circuit containment materials.

Breeder and Power Plant Stress Transients

A common failure sequence for a stressed engineering structure begins with the formation and propagation of a fatigue crack starting at a stress concentration, overlooked in design and/or at a flaw arising from fabrication. Each stress cycle causes an increment of crack growth until the remaining ligament can no longer support the steady load. Large power plant installations are designed and supplied against a "Table of Plant Transients" per year of operation and a "design life" of say 30 years, for normal and upset conditions. For more serious accident conditions, it is necessary to guarantee only safe shut down. In the first instance, the transients are expressed in terms of time, temperature, pressure cycles of the coolant which must by computation and stress analysis be converted to stress cycles in the structural materials and hence into fatigue crack growth rates from flaws of a postulated range of sizes. The transients from power plant include start up and shut down, leak testing, turbine roll tests, load following, and temperature ripple within the limits of the power input control. It remains to be proved whether any type of material could survive the conditions in a fusion breeder blanket with a pulsed mode of power generation in the plasma in addition to those transients common to other types of power plant. This is or certainly will be one of the most critical questions in fusion power generation. It is rendered more difficult by the inevitable radiation embrittlement of structural materials and possibly chemical embrittlement also.

Primary Candidate Alloys

A list of alloys which might be used for first wall and breeder structural applications has been drawn up on an International basis. A collaborative testing programme has also been initiated. In the following paragraphs are summarised the main pros and cons for each alloy type with a final brief mention of more speculative materials.

Ferritic and Martensitic Steels

Chromium-molybdenum steels are used to almost total exclusion of other alloys in fossil fuelled steam raising plant including the superheater sections. For fusion applications it will be desirable to avoid alloying additions and impurities which become transmitted to isotopes which produce hard gamma-rays and have long decay times. The low alloy steels have good resistance to void swelling and loss of creep ductility up to 550°C. The ferritic and martensitic steels have good thermal

conductivity and with chromium and molybdenum additions they have
adequate corrosion resistance in high purity oxygen free water.
Fatigue and "wet fatigue" properties are known for a few alloys
only. The effects of other environments are not known.
Fabrication is well established but welding needs strict control
of composition, welding conditions and post weld heat treatment.
The need for corrosion and compatibility studies in large thermal
gradient loops has already been well recognised.

There are two major problems with ferritic and martensitic
steels. First, they are ferromagnetic, and being situated within
the primary windings of the fusion machine they would suffer
large transient stresses during each power pulse. Secondly the
brittle-ductile transition temperature is raised by fast neutron
irradiation below about 350°C. Thus after irradiation the alloys
could be in a brittle state to well above room temperature.
These are both critical problems which could rule out ferritic
steels for fusion power systems.

Austenitic Steels and Nickel base Alloys

The types of austenitic stainless steels and nickel alloys
used in PWR primary circuits are noted for extremely high
toughness and corrosion resistance in pure water. Stress
corrosion cracking is a well known problem and can be guarded
against. These alloys have rather poor thermal conductivity
which might be a disadvantage for certain pulse length durations.
Up to 300°C radiation effects occur but none are critical.
Effects of helium bubble growth to cause dimensional changes is
not likely to be worse than in other materials. In the range
350°C to 600+°C fast neutron irradiation can cause severe void
swelling and current research aims to reduce this tendency by
careful composition control. Also in the range 550°C-700°C helium
implantation causes a loss in creep ductility and augmentation of
the rate of fatigue crack growth. Here again much work has been
done but not up to the radiation doses that would be relevant to
a fusion power production plant.

For a conservative design the use of austenitic steel up to
300°C has much to commend it, since it also avoids the need for
time dependent stress analysis. The structural materials in
fusion-reactor blankets will experience a much more hostile
environment than is experienced in any fossil-fuel-fired power
plant or fission reactor, giving rise to a correspondingly severe
thermal and mechanical environment. Therefore there is a lot to
be said for a conservative approach which avoids the high
temperature materials problems altogether.

Materials for Future Systems

If, as seems likely, the first wall of NET is run cold, i.e. cooled with non pressurised cold water, then commercial aluminium has a lot to commend it. Good thermal conductivity, good corrosion resistance, low neutron activation, short decay time, low cost and easy to fabricate. There may however be a need to use an unconventional baking system to outgas the components exposed to the plasma.

Commercial aluminium with small addition of nickel is compatible with water and retains useful strength up to about 150°C but suffers severe erosion at higher temperatures if the water flow rate is high. It does not store radiation damage but would swell if much helium is injected. In air, CO_2 and probably other gases it will survive unstressed to about 400°C but helium swelling could then be excessive.

Titanium is used in condensers of conventional steam plant for its excellent resistance to corrosion in rapidly flowing slightly chlorinated seawater at a little over ambient temperature. It is also used highly alloyed for compressor blades in jet engines at temperatures of up to about 550°C for its good mechanical properties and high strength to weight ratio. Some alloy development and testing would be required to find a titanium alloy suitable for the pressurised water tubing of a fusion reactor.

The temperatures in more recent fusion reactor designs have been reduced into a range where conventional materials can be used, and for which there is a far more extensive data base, albeit not all encouraging. Currently vanadium alloys are preferred for their low radioactivation, but lose strength at about 650°C. Vanadium is inherently ductile but it is very reactive chemically and can be embrittled by pick up of nitrogen and other interstitial elements, and by irradiation with fast neutrons. Vanandium alloys have not been used in power plant and it is very doubtful whether they could be suitable.

Apart from choice of materials new to power reactor technology, there is also the possibility of choosing new production routes, hopefully to give improved properties. There is much research currently in the production of refined structures by rapid solidification and production of mechanically dispersion strengthened material from hot pressed mixed powders. Also certain alloys are capable of long range ordering of their constituents giving rise to some interesting properties such as strength at high temperatures and resistance to radiation damage. There has so far, however, been little success in eliminating brittleness from these materials.

The amount of R&D required to bring any one of these materials and/or production routes to a stage where they could be manufactured and used with confidence in the blanket or first wall of a fusion reactor is quite enormous.

3. HIGH TEMPERATURE MATERIALS IN THE SHORT TERM FUTURE

Gas-Cooled Reactors

The development of gas-cooled reactors in Europe has, in the past, included very enterprising and successful joint projects such as the OECD DRAGON project. The interests of the member states have now, however, diverged considerably in this field and further major collaboration is not envisaged.

Sodium-Cooled Fast Breeder Reactors, FBRs

From the very beginning of the development of civil nuclear power in 1946 the need eventually to replace thermal reactors by fast breeder reactors was recognized. Uranium ores of workable enrichment are scarce and are used inefficiently in thermal reactors. Worldwide shortages and price increases were then expected after about 25 years. Since that time the rate of introduction of new thermal reactors has fallen far below the original planned levels, new uranium deposits have been found and technical problems combined with high capital costs have delayed the introduction of FBRs.

Yet the need to introduce FBRs, which extract about 50 times as much energy from a given amount of uranium, remains. To that end a joint European FBR research and development programme was adopted in 1984. Now, in 1989, the European utilities and nuclear design companies have agreed to initiate a design and safety study with a view to building FBR power stations in Europe. The critical problems remain : the need to avoid core melt-down, the need for sodium to water heat exchangers of guaranteed long-term freedom from leaks and the reduction of capital cost. There is still some interest in the low enrichment, large-core, gas-cooled fast reactor of low specific core power density and high-breeding gain. The need to provide high enrichment fuel is avoided and reprocessing would be less frequent. In addition, sodium coolant is not required and the sodium to steam heat exchanger problems are thus avoided. Interest in this system will lie dormant unless the problems with the sodium-cooled system prove to be insuperable.

In the HTM field an important current topic is the development of martensitic steels for application as fuel element wrappers. Maintaining suitable structures and low brittle-ductile transition temperatures in welds is a problem unless weld-free designs are used. Clearly a continuing FBR research and development programme is well justified in view of the future plans for this European industry.

Fusion Systems

There has been much research and much written about the first wall problem of fusion machines, but the flashes of particles and energy from the plasma will affect only a thin layer of material on a support which can be kept at low temperature and carries only trivial principle stresses. The situation is similar, therefore, to a refractories application in, say, a steel melting furnace, where relining and patching must be done during a routine shutdown. For machines of the present and next generation this remains the only high temperature materials problem and a comprehensive research programme is already under way worldwide.

Looking far ahead to the prototype power producing station, i.e. three generations ahead, the high temperature materials problems that must be solved before the prototype can be build can be summarized as follows :

1. Production of structural materials which, after the full reactor lifetime exposure to neutrons, will decay in 100 years to a radiation level suitable for shallow burial. This may involve developing vanadium alloys or steels with different alloying elements and levels of some impurities two orders of magnitude less than the purest commercial alloys in present use, such as high conductivity copper or nickel base gas-turbine alloys.

2. Development of coolant/breeder circuit containment materials and design to guarantee that tritium losses are below 10^{-6} of the inventory with a high efficiency of extraction and recycling.

3. Demonstration that the chosen materials will be totally reliable with respect to conventional power plant conditions, i.e. high temperature, high stress, corrosion and oxidation, table of stress transients such as those found in a PWR, PLUS the stress transients imposed by a pulsed power output and transient magnetic fields.

4. Conditions as in 3 in material having suffered neutron radiation damage a few orders of magnitude greater than the pressure containment circuit materials of a PWR or a fast fission reactor.

5. Active remote maintenance.

It is difficult, in the light of these statements, to recommend a particular materials programme. These critical requirements involving high temperature materials need serious consideration now, even though the full extent of the conditions of exposure will not appear in the next two generations of fusion machines). They would all be avoided if the power output were not in the form of fast neutrons.

Requirements 1 and 2 might be relaxed or circumvented depending on public or regulatory attitudes.

Requirement 3 would be avoided if the output were not pulsed, i.e. if it were constant to $\pm$ 2 % over periods of a few months.

Requirement 4 remains. Only the fuel elements and core internals of a fast reactor are irradiated to anywhere near the same levels, but they are either readily replaced or are not pressure retaining barriers. In addition, the neutron energies are lower, so that rates of helium and hydrogen production are less. The zirconium alloy pressure tubes of SGHW and Candu do form part of the primary pressure retaining circuit and are traversed by fast neutrons passing from the fuel to the moderator. The levels of displacement damage and helium generation are, however, comparatively low, the tubes are of simple shape and are replaceable.

Since the main thrust of the present and near future fusion programme continues to be in plasma physics, the important message from materials experts should be to encourage the physicists to abandon the road to D + T $\rightarrow$ He + n. There are other nuclear reactions which do not involve neutrons. Even so, if pulsed power output is unavoidable, then a fluid heat sink to level out the thermal fluctuations will be required.

4.4. Engines

4.4.1. Aero Gas Turbines

S. Newsam

Rolls-Royce plc

Derby, United Kingdom

1. INTRODUCTION

Aero gas turbines have been a reality for almost half a century, and the major objectives in their design continue to be increased performance, increased thrust to weight ratio, improved fuel efficiency and minimum direct operating costs (DOC). In order to satisfy these demanding requirements continued improvements in both aero-dynamics and materials are required.

The three major materials related drivers are :

1) Reduction in material and fabrication cost. This is required to reduce both engine acquisition costs and life cycle maintenance costs.
2) Lower weight materials to reduce the overall weight of the aircraft and thus improve efficiency.
3) Improvement in physical, thermal and mechanical properties. This will permit higher pressure ratios in the compressor and higher operating temperatures in the turbine, resulting in an overall high cycle efficiency, engine performance and improved specific fuel consumption (SFC).

Early gas turbines of the 1940's and 1950's were constructed primarily of the aluminium and steel alloys of the day (Fig. 1), and had a thrust to weight ratio of less than three, a compression ratio of 4:1 and a turbine entry temperature (TET) of 800°C. Demand for increased engine performance led to the development of high temperature nickel base superalloys and low density titanium alloys, and resulted in the engines of today having thrust to weight ratios of 23:1 and TET 1400°C.

Predicted trends in jet engine material usage

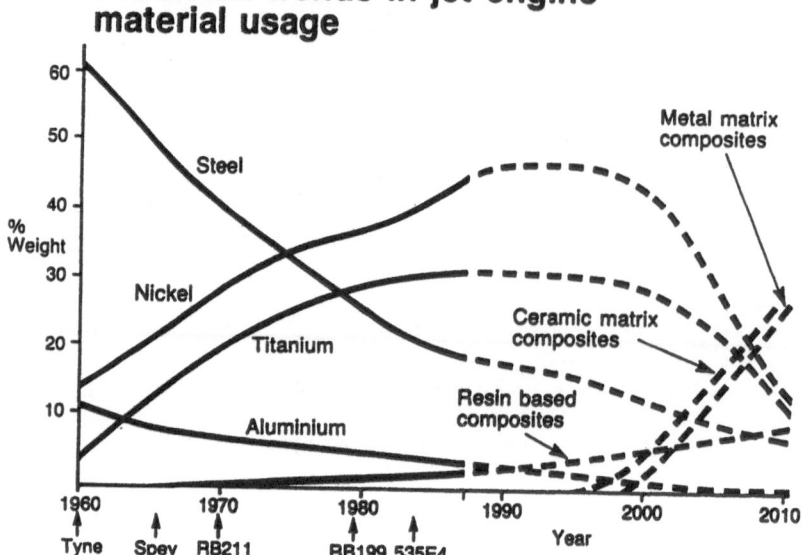

Fig. 1

Additionally SFC has decreased dramatically from 1.2 lb/hr/lb to 0.56 lb/hr/lb as thrust has increased more than ten fold from 5,300 lb to 58,000 lb.

Current engine designs are still based on the traditional metallurgical approach, and continued development of aluminium, titanium and nickel based alloys should result in some limited engine performance improvements. The demanding requirements of higher temperature capability, strength, stiffness and reliability for the engines of the next century, however, cannot be met with such an approach. It is currently estimated that 50% of the capability required for these engines will have to come from new, revolutionary developments in materials and the predicted trend is one away from homogeneous, isotropic metallic materials, to a range of anisotropic non-homogeneous composite systems (Figs. 1 and 2).

This chapter briefly outlines todays engineering requirements and the current material scene, highlights the future requirements which cannot be met by todays materials and discusses the new areas of materials technology which must be developed if future gas turbine engineering needs are to be realised.

Potential applications for composites in future military engines

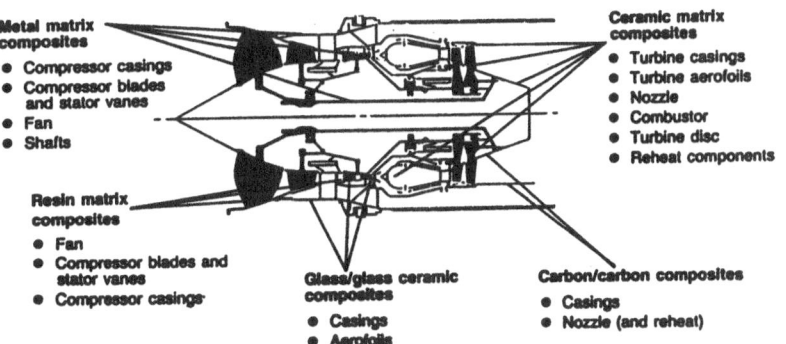

Metal matrix composites
- Compressor casings
- Compressor blades and stator vanes
- Fan
- Shafts

Ceramic matrix composites
- Turbine casings
- Turbine aerofoils
- Nozzle
- Combustor
- Turbine disc
- Reheat components

Resin matrix composites
- Fan
- Compressor blades and stator vanes
- Compressor casings

Glass/glass ceramic composites
- Casings
- Aerofoils

Carbon/carbon composites
- Casings
- Nozzle (and reheat)

Fig. 2

1.1. Military Engines

Current fighter engines are powered by low bypass turbofan engines at speeds up to Mach 2 and by turbojet engines with afterburners at speeds above Mach 3.

The direction to be pursued for the next generation of military engines will be towards higher performance and increased fuel efficiency. This will necessitate a move towards higher fan pressure ratios, higher overall pressure ratios and higher TETs. The objective is to achieve a thrust to weight ratio engine of at least 15:1 (compared to todays value of 7:1), rising to 20:1 by the year 2010. In order to achieve this goal, stoichiometric burning will have to be introduced resulting in TETs approaching 2100°C. An additional target is lower direct operating cost (DOC) via the use of fewer engine components.

1.2. Civil Engines

For civil engines the prime requirement is minimum DOC. The main factors influencing DOC are purchase cost, maintenance costs, component life and fuel consumption, with the latter accounting for about 50 % DOC in todays turbofan engines. Future civil engines will require similar advancements in materials technology for the engine core as required for military engines

in order to permit the use of higher overall pressure ratios and
TETs which should confer a further 5 % reduction in fuel
consumption compared with engines currently under development.

The major advance will however come from the development of
aircraft powered with even higher bypass ratio engines such as
the turboprop (alternatively called the advanced prop fan or
unducted fan), Fig. 3, for aircraft speeds up to Mach 0.7, and
the larger diameter ducted fan to Mach 0.8-0.9. Such engines

Contra-rotating geared propfan

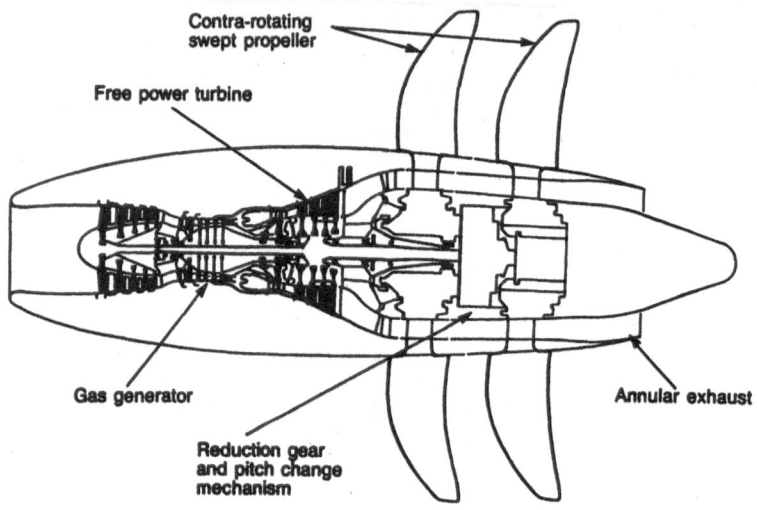

Fig. 3

offer the potential of a further 10-20 % improvement in fuel
consumption depending on the configuration chosen, resulting in a
total potential improvement of between 15 and 25 % when added to
the advances in core engine technology.

It should however be noted that the improvement in fuel
efficiency is offset by higher purchase and maintenance costs
associated with the larger engine size. The case for the new
engine types currently rests on fuel prices about doubling.

1.3. Small Engines

Small gas turbine engines include helicopter, turbopropeller
and missile engines (not discussed here). In these designs, the
requirement for improved SFC and lower weight has to be carefully
balanced against the requirement for low cost and the

manufacturing difficulties associated with the small size. These
requirements lead to turbomachinery operating at very high
rotational speeds (up to 50,000 rpm), requiring the use of high
strength materials in order to minimise the number of components
and resulting cost. In order to improve future efficiency,
increased TETs will be required. Theoretically, this can be
achieved either through the increased use of HP compressor
cooling air (to cool existing metallic components), or via the
use of materials having a higher temperature capability i.e.
ceramic type materials. Due to the manufacturing problems of
introducing fine cooling channels into the very small turbine
blades (typically 25-35 mm) combined with the limited HP
compressor cooling air available in small engines, the attainment
of increased efficiency in the future is only seen to be feasible
via the incorporation of ceramic components. The incorporation
of ceramics into the turbine is estimated to decrease SFC and
increase power output by 6 % and 16 % respectively.

The most serious environmental problems for helicopter
engines are erosion and corrosion. Under normal operating
conditions, dust, salt spray and other debris may be ingested
with the intake air producing severe erosion damage, sometimes
exacerbated by corrosion. Intake particle separators alleviate
the problem but introduce additional weight and some loss in
aerodynamic performance.

1.4. Industrial Gas Turbines

The prime requirement for industrial gas turbines is that of
long life up to 100,000h., maximum efficiency and reliability,
and minimum cost. The potentially more aggressive environments
of industrial units combined with the frequent use of cheaper,
more easily obtainable high sulphur containing fuels,
necessitates the use of inherently more corrosion/erosion
resistant alloys and additional surface protective coatings.
Corrosion resistant superalloys e.g. IN738 and IN792 tend to have
lower strength than the non corrosion resistant varieties. This
necessitates lower rotational speeds and/or reduced TETs when
compared to the aero antecedents from which they are derived.
Corrosion resistance tends to be more problematic than erosion,
as filter units can be incorporated to filter out airborn
contaminants without detracting from overall engine performance.
In general industrial gas turbine usage is continuous and thermal
cycling is not a problem.

Target engine lifetimes are difficult to predict as this
will vary from engine to engine depending on customer usage e.g.
lifetimes of 100,000 hrs are possible using low power ratings and
natural gas, whilst 5-10,000 hrs is more typical for engines run
continuously at maximum power on high sulphur liquid fuel.

Future developments will in general follow proven aeroengine technology, with emphasis on increased corrosion resistant alloys and coatings.

2. ENGINE COMPONENT REQUIREMENTS

In order to define the future objectives for materials operating at higher temperatures in more detail the general engineering requirements will now be considered on a component basis.

2.1. Compressor Blades and Discs

Compressor blades are subject to high rotational speeds, consequently exerting a high centrifugal load on themselves, and on the disc which carries them. The prime requirement for both blade and disc application has therefore been that of high strength and minimum density.

The emergence of titanium as a compressor material in the mid 1950's heralded a new era and enabled enormous strides to be made in aerodynamic, component, cycle and propulsive efficiency. Titanium has exceptional properties of high specific strength, low expansion, corrosion resistance, workability and weldability. Its specific strength, with a density just over half that of steel, is superior to most other structural metals and it is this characteristic, maintained to high temperatures which has resulted in its rapid growth in its use in the compressor stage over the last 35 years. The recent development of the highly alloyed systems, coupled with progress in compressor mechanical design, has enabled titanium alloys to be extended to the rear of the compressor, now delivering a pressure ratio of 36:1 and discharge temperatures in excess of 600°C at peak conditions. Titanium is however a highly reactive material and titanium fires can start when molten titanium comes into contact, with high pressure air under local rub conditions. For this reason titanium stators are generally not used between titanium rotor blades.

The future engineering trend is towards increased pressure ratios and fewer compressor stages. The resulting high rotational speeds and smaller number of parts give rise to increased stress levels and temperatures at each stage. High absolute strength and strain energy absorption are required for blades for impact resistance, particularly for the first stage of the high pressure (HP) compressor, and high absolute stiffness, for aerodynamic efficiency. In addition, high specific strength will be required to minimise centrifugal loads as well as satisfy the general requirement for minimum weight. These improvements must be obtained without deleterious effects on creep and erosion resistance, high cycle and low cycle fatigue properties and vibration characteristics. In addition to the above requirements reduced thermal expansion is required to reduce the thermal stress in the disc bore. Furthermore, improved stiffness is required to resist resonance, carry centrifugal blade loads and transmit torque to drive.

Initial design studies have indicated a 50 % improvement in the specific strength of todays best titanium alloys maintained up to 800°C will be required for HP compressor blades (Fig. 4).

Titanium alloys — Ultimate tensile strength (UTS) v temperature

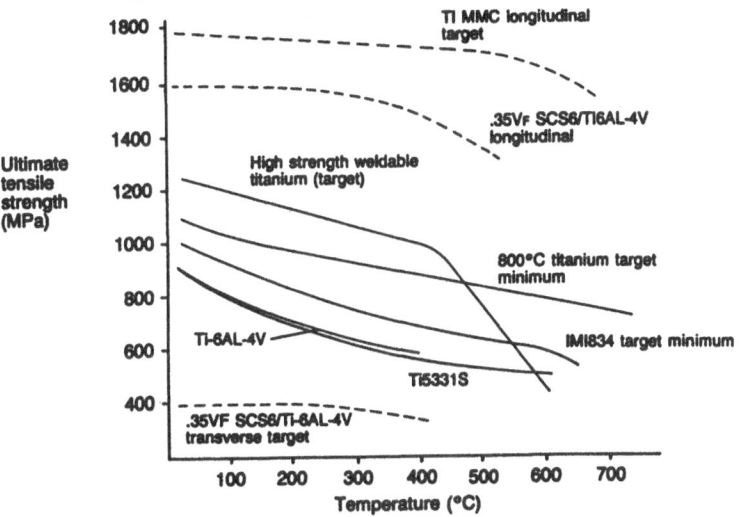

Fig. 4

It is unlikely that this target will be achieved by conventional alloy development or by thermomechanical processing, and in the short term hopes are set on toughened intermetallics (aluminides) and precipitation hardening using rapid solidification techniques. Titanium aluminides, having high temperatures strength, low density and good oxidation resistance to 900°C offer high rewards provided their major disadvantages of reduced ductility and low strength below about 600°C can be overcome. In addition coatings to protect against oxidation, hot salt corrosion and fire risks must be developed.

In the longer term, however, the high stage loadings required by minimum component, high pressure ratio compressors for the year 2010 will be best met by high strength/stiffness, continuously reinforced resin based composites for temperatures

up to 300°C and titanium metal matrix composites and glass/glass
ceramic composites for higher temperatures, possibly to 800°C
(Fig. 5).

Absolute and specific tensile modulus of resin, metal and glass composites

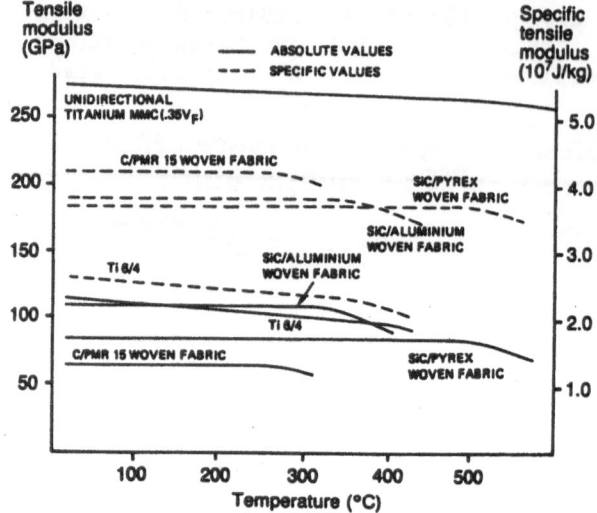

Fig. 5

Temperature limitations and low confidence levels in resin
based composite have, to date, limited their use to non
structural components. Improvements in temperature capability,
advances in behavioural understanding, design rules, predictive
techniques, damage tolerance and impact behaviour will be
required before these materials can be utilised in compressor
aerofoil applications.

Glass and glass ceramic composites using Nicalon SiC fibre
have been investigated extensively over the past 5 years,
primarily in the USA. These materials have a temperature
capability of up to about 1000°C and offer a weight saving of up
to 50 % over conventional titanium alloys. In addition these
composites have the potential to alleviate the problems
associated with titanium fires. Further research into these
systems is required in two basic areas, material behavioural
understanding and fabrication techniques. Conditions governing
fibre/matrix interaction and associated failure modes require
characterising, whilst additional fabrication techniques
(currently limited to the slurry impregnation and hot pressing
route), need to be developed.

A range of metal matrix composites (MMCs) are currently under development. Broadly they can be classified into two categories. The first are composites reinforced by particulates or whiskers. These primarily use aluminium alloys as the matrix although other systems are also being explored. The second group use continuous fibres, mostly SiC, although boron and carbon fibres have been used. The reinforced metals in this case being aluminium, titanium and steel.

For compressor blade applications, the primary requirements of improved specific strength and temperature capability, makes continuous fibre reinforced titanium the only real contender. Continuously reinforced aluminium may find a small applications niche in the IP compressor, however this niche is small and is also being fought for not only by titanium MMCs, but also glass/glass ceramic composites and advanced high temperature resin systems.

The two major problems in the area of continuously reinforced titanium composites are the lack of a suitable fibre and adequate fabrication technology. The current fabrication technology, as previously mentioned, is extremely restrictive and new and innovative approaches perhaps 3D orientated are required. Reinforcement options are currently limited to monofilaments having large minimum bend radii and restricted fabrication routes, or to textile type fibres having inadequate chemical stability, mechanical or thermal properties. A further area of concern in metal matrix composites is the degradation of mechanical and/or thermal properties after thermal cycling. Fibre and matrix coefficients of expansion must be matched, and thermodynamic stability must be controlled if optimum fibre/matrix interface characteristics are to be achieved during fabrication and maintained throughout the component lifetime. In civil applications this could be up to 20,000 hours.

Continuously reinforced MMCs and in particular

Potential free world market for advanced materials (£M)

	1990	2000	2010
C/PMR15	42	255	255
Ti MMC	-	270	1040
Ceramic composites	-	270	1855
Carbon/carbon	-	45	45

Fig. 6

titanium MMCs are one of the most important emerging materials
for future engine use. The potential free world market for the
latter in gas turbine components has recently been estimated at
M£ 270 in the year 2000 rising to M£ 1040 in 2010 (Fig. 6).
Considerable effort is however required to make this market a
reality.

In all cases of fibre reinforcement, major problems of lack
of design technology, fibre availability, predictive behavioural
models, flexible fabrication routes, composite joining techniques
and test methodology (including NDE) will all have to be
overcome. Extensive research effort is clearly required in all
these areas.

2.2. Combustors

The major requirement for civil combustors is that of
extended component lifetime, typically 10-15,000 hrs. Rugged,
long life, cost effective combustors will be required for future
small helicopter engines, whilst improved efficiency via
increased operating temperatures together with long component
life will be essential to the future of advanced military
engines. In order to achieve the long term military goal of a
thrust to weight ratio engine of 20:1 these increased operating
temperatures may be up to 2100°C.

Todays combustor technology utilises welded sheet (nickel)
superalloy fabrication. In order to achieve higher cycle
temperatures it is necessary to burn more fuel which in turn
requires more primary air for combustion. This results in a
reduction of the amount of air remaining to cool the combustor
walls and for reducing the temperature of the combustion gases to
a level which the turbine can accept. Although transpiration and
film cooling technology have improved cooling effectiveness, the
benefits are always offset to some extent by the higher "cooling"
air temperatures arising from increased compression ratios. Thus
increased temperature capability has, and will continue to be a
major requirement for combustor materials, particularly for
military engines.

Thermal barrier coatings (TBCs) have been used in the RB211
combustion chambers for over a decade, doubling combustor life
through component temperature reduction of 50°C or more. Oxide
dispersion strengthened alloys have the potential for operating
at high temperatures. However the unweldable nature of these
materials will necessitate the development of novel fabrication
routes and the poor thermal fatigue characteristics of these
materials will have to be improved. Tile constructions are a
possibility, however the inevitable increased weight of such
designs may not be acceptable in military engines where thrust to

weight ratio is of prime importance. Future military engines, operating under stoichiometric burning conditions will experience combustor temperatures in excess of 2000°C. In addition to oxidation resistance and thermal shock, vibration loading and corrosion resistance will be essential.

The solution to these problems will inevitably involve both design and material innovation. In the near term, the insulating qualities of non-structural ceramics (monolithic and/or composite) should be exploited, probably in tile constructions, whilst fibrous ceramic pressure vessels designed to utilise both the thermal and mechanical advantages of ceramics are targeted for long term applications. Whilst the potential use of these materials is generally accepted, technology development is in its early stages and ceramic combustors are unlikely to become a reality in the next five years. In fact metal fabrications are unlikely to be superceded until material temperature capabilities rise beyond about 1400°C.

Cost reductions are required for small engine applications. These are likely to be achieved through innovative automated manufacturing routes such as metal plasma spraying onto cheap discardable mandrels. Significant effort will be required in this area both to develop the techniques and ensure consistency and repeatability of material properties.

2.3. Turbine Blades and Vanes

Turbine blades operate under the most arduous conditions encountered in the engine. They are subject to large tensile stresses which fall off from root to tip and a temperature profile which increases in the other direction. Blade materials must also satisfy demanding specifications for resistance to creep, low and high cycle fatigue, thermal gradient stresses, hot corrosion and impact damage. In addition, they should ideally have a low density to reduce stress levels both in the blade itself and in the turbine disc and a cost effective processing route.

The importance of TET on engine efficiency is illustrated in Fig. 7 and explains the never ending search for materials with improved temperature capability. The ability to operate turbine blades at temperatures above their melting point has been due to both materials developments and advanced blade cooling concepts. This blade cooling, however has a severe efficiency penalty with up to 20 % of compressor delivery air typically being required in todays large civil engines.

The future trend towards increased engine performance and efficiency will result in higher tip speeds, higher pressure

Effect of cycle pressure and temperature on thermal efficiency

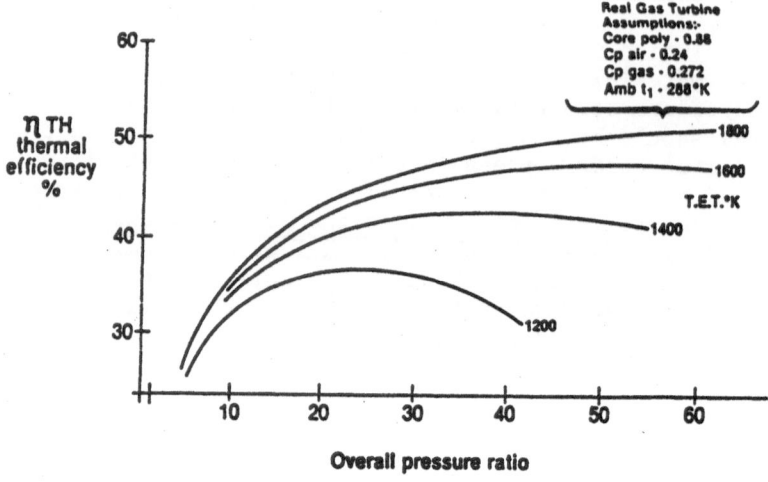

Fig. 7

ratios, higher operating temperatures and increased component life. This will require materials with higher temperature capability and improved specific strength and reliability.

Advances in nickel base alloys together with directional solidification (DS) and single crystal (SX) technology have resulted in temperature capabilities rising ∼300°C over the past 40 years. Further single crystal alloy development is possible, however as current systems are already operating at about 80 % of their melting points, their potential is extremely limited. Oxide Dispersion Alloys (ODS) are stable to higher temperatures than DS or SX alloys. They are however, generally anisotropic, unweldable and have poor thermal fatigue characteristics and limited strength. TBCs are already utilised on turbine stator blade platform reducing metal temperatures by about 50°C and avoiding the need for film cooling. Further developments should permit the utilisation of this technology on aerofoil sections.

Future military engines are targeted to have TETs in excess of 2000°C. If conventional alloys were to be used, they would have to be cooled by over 1000°C. Theoretically this would be possible but it would result in a minimum unacceptable 9 % SFC penalty and 16 % – 20 % reduction in thrust. The amount of HP compressor cooling air required to do this would be > 10 % which could not practically be accepted by the nozzle guide vanes (NGVs) and HP turbine blades. Some degree of cooling is possible, and almost certainly inevitable, but this is unlikely to exceed about 3 % (equivalent to 500-600°C reduction in material temperature). Increasing TET by additional conventional cooling is therefore not a viable option, although some short term gains are predicted from advancements in manufacturing technology which will allow the design of improved cooling systems such as spar/shell or wafer type constructions.

Major improvements in TET will only come from the reduction or elimination of cooling air, with materials operating at temperatures, initially of 1400°C rising to 2100°C by the year 2010. This will lead to the desired improved SFC, specific thrust and reduced core size.

The above requirements can only be satisfied with ceramic type materials. Of the many possible monolithic ceramics silicon nitride (Si_3N_4) and silicon carbide (SiC) have the greatest potential for application in gas turbine engines in components operating at temperatures up to about 1600°C. Materials for use above 1660°C require further definition. These materials are stronger than nickel superalloys above 1000°C, have superior creep strength and oxidation resistance, and are potentially cheaper. In addition their density is less than half that of the superalloys (typically 3.2 g/cm³ compared to 7.9 g/cm³). The Achilles heel of these materials, however is their intrinsic brittleness, flaw sensitivity and consequent lack of reliability.

Before being accepted in the gas turbine these materials must demonstrate a reliability in operation at least as good as the metal components they replace. This can be achieved in one of three ways. The first approach is to learn to live with the brittleness (low K_{1c}) and develop a fundamental understanding of the micromechanics of failure (i.e. flaws and their relation to strength). In this way statistical methods, non destructive evaluation or proof test methodologies can be used to specify design parameters such as strength or component life.

The second approach is to identify the source of strength degrading flaws and then develop improved processing methods to totally eliminate these strength limiting defects. As the critical defect size for ceramics is < 100 μm i.e. some two orders of magnitude less than that for metallic materials at typical operating stresses, this solution is not as easy as it may first appear.

Both of these approaches however rely heavily on the design stress being kept, at all times, below the failure stress. If for any reason the failure stress is exceeded, catastrophic failure will ensue. In meeting current application requirements a basic change in failure mechanism from this flaw sensitive brittle, catastrophic failure to a more forgiving non catastrophic failure is seen to be essential for aero engine operation and desirable for industrial use. This can only be achieved by the third approach; to design ceramic microstructures with improved resistance to fracture, and hence some defect tolerance.

Fibre reinforced ceramic matrix composites have this potential and have consequently recently received a great deal of attention for use in high temperature structural applications. The reason for this interest lying in the assumption that strong ceramic fibres can prevent catastrophic brittle failure in ceramics by providing various energy dissipation processes during a crack advance, and thus provide some defect tolerance. This approach is especially important for applications such as combustors, turbine blades and reheat/exhaust components, where the hostile environment can introduce strength degrading defects, thus negating all the efforts to ensure reliability by identifying or eliminating pre-existing flaws.

Although the feasibility of the fibre reinforced ceramics approach has been demonstrated, however a large amount of effort is required before ceramic composite components become a reality. The major barriers to be overcome with this newly emerging technology are the lack of reinforcing fibres with adequate thermomechanical properties, fibre coating technology, fabrication techniques, behavioural models and as with all composite technology, a complete lack of scientifically based behaviour and design technology.

The availability of suitable fibres is crucial, with the critical requirements being that of high strength and stiffness, low density, small diameter and most importantly thermal stability, both during fabrication and under component operating conditions. It is unlikely that ceramics fibres will be able to provide adequate long term reinforcement at temperatures greater than 1200°C and most are expected to be limited to below 1000°C. Susceptibility to decomposition, oxidation, microstructural instability, accelerated grain growth and creep have been identified as the major obstacles to be overcome. More refractory fibres are coming onto the market and it is not yet clear what impact these will have. However, even higher temperature capability reinforcements will be required if the goal of stoichiometric burning and no, or minimal, cooling is to be achieved.

Fibre coatings are likely to be required for two reasons. Firstly to limit fibre matrix chemical interaction and secondly to act as mechanically weak boundaries for toughening. The long term compatibility of these coatings with both fibres and matrices will also be important in maintaining as-fabricated composite properties.

Fabrication techniques for short fibre/whisker composites have to date been limited almost entirely to hot pressing, whilst various liquid or gaseous techniques e.g. chemical vapour infiltration, reaction bonding, sol-gel and polymer infiltration have been employed for continuously reinforced systems. These techniques, applied to real engine components, are generally at an early stage of development and considerable work will be required to ensure that adequate reliability, reproducibility and surface finish can be achieved in components of complex geometry.

2.4. Turbine Discs

Turbine discs require high tensile strength, to prevent bursting in the event of an overspeed, coupled with good low cycle fatigue life to withstand frequent change in stress accompanying each cycle, defect tolerance and long life.

The future engineering trend towards improved engine performance and efficiency will lead to higher rim speeds, pressure ratios, operating temperatures and improved fatigue life. For military engines this will necessitate the development of materials with increased strengths (up to 1400 MPa for a material density = 7.9 g/cm^3), increased temperature (ca. 800°C) capability and reduced weight.

The likely candidates for such an arduous task are nickel aluminides, nickel metal matrix composites and in the longer term reinforced ceramics. All three areas present challenging problems. As regards nickel metal matrix composites, the availability of suitably refractory fibres, combined with the thermal/chemical stability of the fibre/matrix combination and resistance to thermal cycling and foreign object damage are likely to be the major problems. Limited ductility, adequate strength, metallurgical and dimensional stability, together with adequate environmental surface protection will cause concern with the intermetallics (nickel aluminides), whilst fibre availability, fibre matrix control, adequate strength, life and integrity requirements will seriously hinder the introduction of reinforced ceramics into turbine discs for military applications. The major challenge for civil engine application will be to develop materials capable of operating for long lives, typically 24,000 cycles at temperatures > 700°C. The critical feature for civil engine application will therefore be the optimisation of an alloy system for fatigue resistance. As it is the presence of small scale inclusions (typically 100-200 µm) which currently limits the maximum stress at which the stronger highly alloyed powder products can operate, future effort in this area should be directed towards developing superclean segregation free processing routes.

2.5. Turbine Structure, Reheat and Exhaust Components

As combustion temperatures increase, the temperature requirement of all turbine gas path components will also increase. The development of turbine structures and advanced reheat and exhaust systems requires stiff, lightweight, oxidation resistant materials capable of operating uncooled at temperatures above 1050°C i.e. above the limit of current metal technology, possible up to 2000°C. For turbine casings as with compressor casings, dimensional control (to maintain minimum blade tip clearance) under both varying temperature and load is the most critical requirement. For reheat and exhaust components such as flameholders, jet pipe liners and nozzle petals improved efficiency via weight savings is of primary importance. Likely materials for these components applications are reinforced ceramics (as previously discussed) and carbon/carbon. The latter are a generic class of composites having high specific strength and stiffness, low density (1.7-1.9 g/cm^3) and in an inert atmosphere the ability to operate at temperatures in excess of 2000°C. In air temperatures above 400°C however, oxidation resistant coatings will be required. Simple overlay coatings tend to have higher coefficients of thermal expansion than the substrate carbon/ carbon, and typical engine thermal cycling leads to extensive cracking of the coating. This results in oxygen ingress through the coating, and serious fibre and matrix attack. Mechanical property degradation is severe and failure can be catastrophic. Two approaches have been taken to overcome the problem. The first is via the introduction of viscous glasses which seal the cracks by having appropriate chemical and viscosity characteristics under the prevailing temperature and chemical environment. The second is via the introduction of particulate oxidation inhibitors distributed throughout the matrix and/or fibres themselves, and the open porosity within the carbon matrix. All these features combine to form a very complex, dynamic and interactive oxidation protection system, which tends to work only as an integrated system. Significant research has been devoted to oxidation resistant carbon/carbon composites, mainly in the USA and results are looking encouraging.

However, reliability, reproducibility and scale up to larger more realistic sizes, together with complex component geometries and coating uniformity remain major challenges. In addition to oxidation resistance, low interlaminar shear strength, high processing cost (due to lengthy processing cycles) and a number of component related problems (e.g. erosion, attachments, moisture susceptibility, contact load damage, etc.) must be solved before these material systems become a reality and not just a fantasy material in the eyes of engine designers. The use of carbon/carbon in engine components thus requires a fundamental understanding and control of a large number of underlying variables.

Design studies have indicated a possible 20-50 % weight saving if nickel base alloys were replaced with carbon/carbon in jet pipes and reheat systems. The potential of carbon/carbon (Fig. 2), despite the daunting list of technical problems, is such that it cannot easily be ignored.

3. SUMMARY

Aero gas turbines have been a reality for almost half a century. During that time the use of light alloys and steels has declined dramatically, whilst that of the low density titanium and high temperature nickel superalloys has risen in a similar fashion to the extent that by 1990 these two materials will account for more than 65 % of the engine's weight.

The stage has now been reached, however, where the future engineering requirements of high temperature, strength and stiffness cannot be met by these traditional metallic materials alone and the use of a wide range of composite materials, (resin, metal, glass and ceramic matrix) will be required. Most of these advanced material systems are still in the development stages and need to be properly defined, understood and developed during the next decade. Critical issues for metal, glass and ceramic matrix composites are the lack of reinforcing fibres with adequate thermal stability, inadequate fibre coating technology and control of fibre matrix interface, and as with all composite technology a complete lack of scientifically based material behavioural understanding, innovative designs and predictive component performance techniques.

Resin, metal (in particular titanium) and glass/glass ceramic components have significant potential in the compressor whilst combustor, turbine, reheat and exhaust components are candidates for defect tolerant ceramics and carbon/carbon materials, the obvious limitation with the latter being poor oxidation resistance. Fig. 6 is an estimate of the free world market for engine components in advanced resin, metal, ceramic and carbon/carbon composites and together with Fig. 2 highlights the extent of the applications niche awaiting the introduction of these materials.

The future trend in titanium alloys will be towards

toughened intermetallics, RSP technology and fibre reinforcement, whilst development of superalloys will for discs move towards 'super clean' melting routes and for aerofoils novel manufacturing technology that allows improved cooling systems. This will be augmented by thermal barrier coatings.

Aero engine materials are now moving out of the traditional world of alloy development by relatively simple compositional control, into a more complex environment, where materials are "tailored" or "engineered" to meet specific applications. In moving away from basically isotropic metallic materials to ceramics and anisotropic composites, the current metallic design and manufacturing technologies which have been developed and universally accepted to date will no longer apply. Radical changes in the way engine components are designed and manufactured will inevitably be required. Material and component design, development, fabrication and manufacture are all interdependent and if the long term engineering goals identified are to be achieved, and composites used with confidence, these technologies must be developed in conjunction with the new materials.

The move towards composites will introduce new modes of material behaviour necessitating a change from the earlier empirical approach to material development/component lifing to one based on an understanding of the physical phenomenon involved. This approach will inevitably require sophisticated modelling techniques, innovative component design, improved process monitoring and material inspection techniques. If success is to be achieved, increased and redirected effort will be required and closer cooperation between the engine producer and supplier in all areas of research, development and production will be essential.

4.4.2. Marine Gas Turbines

J.F.G. Condé

Materials Consultant

Broadstone, Dorset, U.K.

1. INTRODUCTION

Marine gas turbines are derived from custom built industrial engines or from aero gas turbines. Conversion to marine use involves modification to design and changes in materials to render the engine more suited to its marine role and able to operate effectively in an aggressive marine environment. Marine gas turbines may be used for a variety of purposes including propulsion engines in ships (principally warships), hovercraft and hydrofoils, pumping duties for gas or oil in bulk carriers, power generation and firefighting duties in ships and similar duties on offshore oil and gas exploration and production platforms.

The characteristics of gas turbines which make them attractive for marine use include high power-to-weight ratio, compactness, quick start-up and fast response to fluctuating power demands. Other advantages include simplicity of operation, suitability for automatic or remote control, and maintenance by replacement on a complete engine or modular basis. In ships, the complete engine may be extracted via the air intake or exhaust gas uptake thus avoiding maintenance under cramped conditions in the ship's engine room. An engine can be replaced in about twenty four hours, if necessary at sea, thus enhancing ship availability and reducing costs. Frequently the use of the marinised aero engine is favoured due to smaller size compared with the custom built industrial engine. Engines may vary in size from 100 h.p. (74.6 Kw) to 30,000 h.p; (22,380 Kw) or greater. They are suitable of operation on a wide range of distillate fuels or gas and have the further advantage that the

power output can normally be uprated without significant changes
in engine size. In general, marinised aero derived gas turbines
tend to be somewhat down-rated compared with the original aero
version of the same engine and hence operate at lower turbine
entry temperatures. Equally conversion of aero type engines to
industrial or marine use lags behind developments in the aero gas
turbine field. Accordingly first stage blade temperatures may be
at least 200°C lower than in state-of-the-art aero gas turbines.
When employed for marine propulsion, particularly in warships,
gas turbines may spend a high proportion of life operating under
part-load conditions and only less than 5% of life at full power.
Thus even engines designed for first stage blade temperatures in
the range 850° to 950°C may operate mainly in the region around
700° to 800°C. In addition many engine types are designed with
first stage turbine blade cooling which inevitably results in
some areas adjacent to the blade root or platform operating in a
similar temperature range, even at full power. This low
temperature region on cooled blades extends further up the mid-
chord of the blade as power is reduced.

The high temperature components of marine gas turbines may
be subject to hot corrosion or sulphidation due to air and fuel
borne seasalt contamination together with sulphur from the fuel.
Diesel fuel or gas oil may contain 0.3 to 1.0% or more sulphur
compared with aviation kerosene with about 0.05% sulphur. 1%
sulphur in the fuel produces about 300 ppm oxides of sulphur in
the combustion gases. In addition the high energy-density
combustion systems normally associated with marinised aero gas
turbines may under fuel rich conditions lead to carbon formation
due to incomplete combustion or pyrolysis of fuel. This
particulate carbon may exacerbate corrosion due to chemical/
electrochemical effects or physical erosion. Where low sulphur
gas is employed as the fuel hot corrosion problems may not occur.
The need to employ a wide range of fuel bunkering facilities
for refuelling of ships worldwide may result in fuels containing
significant vanadium content (up to 10 ppm) being employed in
ship propulsion turbines (e.g. furnace fuel oil, FFO). The
synergistic effects of sulphur, seasalt, vanadium, carbon, etc
may result in hot corrosion damage being enhanced significantly.

Seasalt contamination of the gas turbine can result from
air-borne salt aerosol and also from seawater entrained with the
fuel (from use of seawater displaced fuel tanks or seawater
ballasting). The use of intake air filtration (separators and
knitted mesh filters) can reduce intake salt levels to less than
0.005 to 0.01 ppm and coalescence filters plus centrifuges can
reduce the salt content of fuel to undetectable levels. However,
the compressor section of the gas turbine can act as a further
salt filter, dry salt aerosol passing the intake filter either
eddy impacting onto the compressor blades or wet aerosol drying

out on warm surfaces. Unless removed by cold crank washing of the compressor periodically such salt deposits may break off and re-entrain in the air flow and reach the hot end of the engine.

2. HOT CORROSION

There is substantial agreement that hot corrosion of alloys and coatings is a two stage process (1) and involves an initiation stage or incubation period and a propagation stage. It is also accepted that there are two principal types of hot corrosion known as Type II or the so-called Low Temperature Hot Corrosion (650° to 750°C) and Type I or High Temperature Corrosion (above about 750° to 800°C). The morphologies of the two types are different. Type II corrosion shows localised pitting or nodular attack with a laminar scale and no evidence of discrete sulphides in the metallic substrate but a sulphur enriched region at the scale metal interface. There is normally no significant depletion of major alloying elements (e.g. chromium or cobalt) ahead of the corrosion front. Type I corrosion occurs in a more uniform, broad-front manner with characteristic discrete sulphide phases present in the substrate metal and depletion of alloying elements, such as chromium, ahead of the corrosion front. Both forms of corrosion are considered to occur only in the presence of condensed salt phases arising from contaminants present in the combustion product gas.

The low temperature form at around 700°C is generally agreed to be a form of gas induced acidic fluxing related to the presence of SO_3 and formation of aggressive, low melting point salt eutectics consisting of Na_2SO_4 - $CoSO_4$, Na_2SO_4 - $NiSO_4$ or other similar mixed sulphates.

The high temperature form of corrosion is considered to be produced by acidic or basic fluxing in aggressive salt melts, the acidic form being related to the presence of acidic oxides such as vanadium pentoxide (V_2O_5) derived from the fuel or MoO_3 and WO_3 derived from oxidation of these elements present in the alloy. Basic fluxing is related to the formation of O^{2-} ions which produce basicity leading to solution of the protective oxides (e.g. Al_2O_3 as aluminate).

Detailed discussion of hot corrosion mechanisms and part processes can be found elsewhere (1,3) together with consideration of the influence of chlorides, mechanical and thermal effects on scale integrity, temperature, contaminant quantity, salt chemistry effects, fuel sulphur content, fuel/air ratio effects, heat flux etc. A number of studies have also identified the possibility of inhibiting hot corrosion effects by addition of certain elements such as chromium which produce amphoteric oxides (e.g. Cr_2O_3) or control deposit melting point or friability. In combustion rig studies zinc was found to produce erratic results, varying from alloy to alloy. However, Canadian studies arising from adventitious contamination of fuel with zinc from inorganic zinc fuel tank coatings suggest that with at least two alloys (S816 and AMS5385) zinc may be effective. Such inhibitors may be added as fuel soluble organic compounds (e.g. napthenate) or by other techniques employing oxides.

In general hot corrosion damage is normally confined to blades and nozzle guide vanes (NGV's) and it is unusual for combustion chambers, burners etc to suffer significant damage. The most severe damage is usually found on first stage H.P. turbine blades. This can be explained on the basis of the higher temperature of stators (NGV's) (e.g. 1050°C) and the nature of the alloys from which they are made. Also although NGV's may sustain higher rates of particulate contaminant deposition the residence times may be shorter due to evaporation or greater fluidity leading to re-entrainment.

3. ENVIRONMENTAL EFFECTS ON MECHANICAL PROPERTIES

Laboratory and combustion rig experiments to investigate the behaviour of materials under stress in genuine or simulated corrosive combustion environments have suggested that salt contaminants may enhance creep rates and reduce rupture ductility (1). Sodium sulphate appears to have little effect but sodium chloride has a significant effect and even greater reductions have been found in the presence of chloride/sulphate mixtures or seasalt. Such life reductions are greater than can be accounted for by the loss of load bearing section produced by corrosion. However, there is no clear evidence from engines in marine service that these effects are significant under practical operating conditions.

Fatigue studies carried out under COST 50 Round III demonstrated that fatigue properties may be adversely affected by a corrosive environment and that the fatigue limit can be significantly reduced (1). This is not surprising in view of the nature of certain types of hot corrosion damage which may create stress raisers and initiate cracking. The practical significance of these observations has not been established in engine running.

4. MATERIALS

The development and evolution of the aero gas turbine engine provided the initial and continuing incentive for the development of nickel and cobalt base superalloys. Materials for turbine blades and vanes were optimised to meet the stress, temperature and environmental condition appropriate to the operating

conditions. In general terms the operating temperature of vanes
is about 100°C higher than that of first stage blades and the
stresses in the latter about two to three time greater (2). Both
are exposed to a highly oxidising combustion gas environment.
Normally nickel base alloys are employed for blades and cobalt
base alloys for vanes.

The continuing demand for more efficient aero engines
promoted development of stronger, more creep-resistant nickel
base alloys. This trend resulted in the nickel base alloys
falling broadly into two groups in terms of their oxidation
behaviour (Table 1). Those containing more than about 15%
chromium and less than about 5% aluminium form mainly chromia
(Cr_2O_3) scales and alloys with lower chromium and more than 5%
aluminium form mainly aluminia (Al_2O_3) scales. Certain alloys
are intermediate in behaviour (e.g. Nimonic 105). Cobalt base
alloys do not normally contain aluminium and are chromia formers.
In general Al_2O_3 scales are more resistant than Cr_2O_3 scales to
oxidation at high temperatures in high velocity gases and are
less prone to loss by vaporization. In addition Al_2O_3 thickens
more slowly than Cr_2O_3 (2).

The high aluminium, low chromium nickel base alloys were
quickly found to show poor hot corrosion resistance under marine
gas turbine conditions and about 1970, alloy development for
marine engines diverged from the route followed for high
performance aero gas turbines. The need for enhanced hot
corrosion resistance led to the evolution of a series of higher
chromium alloys of intermediate creep strength. The cobalt
alloys with their higher chromium levels were found to be more
compatible with the marine conditions and so similar major alloy
developments were not required and in certain marine hovercraft
engines vacuum cast cobalt alloy blades were employed
successfully.

When hot corrosion was first encountered in marine engines
the diffusion aluminide coatings which were employed successfully
in aero engines were applied to vanes and blades. No significant
benefits were found with cobalt base alloys but aluminide
coatings on first stage nickel alloy turbine blades provided a
life extension of about 1000 hours. Certain of the improved
coatings developed for aero engines also gave enhanced protection
in marine engines but from about 1975 onwards improved coatings
were developed specifically for use in marine engines.

The most comprehensive information on the hot corrosion
behaviour of alloys and coatings has come from high velocity
combustion rig exposure tests supported by so-called "rainbow"
engine testing of coatings. The cost of operating test bed
engines under realistic conditions together with the time

TABLE 1 (After Ref 1)

EXAMPLES OF SUPERALLOYS USED FOR BLADE AND VANE APPLICATIONS

Alloy (WT.%)	Ni	Co	Cr	Al	Ti	Mo	W	Ta	Nb	C	Others
IN100*	bal	15	10*	5.5	4.7	3				0.18	1V-0.015B-0.6Zr
B1900*	bal	10	8*	6	1	6	0.1	4.3	0.1	0.1	0.015B-0.08Zr
Rene 77	bal	18.5	15	4.25	3.5	5.2				0.15	0.05B-1Fe
Rene 80	bal	9.5	14	3	5	4				0.17	0.015B-0.03Zr
Mar.M.246*	bal	10	9*	5.5	1.5	2.5	10	1.5		0.15	0.015B-0.05Zr
Pd 21*	bal		5.75	5.9		2	10.5		1.5	0.10	0.02B-0.12Zr
INCO 713*	bal		12.5	6.1	0.8	4.2			2	0.12	0.012B-0.10Zr
INCO 738	bal	8.5	16	3.4	3.4	1.75	2.6	1.75	0.9	0.17	0.01B-0.15Zr
IN6201	bal	20	20	2.5	3.6	0.5	1.3	1.5	1.0	0.03	0.8B-0.05Zr
INCO 939		19	22.5	1.9	3.7		2	1.4	1	0.15	0.01B-0.1Zr
Nimonic 105		20	15	4.7	1.2	5				0.08	0.005B
Nimonic 90		18	19.5	1.4	2.4					0.06	1.5Fe
Mar.M.509	10	bal	21.5		0.2		7	3.5		0.6	0.01B-0.5Zr-1Fe
HS 31	10		20		1	2.6	10.7			0.15	
FSX 414	10.5		29.5				7			0.25	0.012B-2Fe-1Mn-1Si
X 40	10		25				7.5			0.5	1.5Fe-0.5Mn-0.5Si
X 45	10		22.5				7			0.25	1Mn-0.01B-2Fe

Al_2O_3-former (IN100* through INCO 939)

Cr_2O_3-former (Nimonic 105 through X 45)

* Alloys with poor hot corrosion resistance

consuming nature of such evaluation procedures makes the use of high velocity combustion rig testing the only viable approach for initial assessment and selection of alloys and coating systems. In practice few rigs have proved capable of simulating the types of corrosion and corrosion morphologies encountered in marine gas turbines. A detailed discussion of the reason for this is beyond the scope of this appraisal but it is related to combustion chemistry and gas velocity (hence residence time), contaminant flux and deposit formation. Most combustion rigs operate at only slightly above ambient pressure, very few high pressure/high velocity rigs having ever been constructed due to cost and complexity. Fortunately low pressure, high velocity, rigs have enabled close simulation of corrosion mechanisms and morphologies in marine gas turbines although corrosion rates may be somewhat accelerated compared with engine experience. High pressure, high velocity, rig studies which closely simulate the fluid flow effects in gas turbines have provided some important pointers to the significance of complete combustion in minimising hot corrosion (4,5) as shown in Fig. 1.

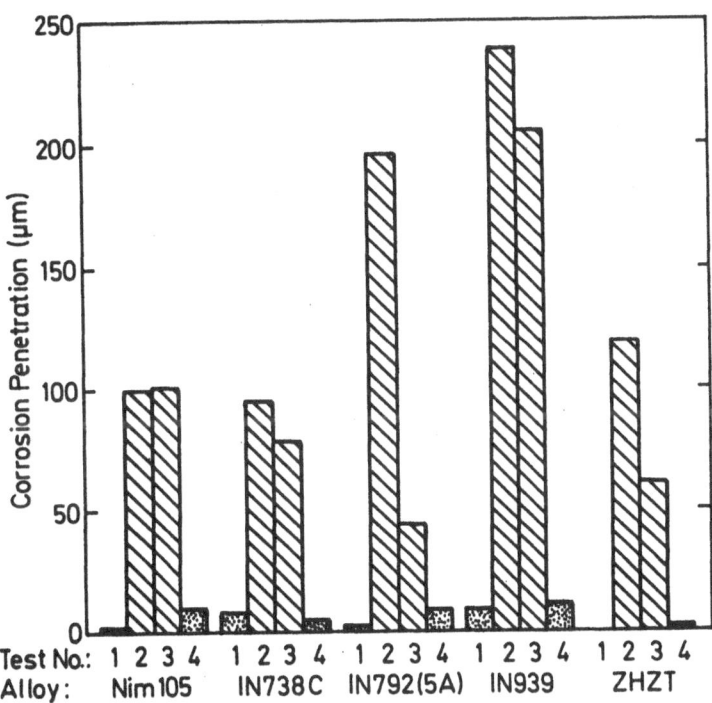

Fig. 1   Results from High Pressure Rig when Using Air-Rich
         Primary Zone Conditions (Tests 1 & 4) and Fuel-Rich
         Conditions (Tests 2 & 3) Ref. 5.

A selection of typical superalloys is shown in Table 1 and an indication of the significant effect of chromium level in Fig. 2. The dependence of hot corrosion on both time and

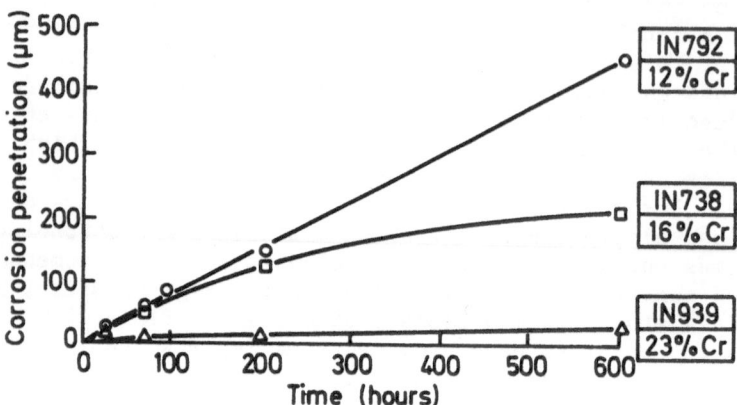

Fig. 2 Effect of Chromium Content on the Corrosion Resistance of Alloys at 750°C

temperature is illustrated in Fig. 3. Corrosion kinetics appear faster at around 700°C although with increasing exposure time break-away corrosion appears to take place at higher temperatures. Hence low operating temperatures around 700°C are not effective in preventing corrosion and in general have the reverse effect. A generally similar situation exists with coatings and in most instances rapid failure ensues at low temperatures in the range 650° to 750°C.

None of the early diffusion aluminide coatings gave substantial improvement in corrosion protection. Subsequently in the mid 1970's a range of overlay coatings were developed, applied principally by vapour deposition (PVD) ion plating, sputtering, arc plasma thermal spray deposition and vacuum or low pressure plasma deposition. In general, overlay coatings were 125 μm thick compared with 40 to 50 μm for the diffusion aluminide coatings. The early overlay coatings were of the M–Cr–Al–Y type where M was Co or Ni and in some instances (Ni + Co). In general these coatings corroded at least as fast as the substrate alloys (Fig. 4), and demonstrated the same pattern of behaviour with respect to Al and Cr content as found for the substrate alloys.

Improved platinum modified aluminide coatings which gave significantly better performance in aero gas turbines especially for low flying aircraft in maritime environments, also performed well under marine conditions at 1000°C, but showed little

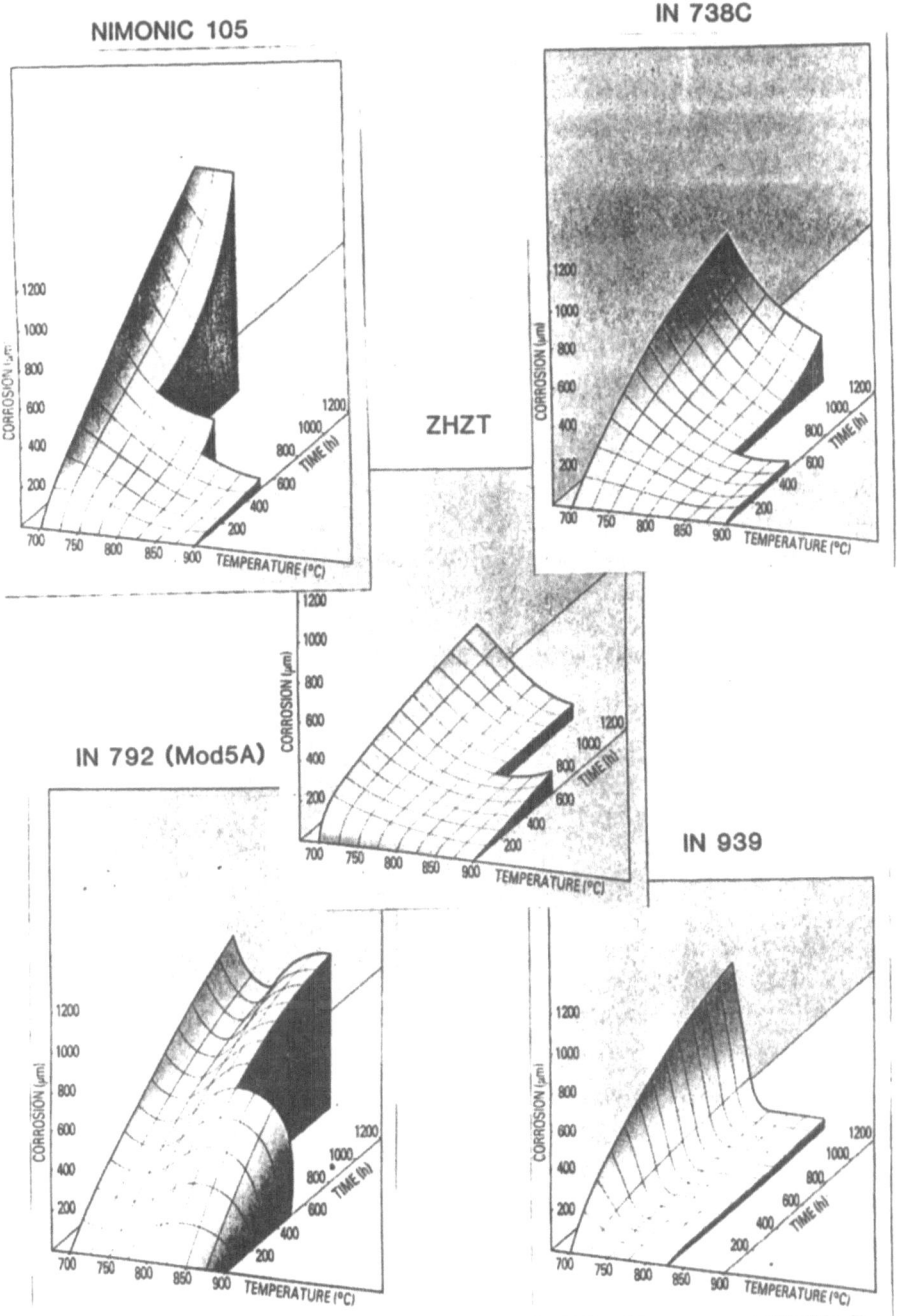

Figure 3. Corrosion—Time—Temperature Profiles for Five Superalloys.

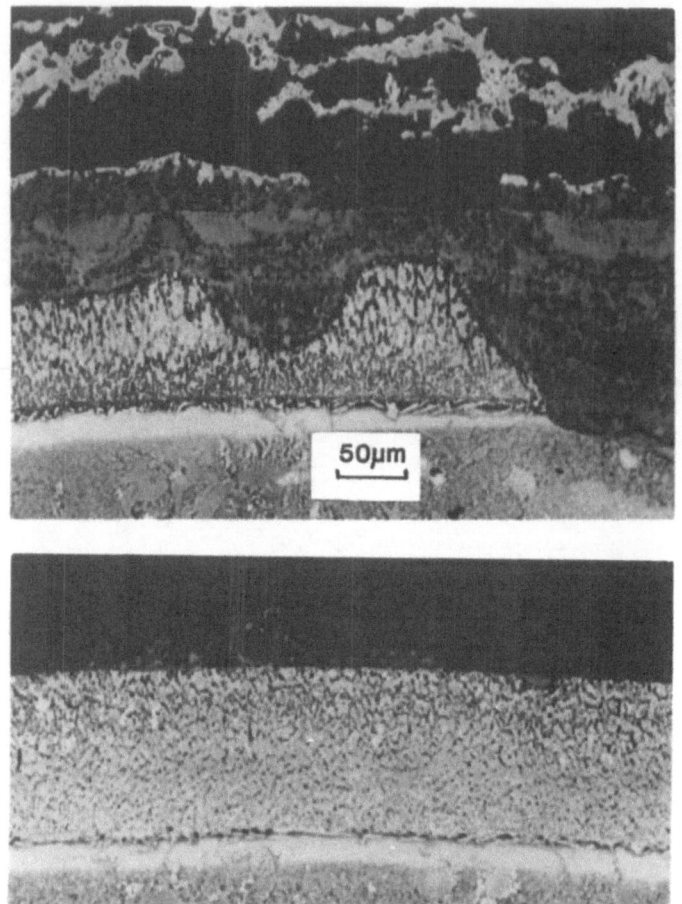

Figure 4 . Corrosion of Co-Cr-Al-Y PVD Overlay Coating
in ARE Low Pressure Rig (a) 200 hours at
750°C (b) 200 hours at 830°C.

improvement at lower temperatures.

Aluminium silicon coating systems were found to have poor resistance at 700°C but to perform better a higher temperatures. Later modifications containing titanium were found to overcome this problem, probably by forming stable compounds limiting diffusion of silicon.

Additions of elements such as zirconium and hafnium have been found to be beneficial in M-Cr-Al type coatings and overlay coatings containing complex additions such as hafnium and platinum have shown enhanced performance. However, to date no coatings systems have been identified which are capable of surviving for more than about 7000 to 10,000 hours.

The problem of low temperature corrosion may be exacerbated by blade cooling as shown in Fig. 5.

4.1. Research and Development Programmes

Significant Government sponsored research programmes were undertaken in the 1970's and early 1980's in the UK and the USA on improved substrate alloys and coatings in support of naval applications. In parallel, in the EEC and Europe under COST 50 extensive programmes were mounted in support of marine and industrial gas turbines. Extensive programmes were also undertaken by engine and superalloy manufacturers and coating vendors.

In the twelve year programme of COST 50 which ended in the early 1980's significant progress in understanding the control of high temperature corrosion and protection in gas turbines, including marine engines was achieved (1). However, in parallel with COST 50, other government sponsored programmes in the EEC and N. America related to marine gas turbine have largely ceased. Whilst certain European engine builders have combustion rig corrosion test facilities there are now no EEC government owned facilities in operation which are capable of simulating the forms of degradation encountered in marine gas turbines.

5. ESSENTIAL MATERIALS RESEARCH AND DEVELOPMENT NEEDS AND PRIORITIES

Current material/coating systems appear unlikely to permit first stage turbine blade lives in excess of 7,000 to 10,000 hours in marine gas turbines. As a matter of basic design philosophy it is usual to employ only blade materials which have inherently good corrosion resistance. It is not considered acceptable to employ a material with appropriate mechanical properties but poor corrosion resistance and then protect it with

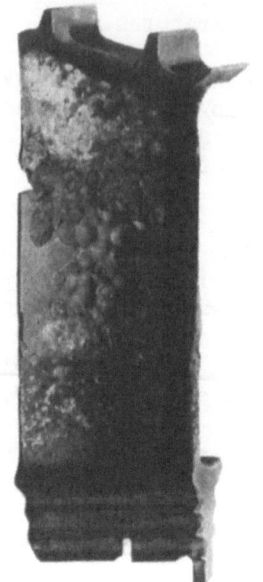

Corrosion of Uncooled Pack Aluminised
Nimonic 105, 1st Stage Turbine Blade.

A

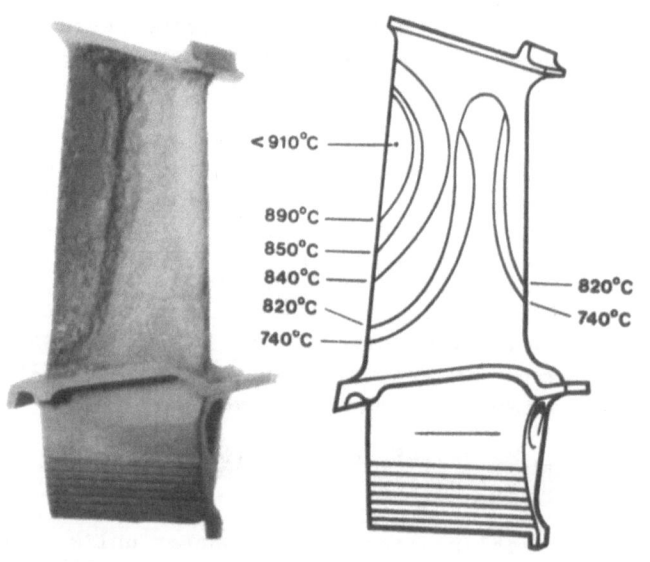

B

Corrosion of Pressure Surface of Cooled HP
Rotor Blade Compared with Temperature
Contours.

Figure 5

a corrosion resistant coating. Both the substrate alloy and the coating must have intrinsic resistance to both Type I and Type II hot corrosion if adequate lives are to be achieved. A life of 100,000 hours would seem a reasonable target for which to aim.

Development of alloys such as IN939 has provided alloys which have high resistance to Type I corrosion but unfortunately not to Type II corrosion. The alloy IN6201, a high boron nickel base alloy, has a good balance of resistance to both Types I and II corrosion but may be somewhat low in ductility. There is therefore a need for an alloy with mechanical properties at least as good as IN738 but with hot corrosion resistance similar to IN6201. In addition to the traditional alloying approach there would appear to be promise in development of oxide dispersion strengthened (ODS) alloys specifically tailored to providing good Type I and Type II corrosion resistance. Incorporation of oxide inhibitors in such alloy may be feasible (e.g. Cr_2O_3).

It is axiomatic that the composition of a diffusion coating must be a function of that of the substrate alloy. The overlay coatings have overcome this problem and have provided a new degree of freedom in enabling free choice of coating composition. The standard M-Cr-Al-Y coatings have good resistance to Type I corrosion and the high chromium types (35% Cr) with reduced aluminium levels also resist Type II corrosion. The question of long term interdiffusion in the coating/substrate system remains since no very long term operating experience exists.

The development of overlay coatings deposited by vacuum plasma, shrouded plasma and electrophoresis and then aluminised or siliconised to enhance oxidation resistance has been pursued actively. Further research and development is required on the complex overlay coatings to establish the optimum levels of alloying elements and the potential long term advantages of surface enrichment.

There is some evidence that ceramic thermal barrier coatings can confer added corrosion protection (6) and in the light of successful use of ceramic coatings on combustion chambers, this approach may be worthy of further investigation. The conventional stabilised zirconia thermal barrier coatings may not be compatible with molten salts since there is some evidence that the stabilising oxide may be leached by sulphate salts. In an early search for ceramics stable in salt melts (7) it was found that calcia and magnesia stablised zirconia were unstable in mixed sulphate melts. Subsequently it has been shown that yttria is also leached from yttria stabilised zirconia. Theoretical studies showed that ceria stablised materials should be resistant. It has been proposed that strain tolerant ceramic coatings have potential for providing enhanced resistance to hot corrosion. Recent studies (8) have also indicated that high chromium, low aluminium bond coats give the best bond coat/ceramic coating system performance.

The position on independently operated evaluation rig facilities is now somewhat unsatisfactory and consideration should be given to a joint EEC facility for this purpose. Whilst the technical advantages of marine gas turbines are clear, the economics are less well defined being dependent on not only fuel oil costs but also through life costs related to corrosion life of materials and coatings. There is thus a strong case for research and development to provide extended life.

REFERENCES

1. A Guide to the Control of High Temperature Corrosion and Protection of Gas Turbines Materials, Duvet-Thual, C., Marbioli, R. and Steinmetz, P., CEC Cost Report EUR 10682 EM (1986)

2. High Temperature Oxidation of Superalloys - High Temperature Alloys for Gas Turbines, Whittle, D.P., Ed. by Coutsouradis, D., et al., Applied Sci. Publ. Ltd., London, 109–123 pp. (1978)

3. Mechanisms of Hot Corrosion - High Temperature Alloys for Gas Turbines 1982, Condé, J.F.G., Erdös, E. and Rahmel, A., Ed. by Brunetaud, R., et al., D. Reidel Pub. Co., Dordrecht, Holland, 99–148 pp. (1982)

4. AMTE Rig Test Data and Their Implications, Taylor, A.F. and Booth, G.C., Private Communication (1984)

5. Evaluation of Corrosion Resistance of Coated Superalloys in Rig Tests, Booth, G.C. and Clarke, R.L., Materials Sci. and Engng., Vol. 2, 272–281 pp., Mar (1986)

6. Behaviour of Plasma - Sprayed Ceramic Thermal - Barrier Coatings for Gas Turbine Applications, Grot, A.S. and Martyn J.K., Ceram. Bull. Am. Ceram. Soc. Vol. 60, No. 8, 807–811 pp. (1981)

7. Unpublished work, Erdös, E.

8. Optimization of the $Ni-Cr-Al-Y/ZrO_2-Y_zO_3$, Stecura, S., Thermal Barrier System, Advanced Ceramic Materials, 1(1) 68–76 pp. (1986)

4.4.3. Turbines for Motor Vehicles

J.F.G. Condé

Materials Consultant

Broadstone, Dorset, U.K.

1. INTRODUCTION

The development of the aero gas turbine in the late 1930's and 1940's was followed by interest in the use of the gas turbine for automotive power. The concept of the turbine powered car is not new and some 300 years ago a model of a turbine powered vehicle was demonstrated in China (1). Numerous patents for similar vehicles were taken out from 1791 onwards but the first serious project on turbine powered motor vehicles was undertaken in 1945. This led to the first demonstration of a turbine powered car on 8 March, 1950. From about 1952 onwards, a large number of companies in the vehicular industry, started programmes aimed at gas turbine propulsion for cars and trucks.

Vehicular turbines need to be simple and relatively cheap. They have a pressure ratio of about 5/1 (cf aero engines; 20 to 30/1) combined with a regenerative heat exchange cycle to recycle waste heat from the exhaust gas and heat the inlet air.

The special attributes of the gas turbine which make it attractive for vehicular propulsion are reliability, cost, performance, clean exhaust, compact size, lower weight and the ability to operate on a wide range of fuels. In addition, the power characteristics are such that only simple gear boxes are required with resultant reductions in all-up weight and cost.

Early metallic gas turbines indicated the need for a regenerative cycle to achieve acceptable fuel consumption and to compete with the high efficiency diesel engine. It was also apparent that higher turbine entry temperatures (TET's) were essential if maximum efficiency was to be achieved (Fig. 1). The increases in fossil fuel oil price during the 1970's, together

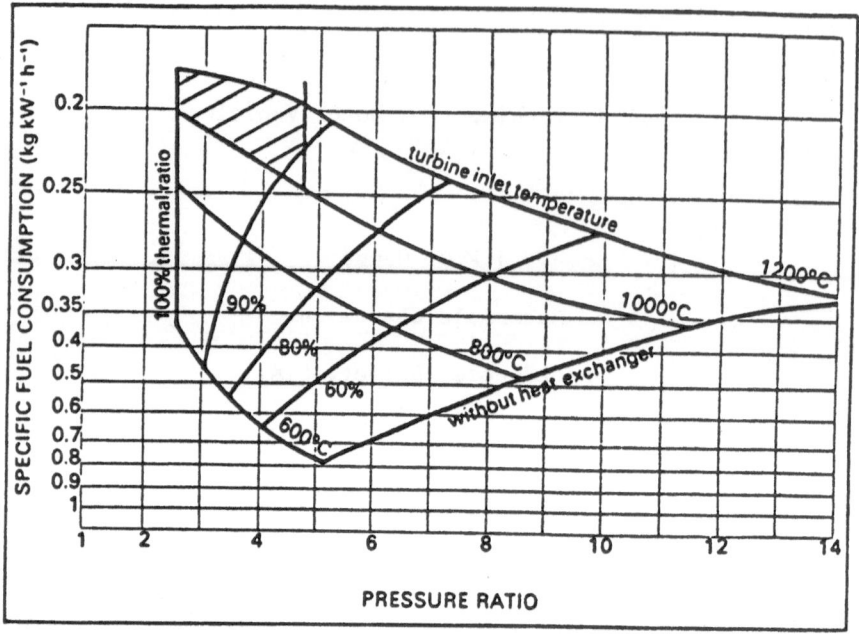

Fig. 1   Parameters Influencing Turbine Efficiency and Showing
Area of Interest for Automotive Engines. (1)

with increasingly severe anti-pollution legislation has made the high efficiency automotive gas turbine an attractive goal and has focussed attention on the application of ceramic materials in the engine in order to achieve high TET's combined with potential cost reductions and reduced consumption of expensive strategic nickel and cobalt base alloys.

2.   MATERIALS

In general metallic vehicular gas turbines are limited to TET's of around 1000°C, even with extensive turbine blade cooling. Increased cooling is counterproductive since what may be gained in higher TET is lost in overall efficiency. The use of uncooled ceramics with their higher melting or dissociation temperatures and superior creep properties, as compared with complex, air-cooled metal components, will permit higher operating temperatures leading to greater thermodynamic efficiency, and lower production costs (2). The hardness and wear resistance of ceramics also offer prospects of enhanced component and bearing durability combined with reduced friction, and greater resistance to erosion by sand and dust particles combined with oxidation and corrosion resistance. Their lower specific gravity offers reduced weight and enhanced power/weight and power density ratios. The reduced inertia in rotating components offers potential for engines with faster response to increase in power demand, rotational speeds and reduced vibration. The enhanced efficiency resulting from the use of ceramics could reduce fuel consumption by around 20% and if low manufacturing costs can be achieved, mass production of vehicular gas turbines will be feasible (2) for cars and trucks.

The principal disadvantage of ceramics, compared with metals, is their intrinsic brittleness. Such low ductility implies low work of fracture to propagate cracks, there being no plastic component to absorb the energy, and high sensitivity to even small flaws. These materials operate better in compression than in tension, and show considerable scatter in mechanical properties.

In recent years some improvement in fracture toughness has been achieved by "transformation toughening" and although these materials have at present severe temperature limitations, they now show three times the fracture toughness of conventional glass and domestic ceramics (3). The moderate fracture toughness of ceramics has attracted attention in the gas turbine principally in the context of failure due to impact damage resulting from ingested material or from combustor generated pyrolytic carbon, a problem in high energy density combustion systems. Vehicular turbines will need intake filters for other reasons and these, combined with rotary regenerators to preheat the air flow will effectively eliminate risk from ingested material. The question of pyrolytic carbon breaking off from deposits on burners and flares remains and can only be overcome by close attention to the combustion system to eliminate the problem.

Early experiments, in Germany, aimed at employing ceramics in gas turbines, involved alumina which was attractive because of its high melting point of around 2049°C. Unfortunately, its high coefficient of thermal expansion results in cracking due to thermal stress at very high temperatures (3) and impurities may enhance creep. The high melting point (2690°C) of zirconia also makes it attractive but even if it is stabilised, cycling above 1000°C can result in failure due to the tetragonal to monoclinic phase transformation. In addition zirconia has a high coefficient of thermal expansion which in high thermal fluxes leads to high levels of thermal stress. Hence it is used for thermal insulation in the form of thin coatings or components comparatively weak and relatively tolerant of cracking (3). The ceramics of principal interest in the gas turbine for high temperature vanes; rotor blades, and combustions components are the silicon carbidés and nitrides; for rotary regenerators, lithium aluminium silicate (LAS, β-spodumene), aluminium silicate (AS) and magnesium aluminium silicate (MAS, cordierite), although silicon carbide and nitride based materials may also be candidates.

The silicon based ceramics have low coefficients of thermal expansion, silicon nitride being about two thirds that of silicon carbide and the elastic modulus of silicon carbide is about 406 GPa whilst silicon nitride is around 290 GPa, so that with the same thermal gradient the thermal stresses will be lower in the nitride ceramic. Conversely because silicon carbide has greater thermal conductivity than silicon nitride (SiC at 20°C, 87 W/mK; Si_3N_4, 20 W/mK), the former develops lower thermal gradients in the same thermal flux (3).

The properties of selected engineering ceramics are shown in Table 1 (4). In general, the silicon carbide and nitride based

TABLE 1.   Properties of some Engineering Ceramics in Comparison with Selected Metals (4).

	Density * ($10^3 kgm^{-3}$)	Bend * strength (MNm^{-2})	Youngs * modulus (GNm^{-2})	Fracture * toughness ($MNm^{3/2}$)	Thermal + expansion ($10^{-6}K^{-1}$)	Thermal + conductivity ($Wm^{-1}K^{-1}$)
Alumina 99%	3.9	400	400	3.0	9.0	24
Aluminium titanate	3.0	40	20		0	1.5
Cordierite (MAS)	2.5	120	110	2.5	2.0	1.5
Silicon carbide						
Reaction bonded	3.1	500	410	4.5	3.8	100
Sintered	3.1	460	400	4.5	4.0	90
Silicon nitride						
Hot pressed	3.2	800	310	6	3.2	20
Reaction bonded	2.5	200	170	3	3.0	12
Sintered	3.2	400–700	250	5	3.4	16
Sialon	3.2	950	290	8	3.1	21
Zirconia						
Plasma sprayed	5.2	6–80	48	2	8.0	1.0
PSZ	5.6	500	205	8	9.5	1.7
TZP (Y)	6.05	1000	210	15	9.0	2.0
Fully stabilised	5.8	180–250	160	4	10.0	2.0
Zirconia toughned alumina	4.1	450	340	8	8.1	23
Nomonic (80A)	8.2		200		13.0	12
Inconel 751				80–100		

* Values at RT.
+ Temperature range 300 - 600 K.

NB.   Values of properties are for guidance only since numerous ceramic grades may exist under one general name.

ceramics do not melt but decompose, the dissociation pressure of silicon nitride reaching one atmosphere at 1900°C and silicon carbide decomposing in the range 2200°C-2500°C. The presence of impurities or elements added deliberately as hot pressing aids, depending on the form in which they are present, may have adverse effects on high temperature creep resistance. If present as glassy grain boundary phases viscous flow will reduce creep strength. This effect can be overcome by altering the composition to induce crystallisation of the glassy phases which helps to prevent viscous flow.

The variations of strength with temperature for some representative types of material is shown in Fig. 2 a (4) in comparison with the metallic material Inconel, and other data in Fig. 2 b (5). The superiority of materials free from hot pressing or sintering aids is clearly apparent at the higher temperatures. Future research should address the influence of sintering aids on ceramic strength.

The silicon based ceramics depend for their oxidation resistance on the formation of a protective film of SiO_2 which when in the crystalline form melts at 1710°C. However, impurities in the ceramics which may be incorporated in the protective film can form a SiO_2 glass with a reduced melting point. Such impurities may also arise from the combustion of impurities/additions in the fuel and whilst purer ceramics may be possible, reducing the level of impurities in the combustion environment is more difficult. Filtration to reduce airborne contamination levels may be feasible at no great cost but any attempt to improve fuel quality (apart from coalescence filtering or centrifuging to remove extrained salt water) may incur a substantial cost penalty as well as creating logistic problems. There is a current trend for impurity levels to rise and, for example, the sulphur levels in one type of gas oil or diesel fuel have risen in the past fifteen years from 0.3 to 1.0%.

Changing sources of crude oil and refinery practise to reduce the petrol fraction have resulted in a gradual change in the cetane value and, in parallel, the composition of the hydrocarbons. Whilst the cetane value appears to be more significant for diesel engines than for the gas turbine the change in composition of the hydrocarbons is significant in enhancing carbon formation in the gas turbine combustion system creating reducing conditions and enhancing corrosion or erosion effects.

The potential mechanisms of corrosion of silicon based ceramics in gas turbines and heat engines have been summarised as follows (5):-

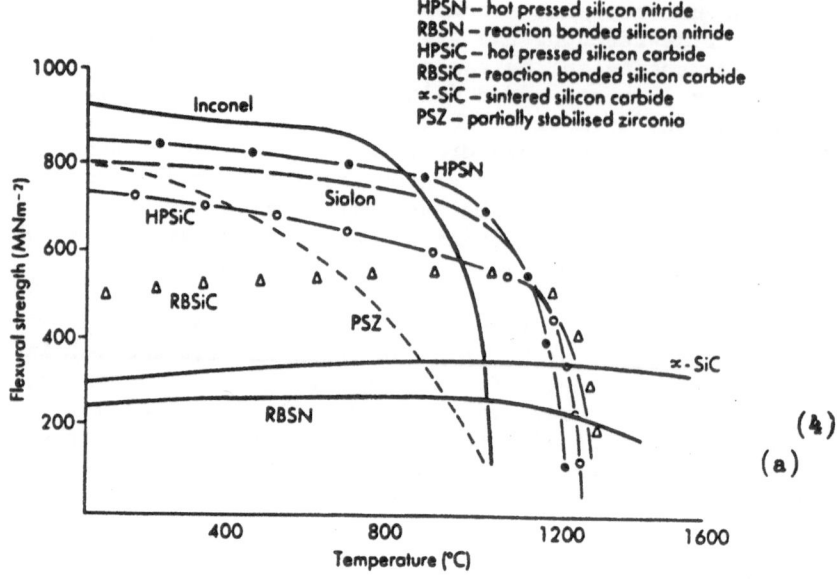

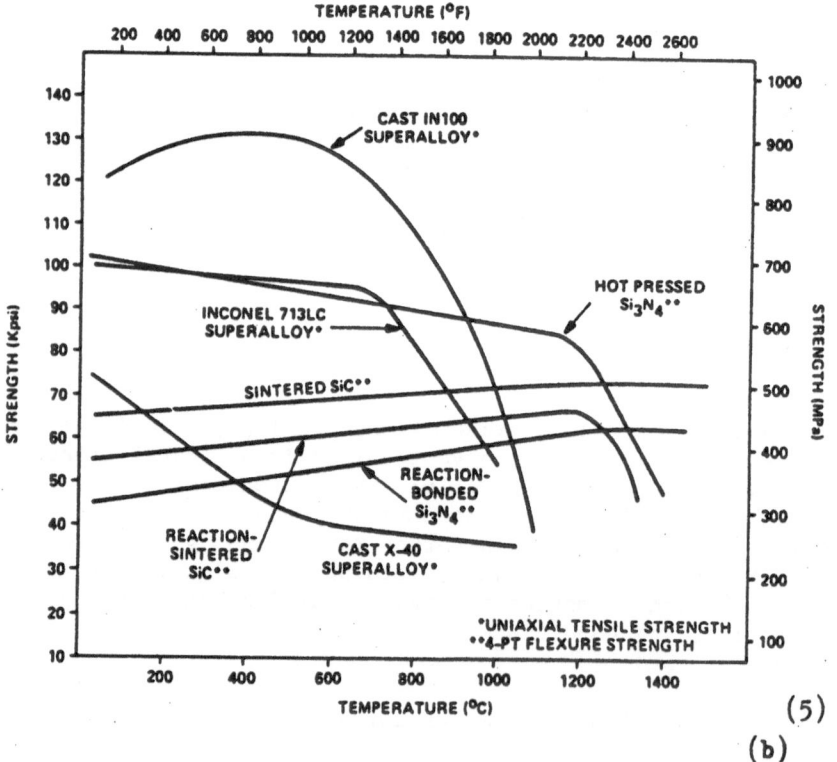

Figure 2 Variation of Properties of Engineering Ceramics with Temperature in Comparison with Superalloys.

Change in the chemistry of the SiO_2 layer, increasing the oxygen diffusion rate.

Bubble formation disrupting the protective SiO_2 layer and allowing increased oxygen access.

Decreased viscosity of the protective layer which is then entrained by the high velocity gas flow.

Formation of a molten slag which can dissolve the ceramic. Localised reducing conditions with reduction of the oxygen partial pressure sufficiently to permit active oxidation.

Formation of new surface flaws e.g. pits or a degraded microstructure with a decrease in strength.

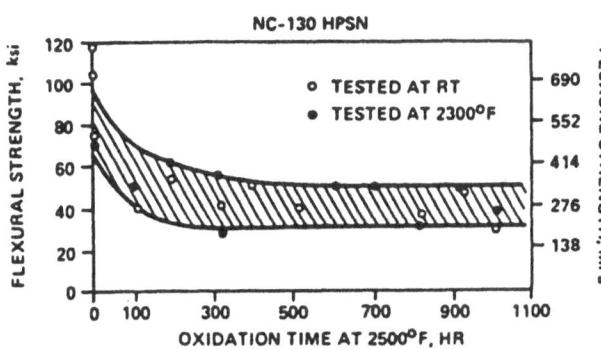

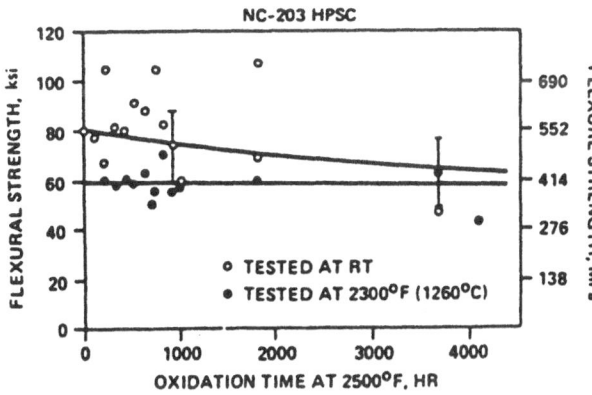

In the past two decades there have been many studies (5-12) of the corrosion behaviour of the silicon based ceramics employing simple laboratory experiments and low and high pressure, high velocity combustion rigs. Temperatures in the range 650°C to around 1400°C have been employed. The overall conclusion is that even in combustion environments heavily contaminated with sodium, vanadium and sulphur the silicon based ceramics are superior to the nickel based super-alloys. Strength reduction can occur under salt-contaminated

Fig. 3   Effect of Exposure of Hot-Pressed Silicon Nitride (HPSN) and Silicon Carbide (HPSC) in a Turbine Environment. (5)

conditions, as shown in Fig. 3 (5) and Table 2 (11). Even simple oxidation will produce reductions in strength and methods for

TABLE 2 <u>Strength of α-SiC Corroded in Na$_2$SO$_4$ (II)</u>

Conditions	Average Strength σ$_f$ MPa	Standard Deviation S MPa	Sample Size N	95% Confidence Interval of σ$_f$ MPa	Δσ$_f$ = σ$_f$ As Received - σ$_f$ as corroded	95% Confidenc Interval of Δσ$_f$ MPa
As Received	409	62	15	433,375	—	—
Burner Rig*	278	58	9	322,233	131	185,77
Furnace+	251	45	8	288,213	158	210,106

* 4 ppm Na; 4 x 10^5 Pa; 1000°C; 13.5 h.

+ 2 - 3 mg/cm^2 Na$_2$SO$_4$; 1 x 10^5 Pa air; 1000°C; 48 h.

inhibiting oxidative strength degradation have been studied (14). This is clearly an area requiring further research and development as well as ways and means for mitigating hot corrosion-induced strength reduction.

The aluminium silicate materials are attractive for rotary regenerators because of low thermal expansion characteristics giving good thermal shock resistance combined with comparatively low elastic modulus (E) and strength. The LAS material has been found to degrade in combustion gas environments due to exchange of Na+ and H +ions in different regions of the heat-exchange matrix leading to local changes in properties with associated distortion and cracking (5). Leaching of the LAS structure to remove lithium prior to application produced aluminium silicate (AS) which had suitable thermal expansion characteristics and did not show distortion or cracking under engine conditions. The temperature limit for AS materials is about 1100°C. MAS has a somewhat higher temperature capability but its larger coefficient of thermal expansion can result in increased thermal stress if the design of the regenerator matrix is not modified to accomodate the different properties.

Cyclic thermal exposure (CTE) testing (2) to simulate an engine maximum acceleration/decleration cycle has demonstrated that neither AS nor MAS materials will withstand an acceptable number of engine accelerations because of the associated temperature peaks. The strength degradation resulting from CTE testing, particularly evident in the MAS materials, is believed to be associated with anisotropic expansion of one or more of the phases present in the ceramic (2). Damage is believed to be a function of grain size and second phase grain boundary constituents. Controlling fabrication to reduce grain size and the form and amount of grain boundary phases has been proposed as a possible approach to improving the CTE characteristics. Other aspects which require further assessment include resistance to mass transfer and corrosion effects associated with exposure to salt contaminants and contact with potential regenerator seal materials such as NiO and CaF_2. Both modulus of rupture (MOR) and the radial compressive strength of the heat-exchange matrix may show significant degradation and further evaluation is required.

Ceramics have good erosion resistance due to their high hardness and in this respect are superior to most metallic materials. Silicon carbide and silicon nitride with Knoop hardnesses of around 2700 and 2200 kg/mm (2) respectively have particularly attractive characteristics. In general, erosion resistance is best with fully dense material, the porous ceramics, such as reaction bonded silicon nitride, being less resistant.

Ceramic materials are also attractive for bearing applications in gas turbines because of their high hardness, ability to take on a good surface finish and their high temperature characteristics. Metallic bearing materials start to lose their hardness and dimensional stability above about 550°C and require lubrication and cooling for reliable long term service. Ceramics can operate effectively in the absence of lubrication and cooling. Silicon nitride and the sialons have been found to give good performance in rolling bearings. Ceramic gas bearings are also an attractive option for rotating bearings in the gas turbine. Both pressure fed, self acting or hybrid type (pressure fed during start-up) bearings are possible and are free from the risk of seizure if "touch-down occurs leading to surface-to-surface contact (a problem with metallic gas lubricated bearings).

3. INTERNATIONAL R & D PROGRAMMES

The world wide effort on ceramics in heat-engines is located principally in the USA, Japan and the EEC countries. The relevant programmes have been reviewed by a number of authors and are summarized in a variety of sources (2, 3, 4, 15, 16, 17). Whilst the Japanese and US programmes on engineering ceramics R&D (17) appear to involve similar expenditure, the US programmes have focussed on engine applications with emphasis on gas turbines. The FRG programme is similarly orientated and, like the US programme, involves heavy government funding. In Japan, whilst the government has led the programme, the bulk of the funding has been by industry and the programme has been spread over a wider range of potential applications. The Japanese programme has been driven primarily by commercial considerations and has emphasised production and applications technology in contrast to the US programmes which, having predominantly military applications, have emphasised the technical goals rather than the economic and production aspects. R & D programmes in the FRG, France, Italy, Sweden and the UK have increased significantly in the past seven years.

4. RESEARCH AND DEVELOPMENT REQUIREMENTS AND PRIORITIES

The principal factors which remain a barrier to application of ceramics in vehicular turbines are said to be cost and reliability (16) hence further R & D will be directed to reduce costs (i.e. cost of powders, forming and consolidation technology, finish machining and quality assurance/inspection). Components are a product of their materials and manufacturing technology and improvements in both are required as well as in probalistic engineering design techniques to achieve enhanced reliability in service.

All of the AGT (Advanced Gas Turbine) programmes from 1965 onwards have led to the conclusion that materials development programmes should concentrate on actual components for engines. It is not possible to progress from optimum test bar properties to equal properties in engine components. The ceramics materials' properties and component design must be based on the requirements of a definite engine environment and hence materials development programmes and basic research must be related to specific engine components. It is axiomatic that success cannot be achieved without the benefit of demonstrator engines which subject components to be most severe duty cycle of the engine (2). Proof testing of components before engine testing is essential to eliminate defective components which lead to costly engine failures and do not further understanding. Simulation of the thermal and stress environment is an essential feature of proof testing. Only components with an acceptably high probability of survival are likely to be productive. To achieve this goal it is necessary to evolve fully effective non-destructive inspection techniques capable of resolving strength/fracture controlling defects.

Existing analytical design methods are capable of designing components which will provide short-term reliability with application of proof testing. However, the design methods need to be enhanced to enable creep, thermal cycling, thermal stress and oxidation/corrosion effects, which are time dependent, to be incorporated. For this purpose further data on these phenomena are required.

The existing and emerging silicon nitride and carbide based

engineering ceramics seem capable of meeting many of the requirements for ceramic gas turbine components. Additional research and development is required on materials for rotary regenerator matrices to enhance temperature capability and environmental compatibility. There may also be reasonable prospects of evolving toughened structural materials which will be more tolerant of defects or inexact design and capable of operating at higher temperatures than the existing zirconia based materials. The scope of fibre reinforcement also requires to be defined in relation to turbine components and the possibility of "pre-stressing" should not be overlooked.

Whilst the use of ceramic (e.g. zirconia based) thermal barrier coatings for the components of metallic air cooled turbines may be realistic for small aircraft auxiliary power units (APU's) this technology is likely to be too expensive for application in vehicular turbines. The use of ceramic components appears to be the best prospect for vehicular use.

A further area where additional research and development is required is in the technology of joining, whether mechanical or by bonding. Seals and bearings also warrant considerable research and development effort. In two areas, rotary regenerators and blade tip seals (shrouds) some success has been achieved with ceramic seals (2, 3). Gas bearings (sialon shells in combination with chromium oxide coated superalloy shafts (3)) have shown promise as well as HPSN and sialon components in roller bearings.

5. REFERENCES

1. Gas Turbines for Land Transport, Penny, R.N., Science Journal, 54-59 pp., April (1970)

2. Ceramic Applications in Turbine Engines, Helms, H.E., Heitman, P.W., Lindgren, L.C. and Thrasher, S.R., Noyes Publications, N.J. USA (1986)

3. The Use of Ceramics for Engines, Godfrey, D.J., Proc. of the Twelth Int. Cont. on Science of Ceramics, Saint-Vencent, Italy

4. Ceramic Components in Automative Applications, Kirk, J.N., Metals and Materials, Vol. 3. No. 11, 647-652 pp., November (1987)

5. Modern Ceramic Engineering Properties, Processing and Use in Design, Richerson, D.W., Marcel Dekker Inc., N.Y. (1982)

6. Corrosion Behaviour of Silicon Nitride and Silicon Carbide
    in Turbine Atmospheres, Singhal, S.C., 1972 Tri-Service
    Conference on Corrosion MCIC 73 - 19, 245-250 pp. (1973)

7. Oxidation and Hot Corrosion Behaviour of Si_3N_4 and Sialon,
    Schlichting, J., Nitrogen Ceramics, Ed. Riley, F.L., Pub.
    Noordholdt Leyden, 627-634 pp. (1978)

8. Molten Salt Corrosion of SiC and Si_3N_4 Ceramics, Tressler,
    R.E., Meiser, M.D. and Yannshonis, J. Am. Ceramic Soc. Vol.
    59 No. 5 - 6, 278-279 pp. (1976)

9. Corrosion of Silicon Carbide in Gases and Alkaline Melts,
    McKee Out and Chatterji D., J. Am. Ceramic Soc. Vol. 59
    No. 5 - 6, 441-444 pp. (1976)

10. Hot Corrosion of Sintered and SiC at 1000°C, Jacobson, N.S.
    and Smialek, J.L., J. Am. Ceramic Soc. Vol. 68 NO. 8,
    432-439 pp. (1985)

11. Burner Rig Corrosion of SiC at 1000°C, Jacobson, N.S.,
    Stearns, C.A. and Smialek, J.L., Advanced Ceramic Materials
    1, (2), 154-161 pp. (1986)

12. Ceramic Heat Exchanger Concepts and Materials Technology,
    Bliem, C., et al., Noyes Publications, N.J. (1985)

13. Brittle Materials Design - High Temperature Gas Turbine -
    Materials Technology, Miller, D.G., et al., AMMRC CTR - 76 -
    32, Vol. 4 (1976)

14. Private Communication, Godfrey, D.J.

15. Progress in Nitrogen Ceramics 1983, Riley, F.L., Ed.,
    Martinus Nijhogg Publishers, Boston/The Hague/Dordrecht/
    Lancaster, Published in Co-operation with NATO Scientific
    Affairs Division.

16. Ceramic Materials and Components for Engines, Bunk, W. and
    Hausner, H., Eds., Proc. of the Second International
    Symposium, Lübeck-Travelmünde, FRG, Verslag Deutsche
    Keramische Gesellschaft, 1023-1034 pp., April 14-17 (1986)

17. State of the Art Report on Engineering Ceramics as Applied
    to Reciprocating Engines, McClintock, A.L.M., Kannan, K.R.
    and Probert, C., Department of Trade and Industry, National
    Engineering Laboratory, June (1987)

4.4.4. Reciprocating Engines - Diesel - Otto

D.J. Godfrey

Admiralty Research Establishment

Holton Heath, U.K.

1. INTRODUCTION

Reciprocating engine development has always been strongly dependent upon metallic materials. In the early years of development of spark ignition (Otto) and compression ignition (Diesel) engines the availability of cast iron and steel products with suitable properties and adequate fabricability and cost was crucial, and in later years the advantages of lightweight and castable aluminium alloys have allowed improvements in engine performance and cost. Only in the case of spark ignition engines was the need for a ceramic material critical, and the development first of clay-derived aluminosilicate and then of alumina ceramic sparking plug insulator components is a fascinating story in which resistance to mechanical and thermal stress, metal/ceramic joining technology, and material and processing cost reduction played major roles.

The development of strong silicon nitride ceramics, with good heat resistance and thermal shock behaviour raised the possibility that they might be used in reciprocating engines (1), since even the lower strength reaction bonded form of silicon nitride could survive in both spark ignition and diesel engines (2) (3) (4). Survival having been demonstrated, albeit somewhat marginally in highly rated diesel engines (4), much thought has been exercised as to how the use of ceramics could be exploited with advantage in reciprocating engines.

In recent years in parallel with development of ceramics, metal matrix composites (MMC's) have been developed into useful engineering materials and some forms are already finding applications in diesel engines (5). One particular manufacturer is already producing some 100,000 reinforced pistons per annum for automotive diesel engines. These are made from short fibre reinforced aluminium alloy formed by squeeze casting employing ceramic fibre preforms. It is claimed that selective reinforcement of the piston crown and ring groove to enhance thermal fatigue, wear resistance and dimensional stability has enabled the more expensive Ni-Resist ring insert to be replaced at no extra cost.

2.   MATERIALS

2.1. Metal Matrix Composites

In the automotive engine field of reciprocating engines, cost considerations are preeminent and it is claimed that add-on values are generally limited to £1/kg (5). The short to medium term future suggests expansion of the application of MMC's based on the already successful and cost effective application of short fibre reinforcement to pistons. Conventional aluminium silicon alloys lack the necessary mechanical properties to make high performance pistons possessing the required long life characteristics. Short ceramic fibre reinforcement can enhance high temperature performance, reduce thermal expansion coefficient enabling a closer fit to the cylinder bore with attendant improvement in fuel consumption and wear resistance (5). Other components where MMC's may find application include conrods, gudgeon pins, cylinder liners and possibly clutch plates, brake discs, turbo compressor vanes and selectively reinforced housings.

2.2. Ceramics

In the longer term future it seems likely that ceramics, either monolithic, reinforced (ceramic matrix composites) or coatings will become of growing importance in reciprocating engines. The potential benefits can be grouped into categories on the basis of the five different characteristics of ceramics as compared with metals:-

(a)   Refractoriness

The main potential applications are to replace more expensive heat resistant metals, used in key components in highly rated engines, where thermal loadings prove to be excessive for cast iron, steel or aluminium components, and thermal stress cracking, creep, partial melting and local oxidation problems are severe. The use of ceramics in such applications has been little explored, although thermally sprayed or sputtered zirconia coatings show promise. Coatings have the advantage that the special fixture which a monolithic ceramic component would require is not needed, however adhesion of the coating may prove to be inadequate and high bond strengths are not readily achievable. In addition highly dense coatings are less flexible and therefore less resistant to thermal stresses, and experience with thick coatings has shown that the adhesion problem becomes more difficult as the coating thickness exceeds about 0.3 to 1mm. Considerable progress has been made in the evolution of well adherent thin zirconia coatings for thermal protection of superalloy gas turbine combustors and blades (1), for thermal and corrosion control in diesel engines, and to deal with the corrosive effects from the combustion products from low grade fuels, some of which can degrade zirconia ceramics. Apart from zirconia, there has been some work on calcium titanate coatings, but alternative materials have received little attention since there are few obvious candidates with both refractoriness and good thermal expansivity match to metal substrates.

Although coatings developed for superalloys have been applied with some success to iron and aluminium alloys in diesel applications, expansion mismatch (especially with aluminium) is a serious difficulty. A survey of suitable alternatives to zirconia followed by exploratory work on coating deposition and performance evaluation might yield major advances. Adhesion might be improved by the development of innovatory interface texturing procedures, as an alternative to the multilayer metal ceramic mixture "bond coat" approach.

The attachment of thick layers might also be feasible if a suitable metal or ceramic interlayer adhesive bond system could be developed, as for valve technology for alumina (molybdenum-manganese-glass and solder glasses), and such techniques might be applicable to monolithic ceramic shields, as an alternative to clamping by flexible or tension metal fasteners. The attachment of ceramic caps to aluminium pistons has proved to be a difficult engineering problem, and protracted engineering development has proved necessary to overcome the complex, interaction of thermal expansion mismatch stresses, thermal gradient stresses, loosening due to metal accommodation under stress and metal oxidation growth.

(b) Thermal insulation

A further aspect of the use of ceramics as heat barriers, has been the possibility that thermal insulation could be used advantageously. Both zirconia and glass ceramics are ceramics which have very low thermal conductivity and useful mechanical strength. Silica and mixed oxide glasses also have low conductivities, but their lower strengths, especially at elevated temperatures, have discouraged their investigation for use in reciprocating engines.

In the last decade there has been intense interest in the possibility that insulation of the regions of the reciprocating engine containing hot gas might improve the efficiency of the engine. A considerable proportion of the heat energy in the fuel flows into the engine structure, ranging from 1/3 to 1/5, varying with the engine size, design, and operating speed and power. The surfaces through which the majority of the heat transfer takes place are those of the piston crown and cylinder head; as the cylinder bore is progressively exposed during the expansion phase heat transfer to it from the gas falls rapidly and its heat transfer role is rather minor.

Hopes that insulation would lead to better engine performance have not so far been justified in terms of engine efficiency. Computer simulations show that the beneficial effects of increased temperatures are offset by the increased temperature of combustion space restricting volumetric capacity because the induced mixture becomes hotter and less can therefore enter before ignition takes place. The role of the laminar boundary and radiation in heat transfer are also complex and difficult to predict. Engine testing has so far not demonstrated any significant gain in thermodynamic efficiency, although there are indications that some kinds of unwanted pollutant emissions may be reduced by the hotter conditions, and that noise may be affected beneficially in some engines.

There has been much controversy about the effects of insulation on thermal efficiency, and this has tended to obscure the positive gains in overall engine efficiency by the reduction in the engine's cooling needs when heat flux is shifted from the engine structure to the exhaust gas. The cooling system absorbs a not inconsiderable fraction of the engine power, is a small but significant part of the vehicle's cost, and adds to its weight, size and aerodynamic drag. It appears that several percent improvement in overall vehicle efficiency is possible when even ceramic coatings are used for insulation. Insulation also aids in the application of the 'precision' cooling concept to engines; i.e. cooling applied differently in different engine regions so that only as much as is necessary is provided.

Insulation is thought by some engine specialists to be potentially beneficial in engine combustion chambers, where it might decrease engine warm-up time and also modify combustion processes and products beneficially, a 10% improvement in output has been reported (6) also from a ceramic 'hot plug chamber, and noise reduction of 10% was also reported (7). Work with an all ceramic swirl chamber was described by Sakurai and Matsuoka (8) in which the particulate level in the exhaust fell by two thirds.

Rather than zirconia, aluminium titanate Al_2TiO_5 has been the material most favoured for this application, since it has a combination of low thermal conductivity and low thermal expansion coefficient, and consequently resists thermal stress quite well. It has rather low strength (ca. 30 MPa), but is a material which after some microcracking may still be functional, especially if held in place compressively. The possibility of a material improvement by microstructural and compositional control is actively being investigated (9); dispersant transformation and fibre toughening merit investigation.

Silicon nitride has a relatively low thermal expansion coefficient which enables it to resist thermal stresses quite well, but its thermal conductivity is not particularly low, and is similar to that of many superalloys (at room temperature). Godfrey has developed a new type of insulating material (9), which by reaction bonding in nitrogen can be produced without sintering shrinkage. This material has the thermal diffusivity of zirconia, and at least twice the strength of aluminium titanate ceramics, and the considerable economic advantage of near-zero dimensional change when consolidated by nitridation. There is further room for innovatory thinking in devising better materials for insulation. However, all ceramic insulating materials will have to compete with the alternative stratagem of providing an air space in the metallic component, which is an effective means of insulation, provided the hot metallic face does not fail thermally, and that increased cost of the more complex component with an air space is acceptable.

(c) Wear resistance

Ceramics are also potentially usable as wear-resistant components, because of their high hardness and strength, especially at elevated temperatures. Wear resistance is a complex property and varies with loading intensity and cycle character in varying load applications, and also even with ceramics with chemical effects from the environment, so that behaviour is difficult to predict, and practical tests are presently the best way to rank materials. Alumina, zirconia-toughened alumina, zirconia and silicon nitride and carbide are all currently being investigated for applications such as tappet faces, push rod ends and valves, but no material has yet emerged as the best for all tribological applications. Research aimed at understanding the tribological behaviour of ceramics has not yet produced a clear and unified theoretical basis, although there are indications that its complexities are not insoluble, and further research is clearly desirable. Boron carbide might also be a useful hard material but is expensive with costly processing needs. Methods to develop inexpensive means of producing coatings could usefully be investigated. The possibility of using thick diamond-like carbon coatings made by plasma activated vapour deposition merits attention. Titanium carbide coatings have become useful tribological materials, although it is still difficult to predict whether they will perform well in a new application. Thermally sprayed oxide coatings are known to perform well in some tribological applications (e.g. alumina/ titania and chromium oxide), and could usefully be investigated for engine applications. The treatment of iron cylinders with silicon carbide powder, which becomes embedded in the cast iron, has proved to be an effective means of reducing cylinder liner wear in diesel engines, and the principle of improving wear resistance by forcing in a hard ceramic particulate material might be usefully investigated more widely. Ion implantation to produce hard surfaces is already established with ceramics, but requires vacuum chamber processing which is costly, although the ability to thereby modify corrosion behaviour might be useful in oxidation, corrosion or tribological environment degradation situations. However, the survival of the implanted layer under severe corrosion may be doubtful.

(d) Corrosion resistance

Ceramics also might find applications because of their corrosion resistance, especially with the tendancy in large size diesel engines to use low grade fuels whose products of combustion are severly corrosive to many metallic materials.

Although there has been a certain amount of research on the corrosion of simple ceramics by ash products, there has been little published work on ceramic 'alloying' to produce more resistant materials. Sulphur oxides and H_2SO_4 attack (lithium leads) of lithium aluminium silicate glass ceramics proved to be a serious problem in long term durability studies with ceramic rotary regenerator cores for gas turbine efficiency improvement, although magnesium aluminium silicate (Cordierite) is more resistant to combustion products from sulphur containing fuels. The corrosion resistance of ceramics also might be important in exhaust system applications, although economic considerations already prevent the large scale use of stainless steel or ceramic enamel coating for automobile exhausts for relatively low cost mass produced vehicles.

(e) Low density

The low density of ceramics as compared to metals offers some potential improvements in vehicle performance e.g. for high performance car engines. Wide interest in turbocharger applications has been generated by the recent use of a sintered silicon nitride turbo rotor in sports cars. The difficult problem of ceramic attachment to the steel rotor shaft was solved by a complex multilayer (Si_3N_4/Ni/W/Ni/ Steel with Cu/Ag(Ti) braze) attachment, but it is probable that defect control in the rotor has been a serious economic problem in reducing the manufacturing yield of acceptable quality rotors. The lower density of silicon nitride (3.2) permits more rapid acceleration of the vehicle. The lower density of other ceramic engine components could reduce overall engine weight, and as with the reduction of cooling system size by ceramic insulation, improve acceleration efficiency.

Electrical properties

Ceramics are used or are being considered for a wide variety of other applications in automobile engines (11), many of which are of an electronic or sensor character, e.g. coolant, inlet air and oil temperature, oxygen (combustion) and exhaust gas, fuel level, knock, ignition timing and crank angle/revolution, rain drop and icing, and road clearance and near obstacle sensor applications. Electronic applications include fluorescent, LED and electroluminscent displays, automatic lighting control, IC substrate and condenser warmer and window heater coatings, alarm buzzer, blower motor varistor and diesel fuel, automatic choke, cold mixture heaters and glow plugs. Exhaust purification catalyst and catalyst substrate applications are also important, and can involve the production of honeycomb, high surface area, ceramics. There are many research opportunities in these areas of consumer electronics applied to automobiles and their engine optimisation and control systems, both for new types of application, improved performance and material and manufacturing cost reduction.

Future Development Trends and Research Needs

Metal Matrix Composites

Almost all of the important automotive engine and associated component manufacturers are alert to the potential benefits of MMC's and are actively pursuing R+D as well as limited production of components. Development trends will be oriented to cost effective and competitive production of components as well as improved performance. It seems likely that for this cost conscious field emphasis will be on particulate and short fibre reinforced materials which, for the most part, can be manufactured directly or by adaptation of existing metal manufacturing technology. Attempts to use long or continuous fibre reinforcement have not so far proved cost effective in automotove components and seem unlikely to make headway until fibre costs fall through quantity production and manufacturing technology for long fibre reinforced components is simplified.

Significant scope exists for development of new or improved short and continuous fibre materials by novel routes aimed at reducing production costs. The use of alternative cheaper particulate reinforcements also exist, including possibly the incorporation of solid lubricants. There is also need to explore alternative matrices such as magnesium alloys which may offer advantages in weight saving and ease of manufacture. The use of novel surface treatment and coating processes on fibres to improve fibre-matrix compatibility in some instances may also result in improved properties and performance. Whilst for higher temperature applications there is a need for fibres having better mechanical properties at high temperatures, the existing fibres will probably be adequate for foreseeable developments in internal combustion engines. The damping properties of MMC's are in advance of those of metallic materials but the precise level of advantage needs to be quantified.

Summarising, there is a need for R + D in the following areas (see also section 3.3.2.) specifically in relation to applications in reciprocating engines:-

a)   Lower cost short and continuous fibre reinforcement, possibly having improved high temperature properties.
b)   Improved manufacturing technology for components to provide cost effective and efficient production methods leading to near net shape components with minimum finish machining requirements.
c)   Improved understanding of design in order to maximise the benefits of use of MMC's in components.
d)   Alternative aluminium and/or magnesium alloy matrices.
e)   Surface treatment of fibres to improve fibre/matrix compatability (i.e. wetting, stability etc)

Ceramics

Large research programmes exist in the USA (ca $ 50M/year) and Japan (16-40 M $) with the latter seemingly more centred in manufacturing industry, but both having considerable government support. In Europe, Germany has had an important national programme for some years (1974 onwards; $ 15M 1983 - 1986) and in the UK a government cooperation programme with engine and ceramic manufacturing firms has been operative for two years (CARE Programme $2M/year). Important research efforts in France, Italy and Sweden also exist. The spectrum of R + D world-wide has been well described elsewhere (12).

An outstanding shortfall with currently available ceramics is their lack of toughness, due to the ease with which cracks are propagated to allow the catastrophic failure of components. Some improvements in toughness have been made, by for example, transformation toughened dispersion materials, principally (zirconia either pure or in alumina, mullite or glass ceramics) and from particulate, microcrack and fibre dispersion approaches, but all the improvements are so far insufficient for major increases in component reliability to be assured. Whilst the properties of these toughened ceramics need to be optimised, it should not be forgotten that composite or dispersion fabrication processes are expensive, and research aimed at processing cost reduction is very important. Whilst toughness is of paramount importance, expectations for rapid and considerable success are low.

Recent development programmes have highlighted the problems of producing reliable components. In the USA and Germany the emphasis of government sponsored programmes has accordingly shifted from engineering evaluation to reliably high strength ceramic manufacture. In this context basic research on ceramic processing to find methods of consolidation which minimise defect size is therefore highly relevant. Methods of producing ceramics with better corrosion resistance to fuel impurities also deserve attention.

Whilst ceramic coatings are finding more ready acceptance in engine applications than monolithic ceramics, a very serious problem in assuring that they are reliably adhesive remains. More progress on coating and interlayer formulation is urgently needed, particularly the technological pressures are for thicker coatings, where adhesion becomes more difficult. New coating systems merit some attention, and coatings for aluminium alloys may benefit from new innovatory thinking.

Aluminium titanate combustion chambers appear to offer some promise, but work on material optimisation and on competitive materials (such as reaction bonded silicon nitride/zirconia composites) seems probable.

Improved glass ceramics having better high temperature strength, seem likely to be evolved, such as materials based on barium e.g. the Celsian family. Improved strength silicon carbide will make this material more competitive with sintered silicon nitride and Sialon materials, whilst the organometallic polymer route, may yield not only improved silicon carbide materials, but new forms of coating containing silicon nitride, titanium carbide and other ceramics. Composite materials with improved toughness may appear, with acceptable processing cost. Carbon can now be produced as a coating ("diamond-like carbon") with very high hardness, and such materials could have an important effect on tribological performance if they could be produced by low cost methods and in appreciably greater thicknesses.

For ceramic materials therefore R + D needs may be summarised to include:-

1) Tougher ceramics
2) Better strength and reliability in ceramics made by low cost routes
3) Better adhesion for coatings
4) Research on optimising and understanding the wear performance of ceramics for engine applications
5) Development of materials with improved resistance to fuel impurities
6) Better understanding of engineering methods for fixing ceramics to metals
7) Low cost effective methods for non-destructive evaluation of components.

REFERENCES

1. Survey of the Technological Requirements for High Temperature Materials R + D, Section 1: Diesel Engines., Ed. Timoney, S.G., EUR7660EN, CEC, JRC, Petten (1983)

2. Ceramics in Severe Environments, Godfrey, D.J. and May, E.R.W., Materials Science Research Vol. 5, Ed. Kregel and Palmow, Plenum Press, NY, 149-162 pp. (1971)

3. Silicon Nitride Ceramics for Engineering Applications, Godfrey, D.J., SAE740236, Trans SAE, 83, 1036-1045 pp. (1974)

4. Ceramics for High Performance Applications III, Godfrey, D.J., et al., Ed. Lenoe, E.M., et al., Plenum Press, NY, 81-99 pp. (1983)

5. Metal Matrix Composites - Applications and Prospects, Trumper, R.L., Metals and Materials, 662-667 pp., November (1987)

6. SAE Paper 850523, Kamiya et al. (1985)

7. SAE Paper 840426, Matsuoka (1984)

8. SAE Paper 861407, Sakurai and Matsuoka (1986)

9. Ceram. Trans. J., Kajiwara Brit., 86, 77 (1987)

10. Patent Application, Godfrey, D.J.

11. Ad. Ceramic Materials, Taguchi, M., 2, (4), 754 pp. (1987)

12. State of the Art Report on Engineering Ceramics as Applied to Reciprocating Engines, McClintock, A.L.M., Kannan, K.R. and Probert, C., DTI, NEL, June (1987)

4.5. Iron and Steel Production

Dr. Ir. J.T. Van Konijnenburg

Structural Ceramics – Hoogovens

Ijmuiden, The Netherlands

1. INTRODUCTION

The most common route for iron- and steel making is the blast furnace oxygen steelprocess route. To achieve low production costs large capacity furnaces have become the norm. The disadvantage, however, is a loss in flexibility. About 80 % of all steel is produced in this way; the remainder is produced by melting scrap or pre-reduced iron ore in an electric arc furnace (EAF).

The principal and relevant installations of the blast furnace – oxygen steelmaking route are : sintering machines, pelletizing plants, coke furnaces, blast furnaces, hot blast stoves, torpedo cars and mixers, hot metal ladles, converters (e.g. BOF), steel ladles, vacuum degassers, tundishes of continuous casting installations.

For the EAF-route they are : a reduction furnace for pelletized ore, the EAF and subsequent equipment similar to that of the blast furnace-BOF-route for handling liquid steel. Important high temperature equipment used in the rolling of steel, includes reheating furnaces and annealing furnaces.

In most of these installations process temperatures exceed 1000°C and in some instances are as high as 1700°C. The common configuration is that a steel shell or a steel harness supports a refractory lining which contains the materials to be processed. The refractory lining protects the steel structure of the furnace or vessel from direct contact with the process and it acts as an insulator to prevent the steel structure from overheating. In this account only the lining materials will be considered.

The iron- and steel industry is by far the largest consumer of refractory materials (for a survey of refractory materials see Section 2.4.2.). A great variety of materials is in use. The materials used are mainly oxides and combinations of oxides, but also carbon and silicon carbide have found application. In the last decades the quality of refractory materials has been developed significantly. The purity of the raw materials used has increased and in many instances synthetic raw materials which have been chemically processed are used. Although fired bricks are still widely used, unfired chemically bonded bricks (binders are e.g. phosphate, tar, resin) have found extensive application. The introduction of carbon or graphite in bricks to increase the resistance to slags has found widespread acceptance.

Mixtures of different components are used to tailor the properties of bricks. Generally in the production of refractory bricks properties are closely controlled and adapted to the specific needs of their application. Finally it should be mentioned that unshaped refractory material are increasingly being used as castable, gunning or ramming material.

The specific consumption of refractory materials in iron- and steelmaking has steadily decreased, due to changes and improvements in processing techniques (e.g. the decline of the open hearth process and the rapid rise of the oxygen steelmaking processes, the disappearance of pit furnaces with the advent of continuous casting) but also due to the improvement in quality of these materials. New processes such as continuous casting or vacuum treatment, however, created demand for new types of refractories. The higher costs for the more advanced refractory materials prompted the iron- and steelmakers to increase the performance of the linings by a concerted effort to improve lining construction and installation, development of intermediate repair methods and process adaptations. Remarkable results have been obtained. Blast furnace campaigns of 8 years have been registered and converter campaigns of several thousands of heats are no longer exceptions. These long campaigns not only lead to a low specific consumption of refractory material but they also enhance the availability of the installation for production.

After the energy crisis the awareness of saving money by decreasing energy losses has become manifest in the iron- and steel industry. Reheating furnaces are an example where an external layer of insulating refractory material was applied successfully. Not always, however, is better insulation a good solution because it generally involves higher temperatures at the hot face of the lining which may lead to faster wear. In some instances forced cooling of the lining is used advisedly to increase lining life, e.g. in blast furnaces. The most extreme example, however, is the use of water cooled side walls for the EAF which has virtually eliminated the refractory lining.

Some refractories used in iron- and steelmaking are not used for containing the process and reducing heat loss, but have a more functional application, e.g. ceramic burners in hot blast stoves, checker bricks in regenerators, gas injection bricks in converters and ladles, tubes and nozzles in continuous casting. Although some of these materials are advanced and expensive materials they can be classified as refractory materials. In recent years advanced ceramics also described as engineering ceramics or technical ceramics have also found application in iron- and steelmaking. Leaving aside low temperature applications (e.g. alumina tiles for abrasive conditions) these materials have been used for breakrings in horizontal continuous casting machines, radiant tubes and skidbuttons in reheating furnaces. At present their use is still limited but more applications may develop. A survey of applications is given by H. Fukuoka (1).

2. INSTALLATIONS

2.1. Coke ovens

Coke furnace batteries are huge, complex structures of refractory material, one unit with an annual production of 1 million tonnes of coke containing 16.000 tonnes of silica and chamotte bricks. Silica is universally used as the construction material for the coke and combustion chambers because of its peculiar expansion behaviour. When heated a silica brick shows a very pronounced expansion of about 1,2 % between room temperature and 300°C, but once this has been accomplished the dimensions remain stable when the temperature is increased further. The coking process is a cyclic process with a cycle time of 16 to 24 hours. This means that in a period of 30 years, which is the lifetime of a battery, silica bricks will undergo 10.000 to 15.000 cycles. Due to the weak expansion in the temperature interval at the cycle, silica bricks can survive this treatment. Also, the weight of the larry car that travels on the top of the battery creates a considerable, variable tension in the ceramic structure. The long term effects of thermal cycling and of variable mechanical loads is not fully understood for refractory materials in general and for coke furnace silica bricks in particular.

Silica bricks, however, combine good strength, refractoriness, resistance to shrinkage and resistance to high temperature cycling; they generally have a good performance in coke furnaces. Stability problems in coke furnaces (with wall lengths up to 16 meters and wall height up to 8 meters) must be attributed generally to constructional deficiencies. Chemical attack on the refractory materials of the coke chambers and the heating chambers is not an important factor. However, better

thermal conductivity of the silica wall bricks would be
desirable, allowing lower flame temperatures. Efforts have been
made to obtain this goal without sacrificing the essential
properties of silica bricks. Bricks of higher density have been
proposed as well as bricks with additions of SiC. Success,
however, has been limited.

2.2. Hot Blast Stoves

Also in hot blast stoves the combination of properties of
silica bricks (good strength, low creep, low expansion at high
temperature) have led to their application, in dome, in the upper
parts of wall, burnershaft and chequers. Also, high alumina
bricks with 70-80 % Al_2O_3 are used. The constructional stability
that is achieved with silica, once the stove is heated up, gives
this material an edge over others for those parts that undergo
the largest temperature fluctuations. Dome temperatures between
1500 and 1600°C are possible, corresponding to blast temperatures
of 1400°C. Use of materials such as magnesia and mullite have
been considered for chequer materials because of their greater
density and consequently higher thermal capacity. However, for
both technical and economical reasons this was not feasible.

In conclusion it can be stated that modern high temperature
hot blast stoves are mature from the point of view of construction and refractory materials. Because of the high investment
involved it is tempting to explore the possibilities of high
temperature recuperators instead of regenerators. A recuperator
does not need the on- and off operation of a regenerator. The
required high temperature of the hot blast rules out all materials except ceramics. Efforts are being made in several countries to develop ceramic recuperators for different applications.
Problems pertaining to materials properties and construction have
to be solved, however, before such a recuperator could compete
with the hot blast stove.

2.3. Blast Furnace

In a modern blast furnace some 10 zones can be distinguished
in which the conditions to which the refractory lining is
exposed, differ considerably. All modes of attack are present,
such as high temperature and thermal cycling, pressure and
mechanical effects, chemical reaction by gas, slag and metal and
physical attack by dissolution.

The designer of the refractory lining tries to cope with
this multitude of factors and this is reflected by a design which
shows more variation in refractory materials used and in
constructional details than in all other installations in the
iron- and steel industry. Table 1 lists the different zones, the

AREA	PRINCIPAL ATTACK PHENOMENA	RESULTING DAMAGE	HOOGOVENS INTEGRATED COOLING/REFRACTORY DESIGN	SELECTED CONSTRUCTION
UPPER STACK	ABRASION MEDIUM TEMPERATURE FLUCTUATIONS IMPACT	ABRASIVE WEAR LOSS OF BRICKS	HIGH ABRASION RESISTANCE PLATE COOLER SPACING FOR ADEQUATE SUPPORT OF REFRACTORY PROFILE FIXATION	SiC WEAR SURFACE ALUMINA BRICK BACKING (42-44%) Al_2O_3
MIDDLE STACK	ABRASION HEAVY/MEDIUM TEMPERATURE FLUCTUATIONS GAS EROSION OXYDATION ALKALI ATTACK	ABRASIVE WEAR SPALLING WEAR DETERIORATION	COOLING WATER DESIGN (Q, V) TO COPE WITH LOCAL PEAK LOADS MEDIUM DENSE TO DENSE PLATE COOLER SPACING HIGH/MEDIUM SPALLING RESISTANCE MEDIUM/HIGH OXIDATION RESISTANCE MEDIUM/HIGH ABRASION RESISTANCE	SiC + GRAPHITE GRADUALLY DECREASING SiC - GRAPHITE SURFACE AREA RATIO SPALLING RESISTANT LINING CONSTRUCTION
LOWER STACK	HEAVY TEMP. FLUCTUATIONS EROSION BY GAS JETS ABRASION ALKALI OXIDATION THERMAL FATIGUE	SEVERE SPALLING WEAR DETERIORATION SHELL DAMAGE AND CRACKS	COOLING WATER DESIGN (Q, V) TO COPE WITH EXTREME PEAK LOADS VERY HIGH SPALLING RESISTANCE MEDIUM ABRASION RESISTANCE HIGH CHEMICAL RESISTANCE PROPER EXPANSION ALLOWANCE	DENSE PLATE COOLER SPACING SiC + GRAPHITE SPALLING RESISTANT LINING CONSTRUCTION PANEL CONSTRUCTION
BELLY	MEDIUM TEMPERATURE FLUCTUATIONS ALKALI OXIDATION ABRASION AND GAS EROSION HIGH TEMPERATURE	SPALLING DETERIORATION WEAR	COOLING WATER DESIGN (Q, V) TO COPE WITH LOCAL PEAK LOADS RESISTANCE AGAINST GAS EROSION MEDIUM DENSE PLATE COOLER SPACING MEDIUM SPALLING RESISTANCE HIGH SLAG/CHEMICAL RESISTANCE	SiC + GRAPHITE SPALLING RESISTANT LINING CONSTRUCTION PANEL CONSTRUCTION
BOSH	HIGH TEMPERATURE SLAG ATTACK ALKALI MEDIUM TEMP. FLUCTUATIONS ABRASION	STRESS CRACKING DETERIORATION, WEAR SPALLING WEAR	COOLING WATER DESIGN TO COPE WITH MEDIUM HEAT LOAD MEDIUM SPALLING RESISTANCE LOW ABRASION RESISTANCE HIGH TEMPERATURE RESISTANCE HIGH CHEMICAL RESISTANCE	ALTERNATING RINGS OF SEMI-GRAPHITE AND GRAPHITE EXPANSION ALLOWANCE BY PANEL CONSTRUCTION
RACEWAY AND TUYERE ZONE	VERY HIGH TEMPERATURE TEMPERATURE FLUCTUATIONS OXIDATION (WATER AND O_2) SLAG ATTACK EROSION DAMAGE FROM SCABS	STRESS CRACKING AND WEAR SPALLING DETERIORATION WEAR LOSS OF PLATECOOLERS AND TUYERES BREAK OUT RISK	HIGH TEMPERATURE RESISTANCE STRESS REDUCTION SPALLING RESISTANCE SLAG AND CHEMICAL RESISTANCE SPALLING RESISTANT TUYERE BLOCKS OXIDATION RESISTANCE	SPECIAL DENSIFIED GRAPHITES OXIDATION RESISTANCE IS NOT THE FIRST PRIORITY
HEARTH	OXIDATION (WATER) Zn + ALKALI SLAG ATTACK HIGH TEMPERATURE EROSION FROM HOT LIQUIDS	WEAR DETERIORATION STRESS BUILD UP AND CRACKING BREAK OUT RISK	STRESS ELIMINATION OXYDATION PROTECTION CHEMICAL RESISTANCE EXPANSION ALLOWANCE	DENSE SEMI-GRAPHITE REFRACTORY DESIGN WITH METAL GAS SEALS
IRON NOTCH	HEAVY TEMP. FLUCTUATIONS ERIOSION (SLAG AND IRON) Zn + ALKALI ATTACK GAS ATTACK OXIDATION (WATER ATTACK)	SPALLING TAPHOLE WEAR DETERIORATION WEAR AND DETERIORATION	SPALLING RESISTANT TAPHOLE BLOCK OXIDATION PROTECTION (H_2O) CHEMICAL RESISTANCE GAS LEAKAGE REDUCTION	SEMI GRAPHITE DOOR GRAPHITE TAPHOLE CYLINDER ALUMINA INSULATION BRICK
BOTTOM HEARTH CONNECTION	HIGH TEMPERATURE Zn + ALKALI HOT METAL EROSION SLAG ATTACK FERROSTATIC PRESSURE	STRESS BUILD-UP AND CRACKING DETERIORATION INVERTED MUSHROOM EFFECT IRON PENETRATION IN PORES SALAMANDER FORMATION	STRESS REDUCTION BY : 1 - EXPANSION ALLOWANCE 2 - PROPER COOLING INCREASED CHEMICAL RESISTANCE SELECTED THERMAL CONDUCTIVITY RESULTING IN LOCAL FREEZING	SEMI GRAPHITE COOLED BY GRAPHITE RESULTING IN SUPPRESSING OF INVERTED MUSHROOM EFFECT
BOTTOM	HIGH TEMPERATURE SLAG ATTACK Zn + ALKALI + Pb FERROSTATIC PRESSURE SOLUTION IN IRON EROSION FROM HOT METAL	STRESS BUILD UP AND CRACKING DETERIORATION SALAMANDER FORMATION	STRESS ELIMINATION BY : 1 - EXPANSION ALLOWANCE 2 - PROPER COOLING 3 - SANDWICH CONSTRUCTION INCREASED CHEMICAL RESISTANCE THERMAL CONDUCTIVITY BASED ON ISOTHERM CALCULATIONS	SANDWICH CONSTRUCTION HIGH ALUMINA TOP SEMI GRAPHITE GRAPHITE CARBON GRAPHITE

Table 1: Blast furnace zones, principal attack phenomena and main construction features

different attack phenomena, the kind of damage and the main construction features and materials. It must be emphasized that there exist differences in types of lining depending on the size of the furnace, the process conditions and other local factors and on the philosophy of those responsible for the construction of the lining.

The general design criteria for a blast furnace lining are :

- selection of refractories for each zone to cope with the different mechanisms of attack and eventual combination of different refractories;
- adequate cooling of refractories;
- adequate expansion allowance.

Two important categories of refractory materials for the modern blast furnace are graphite and silicon carbide. Graphite is very effective when properly cooled. Because cooling of the blast furnace lining by plates, staves or shellcooling is essential, graphite plays an important role. It has some drawbacks, however. It is soft and oxidation prone and, because it often contains a certain iron content, may suffer from carbon monoxide disintegration due to carbon deposition in the pore structure. The use of purer, denser material gives a notable improvement, but makes the product much more expensive. Top grades are repeatedly impregnated with tar and refired in order to eliminate porosity.

Silicon carbide bricks have increasingly found application in the blast furnace lining. They combine high strength and high thermal conductivity and therefore have good thermal shock resistance. Also resistance to corrosion by acid slags, alkali vapor, zinc vapor and oxidizing gases is good. These properties have made silicon carbide bricks excellently suited for application in the most demanding locations of the blast furnace shaft. The main research goal in developing the quality of these bricks remains to determination of the best bonding system for the SiC-grains. The bonding phase can be SiC itself, but also silicon nitride or sialon. Practical tests are being carried out to find the answer. A wall lining in which cooling elements, graphite bricks and silicon carbide bricks are integrated is presently the best solution to cope with aggression by a complex process.

In the hearth, liquid metal, liquid slag, zinc and alkalies are the factors to be dealt with. One concept is to make a crucible of high thermal conductivity, from carbon and graphite materials, that is well cooled and on which iron and slag form a solidified layer. The design must guarantee an adequate heat flow and high thermal conductivity, from carbon and graphite materials, that is cooled well and on which iron and slag form a solidified layer. The design must guarantee an adequate heat flow and high thermal conductivity of the graphite and semi-graphite material employed is essential. Low porosity is important for this property and also to prevent penetration of iron into the pores.

Another approach to the construction of the hearth and bottom of the blast furnace is to use high grade oxide refractory materials, backed up by graphite layers, and hearth and bottom cooling. The demands on these oxide materials are high, refractoriness, thermodynamic stability, resistance to corrosion and resistance to thermal cycling must be optimal. With temperature cycling thermal expansion is a critical property. Picochromite is being studied as a potentially feasible material.

In the taphole of a blast furnace the eroding effect of the hot metal stream is severe. A good solution has been found in using blocks of high grade graphite. High strength and high density are prerequisites and manufacturers try to achieve this by impregnating graphite blocks with pitch and refiring them, sometimes repeatedly.

2.4. Hot Metal Transport

The commonest way to transport hot metal from the blast furnace to the converters of the oxygen steelmaking shop is by torpedo car and, subsequently, by hot metal ladle. A substantial number of plants, however, still employ mixers. They transport the hot metal by ladle to the mixer (which can contain several torpedo ladle charges), and they use a ladle to bring the hot metal from the mixer to the converter. In some cases hot metal is transported by ladle directly from blast furnace to converter.

During the last 10 to 15 years the practice of pre-treating the hot metal has been developed, in particular the process of desulphurization. This has increased the load on the linings of ladles and torpedo cars, due to increased temperature, longer residence time, mechanical influence due to gas injection and corrosion by aggressive slags.

Fireclay bricks have long been used as the lining material of torpedo cars. But with more aggressive circumstances, fired high alumina bricks, based on bauxite and andalusite became a cost effective alternative. The injection of basic components into the hot metal made the slags very aggressive and refractories engineers looked for even more resistant bricks. Alumina-chrome bricks, basic bricks and in the last years also bricks made from Al_2O_3-SiC-graphite are being used. Particularly the latter type is expected to become generally used for the parts of the lining subject to severest attack. Consideration could be given to using magnesia-graphite bricks, which are well established in converters, for torpedo car linings. This would require development work to adapt this type of brick to the specific demands of this application. With the better performance of the new lining materials the need for gunning repairs has decreased. A vulnerable area that still needs a lot of attention is the mouth of the torpedo car. Often a monolithic lining is used and intermediate repairs are necessary. These repairs imply cooling off the lining and decreased availability of the installation.

Fireclay bricks still are the most widely used lining materials for hot metal transport ladles. In those ladles in which desulphurization is carried out the performance of fireclay bricks may be inadequate and bricks of higher alumina content are used. For slaglines, bricks of even higher quality such as Al_2O_3-SiC-C are sometimes used.

2.5. Converters

Oxygen converter processes exist with varying modes of oxygen injection, through a lance, through the bottom or both. A widespread operating practice is topblowing with bottom stirring with inert gas. The converter process has also been influenced by the introduction of continuous casting and by vacuum treatment of the steel after tapping the steel from the converter. The operating conditions of oxygen converters have become more severe because of increased stirring and higher tapping temperatures. On the other hand, hot metal pretreatment to remove sulphur, phosphorus and silicon, and ladle metallurgy have taken some of the burden off the converter process. A reduced slag volume, a lower iron content of the slag and a shorter process cycle are favorable for the performance of the refractory lining. In the last decade important developments have taken place in converter refractory materials and converter campaigns of many thousands of heats have been achieved. In Europe there has been an evolution from tar-bonded doloma as the main lining material to tar-bonded magnesia. In recent years magnesia-graphite bricks, often with a resin bond, have been introduced especially for those parts of the lining that are exposed to particularly aggressive attack such as slag lines and around tuyères and gas stirring bricks.

Because of the ratio in price between tar-bonded doloma bricks, tar-bonded magnesia bricks and magnesia graphite bricks is roughly 1 : 2,5 : 5, each type of brick has its place in lining design; each steel-plant has its own design adapted to the local process situation. When converter availability is not a constraint, as often is the case with less than full capacity production, steel plant managers tend to use the cheaper brick qualities, but with high production levels the expensive linings become cost effective.

With natural raw materials such as doloma and magnesia produced from mineral deposits, the purity of the refractory grain that can be obtained is limited. This has led to the development of synthetically produced magnesia from seawater or brines. Magnesia with 98, 99 or 99$^+$ % MgO has become available in quantity for brick production.

The introduction of graphite in magnesia bricks has boosted the performance of these bricks by tens of percent. It was found that flake graphite, produced from natural rock deposits, showed the best performance, due to its good oxidation resistance. Synthetic alternatives have not so far been able to match the quality of the natural product. Addition of metal powders (Al, Si, Mg) has improved remarkably the performance of magnesia-graphite bricks. Their effect on the properties of the brick is complex. An important fourth component is the organic binder. Phenolic resins have shown advantages over pitch. Further development work could still improve this type of brick.

An important tool in controlling lining wear is repair by gunning. It enables the operator to balance the lining by strengthening weak spots in the lining either preventing or by repairing holes that have developed. The disadvantages of this method are obvious, operation and gunning material are costly and it decreases the availability of the installation. To be effective in actually prolonging campaign life it is necessary to give a treatment every few heats. Instead of wet-gunning a more durable repair has been sought in the development of flame gunning. Equipment and process are much more costly than wet-gunning and the results have generally not been convincing.

The taphole of a converter is an area of high wear. Periodic repairs are a necessity. Rings of high grade fired magnesia, pitch impregnated are used as replacements in a worn taphole. It remains desirable to improve material quality and repair technique in order to limit as much as possible the frequency of repairs.

Gas injection bricks in converter bottoms are standard. According to the process, different designs are used. A prime difference is whether oxygen containing gas is injected or inert gas. In the case of oxygen, annular designs with an enveloppe of protective gas are used. For inert gas bricks with pipes, capillaries or slits are used. The converter operator has to pay great attention to these bricks and their immediate vicinity and control wear by proper slag practice and gunning. A more resistant brick would be welcome. It is necessary to investigate this area in more detail in the near future.

2.6. Argon-Oxygen Decarburisation Converters

AOD-converters which produce the bulk of stainless steel are relatively large consumers of refractory material, the specific consumption in Europe lying between 10 and 15 kg per tonne of steel. In Europe doloma is the refractory material used, both as fired bricks and pitch-bonded bricks. In Japan much more expensive magnesia-chrome bricks are used, the specific consumption is much lower, but the cost per tonne of steel will be higher. If in Europe more severe operating conditions must be met or longer campaigns were required, magnesia-chrome bricks offer the possibility of achieving these. A range of qualities, direct bonded, semi-rebonded and rebonded magnesia-chrome bricks are available.

2.7. Electric Arc Furnaces

EAF's produce a sizeable part of the total steel production (about 20-25 %) and this process is considered likely to remain one of the prominent steelmaking routes. The ultra high power (UHP) operation of EAF's is possible by improvements in the quality of the graphite electrodes and in the refractory lining. Electrodes with high electrical conductivity have been developed. Electrode consumption is an important cost factor in EAF-steelmaking. One of the attempts to prolonge the life of electrodes is the development of coatings that retard oxidation of the graphite. To cope with the very high temperatures and radiation from the arc, the refractory lining above bath level has, in many EAF's in Europe and in Japan, been replaced by water cooled panels. It is expected that this trend will continue and lead to decreased consumption of refractory material.

2.8. Steel Teeming Ladles

An important feature to mention about steel teeming ladles is that metallurgical treatment of the steel in the ladle is expanding. This implies that the load on the refractory lining materials is increased. Secondary steelmaking processes comprise vacuum treatment, gas stirring, ladle furnace treatment etc. and are associated with higher steel temperatures and longer resistance times. Also the SiO_2-rich lining materials affect the quality of high grade steels unfavourably and reduce the effect of desulphurisation treatment. Because of this the traditional ladle lining materials such as sand, firclay and pyrophyllite are being replaced by more resistant materials. In Japan zircon silicate bricks and mixes have been developed. In Europe and the US high-alumina bricks have found application and in Europe, also doloma bricks. Magnesia chrome bricks are often used in the US. To a lesser extent magnesia, magnesia-graphite, silicon carbide and zircon-alumina bricks are being used. Basic and high-alumina linings require careful preheating and maintaining at temperature. Therefore sliding gates and heating equipment must be installed.

In recent years the application of castable linings has become popular, especially in Japan. The main advantages, which this method shares with monolithic sand and zircon linings, is the case and speed of installation. Low cement castables with high alumina grain are giving a good performance. Summarizing, it can be said that there is a great diversity of steel ladle linings of both monolithic and brick type. In Europe the sand lining installed with a slinger, is still popular but high alumina and especially doloma brick linings are making inroads. It is to be expected that the increase in continuous casting and secondary treatments and the urge to make cleaner steel will lead to increased use of high grade lining materials such as doloma and high alumina bricks. On the other hand, the labour saving castable linings will be attractive alternatives but more development work will have to be done in Europe.

By now, the transition from stopper rod to sliding or rotating gates has been completed in modern steelworks. These gate systems are arrays of refractory parts and mechanical components. The crucial refractory parts are the plates, which by sliding or rotating against each other, open or close the bore that permits steel flow. The most popular material is high alumina, corundum crystals in a mullite matrix.

Quality improvements that have been made include pitch impregnation and the introduction of graphite. Avoiding smoke formation (pitch) is essential.

The performance of a pair of plates is much lower than the life of the ladle lining, which means that plates have to be replaced frequently. Extended life would be desirable.

Magnesia plates are more resistant to slag and metal than hgh alumina plates, but their high thermal expansion makes them vulnerable to thermal shock. They are used mainly for special steels in small sizes. Zirconia plates are known to give excellent performance but are too expensive for general use.

Gas injection through special bricks in the bottom or the wall of a ladle is widely practised, the main purpose being to homogenize the steel in the ladle. The bricks are made of MgO or Al_2O_3 and wrapped in a steel casing. Porous material that gives off fine bubbles to the steel is common, but also dense material is used. In that case injected gas passes between brick and steel shell. Gas injection bricks and their immediate surroundings are areas of preferential wear. More resistant bricks which last as long as the ladle bottom are needed.

2.9. Continuous Casting Systems

Over the last decade continuous casting has replaced the traditional block teeming in most integrated steel mills. By continuous casting slabs, blooms and billets can be produced avoiding soaking of semi-finished products in soaking pits. Using this technique energy is saved and the production route is shortened.

Refractories are only used in the tundish. In the tundish bricks, castables or special linings slabs are used. The main area of development in the tundish is at the moment the development of baffle walls, which are used to avoid undesirable vortex formation. In these walls and in the nozzles, experiments are being carried out with ceramic filtering systems.

For special steels the horizontal continuous caster is under development. In this case the connection between the tundish and the metal mould is crucial. For this purpose breaking rings are under development nowadays made out of boron nitride or composite materials.

2.10. Steel Mills

In the rolling mills processes are carried out at such temperature levels that no severe problems occur for the materials involved.

3. DEVELOPMENT OF NEW PROCESSES

3.1. Ironmaking

New processes for iron making from iron ore are under development. The aim is to develop processes which are more flexible, use less energy, show less environmental problems and are less capital intensive. Several steelmaking companies and engineering firms are now developing cokeless iron production units. The most advanced process is the cokeless ironmaking KR-process which has been developed by Korf Engineering in West Germany (2, 3). Demonstration plants are now under construction in South Africa and the United States. In Western Europe and Japan related processes are also under development. All the processes have one problem in common, this being the refractory lining to be used in the reduction and smelting vessel. For the time being $MgO-Cr_2O_3-C$ bricks seem to have the best chance. Thorough investigations are needed to find the best possible combinations of material and construction.

3.2. Steelmaking

In Brazil a new type of converter is being introduced. This little converter is called the energy optimizing furnace (EOF) (4). The EOF process is a mixture of oxygen and open hearth steelmaking, but built in a furnace similar to an electric arc furnace. The refractories and cooling used, are very similar to these of electric arc furnaces.

In the area of electric steelmaking the development of DC-current arc furnaces and plasma arc furnaces is of importance. In both cases the process conditions are more severe for the refractory bottom lining.

3.3. Steelcasting

As mentioned earlier, continuous casting of steel is becoming more and more important. Also dramatic changes can be observed in continuous casting techniques. First of all the introduction of horizontal billet casting. In this case a large scale industrial breakthrough is held up since the connection between the tundish and the mould, the breakring, is not sufficiently efficient. This piece of advanced ceramics is either too fragile or too expensive. A cheap and reliable product needs to be developed.

Other developments show casters producing very thin slabs, which can be rolled in much simpler wide stripmills.

This development indicates that large reheating pusher furnaces will become superfluous in the near future.

3.4. Steel Rolling Mills and Coatings Facilities

As mentioned above the rolling mills will become simpler when thinner material can be cast. This gives way to processes at lower temperatures. In the cold strip mill a shift from batch annealing to continuous annealing is well underway. The continuous annealing lines are very complex. An important development is the introduction of ceramic radiant tubes to heat the strip. Development in this area is gaining attention.

4. FINAL REMARKS

Modern steelmaking is heading towards relatively small and flexible integrated units using a smaller number of process steps, avoiding a large number of installations and furnaces. This trend will diminish the refractory consumption per tonne of steel. On the other hand the refractories needed for the new installations need well defined properties and high quality.

The new processes can only be developed when related research is carried out on the refractories and advanced ceramics which are required. It will be difficult for the declining refractory industry to carry out this research, and the assistance of larger industries and research institutes will be needed.

5. REFERENCES

(1) Fukuoka, H., "Fine Ceramics for Future Creation", Japan Fine Ceramics Association, Annual Report for Overseas Readers (1986), pp. 4-52.

(2) Korf, W., Wervelbed directe reductie EGKS studieberich, (1981).

(3) Papst, G., "The KR-Process, a Cheap Basis for Hot Metal Production", Lecture Metec (1984).

(4) Weber, R., Wells, W., Stand des EOF - Stahlherstellungs-verfahrends, Stahl und Eisen, 103 (1983), pp. 127-1208.

4.6. Materials used in the Processing of Superalloys

Dr. J. Morlet

Imphy S.A.

Paris, France

The compositions and some of the properties of superalloys create, for the processing of these materials, unusual conditions in terms of chemical reactivity, hot temperature characteristics, machiniability etc. The operating conditions and the quality requirements have a direct impact on exceptional requirements for the manufacturing process. All this should tend to create a special technology with respect to equipment and also materials which contribute to the processing, such as furnace refractories, tools and dies. On the other hand the quantities involved are not large enough to justify systematically a specific technology and except for a few examples the materials which are used are essentially adaptations of existing products by selection of sources or minor adjustments of materials used also by other technologies. The following note makes a brief survey of the main materials in use and their potential for improvement.

In the manufacturing process of superalloys through vacuum induction melting, the lining of the furnace is part of the system in the same way as the atmosphere. Reactions take place during melting, refining and final deoxidation and the wear of the lining is one of the limiting factors to the possibilities to the furnace. Those limitations concern the bath temperature, the vacuum level and the holding time at high temperature and under vacuum. This partially contributes to the reputation of the VIM as a poor chemical reactor compared with other processes used in metallurgy. The cost of the lining is also one of the major constituents of the melting cost.

Since every ceramic lining more or less reacts with the molten superalloys (the use of pure Ca O lining is only theoretical) all industrial linings represent a compromise between, chemical inertness, mechanical and wear resistance, cost and handling ability. Compositions mostly used are magnesia based on a 90-10 type available now as brick or ramming powder. For some Ni based alloy compositions a spinel type lining is 70 % $Al_2 O_3$ - 30 % Mg O is also used. Refractory makers tend to hold to these compositions and improvements are related to:

- minor elements in the chemical composition. This concerns mainly a decrease in the content of oxides which have higher reactivity with the melt, essentially Si O_2 and iron oxide. Chemical interaction of the melt with the lining usually correlates with wetting and penetration. This can be evaluated by examining the lining after a malting campaign or by a laboratory test such as the sessile drop method.

- a selective particle size distribution, a better blending and mixing technique and firing sequence to get a more compact material.

- an elaborate shape of bricks allowing to decrease or even avoid the use of cements.

- ramming powders more tolerant to dimensional variations to allow a mixed mounting with bricks and eventually a porous plug. For these materials a slow sintering process is wished.

    Among other auxilliary equipment using refractory materials in the manufacturing of superalloys, the newer technologies involve:

- porous plugs which have to behave in the same way as the lining in terms of wear (25 to 30 heats) while maintaining their porosity. Much progress has still to be made not only in plug manufacture but also in the whole injection technology.

- the filters (porous or extruded parts). Although the concept is particularly appealing, the technology so far remains unsatisfactory as yet. While the technique is particularly efficient to entrap solid particles in the liquid bath, it is difficult to cast large electrodes at a steady rate and many foamy materials remain too brittle with the danger of pollution of the ingots by extracted filter particles.

- nozzles for inert gas atomization. These components are exposed to a long contact with liquid metal flowing at high speed. The cleanliness requirements for the powder imposes a perfect integrity of the nozzle all along the atomization process. Several processes have been proposed to manufacture these critical parts including a final hot isostatic pressing.

INSULATING MATERIALS

The narrow hot working range of superalloys requires special precautions to minimize heat losses during the hot working sequence. Some years ago asbestos was used but the health regulations concerning this material have modified the technology.

For hot topping of ingots or electrodes, since active hot topping has often been replaced by inert systems, materials now belong to the same type of insulating product developed to improve furnace efficiency. In this case, the purpose is not the integrity of the ingot but the necessity to create a compact zone to be able to weld the stub for VAR or ESR remelting. The hot topping rings are made of mineral wools developed in the last decade to make light highly efficient insulating linings essentially for heat treatment furnaces.

The same types of materials are used as "blankets" to minimize heat losses during forging. They are used in combination with lubricants like glass weaving. All together these developments may halve the required number of reheating sequences.

DIES AND TOOLS

The high temperature characteristics of superalloys impose severe conditions on forging and extrusion tools. Regular hot die steels are used but in some cases it is nessesary to manufacture inserts in high temperature alloys such as A 286, Refractaloy 27 or Inco 718. Ceramic coated dies are also used.

For extrusion of superalloys and mostly consolidation of prealloyed powder, the stress levels on the tools are such that Maraging or 35 NCD 16 steel are required for the rams.

The classical closed die forging operations are carried out with the standard die steels but the tool life is often short. The isothermal forging of superalloys requires special molybdenum base alloys (TZM) which makes tools expensive.

Experiments to find a cheaper solutions have been so far unsuccessful. IN 100 grades are still required for isothermal forging of titanium alloys.

Cutting speeds are low and tool wear rapid when machining is carried out with standard grades of high speed tool steel. The powder route enables a finer structure and novel composition for tools to be developed.

Among the successful compositions proposed lately are

	C	Co	W	V	Cr	Mn
A	2,3	11	9	6	4	5
B	2	15	9	5	3,5	3

For finishing operations ceramic tools are now available, some being reinforced with whiskers. This changes dramatically the machining conditions: cutting speeds can be seven times higher and material removal rate four times greater.

4.7. Materials for Sensors to be used

at High Temperature

Dr. O. De Pous

Eniricerche - Monterondo

Italy

1. INTRODUCTION

The direct monitoring of industrial processes, using on-line computer control technology, currently represents a significant economic opportunity, particularly for processes operating at high temperature. Information on direct inputs such as temperature, gas composition, pressure, etc., has to be fed to the control system, using various sensing devices.

The main requirements of these sensing devices are their sensitivity (selectivity) and their infallibility when operating in harsh environments. This can be achieved by using high performance materials, commonly used for their structural properties as well as their functional characteristics.

This document is mainly devoted to the needs of R and D in the field of materials which require sensors working at elevated temperatures 200°C-2000°C.

2. TEMPERATURE CONTROL

For the determination of temperatures we can use:

- thermocouples
- thermistances
- radiation thermometers
- differential extensiometers

2.1. Thermocouples

For intermediate temperature ranges (500°C to 1200°C) chromel/alumel thermocouples can be used in oxidizing as well as reducing atmospheres, without specific problems.

At elevated temperatures > 1200°C a platinum/platinum rhodium system is used and in order to prevent rapid deterioration of the metal wires, high purity alloys are employed. Unfortunately, high purity metals exhibit significant grain growth at elevated temperatures with consequent deterioration of mechanical properties and progressive weakness in the welded area.

To overcome the grain growth problem, ceramic particulates, such as thoria, are dispersed within the metal, which also improves resistance to abrasion and rigidity of the wire.

A second problem associated with the use of platinum base alloys results from the significant volatility of platinum (as platinum oxide). As a consequence, and in particular under alternating oxiding/reducing atmospheres, the cross section of the wire decreases and the wire becomes susceptible to deformation.

The same problems apply for tungsten/tungsten-rhenium thermocouples. In fact, these thermocouples are not only sensitive to traces of oxygen, but also to atmospheres containing hydrocarbons and nitriding components.

It should be noted that the problems reported above are aggravated when thermocouples are used under gas pressure. This is particularly the case in recently developed hot isostatic pressing (HIP) facilities for which carbon or tungsten based heaters are used. In fact, even under argon pressure, a significant contamination of the thermocouples can be observed, due to the deterioration of the heating element, or the decomposition of the load present in the furnace. For example, tungsten and rhenium are sensitive to oxygen, carbon, carbon monoxide or nitrogen contamination. Carbon thermocouples (carbon and carbon doped with boron), used, for example, around 1800°C – 2000°C for the drawing of optical fibres, are sensitive to oxygen and nitrogen and also to molybdenum and tungsten. Platinum thermocouples can be poisoned by silicon, particularly when carbonizing atmospheres are present (sintering of Si_3N_4, SiALON).

In the design of sheathed thermocouples (Fig. 1), several insulating materials can be used. For a perfect fit of the interstices, monospherical powders produced by sol-gel or

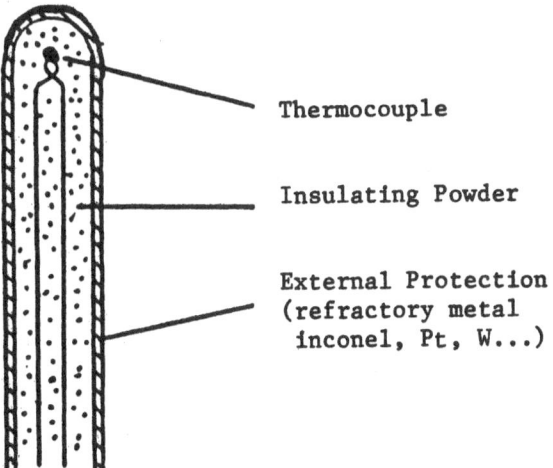

Fig. 1.   Sheathed Thermocouple

equivalent technology (Laser, colloid chemistry) could be of
interest.

The most commonly used insulating materials are:

- Alumina powder. The temperature of application is limited to
  1400°C. If this temperature is exceeded, a significant
  sintering of the powder can be observed.
- Boron nitride. This material is expensive and contains B_2O_3
  which could result in the formation of glassy phase and, at
  high temperatures, volatilisation of different boron oxides.
- Aluminium nitride. This powder has recently become available
  on the market and is of great interest as its thermal
  conductivity is approximately three to four times higher than
  that of alumina. Using aluminium nitride as insulating
  material should enable an improvement in the response time of
  the thermocouple.
- Beryllia. This, in principle, is the best insulating material
  for elevated temperatures. There are, at present, some
  problems concerning environmental safety, but it remains a
  good candidate for thermocouples used in nuclear power plant
  technology.
- Magnesia. This remains the most commonly used insulating
  material because it is cheap and can be used at temperatures
  as high as 1800°C.

For sheathed thermocouples, the temperature limit is
indicated by the materials used for the protection. The
protection can be particularly exposed to corrosion and erosion
effects and the recrystallization at elevated temperature causes

some brittleness.

According to industry, the problems to be faced today are:

a) The requirement for thermocouples to work for long periods in air > 1800°C.

At such temperatures, engineers are looking on ceramic as the most suitable external protection. Alumina, silicon carbide, silicon nitride, aluminium nitride and zirconia protections are possible candidates. Unfortunately, silicon nitride, aluminium nitride and carbide are not resistant in air at temperatures over 1400°C and 1600°C respectively, and zirconia becomes pervious to oxygen at elevated temperatures. In extreme cases, ceramic protection consisting of beryllia may be considered.

b) The second requirement concerns the progressive drift of signals for systems operating for weeks and months on end. This is the case, for example, for creep resistance measurements as well as safety control systems for petrochemical plants or nuclear plants.

To improve the thermal stability of these systems, it is necessary to have a better understanding of the mechanism of degradation of thermoelectric response, especially in terms of interdiffusion, gas dissolution, etc.

2.2. Thermistances

The measuring of the temperature within the engine technology is also an important task. Information on the air intake temperature can be helpful for the optimization of the combustion system (pollution level versus fuel consumption). The new materials used for such purposes are negative temperature coefficient thermistances (nickel manganese oxide). Thermistances are small and cheap but suffer from a lack of reproducibility, non-linearity and stability. In fact, the active part of the sensor should be well protected from external contamination.

2.3. Radiation thermometry

This technology is used for extreme temperatures, i.e. 1600°C to 2500°C, or when the presence of electromagnetic fields represent significant errors in the signals provided by conventional thermocouples. The main problem concerns the difficulty in obtaining exact indications of the temperature. Systematic or erratic errors in temperature measurements could result from:

- variation of emittance from the surface (in particular for liquid metals)
- parasitic reflectance from the hot furnace walls
- the difficulty of accurately controlling the temperature within a wide range of values
- the presence of dust in the sight path as a consequence of high absorption and re-emittance of radiation (light emitted by the hot dust)
- problems resulting from the small size of the target area
- the selective absorption of gases
- the deposition of an absorbing, thin film on the surface of the refractory, especially in the case of furnaces operating under vacuum
- temperature instability, especially in the case of furnaces equipped with gas or oil burners.

The solution to these problems is the use of a complex multi-wavelength system.

For a complex multi-wavelength system, the concept of the measurement of temperature is the comparison of infrared radiation levels measured at two selected wavelengths. As this does not represent an absolute determination of the level of radiation at a single wavelength, the measurement is immune from errors caused by emittance changes of the target materials, bursts of steam, clouds of dust, etc., in the sight path. In order to obtain maximum accuracy in the measurement of the temperature, the materials used for the manufacture of the window have to be chosen carefully. The following should be considered:

- the availability of transparent materials for the manufacture of a non-absorbing window operating at elevated temperatures, i.e., 1500°C - 2000°C
- comprehensive information on the emissivity of different refractory materials (with different surface finishes), Al_2O_3, ZrO_2, chromium carbon oxides
- the manufacture of special optical fibres operating in the IR range (fluorinated glass)

2.4. Extensometry

Such a system could be used at elevated temperatures with possibly a minimum of interaction with the surrounding atmosphere. The use of a sophisticatd displacement device is necessary, in order to detect the differences in thermal expansion. If the sensitivity is low, the reliability remains high.

In this field, the development of new materials, in particular metal oxides, with a high anisotropy of thermal expansion should represent a significant break-through for high temperature measurement in gas furnace.

Few attempts have been made to use materials with different thermal expansion. For example, at elevated temperature it is possible to make use of the benefit of the large difference of thermal expansion of pyrolytic carbon in orthogonal and parallel direction to the preferentially oriented graphite plane. At 800°C thermal expansion is respectively $30 \times 10^{-6}/°C^{-1}$ and $3 \times 10^{-6}/°C^{-1}$. Example of sensor design is shown in Fig. 2.

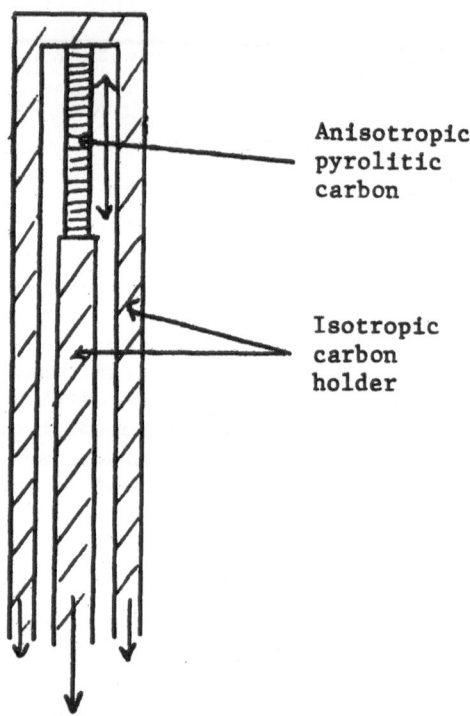

Anisotropic pyrolitic carbon

Isotropic carbon holder

Fig. 2.   Extensometer for the determination of elevated temperatures

3.   GAS COMPOSITION

Many devices have been proposed for the determination of the composition of gas mixtures at room temperature or for medium range temperatures (300°C). The main problem of this type of sensor is the absence of selectivity. However, selective gas sensors can be manufactured using a solid electrolyte. Typical examples of solid electrolytes are stabilized ZrO_2 for oxygen, LaF_3/BaF_2 for fluorine, ionic metal hydrides and protonically doped beta-alumina for hydrogen.

Examples of applications are also the determination of gas diluted in molten media such as liquid metals e.g., steels, copper etc., or molten salts e.g., liquid electrolyte in the production of aluminium. Research efforts should be devoted to the stability of solid electrolytes selected for the manufacture of the sensor. In fact, at very low concentrations and very high temperatures, the solid electrolyte composition diverges from stoichiometry. Hence, the membrane becomes weak as a consequence of an increased level of voids in the anion network and the signal output does not conform to theoretical values forecast by the Nernst equation:

$$\text{Sensor output} = \frac{RT}{4F} \ln \frac{P_{gas}}{P_{ref}}$$

For example, this is the case for sensors used for the on-line determination of oxygen levels below 10 ppm in liquid steel produced by continuous casting.

This type of sensor can also be used to determine the composition of gas mixtures. In such cases, electrodes (commonly made of platinum) are deposited on each side of the solid electrolyte membrane to collect the electric signal which is a logarithmic function of the oxygen concentration (Fig. 3 & 4).

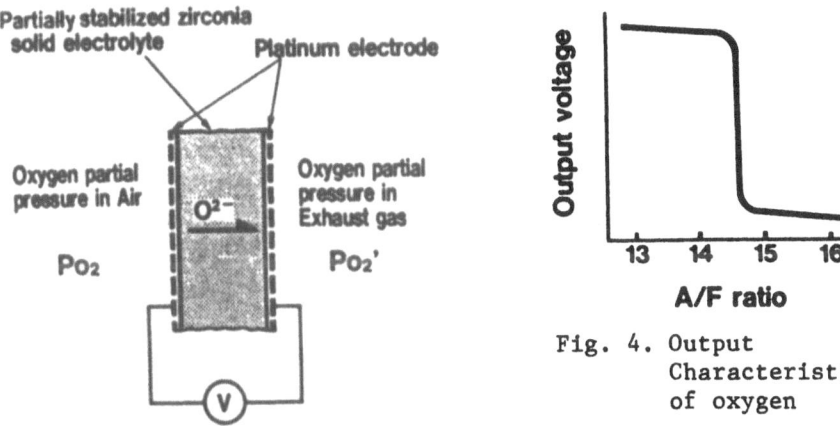

Fig. 4. Output
Characteristics
of oxygen

Fig. 3. Schematic representation of
conventional oxygen sensor

Such devices, for example, are used for the determination of the stoichiometry of the exhaust mixture from combustion systems. Unfortunately, porous electrodes made of platinum are sensitive to poisoning (such as any platinum catalyst), to sintering (when they are exposed to excessive temperature) and they are not resistant to gas flow erosion.

486

A specific requirement is the manufacture of metal oxide (carbide/nitride) electrodes. This approach which has been the aim of several studies concerning the development of fuel cell technology, can also be used for the manufacture of gas sensors. It would be useful to develop special coating technologies for the deposition of such non-metallic electrodes by, for example, electron gun vacuum deposition or RF sputtering technology, and to analyse, on a more fundamental basis, the interface between the non-metallic electrodes and the solid electrolytes.

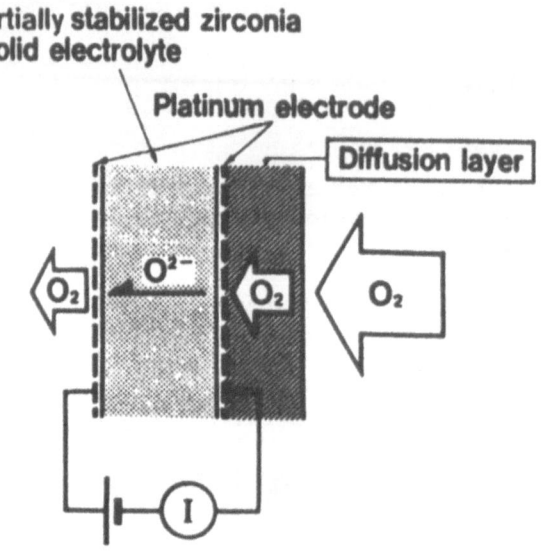

A need for the manufacture of a sensor with linear response over a wide range of oxygen concentrations has been identified, in particular for the control of automobile engines. The basic concept concerns the use of a solid electrolyte as a pumping system and to measure the gas diffusivity through a porous membrane or through well calibrated orifice (Fig. 5 & 6).

Fig. 5. Schematic representation of Lean Mixture Sensor

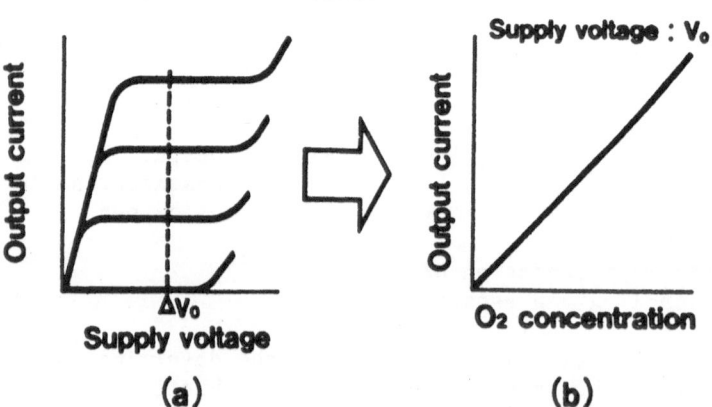

Fig. 6. Characteristics of Lean Mixture Sensor

The use of these systems presents us with the same problems concerning the selection of electrode materials. In this case also, the use of metal oxide electro-conductive electrodes such as perovskite will be of interest.

In an alternative technology, semiconductor metal oxides such as TiO_2 and doped SnO_2, can be used as sensitive material. The problems relating to the use of these semiconductors concern the signal dependence of temperature, and the fact that such systems cannot be used permanently in a range of oxygen concentrations as a great reduction in the oxide is observed ($Ti^{+4} - Ti^{+3} - Ti^{+2}$), (Fig. 7).

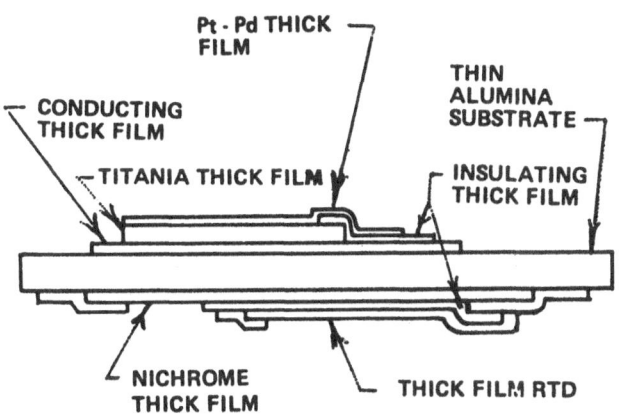

Fig. 7. Schematic of thick film titania oxygen sensor

Systems such as doped silicon carbide or other refractory elements should be considered for very low oxygen concentrations.

In both cases, the interface between the metal connection (wire, thin or thick film metal layer) and the active materials TiO_2 would justify a more in-depth analysis of interface problems because complex diffusion mechanisms can generate erroneous output.

4. PRESSURE AND ACCELERATION SENSORS

Amongst the different fields of application, we would like to report on some recent developments which have taken place within the automotive industry.

The first development concerns the replacement of the crank angle sensor by a combustion pressure sensor. This is needed to control the ignition timing and the fuel mixture.

This sensor is made from PZT piezo ceramic and is inserted withing the cavity of the spark plug (Fig. 8). The difficulty lies in estimating the absolute pressure. Even under normal circumstances, some fluctuations are observed which can be associated with the fluctuations in the temperature of the spark plug itself. There is a need to develop materials of piezo composition with the lowest possible temperature dependence. The use of thermocouples for automatic corrections of the sensor's response is not suitable due to the slow response time of this type of thermal sensor. Small, thin or thick platinum film or nickel probe should be more appropriate for this type of temperature compensation.

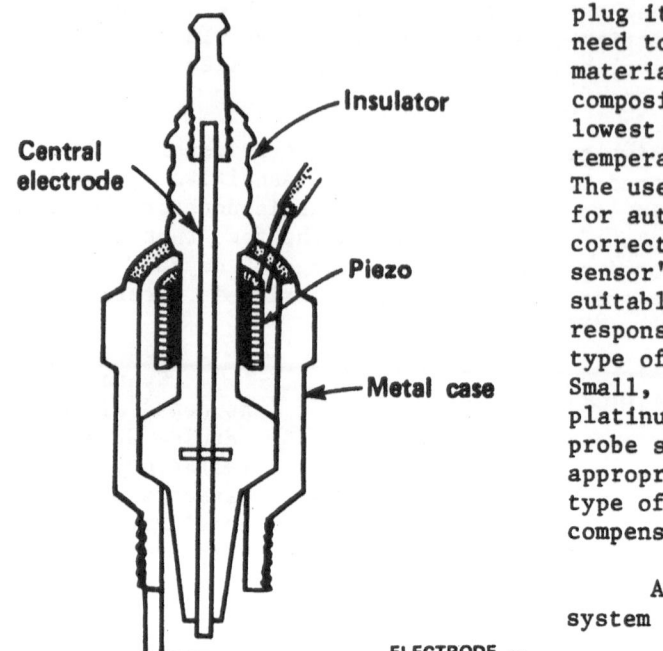

Fig. 8. Cross-section of spark-plug containing internal sensor

sensor, also based on the piezo-electric response of ceramic materials.(Fig. 9). In fact, this sensor does not, as is the case with the previous one, provide information on the pressure but on the presence of knocking instead. This sensor is made of poled material (barium titanate, monolithic or multi-layered) which generates a voltage when the sensor mass is subject to force during the acceleration

A far more simple system is a knock

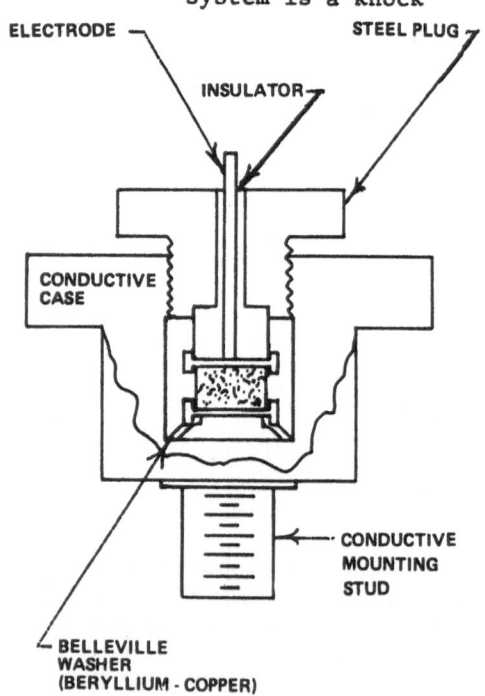

Fig. 9. Schematic of Piezoceramic knock sensor

characteristic of the knock. The amplitude is proportional to acceleration and consequently to the magnitude of the knock condition.

Equivalent systems can be made with a conventional pressure gauge (silicon micromachining techniques). The selection between piezo and silicon technology will be made based on the lowest temperature dependence of either device.

5. CONCLUSIONS

A multipluridisciplinary approach, including physics, chemical, electronic, surface and materials sciences, is required, for the development of a suitable sensor. Specific problems are pertinent to the use of sensors at high temperatures. Those problems relate to the thermal stability of the materials from which the sensor is made, the effect of the temperature on the response of the sensor and the interaction between external electro-magnetic field and the sensor itself. In addition, the selectivity of sensors vanishes at extremely high temperatures and controlled or uncontrolled fluctuations are often encountered. It must also be noted that apart from the problem with high temperatures, sensors are usually used in harsh environments and are consequently subjected also to corrosion, erosion, dust and contamination.

Another general comment concerns the aggravation of the time dependence resulting from either a degradation of the materials, or a slow migration of atoms which cause a change in the micro-structure of the materials.

Finally, one could summarise the research requirements as follows:

- research on the mechanism of diffusion at high temperatures on metal/metal, metal/ceramic, gas/metal, gas/ceramic
- research on re-crystallization, sintering mechanisms and creep resistance of materials
- research on electric contacts (and their passivation) to be used at elevated temperatures (400°C - 700°C)
- research on materials suitable for the electro-magnetic shielding of sensors at high temperatures
- research on the parameters influencing the stability of the temperature dependence of thermistors, advanced piezo materials
- research on the self-correction of temperature dependence employing integrated thermistors.

It should be noted that apart from the use of electro-physical or electro-chemical sensors to avoid some of the problems reported here, an innovative way in which to achieve this is the use of electro-optic devices. We believe that specific materials need to be developed for both the light emission or detection and the signal transfer (optical fibres, micro lenses suitable for high temperature application).

With regard to cost reduction, it should be of interest to consider the new opportunities resulting from the possibility of thin or thick film technology applicable at medium or high temperatures (passivation layer).

Amongst the problems to be solved, we should like to mention the need for a highly reliable sensor for the selective detection of NO_x in exhaust gases and for the control of emission of solid micro particles.

Generally speaking, we can forecast that gas sensors for the control of static combustion systems (domestic or industrial) will represent one of the most promising areas. In fact, it could be compared to that of the automotive application of a few years ago, for the control of pollution levels and the ability to keep fuel consumption down to its lowest possible level.

4.8. Materials for Furnaces

Mr. D. Brun

Stein Heurtey

B.P. 69, 91002 Evry Cedex, France

1. FURNACE EVOLUTION

During the last decades the development of thermal equipment has been directed toward:

- reducing inertia during transient periods,
- transfer with a qualitative and quantitative accuracy the heat required by the load within a minimum time,
- automatic control by models operating in real time, which implies a minimum alteration of the physical characteristics of the furnace components.

A furnace or a thermal equipment in operation may schematically be split up as follows:

- a load M which must either go through a precise cycle or be reheated from a temperature θ_0 to a temperature θ_1, thus receiving such a heat amount $Q = M\, Cp\, (\theta_1 - \theta_0)$; where M is the weight and Cp the specific heat;
- a load support or container which is a source of heat losses and whose influence on the load thermal profile must be minimized;
- an enclosure around the load aimed at reducing the heat losses and the influence of the outer atmosphere;
- a source which has to supply a sufficient energy to heat the load, its support and the enclosure. That energy supply must therefore be: $Q' = Q + losses$.

2. HEAT TRANSFERS

The load heat up speeds can be intentionally limited to avoid localized overheatings, damages or deformations of treated parts due to an excessive ratio between external heat flux compared to thermal conductivity but in most cases those limitations are imposed by the capacity of the heating source.

Apart from direct heating of the load by Joule effect, conduction or induction, two modes of heat transfer are possible, either seperately or jointly: radiation, convection.

2.1. Radiation

Radiation is appropriate to the whole temperature range, from low temperatures (some tens of degrees): infrared radiation for paint baking, to the high temperatures (some thousands of degrees): refractory ceramic melting.

The heat fluxes transferred depend on the physical characteristics of the product treated and of the emitter, in particular emissivity ε, as:

$$Q' + q \mu S (\theta_s^4 - \theta_m^4) t$$

where σ = mutual radiation coefficient
μ = form factor
θ_s = temperature of emitter source
θ_m = temperature of the load or receiver.

In most industrial applications, σ is determined by the load, and when μ is optimised the rate of heating is limited by the maximum admissible temperature of the source.

2.1.1. COMBUSTION

In the case of transfers by radiation from a flame, combustion gases can reach temperatures ranging from 1800 to 2800°C insofar as air is preheated or oxygen doped.

Compared with the present capability of some tens of $kW.m^{-2}$, fluxes ranging about hundreds of $kW.m^{-2}$ could be envisaged in the future for processes in which the load must reach 1000–1600°C. The rate being limited by the economics of burner lifetime and refractories maximum allowable operating temperature.

The materials must be designed to withstand chemical phenomena, viz. possible carburizing due to the presence of free carbon at injector nozzle, oxidizing due to free oxygen strongly reactive at the root of the flames, as well as mechanical phenomena, viz. creep, grain growth, crystal structure changes giving cause for fissuring and dislocation due to repeated thermal shock. Added to this damage to injectors or baffles is the progressive deterioration of the quarls or refractory blocks containing the flame where fracture and fissures allow the combusting gases to flow toward the platework and the insulating material anchors. Rapid on-off furnace cycling exacerbates the problem.

One possible solution is offered by the development of components in silicon carbide but much is left to be done to fabricate components with sufficient reliability and at an appropriate cost for complex shape.

The case of confined combustions, i.e. in isolated atmospheres is even more difficult. Combustion takes place inside a tube which is heated up to an adequate temperature and consequently radiates toward the load. The flux emitted by a tube unit area is therefore directly related to the temperature limit of the materials used for the casing and the inner components during a long operating period.

For the time being, with the metallic alloys used in current processes such as heat treatments of steel parts or strips, it is barely possible to reach 1100°C. Accordingly, for load target temperatures of 1000–1050°C the fluxes are, at best, 19 kW.m^{-2} and require excessive transfer times compared to the possibilities offered by radiant tubes with skin temperatures approaching the combustion gas temperatures. With an increase of 300°C of tube temperature it is possible to multiply those fluxes by 2 to 5.

Because of the oxidizing atmospheres inside the tubes, and the reducing, nitriding or carburizing atmospheres inside the treatment enclosure, it is even more difficult than with the open atmosphere to choose the materials with the appropriate life-time. Current research involving non-metallic carbides or oxides address the problem but there still remains substantial progress to be made in that field: maximum tube length, tightness of U or W tubes,...

The trend toward a combination of metallic and ceramic elements could also be an answer if the linkage and differential expansion problems were solved.

2.1.2. Electrical Resistors

High temperature electrical resistors provide the best hope for heating in controlled atmospheres. Even though radiating elements capable of withstanding operating temperatures above 1100°C do exist and work, their utilization is nevertheless limited and reserved for processes with a high added value. Also, because of their lower mechanical strength and their high-temperature brittleness, these materials cannot be used in furnaces where charging or discharging operations may cause vibrations and shocks. In addition to the high material cost, there is the price of a complex electric power control and regulation in order to take charge of the large variations of material's resistivity from ambiant to 1200/1400°C.

On top of all those limitations, the problem of how to fit the resistors through the enclosure walls still remains to be solved. The materials used must at the same time function as perfect insulators, have mechanical strength to support their own weight and the heating element expansion, and frequently as perfectly tight seals between the enclosures and the outer atmosphere.

2.2. Convection

2.2.1. Forced Convection

The second transfer mode requires a heat transfer medium. Here the fluxes depend on the load geometry, in particular the ratio between surface and mass, and on the product of the exchange coefficient, itself a function of the medium velocity, and the temperature difference between the medium and the load. The supply from the source can be written as follows:

$$Q' = h \, S \, (\theta_s - \theta_m) \, t$$

where h is a function of the velocity and of the physical characteristics of the medium at the temperature θ_s.

As against the radiation method, the transfer mode is independent of the load emissivity but is dependent on load surface characteristics. Thus it is possible to reheat by convection, either homogeneously or locally, all substances, including those which behave as reflectors of radiated energy.

Increasingly hydrodynamic-thermal modelling is employed to anticipate the heat transfer between media and of the load. It is hence possible to design furnaces in which the velocity and the behaviour of the media are controlled in order to obtain a preset heating profile.

Strong fluxes can be taken into consideration, including in a gaseous medium where coefficients varying from 30 to

$400 \text{ W.m}^{-2}.°\text{C}^{-1}$ are used, provided the processes remain within the average temperature limits, i.e. from ambient temperature up to 600/700°C. This is the current limit on the industrial scale for fans with low and mean flow rates ranging from 1,000 to 10,000 $\text{m}^3.\text{h}^{-1}$, to supply pressures necessary to obtain the velocities of 100 to 200 m.s^{-1}, in order to reach the above mentioned exchange coefficients.

Designing fans with ceramic wheels and casings to convey gases between 800 and 1200 or 1300°C would not only result in a substantial reduction of the heating time but also lead to an improvement in control precision of temperature profiles of the load and to increased efficiency through a more extensive use of convection preheating and homogeneizing.

At these temperature levels the thermal fluxes reached by forced convection could be 1 to 10 times as high as those obtained by radiation, for the same temperature difference or gradient between environment and product which radiation coefficient is low.

2.2.2. Heat Recovery

Combustion is associated with the notion of open loop. The continuous supply of oxidant and combustible agents implies the extraction of an equivalent mass from the process. The temperature of flue gases extracted is a direct function of the transfer temperature level. The user is therefore often faced with a problem of economical exploitation/recovery of flue gas latent heat. As soon as the process temperatures reach or exceed 1000°C there arises the question of thermal erosion wear and of oxidation resistance of the made use of materials. The thermal, chemical and mechanical problems are of the same nature as those mentioned for the combustion equipment, albeit with higher flux densities.

3. ENCLOSURE INERTIA

Any research into temperature variation of the load must take into account the enthalpic variation of the enclosure and load supports. Any attempt at optimizing the reheating cycles must face the question of the thermal behaviour of the containment.

A first improvement consists in decreasing the weight of the enclosure through the use of fibrous materials with low density and low conductivity. The inertia thus reduced can be significant. Further advantage can be gained by the use of insulating materials withstanding higher operating temperatures (above 1400/1500°C) as well as the thermal, mechanical and chemical attack of the flame.

Advances are still to be made with fibrous materials used in the form of felts or oriented fibre blankets. It could be interesting, in order to obtain a shorter response time, to cover fibrous materials of very low equivalent conductivity with a thin superficial layer notably more conductive. That layer, when exposed to the fluxes, would instantaneously and almost homogeneously reach a balance between the source and the load temperature, requiring only a low enthalpic supply. The superficial layer would radiate the energy received from the source onto the load with the maximum efficiency. The inevitably long time necessary for the extra-insulating sub-layers to reach their balance would then hardly influence the response times of the enclosure during the transient periods.

The load supports which are part of the enclosure are often one of the main sources of heat losses, even effecting production costs. There is, therefore, a requirement for support systems of high mechanical strength with minimum density and thermal conductivity.

If the load is not chemically inert at the operating temperatures considered, the materials must not only withstand mechanical stresses but also corrosion due to the furnace atmosphere as well as to the load itself.

Numerous tests have been performed with ceramic/metallic combinations and with ceramic alone. Provided there are no mechanical shocks during the process operations, some solutions - ceramic coating of cast parts - would apparently be satisfactory, but their implementation is complex because of the precautions to be taken during variations of temperature and of the linkage design. Non metallic oxides offer future hope, but problems of component production and mechanical characteristics versus time and temperature do not make them an economic proposition at the present time.

5. Optimisation of Components

5.1. Testing Metrology

Malcolm S. Loveday

Division of Materials Applications, National Physical Laboratory

Teddington, TW11 0LW, United Kingdom.

1. INTRODUCTION

Testing Metrology is a generic name used to describe the combination of measurement techniques of sufficient accuracy necessary to determine materials property data. The appropriate accuracy is to some extent dependent upon the purpose for which the data is required, since the level of precision sufficient for product release may be inadequate for determining reliable design data for high technology plant operating in a hostile high temperature environment. Relatively simple product release tests, such as hardness, impact strength, tensile testing (RT and HT) or even the simple one hundred hour stress rupture test, do not provide the design engineer or the plant operator with the necessary data to safely design or predict remnant life of components. Sophisticated testing techniques, such as tests to provide multiaxial creep data, low cycle fatigue, with or without combined creep dwell cycles, or creep crack growth measurements are becoming more important. Despite an improved understanding of the behaviour of materials, in some circumstances it is still regarded as necessary to resort to model component testings (1) or to test full size samples for which testing nuclear containment pressure vessels has provided the impetus to develop 100 MN testing facilities - see Fig. 1 (2). This latter machine will be capable of testing specimens up to 2.5 m in width, under uni-axial or bi-axial stressing conditions, up to approximately 300°C. The specific subject of high temperature testing metrology has been promoted in the UK during the last few years by the activities of the High Temperature Mechanical Testing Committee, resulting in a number of publications covering a broad range of testing techniques (1, 3 to 7).

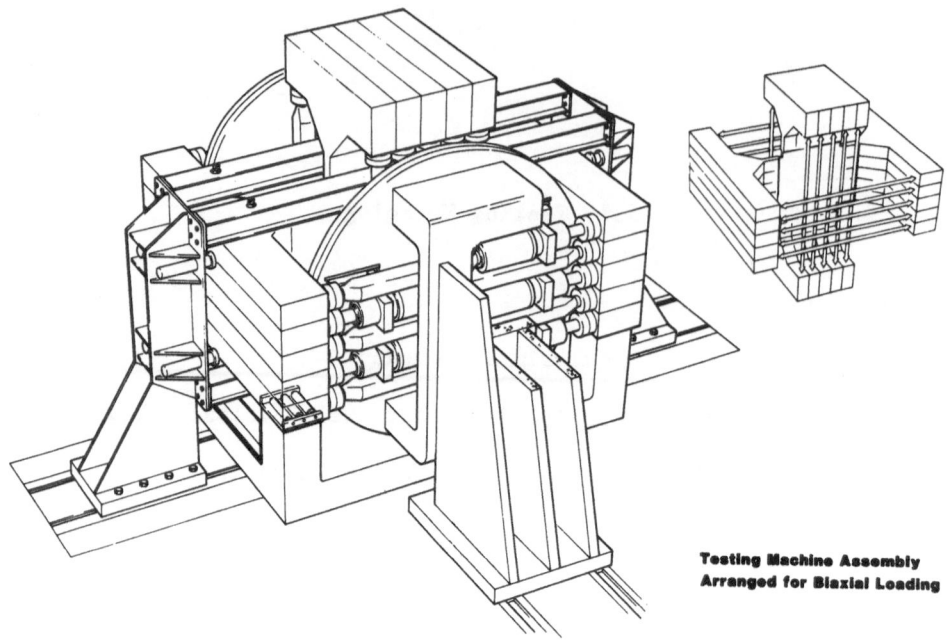

Testing Machine Assembly
Arranged for Biaxial Loading

Fig. 1. 100 MN structural features testing facility, under
construction at Risely Nuclear Laboratories (courtesy of
UKAEA).

Considerable advances have been made in improving testing
techniques to provide design engineers with more accurate and
relevant data than had previously been possible. Widespread
adoption of quality assurance schemes, following BS 5750 and
ISO 9000, are improving reliability of products. In the UK,
accreditation of testing laboratories is organised by NATLAS
(National Testing Laboratory Accreditation Scheme) and
administrated by NAMAS (National Measurement Accreditation
Service) giving with traceability of calibration via BCS
(British Calibration Service) approved contractors to provide the
documented traceability through a hierarchical chain to the
national primary standards. Similar schemes are operated in many
countries and some international agreements already exist to
provide reciprocal recognition of accreditation and thus help to
remove trade barriers.

Regrettably all these advances are of little use if the

testing standards on which they are based do not specify
adequately the accuracy in tolerances in the testing parameters,
which are easily achievable using modern technology. Indeed it
is of some concern that recent developments in the standards
field have been counter-productive to improved testing practice.
This apparent paradox is a consequence of the relationship
between the national standards organisations, e.g. BSI, ASTM,
AFNOR, DIN, JIS, etc. and their participation on an international
level at ISO (International Standards Organisation) or in the
recent discussions taking place within the EEC on the status of
European standards. It has now been agreed by the Heads of
States within the EEC that new European Standards will supersede
the equivalent national standards, which will be withdrawn within
a few months of publication of the new European standards.
Unfortunately International Standards tend to have wider
tolerances in the testing parameters since they are based on a
consensus of the various National Standards and thus to some
extent the present accuracy currently conferred by adherance to
individual National Standards will be lost when they become
superseded by the new European Standards, especially if they are
based on existing International Standards.

Finally it is perhaps surprising that in most countries some
of the most widely used materials property data used by
generations of engineers were acquired without any form of
specific code of practice or testing standard dictating the
manner in which the data is measured. Two examples of material
data in this category are Young's Modulus and Coefficients of
Thermal Expansion; values of such parameters are so often taken
for granted so that the question of their provenance is rarely
given a second thought.

2. CURRENT TESTING METROLOGY

The wide gamut of testing now required to determine a
material's properties may be appreciated by examination of
Table 1. Up-to-date reviews on most of the conventional testing
techniques are available in the publications cited (1, 3 to 5),
and the present state of the art on high temperature strain
measurement (8) and fatigue (9) are given elsewhere and the topic
has been recently reviewed (7). Requirements for testing above
1000°C are growing. New testing machines are now available
commercially for testing ceramics, see Fig. 2, and a recent
conference on Mechanical Testing of Engineering Ceramics at High
Temperature was well supported (17). A detailed comparison
between the tolerances in the testing parameters for tensile and
creep testing, specified in the new ISO standards and the
equivalent British standards has been presented elsewhere (18).

Hot hardness testing can usefully be exploited for

High temperature test description	Material Property	Symbol	Current Testing Standards	Associated Calibration Standards	Comments
Tensile	Young's Modulus Upper and Lower Yield Strengths Proof Strength Tensile Strength Fracture Elongation (Ductility) Reduction in Area	E R_{eH} & R_{eL} R_e R_p R_m A_t Z	BS 3688 ASTM E21 DIN 50145 ISO DP783	BS 1610 - Load ISO BS 3846 Extensometers	Compare with room temp standards:- BS 18 ISO 6892
Creep (Uniaxial)	Rupture lifetime, Creep Rate, Ductility Creep Crack Growth	t_f f	BS 3500 ISO R203 ISO R206 ASTM E139 DIN 50118	BS 3846 Extensometers	Also shape of curve, primary, second and tertiary
High Cycle Fatigue	No of cycles to failure S-N curve Fatigue Crack Growth	N_f ±	BS 3518 ISO R373 ASTM E647	BSI DD2	Load Control
Low Cycle Fatigue	No of cycles to failure S-N curve	N_f, T, or IN P e	ASTM E606 HTMT Committee Code of Practice (NPL)	BS 1610, Load, BS 3846 Extensometry BS 3500 Temperature limits	Strain or Load Control
Fracture Toughness	Plane strain fracture toughness Stress intensity Contour intergral Crack Growth Measurement	K_{1C} K C^* -	BS 5447 ASTM E399		Tests based on RT standards

Test	Description	Symbol	Standards	Standards	Notes
Impact	Absorbed energy Transition temperature	J T_T	BS 131 ASTM E23	BS 1610 Load	Charpy Isod New Precision Impact Testing Standard now drafted (BS)
Instrumented Impact	Absorbed energy (Area under load versus displacement trace)	J	No standards	BS 1610 Load E23	Charpy
Hardness	Brinell Rockwell Rockwell (N & T) Vickers	HB HR HR HV	BS 240 Pt 1 ISO R191 BS 891 ISO R80 BS 4175 ISO R1079 BS 427 ISO R192	BS 240 Pt 2 ISO R156 BS 891 Pt 2 BS 4175 Pt 2 ISO R1355 BS 427 Pt 2	
Wear	Pin on wheel method, Depth of groove for given number of revolutions		ASTM B611		
Thermal Expansion	Coefficient of Expansion		No standards		
Youngs Modulus		E	ASTM E231		Static Methods only

Table 1   Summary of high temperature tests and associated standards.

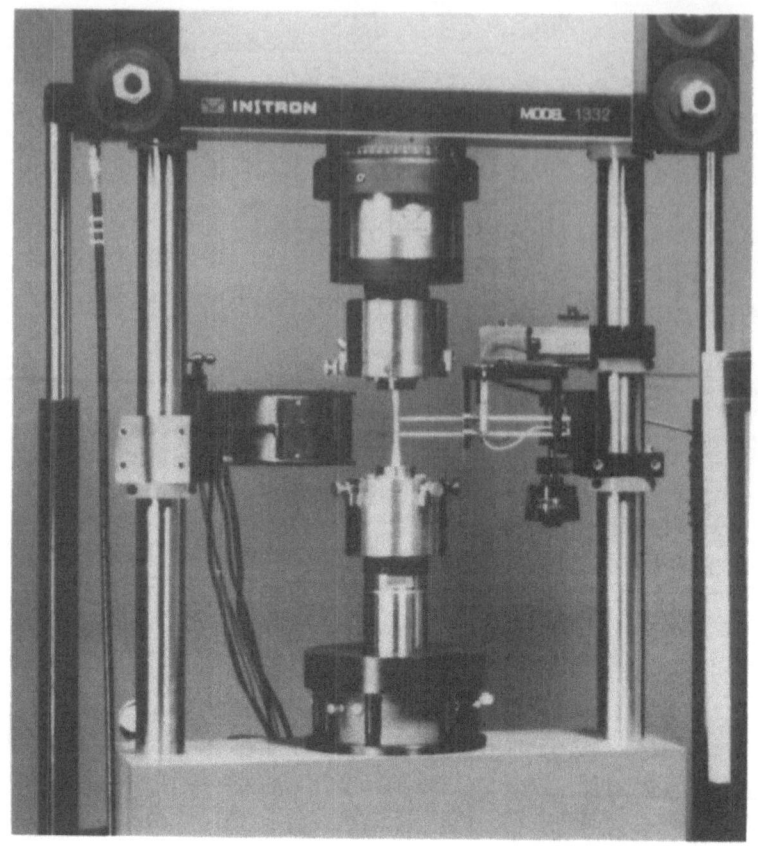

Fig. 2. Testing machine used to determine the tension/compression
cyclic fatigue properties of ceramic materials at
temperatures up to 1500°C. (photo courtesy of Instron
Ltd).

determining the availability of hard materials to retain rigidity
at high temperatures (10), and the hardness value can show a
linear correlation with creep strength. Hot wear testing is also
now gaining credence as a useful test method ((11-14) - see
Fig. 3). In the space available it is impossible to review in
detail all the recent developments in testing metrology that have
taken place over the past few years. However, the publications
cited will provide a good understanding of the subject.

3. FUTURE NEEDS IN TESTING METROLOGY

The future needs may be summarised under two headings :
firstly, General Requirements, which cover a variety of topics

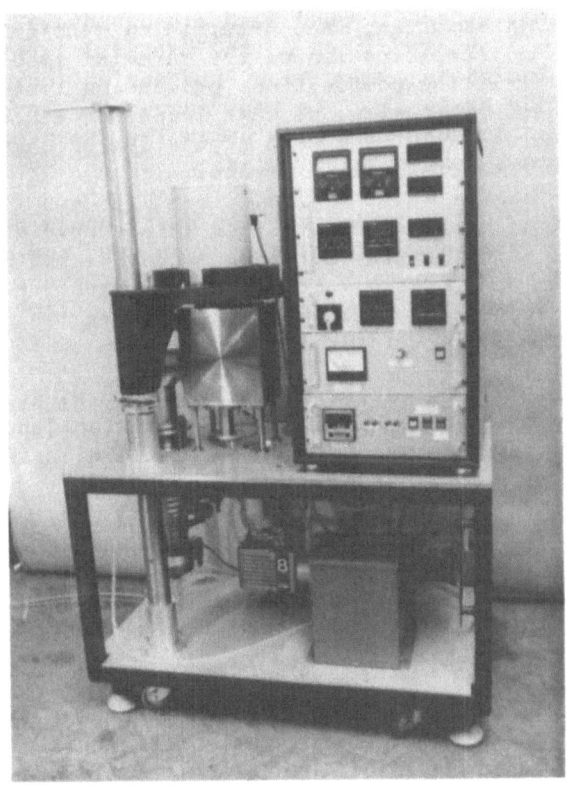

Fig. 3. High temperature wear testing facility for use up to
1000°C, developed at National Physical Laboratory (photo
courtesy of Dr M Gee (11) Crown Copyright).

which, although not necessarily regarded as main line metrology,
nevertheless need to be addressed if there are any benefits to be
accrued from the topics listed under the second heading of
specific future needs in High Temperature Testing Metrology.

3.1. General Requirements

3.1. <u>General Requirements</u>

1. The benefits offered by improvements in testing accuracy
   resulting from advances in the design of testing machines and
   in the use of computer or microprocessor controllers will only
   be exploited provided there are commensurate improvements in
   the specified tolerances in the relevant European and
   International Standards. At present the trend is a broadening
   of the permitted tolerances in testing conditions to
   accommodate disparities in the various national standards.
   This clearly will have the detrimental effect of widening the
   scatter bands on internationally collated data bases of high
   temperature materials properties.

2. It is vitally important when determining materials properties that the full specification of the material is simultaneously recorded, including composition, processing route, heat treatment and grain size, so that inappropriate comparisons are not subsequently made with nominally the same material but in a different metallurgical state.

3. The format of the recorded testing data should not only be compatible with computer database archives, but must also provide the appropriate information to interface with computer predictive modelling packages, e.g. Crispen (15) or the θ-projection concept (16).

4. There needs to be greater centralised co-ordination of data, and in particular data bases should be established for the various nickel-base superalloys, titanium alloys, metal matrix composites and engineering ceramics.

5. All test data should be subject to independent scrutiny before being fed into international databases. Test data should only be acceptable from fully accredited laboratories assessed under schemes administered by NAMAS in the UK, or other equivalent recognised bodies.

6. A wider range of fully characterised reference materials is needed, both to assist accreditation of laboratories and to verify compatibility of international testing facilities.

7. New Standards or Codes of High Temperature Testing Practice are required in the following areas :

   a. Creep testing of internally pressurised tubes.
   b. Creep testing of circumferentially notched specimens used to generate triaxial stress state creep data.
   c. Torsion creep testing.
   d. Instrumented impact testing.
   e. Hot hardness.
   f. Wear testing.
   g. Dynamic calibration of
      (i) Load cells
      (ii) Extensometers.
      used for fatigue testing.
   h. Measurement of crack growth under creep and fatigue conditions.
   i. Methods of validating testing machine software.
   j. Determination of Young's modulus.
   k. Thermal expansion measurements.

3.2. Specific Future Needs in High Temperature Testing Metrology

1. There is an immediate need for tensile and creep testing
   facilities capable of producing reliable data at temperature
   up to approximately 1300°C. This would be of particular
   interest for oxide dispersion strengthened (ODS) alloys and
   thermal barrier coated superalloys.

   In this case the main problems to be solved by testing
   metrologies in order to achieve reliable test results are

   a. the methods of gripping the specimens in pull rod
      adaptators which should preferably be reusable,
   b. the manufacture of reasonably priced pull rods,
   c. the development of reliable extensometry for operating up
      to 1300°C for measuring strain directly from the specimen
      gauge length.

   The solution of the last item may already exist in the form of
   various types of extensometers using ceramic limbs and heads.
   The problem of heating the specimen to the desired temperature
   is not too difficult since it is well within the range of
   conventional muffle furnaces wound with Platinum, or furnaces
   with Super Kanthal (molybdenum disilicide) elements.

2. Facilities are required for providing good engineering design
   data on ceramic materials, in particular uniaxial tensile
   creep and fatigue facilities with provision for eliminating
   off-axis loading. In practical terms it is foreseen that
   facilities will rapidly become available for testing up to
   approximately 1600°C, with the temperature increasing to about
   2000°C before the end of the present millennium.

3. Ceramic materials like many advanced materials such as metal
   matrix composites, exhibit low fracture ductilities. Although
   the latter alloys will see service at much lower temperatures,
   both materials impose severe demands on metrologists to
   develop accurate extensometry capable of working reliably over
   the required temperature range.

4. Inaccuracy of temperature measurement is probably the most
   important single factor contributing to errors in the
   determination of high temperature material properties,
   espacially in long-term creep testing. In particular, noble
   metal thermcouples are prone to drift, due to diffusion of the
   alloying elements across thermocouple junctions. Although
   some data exists on the magnitude of draft for various types
   of thermocouple up to 1000°C, there is no published draft data
   for higher temperatures and such information is clearly
   required if reliable test data above 1000°C is to be obtained.

5. Although most of the high temperature testing standards recommend that extensometers should be calibrated at the temperature that they are to be used, in reality virtually no calibration facilities exist which permit compliance with that recommendation. There is clearly a need, either for a wide-spread improvement in the high temperature calibration facilities, or for a reputable national standards laboratory to establish a high temperature metrological primary calibration facility, probably utilising a laser interferometer, which should be used to validate the continued justification of room temperature calibration of high temperature extensometry.

6. Considerable progress has recently been made in providing a viable means for dynamic calibration of load cells used in fatigue testing machines, as specified in a new draft British Standard. The use of a suitable transfer standard for the dynamic calibration of machines used for HT high or low cycle fatigue testing will become mandatory, and will hopefully eliminate some of the scatter in data collected from the laboratory into comparison exercises.

7. Similarly the dynamic calibration of extensometers used in HT fatigue is a topic which deserves attention. Again, there is a need for a primary calibration facility to be established at a national standards laboratory, and for suitable means to provide secondary transfer calibration facilities capable of being used for calibrating the extensometers attached to testing machines, thus providing traceability to the primary standards.

8. At higher temperatures, pyrometers become increasingly appropriate for temperature measurement, and advances in automatic control, stability and calibration are likely to ensure their extended application for temperature control. Again, demands will be placed on the metrologist to ensure that such devices can be accurately calibrated in situ to minimise errors.

9. In the field of creep testing there will be growing demand to measure creep properties under constant stress rather than constant load, conditions particularly for providing data to verify theoretical models of creep behaviour. However, it should be appreciated that although there is a considerable difference between constant load and constant stress creep behaviour for simple solid solution alloys and pure metals and most steels, that is not the case for many advanced engineering materials such as nickel-base superalloys. In the latter materials, tertiary creep is mainly due to changes in their metallurgical state, rather than being due to the reduction in cross-sectional area of the tensile sample as

deformation progresses. Thus careful consideration should be given before infesting in expensive servo-controlled feedback machines, or even the simpler cam lever creep machines in order to achieve constant stress capabilities.

10. In the field of tensile testing, only data obtained under prescribed __strain rate__ conditions, measured from HT extensometers attached to the specimen, should be regarded as sufficiently accurate for design databases. Similarly, on modern tensile testing machines yield stress, proof stress and Young's Modulus are calculated by microprocessor control systems and the results printed automatically. It therefore becomes necessary to validate the machine's internal software, and calibrating authorities will require independent means of verifying that the values calculated are correct.

11. In the field of impact testing it is foreseen that instrumented impact testing will come to the fore, particularly since with such equipment it is thought possible to differentiate between crack initiation and crack propagation in the material being tested.

4. CONCLUDING REMARKS

High temperature testing must be seen as being complementary to a soundly based scientific understanding of materials behaviour, and testing will clearly remain vitally important for quality insurance, product release and for providing design data.

The widening of the accepted tolerances in the parameters specifying testing conditions, accepted in the recent ISO Standards for room temperature and elevated temperature tensile testing and uniaxial creep testing as compared to the tighter tolerances specified in the National Standards of the individual countries, will inevitably lead to an increase in the scatter in the measured material's properties determined using the ISO criteria. This increase in scatter of materials properties will inevitably be embodied into European data bases as the ISO Standards become adopted as the new Euronorm Standards. It is highly unlikely that the tolerances in the testing parameter will be tightened unless accredited data is forthcoming to demonstrate the penalties of adopting the wider tolerances in testing parameters, and to conclusively show what are both practical and realistic acceptable tolerances. Clearly there is a need for such reliable accredited data to be accumulated within the framework of the European Community.

The ever increasing demand to raise the operating temperature of advanced gas turbines has presented the testing metrologist a considerable challenge. The need exists now for reliable tensile and long term creep measurements on metallic materials up to 1300°C, and on ceramic based materials up to 1600°C, with the latter requirement rising to over 2000°C before the year AD 2000. Totally reliable solutions have yet to be found to the problems of 1) achieving and maintaining reliable specimen temperatures; 2) of gripping test pieces in order to sustain the applied loads; and 3) of measuring deformation from the test piece gauge length in a reliable manner.

Although the practical need will still exist within each technologically advanced country for calibrating authorities to provide primary calibration of force proving devices and of extensometer calibrators for use in verifying testing machines, a case may be strongly argued, mainly on economic grounds, for carrying out certain primary calibration activities at only one member state of the EEC. Indeed this philosophy has already been adopted within the EUROMET framework for primary metrological activities supporting the determination of fundamental constants. In the field of determining the properties of materials is already proposed within the EEC to create a traceability route for hardness determination to a single standards laboratory in Italy. It would likewise be sensible to allocate the tasks to various competent standards laboratories of 1) determining the necessity of calibrating creep extensometers at elevated temperatures; 2) of dynamic calibration of load and of strain measurement; 3) the evaluation of long term drift of thermocouples at temperatures greater than 1000°C; and 4) for validating the software used for determining materials properties in advanced testing machines.

Progress needs to be made towards establishing a European agency for accrediting testing laboratories, with an eventual aim that only data from accredited sources can be fed into European data bases. A full portfolio of suitable reference materials for proficiency testing over a wide range of temperatures will also be a useful asset for accreditation purposes.

It should be recognised that one reliable test data point obtained under conditions of known prescribe accuracy is of more value than several test results of unknown accuracy.

5. ACKNOWLEDGEMENT

The author acknowledges useful informal discussions with colleagues during the preparation of this paper. However the contents are the personel views of the author and are not necessarily the official policy of NPL.

6. REFERENCES

(1)   "Techniques for Multiaxial Creep Testing", ed. Gooch, D.J.
      and How, I.M., Chs. 17-19, pp. 305-356, pub. Elsevier
      Applied Science (1986).

(2)   "Advances in Large Scale Testing", Metals and Materials,
      3(8), pp. 436 (1987).

(3)   "Measurement of High Temperature Mechanical Properties of
      Materials", eds. Loveday, M.S., Day, M.F. and Dyson, B.F.,
      pub. HMSO (1982).

(4)   "Techniques for High Temperature Fatigue Testing", ed.
      Summer, G. and Livesey, V.B., pub. Alsevier Applied Science
      (1985).

(5)   "A Code of Practice for Constant Amplitude Low Cycle Fatigue
      Testing at Elevated Temperature", pub. National Physical
      Laboratory (1986).

(6)   "A Code of Practice for the Use of High Temperature
      Nickel-Chrome base Alloy Creep Extensometers", Loveday, M.S.
      and Gibbons, T.B., pub. National Physical Laboratory (1987).

(7)   "Testing > 1000°C", Loveday, M.S. and Evans, R.B., In Press
      (1987).

(8)   "High Temperature Strain Measurement", ed. Hurst, R.C. et
      al., pub. Elsevier Applied Science (1986).

(9)   "High Temperature Fatigue", ed. Skelton, R.P., Applied
      Science (1983).

(10) "Mechanical Testing of Hard Materials", Almond, E.A.,
      Roebuck, B. and Gee, M., Institute of Physics Conference
      Series No. 75, Ch. 2, pp. 155-157 (1986).

(11) "A High Temperature Wear Testing Machine", Matharn, C.S. and
      Gee, M., J. Phys. E. (to be published) (1987).

(12) "Tribology of Selected Ceramics at Temperatures to 900°C",
      Sliney, H.E., Jacobson, T.P., Deadmore, D. and Miysohi, K.,
      NASA Tec. Mem 87267, Presented at 10th Conf. on Comp. and
      Adv. Ceramic Materials, American Ceramic Soc. Florida,
      January 1986.

(13) "A High Temperature Test Rig for Sliding and Rolling Wear",
      Hammarsten, A. and Hogmark, S., Wear 115, pp. 139-150
      (1987).

(14) "Friction and Wear Behaviour of Ion Beam Modified Ceramics", Lankford, J., Wei, W. and Kossowsky, E., J. Mat. Sci., <u>22</u>, pp. 2069-2078 (1987).

(15) "Modelling Creep for Engineering Design", Ion, J.C., Barbosa, A., Ashby, M.F., Dyson, B.F. and McLean, M., NPL Report DMA(A)115 (1986).

(16) "Creep of Metals and Alloys", Evans, R.W. and Willshire, B., Book No. 304, Institute of Metals (1985).

(17) "Mechanical Testing of Engineering Ceramics at High Temperature" Ed. Dyson, B.F., Lohr, R.D. and Morrell, R., publ. Elsevier Applied Science (in press).

(18) "Standardisation of Mechanical Testing and Quality Control" Loveday, M.S. and Morrell, R., "Mechanical Testing and Engineering Ceramics at High Temperature", Ed. Dyson, B.F., Lohr, R.D. and Morrell, R., pub. Elsevier Applied Science (in press).

5.2. Prior Inspection/NDE

W.N. Reynolds

National NDT Centre

Harwell Laboratory, UKAEA, England

1. INTRODUCTION

High temperature materials and processes have been used industrially for many years. High quality grades of graphite, for example, are produced for nuclear and other purposes at temperatures around 3,000 K. Existing methods of nondestructive examination of such refractory materials by ultrasonic, radiographic, thermal, and other methods are reasonably adequate. The applications and general limitations of these techniques have been assessed in recent publications[1,2].

New problems arise when high temperature materials are required to retain high strength, stiffness and integrity under operating conditions where stress, fatigue, corrosion or impact are to be withstood. There is a wide range of diverse materials now emerging from development and virtually all of them raise demands which cannot be met by current NDT technology. The new applications arise largely in the fields of transport, aircraft and aerospace where dead weight must be minimised. It is often found that the resolving power, energy or frequency range of existing methods, which have been optimised to deal with steels or other conventional materials, are ineffective with the new materials. Thus existing methods require some adaptation : on the other hand, new techniques based on different principles but suitable for strong light section items may be demanded.

In general advanced materials and their applications are of many different kinds, each raising its own particular problems. There is, consequently no panacea in this field and an range of different developments is necessary.

2. MATERIALS TO BE TESTED

The main high temperature materials areas to covered are as follows :

1. Refractory metals and alloys, mainly nickel and titanium based.

2. Ceramics and ceramic composites.

3. Thermal barrier and tribological coatings.

4. Ceramic particles and fibre reinforced metals.

5. Carbon fibre reinforced carbon.

6. Inspection of bonds and joints involving any of these materials.

All of these materials raise special NDT problems at present unsolved. The methods to be developed must have adequate sensitivity and resolution in each case, and many problems call for rapid or non-contacting scans. The cost of the installation or of inspection of any particular item must fall within acceptable limits.

2.1. Refractory Metals

Items up to 15 mm in thickness such as gas turbine blades can be examined for cracks, porosity and inclusions by microfocus projection radiography or by neutron radiography. New problems arise in heavier section materials such as turbine discs and work is in hand to develop suitable ultrasonic tests. Owing to the comparatively high degree of crystal anisotropy and the variability sometimes experienced in grain size distribution, conventional ultrasonics suffers from high noise levels.

Much current research is aimed at overcoming this problem by the use of suitable computer software in conjunction with wider frequency band ultrasonic probes and scanning techniques. The situation is not yet satisfactory, and much remains to be done in this area.

2.2. Ceramics and Ceramic Composites

NDE of these materials has attracted a great deal of interest in recent years. Conventional equipment designed to deal with steels and other structural metals is not very effective for most of the problems that arise. Ceramics are commonly characterised by : comparatively high hardness, stiffness and strength; low toughness or resistance to fracture and low density; very low electrical and thermal conductivity. It is therefore necessary to investigate what modifications are required to existing equipment by way of changes in operating energies, frequencies, etc. and also to consider what new methods might be profitably introduced. A related problem is that of inspecting unfired mouldings, which are extremely fragile and can rapidly disintegrate in contact with fluids such as acoustic coupling media.

a. Radiography

In radiographic studies of ceramics, there are several essential differences with common practice required to maximise contrast and resolution. First, fairly low energies of, say, 25-80 keV must be available. Second, the focal spot size of the source must be as small as practicable to minimise geometrical unsharpness and the intensity must be as high as possible to minimise exposure time. Third, a projection system should normally be used whereby an enlarged shadowgraph of the ceramic item is produced on a recording screen situated at a distance 3 to 15 times that of the object from the source. The major benefit of such an arrangement is that most of the scattered photons are deflected completely out of the field of view and do not contribute to the background noise. Thus constrast and resolution are enhanced.

A number of X-ray sets with focal spot sizes of 10-20 µm are now commercially available. The results obtainable with such a system were described by Reynolds and Smith (3) and have since been confirmed and extended by a number of other groups (e.g. Goebbels and Reinhold (4)). They are summarised in Table I.

Radiographic detection of cracks and delaminations requires that the plane of the defect be aligned fairly closely with the direction of the radiation. In order to view an object from all required directions, systems of real-time and tomographic radiography are needed. Those which have been demonstrated so far have spatial resolving powers in the range 100 µm - 1 mm. Improving techniques with better sources, digital recording and fast image processing should improve this performance.

b. Ultrasonics

Ultrasonic C-scan with plane 25 MHz probes has produced useful results (see Table I). Further progress is required, however, in terms of higher resolution, speed of scanning and the analysis of surface features. Many surface breaking cracks can be detected by the use of fluorescent dye penetrants, but some items, particularly those which have been machined, exhibit

TABLE 1

Inclusions detected in hot pressed (HPSN) and Reaction bonded (RBSN) Silicon Nitride, according to ref. 3.

	Defect	25 μm	125 μm	250 μm	510 μm
HPSN	Fe	X,u	X,u	X,u	X,u
	WC	X	X	X,u	–
	Si	N	X	X	X
	SiC	N	N	N	X
	BN	X,u	X,u	X,u	X,u
	C	N	N	X	X,u
RBSN	Fe	–	X,u	X,u	X,u
	Si	–	N	X	X,u
	SiC	–	N	X	X,u
	C	–	u	X	X
	Low density	–	N	–	X
	Pores	–	X	X,u	X

X defects detected by 80 keV projections radiography
u defects detected by 25 MHz ultrasonic C-scan
N defects not detected
– defects not present

surface pits, gouges and scratches which are potentially serious
if flexural stresses are applied, for example. Acoustic
microscopy in the GHz range has revealed many aspects of surface
and near-surface structure and is therefore of interest in
monitoring constitution on a sampling basis, but is too slow to
apply as routine whole-body NDT for engine parts. However, a
number of intermediate techniques are now being investigated.

Of immediate interest is the availability of rapid scanning
systems using focussed probes of 50-100 MHz. These may be used in
various modes, including B-scan, C-scan or time-of-flight, with
digitised recording and data processing. Furthermore, if the
probes used have a fairly wide numerical aperture, i.e. if a good
range of angles of incidence from the coupling medium to the
ceramic surface is available, surface waves are excited and may
be used to monitor surface defects. In this mode the transducer
is focussed just below the ceramic surface. Alternatively, by
the employment of an acoustic lens of low numerical aperture
(larger f-number) a C-scan for defects at a chosen level below
the surface of the artefact may be produced.

An alternative approach, still requiring detailed
evaluation, is the use of the scanning laser acoustic microscope
(SLAM). In this instrument, a selected volume of the material is
uniformly isonified (illuminated) by a suitable extended source.
The opposite face is scanned at TV frame rates by a laser spot,
the reflections of which produced an image related to the
internal structure of the material. The images so obtained are
not always easy to interpret, however, and the application
requires further evaluation.

c. Thermography

PVT (Pulse Video Thermography) is a recent development (5)
which provides a rapid non-contacting method of detecting
delaminations in solid materials. The detector is a TV-video
imager for the 10 μm infra-red wave band which provides a
complete scan of the field of view every 20 ms. A chosen heat
input is supplied from a controlled source such as a quartz-xenon
flash tube, which can deliver up to 5 J cm^{-2} in 5-10 ms to an
exposed surface. The resulting thermal history of that surface,
or of the opposite surface is then recorded by means of a high
quality VCR, and, on play-back reveals areas of higher of lower
thermal diffusivity. It is therefore an effective means of
detecting sub-surface defects (provided their depth is not
greater than about twice their individual diameter), and
variations in porosity content or density. It is well suited to
the examination of unfired mouldings.

A comparative example of the three techniques is given in Fig. 1. Fig. 2 is an infra-red micrograph of a ceramic surface. Note the high contrast of the surface defects.

2.3. Thermal Barrier Coatings

A possible method of exploiting the advantages of thermal insulators in engines is to use them as coatings on metals. Thermal barrier coatings in the thickness range 100 μm to 3 mm are now produced by methods such as flame or plasma spraying. The inspection of such coatings raises formidable new problems, and the situation has recently been reviewed in some detail (6). The immediate problems are the measurement of thickness, porosity and constitution of coatings (many of which include several distinct layers of different materials) and the quality of the bonding to the substrate. Methods available are as follows :

a. Radiation back-scatter

Existing X-ray and beta back-scatter gauges can interrogate layers up to a thickness of 0.5 mm. Higher energy gamma sources could be adapted for thicker layers.

b. Eddy gauges are available for the measurement of coating thicknesses up to 1 mm or so. Many developments are possible in this technique.

c. Thermal and Opto-thermal devices are already available and could be exploited in several ways. A single point or small area may be illuminated by a laser beam, which can itself be CW (flying spot), chopped or pulsed. The thermal response of the surface is sensitive to the local material characteristics and particular modes of operation can be selected to measure coating thickness, porosity or adhesion. For some purposes, thermography (PVT) can be used to scan a selected area in a very short time. Fig. 3 shows an artificially disbonded area in a plasma sprayed surface about 60 mm in diameter inspected by this method.

d. Ultrasonics

The inspection of coatings by ultrasonics methods necessarily requires the use of high frequences. Many coatings have a disordered structure which produces a high level of scattering, so that useful signals are not obtained. Some success has, however, been reported in the use of 50 MHz scanning probes for the study of alumina coatings.

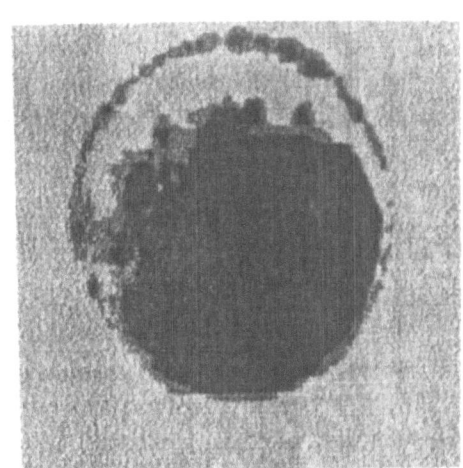

(a)

FIGURE 1:

Part fired
10 mm disc
of silicon
nitride with
an internal
delamination
as recorded by :

a. Ultrasonic
   C-scan

b. High definition
   X-radiography

c. Pulse Video
   Thermography
   (PVT).

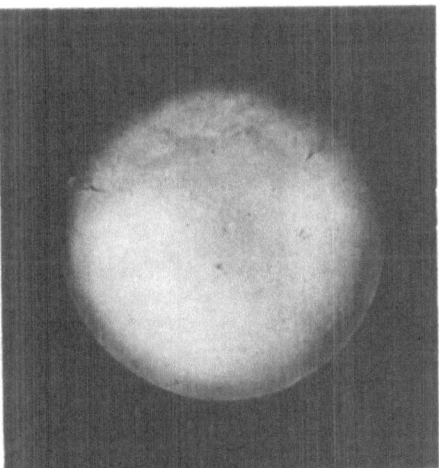

(b)

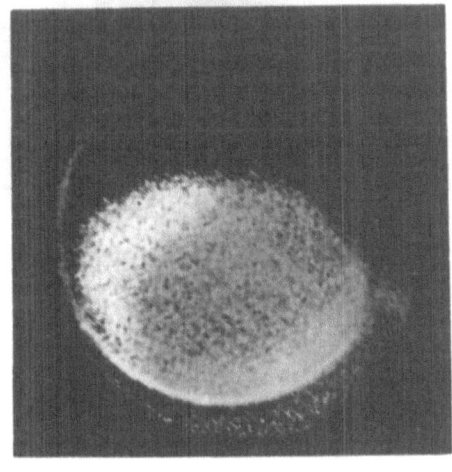

(c)

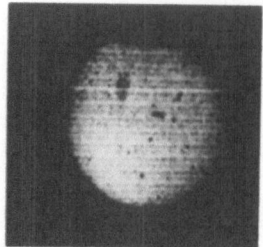

FIGURE 2 : Infra-red micrograph of silicon nitride
surface showing characteristic defects.

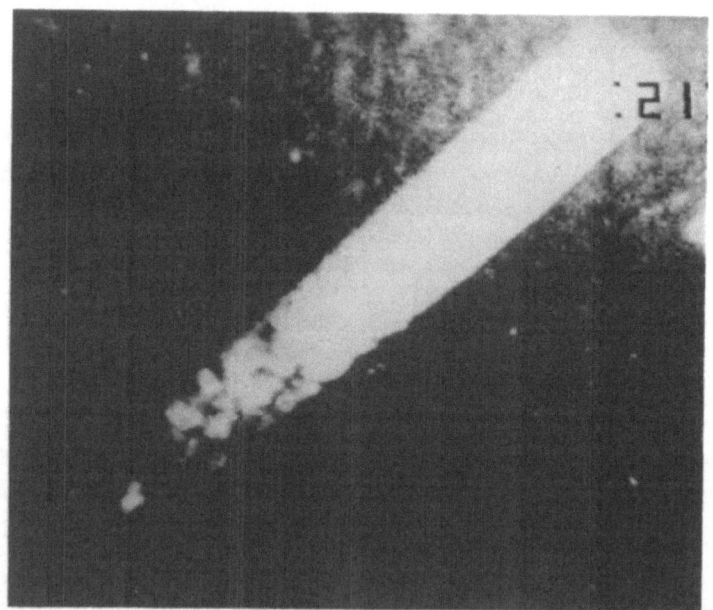

FIGURE 3 : Plasma-sprayed steel surface with artificially
disbonded area (light) viewed by single-sided PVT.

2.4. Metal/Ceramic Composites

The addition of suitable amounts of strong ceramic fibres, whiskers or particles to metals can produce considerable improvements in elastic modulus and strength, particularly at higher temperatues. There is thus considerable interest in the development of Metal Matrix Composites (MMC's) in this field. Once again, new problems of materials quality control are raised. The fibres and fillers used are often of such small dimensions (ca. 1 μm) that they are very difficult to detect individually. Nevertheless, the velocity and attenuation of ultrasonic waves should be detectably altered by any appreciable concentration, and variations in concentration should be detectable radiographically.

Successful results for aluminium matrix materials have been described by Blessing and Elban (7) but radiographic work has hitherto been sparse and inconclusive.

2.5. Carbon Fibre Reinforced Carbon

This is a difficult material to test. It cannot be exposed to coupling liquids, dye penetrants or radiographic enhancement penetrants. Its thermal anisotropy renders it unsuitable for PVT except in thin sections. An ultrasonic scanning test might be developed for items which could be coated in a removable waterproof plastic coating.

2.6. Bond Testing

A range of ultrasonic tests is available for the inspection of welded, brazed, soldered or adhesively bonded joints. To inspect joints between items for high temperature work will required the optimisation of special systems for specific purposes, e.g. ceramic-metal bonds. PVT systems, whereby the migration of heat across such boundaries is monitored after a suitable heat pulse has been injected, should also find application.

3. INDUSTRIAL REQUIREMENTS

It is a general principle that any inspection systems should cause minimum added cost and production time consistent with the required quality and reliability of the product. Thus essentially slow processes such as microscopy can be undertaken only on a sampling basis, and any comprehensive tests must be rapid, inexpensive and, in practice, usually non contacting. It should be said, however, that it is the inspection cost per item which must be low - the installation itself, if well designed and technically effective can soon be cost effective.

The current industrial situation of high temperature
materials is that many products for low critically applications
are being developed, notably in the field of monolithic ceramics
and coatings, without any real provision for NDE. The
applications of these materials in critical areas await the
development of confidence in quality and reliability. It is here
that proven techniques of NDT have a major part to play.

In the light of the current stage of development of the
various materials, attention should be focussed on the types of
material in the following order.

a. Refractory metals
b. Thermal barrier coatings
c. Monolithic ceramics
d. Bonding techniques
e. Composite ceramics
f. Metal matrix composites.

4. RESEARCH NEEDS AND PRIORITIES

In view of the technical factors discussed in this section;
particularly the wide diversity of the problems raised by the
different classes of material, research priorities should be
ordered as follows :

a. Rapid analysis of ultrasonic characteristics of nickel and
   titanium based materials.

b. New methods, including thermal, eddy current and radiation
   back-scatter techniques for the assessment of thermal barrier
   coatings.

c. Rapid high resolution scanning methods for the detection of
   defects in high strength ceramics. This should include both
   radiographic and ultrasonic techniques.

d. Methods for assessing unfired ceramic mouldings.

e. Fundamental studies of such materials as metal matrix and
   composites and carbon fibre reinforced carbons.

5. REFERENCES

1. Quality Technology Handbook, 4th Edition, ed. by Sharpe, R.S.,
   West, J., Dean, D.S., Tyler, D.A. and Coole, H.A. Butterworth
   1984.

2. Capabilities and Limitations of NDT, ed. Hanstead, P.D. To be
   published by Brit. Inst. NDT 1987.

3. Reynolds, W.N. and Smith, R.L.   Brit. J. NDT, Vol. 24,
   pp. 145, 1982.

4. Geobbels, K. and Reinhold, A., Proc. 2nd Int. Symposium on
   Ceramic Materials for Engines, Lübeck–Travemünde (FRG) 1986,
   ed. Burk, W. and Hausner, H., Verlag Keramische Gesellschaft.

5. Reynolds, W.N. and Wells, G.M.. Brit. J. NDT 26, 40, 1984.

6. Reynolds, W.N., NDT International, 20, 153, 1987.

7. Blessing, G.V. and Elban, W.L. J. Appl. Mechanics, 48, 65,
   1981.

5.3. Design of Materials for Components

5.3.1. Basic Principles for Modelling of Deformation and Rupture

Prof. D.R. Hayhurst

Department of Mechanical Engineering

Sheffield University, U.K.

SUMMARY

The paper reviews the state-of-the-art in constitutive modelling of deformation and rupture in high-temperature creep under multi-axial states of stress. Two formulations are considered; the materials science and the phenomenological or state variable; and the equivalence of the two approaches is demonstrated. The use of the constitutive equations in the Finite Element numerical technique, to predict component behaviour is demonstrated for a wide range of geometrics. Finally, the impact of developments, in computer technology and in Computer Aided Design techniques, upon the use of materials data and models is discussed in terms of future engineering design practice.

1. INTRODUCTION

When metals operate for long periods, typically 10^5h, at temperatures in excess of $0.4 \, T_m$, where T_m is the melting temperature of the base metal in degrees Kelvin, and at stresses below one half of the uni-axial yield stress consideration must be given to the possibility that the lifetimes of structural components may be limited either by excessive creep deformation or by failure due to creep rupture. In structural components it is necessary to perform calculations which take account of the effects of two aspects of material behaviour. These are: the increased strain rates which occur during tertiary creep and the form of the multi-axial stress rupture criterion of the material.

Before the significance of these phenomena can be assessed in terms of the overall structural performance it is first necessary to develop constitutive equations which accurately describe the creep deformation of the material and to develop techniques for structural analysis which allow one to model the progressive material degeneration which occurs during creep rupture.

In the development of constitutive equations two approaches have been used. The first is the so-called phenomenological method {1,2} which has been shown to be equivalent to a single state damage variable theory {3}. The second is the approach of the materials scientist in which the physics of the microstructural processes are described either by experimental measurement {4,5} or by the usage of suitable material models {6,7}. The effectiveness of both approaches has to be judged by their ability to describe the strain rates and lifetimes measured in long-term tests under both uni-axial and multi-axial stress.

Given an accurate description of the material behaviour and validated methods for the numerical analysis of structures, they can be used to identify and to study the physical features which control the creep rupture of structures. The types of structures in which creep rupture is important, and is likely to provide a limitation on service life-times, are those where stress concentrations occur due to changes in geometrical form: e.g. at holes, notches, fillet radii and at sharp defects or cracks present within the material.

The purpose of the paper is threefold. Firstly, to describe the theory of creep continuum damage mechanics; and, the development of constitutive equations for both creep deformation and rupture. Secondly, to show how the constitutive equations can be used with Finite Element numerical techniques to predict the behaviour of the range of component types given in the previous paragraph. Thirdly, to discuss the likely development of Computer Aided Design techniques; of computer science; and, to discuss the consequent materials modelling – data requirements.

2. MATERIAL MODELLING

2.1. Constitutive equations for steady load and temperature

Constitutive equations have been developed for primary and secondary creep deformation under both uni-axial and multi-axial stress conditions. The equations are well proven, particularly for steady-load conditions, and, here it will be shown how uni-axial primary-secondary-tertiary creep behaviour can be described using a single state damage variable theory and how this theory can be generalised for steady multi-axial stress conditions.

2.1.(a) Primary-secondary creep

The uni-axial creep strain rate is usually described by Norton's law,

$$\dot{v}/\dot{v}_o = (\sigma/\sigma_o)^n K(t) ,$$ (1)

where n is a material constant, $\dot{v}_o$ is the uni-axial strain rate due to the stress σ_o and $K(t)$ is a function of time selected to describe the decrease in strain rate with time.

The generalisation of equation (1) to multi-axial stresses has been carried out by Odqvist (8) and results in the following equation

$$\dot{v}_{ij}/\dot{v}_o = (3/2) \; (\sigma_e/\sigma_o)^{n-1} (S_{ij}/\sigma_o) K(t)$$ (2)

where $S_{ij} = \sigma_{ij} - \delta_{ij}\sigma_{kk}/3$. Both equations (1) and (2) have been verified by the test results of Johnson et al (9).

2.1.(b) State variable description for uni-axial continuum damage

In a uni-axial creep test the creep rate increases from the steady-state, or secondary creep, value as damage causes an advance into the tertiary portion of the curve. The steady-state creep rate can be expressed as a function of the applied stress σ alone. In order to account for the increase in strain rate during a constant stress test it is necessary to introduce a new variable into the strain equation. Since the increase in strain rate is the result of a damage process the variable ω introduced is referred to as the damage state variable. The strain rate equation then takes the form

$$dv/dt = f(\sigma,\omega),$$ (3)

where f is a function which has yet to be defined. However, in the undamaged state when $\omega = 0$ the equation must reduce to the form observed for steady value of ω as well as the stress σ be known and consequently an equation must be introduced which defines the growth of the damage state variable. The assumption that the damage rate depends both on the current state of stress and damage gives the equation

$$d\sigma/dt = g(\sigma,\omega) .$$ (4)

The problem is how to find the functions f and g in a systematic manner. This may be achieved by following a testing procedure first suggested by Leckie and Hayhurst (3), which makes use of the results of tests with step changes in stress, and described in detail elsewhere {10}.

These procedures are applicable provided that the state of damage can be described by a single damage parameter. If a single state parameter suffices it might still be necessary to modify the equations (3) and (4) for loading and unloading structures. This is certainly the case for creep deformations in the primary region when it is found that the growth laws of dislocation density differ for loading and unloading. This point has been discussed by Leckie and Ponter (11).

2.1.(c) Rabotnov-Kachanov equations for uni-axial creep – A phenomenological approach

Unfortunately it is difficult and time consuming to conduct experiments described in the previous section and the variable stress experiments which have been performed are limited to the prediction of rupture life under cyclic conditions of loading. Faced with this difficulty Rabotnov (12) proposed modifications of constitutive equations first suggested by Kachanov (13) which described existing experimental results. For uni-axial stress tests the growth laws are assumed to have the simple form:

$$\dot{v}/\dot{v}_o = (\sigma/\sigma_o)^n K(t)/(1-\omega)^{n^{\cdot}} \text{, and} \tag{5}$$

$$\dot{\omega}/\dot{\omega}_o = (\sigma/\sigma_o)^{\chi}/(1+\phi)(1-\omega)^{\phi} \text{,} \tag{6}$$

where χ, ϕ and $\dot{\omega}_o$ are constants. In a constant stress test the value of ω monotonically increases from zero to a value of unity at failure. When $\omega = 0$ equation (5) reduces to Norton's law, given by equation (1), and when $\omega = 1$ the strain rate is infinite. Integration of equation (6) between the limits $\omega = 0$, at $t = 0$, and $\omega = 1$, at $t = t_f$, yields the experimentally observed expression for the rupture lifetime

$$t_f = 1/\dot{\omega}_o (\sigma/\sigma_o)^{\chi} \text{.} \tag{7}$$

Equations (5), (6) and (7) have been shown (14) to accurately describe uni-axial creep behaviour.

2.1.(d) Generalisation of state variable description

The generalisation of equations (5) and (6) for multi-axial stresses has been achieved by making the assumtion that the influence of continuum damage on the deformation rate processes is scalar in character .
Equation (2) can then be written as:

$$\dot{v}_{ij}/\dot{v}_o = (3/2)(\sigma_e/\sigma_o)^{n-1}(S_{ij}/\sigma_o)K(t)/(1-\omega)^n. \tag{8}$$

In the generalisation of equation (6) the stress-state effects observed from the isochronous loci must be reflected; this is achieved by the introduction of the homogenous stress function $\Delta(\sigma_{ij}/\sigma_o)$ as follows:

$$\dot{\omega}/\dot{\omega}_o = \Delta^X(\sigma_{ij}/\sigma_o)/(1+\phi)(1-\omega)^{\phi} \; ; \qquad (9)$$

for copper $\Delta(\sigma_{ij}/\sigma_o) = \sigma_1/\sigma_o$ and for aluminium alloys $\Delta(\sigma_{ij}/\sigma_o) = \sigma_e/\sigma_o$. Integration of equation (9) for the conditions $\omega = 0$, $t = 0$ and $\omega = 1$, $t = t_F$ yields the experimentally observed result $t_f = 1/\dot{\omega}_o = \Delta^X(\sigma_{ij}/\sigma_o)$, which may be normalised to give

$$t_f/t_o = 1/\Delta^X(\sigma_{ij}/\sigma_o) \; ; \qquad (10)$$

substitution of $t_f = t_o$ gives the equation of the isochronous surface $\Delta(\sigma_{ij}/\sigma_o) = 1$.

Equation (8) has been verified using the results of Johnson et al (9) for tension-torsion tests. If in such a test, γ is the shear strain, v is the axial strain and γ_f and v_f denote their values at failure, then is may be shown from equation (8) that the ratio $(\gamma/\gamma_f)(v/v_f)$ must remain constant throughout the test. The experimental verification of this may be seen in Fig. 1, where the results of the tests by Johnson et al on copper and aluminium have been plotted. The implication of this result is that the state variable ω has a scalar interpretation in the strain rate equation. In addition equation (9) is capable of describing the multi-axial rupture behaviour.

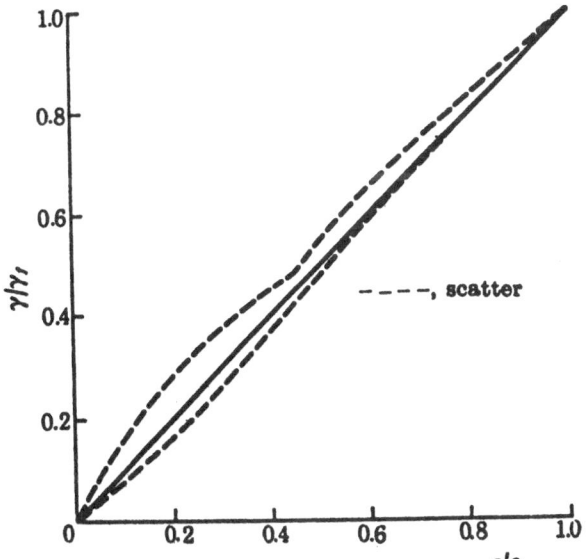

Fig. 1.   Variation of axial and shear creep strains for copper and aluminium alloys

2.1.(e) Isochronous rupture surface

The concept of an isochronous rupture surface has been introduced in section 2.1.(d) where the dependence of the damage growth rate upon stress state is expressed by equation (9). A review of bi-axial creep rupture behaviour, carried out by Hayhurst (15), indicates that a general representation may be used of the form

$$t = h(J_1, J_2, J_3) , \tag{11}$$

where J_1, J_2, J_3 are the invariants of the stress tensor and h is a homogeneous algebraic function of degree - χ in stress. Hayhurst (15) has shown that it is necessary to replace J_3 by the maximum principal tension stress σ_1 and, that the behaviour is accurately described by:

$$t = M\{\alpha\sigma_1 + \beta J_1 + \gamma \, (J_2')^{1/2}\} \tag{12}$$

where M is a constant independent of stress, α, β and γ are constants, and J_1 and J_2' in terms of principal stresses are given by

$$J_1 = \sigma_1 + \sigma_2 + \sigma_3, \; J_2' = \frac{1}{6} \{(\sigma_1 - \sigma_2)^2 + (\sigma_2 - \sigma_3)^2 + (\sigma_3 - \sigma_1)^2\}.$$

In practice the β-term is usually small and the behaviour of most materials can be represented using $\beta = 0$. Under these conditions $\alpha + \gamma = 1$ and equation (12) can be written as

$$t = M \{\alpha\sigma_1 + (1-\alpha)J_2'^{(1/2)}\}^{-\chi},$$

or, to a close approximation by

$$t = M \{\alpha\sigma_1 + (1-\alpha)\sigma_e\}^{-\chi} , \tag{13}$$

where σ_e is the effective stress. The term in brackets corresponds to the function $\Delta(\sigma_{ij})$ introduced in section 2.1.(d). The corresponding damage rate equation is

$$\dot{\omega} = \{\alpha\sigma_1 + (1-\alpha)\sigma_e\}^{\chi}/(M(1+\phi)(1-\omega)^{\phi}). \tag{14}$$

The material constants M, ϕ and χ may be determined from the results of uni-axial tests but the determination of the constant α requires the results of rupture tests to be obtained under two different states of stress. The tests usually carried out to determine α are uni-axial and pure shear tests ($\sigma_1 = -\sigma_2$, $\sigma_3 = 0$). From (14) the isochronous rupture surface can be constructed in stress space, examples will be given in following sections.

2.2. The approach of the materials scientist

In this section an attempt is made to present the approach of the materials scientist and to identify those features which are common with the single state damage variable approach described earlier. A general formulation for the growth of damage at grain boundaries has been given by Ashby and Raj (16). The growth of damage is expressed in terms of two mechanisms. One mechanism, referred to as nucleation, gives a measure of the rate at which grain boundary voids are formed. The other mechanism, referred to as growth, gives a measure of the rate of growth of void size. Various studies have been made with the objective of the formulation of equations which define the nucleation and void growth rates. Two particular examples are now discussed against the background of the phenomenological procedures already outlined.

2.2.(a) The Greenwood equations

Greenwood (5) has studied the growth of creep damage in copper at a temperature of 500°C. The tests were performed under conditions of constant uni-axial stress and measurements made of the number and size of voids. No attempt was made to study the effect of damage on strain rate although it appears that the tertiary strains were little greater than those predicted by steady state theory.

Suppose that a uni-axial test is conducted at constant stress σ_o. Let the rupture time be t_o when the strain is v_o, the hole density n_o and the average volume of the holes ν_o. With reference to these physical values the growth equations for a uni-axial stress σ are

$$d(n/n_o)/dt = (\sigma/\sigma_o)^2 d(v/\dot{v}_o)/dt \ ;$$

$$d(\bar{\nu}/\bar{\nu}_o)/dt = (\sigma/\sigma_o)/t_o \ ;$$

$$d(v/v_o)/dt = (\sigma/\sigma_o)^5/t_o \ . \tag{15a-c}$$

In the first of these equations the nucleation rate is proportional to the strain rate but is, in addition, proportional to the square of the applied stress. In the second the volume rate of the voids is assumed to be controlled by a diffusion process which is proportional to the stress. The third equation illustrates that the strain rate is proportional to the fifth power of applied stress and the effect of damage on strain rate is neglected.

The damage is defined as

$$\omega = n\bar{\nu}^{-2/3}/n_o \bar{\nu}_o^{-2/3} = A/A_o \ , \tag{16}$$

and the rupture condition as $\omega = 1$. This is the same condition as proposed by Greenwood (5) who used the concept of a critical area fraction at failure. For multi-axial states of stress Leckie and Hayhurst (17) suggested that equations (15) and (16) take the form

$$d(n/n_o)/dt = (\sigma_1/\sigma_o)^2 d(v_e/v_o)/dt \; ;$$

$$d(v/v_o)/dt = (\sigma_1/\sigma_o)/t_o \; ;$$

$$d(v_e/v_o)/dt = (\sigma_e/\sigma_o)^5/t_o \; . \qquad (17a\text{-}c)$$

and the damage is $\qquad \omega = n\bar{v}^{-2/3}/n_o \bar{v}_o^{-2/3} \; .$

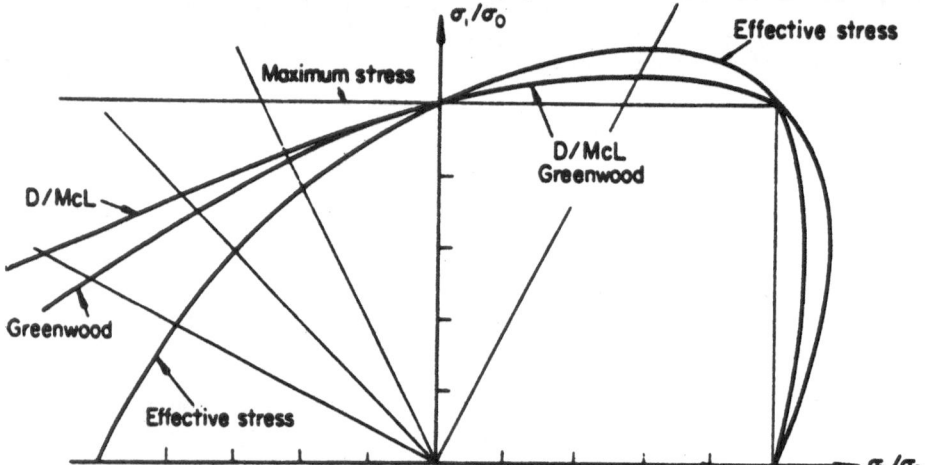

Fig. 2   Predictions of isochronous rupture loci

Integration of the above equations, for the rupture condition $\omega = 1$, gives the isochronous surface for rupture time t_o:

$$\Delta(\sigma_{ij}/\sigma_o) = (\sigma_1/\sigma_o)^{8/23} (\sigma_e/\sigma_o)^{15/23} = 1. \qquad (18)$$

The corresponding isochronous rupture surface is shown in Fig. 2. For constant stress, σ, the equations are easily re-expressed to give the following expression for damage rate:

$$d\omega/d(t/t_o) = (5 \, \Delta^{23/5} (\sigma_{ij}/\sigma_o)/3)\omega^{2/5} \; . \qquad (19)$$

For the uni-axial case, $\sigma = \sigma_o$, the variation of ω with time is given in Fig. 3. If now the damage rate equation is written in the following particular Rabotnov-Kachanov form

$$d\omega/d(t/t_o) = 2 \, \Delta^{23/5} (\sigma_{ij}/\sigma_o)/3(1-\omega)^{1/2} \; . \qquad (20)$$

with the rupture condition $\omega = 1$,
it is found that this gives the damage-time graph shown in
Fig. 3. It can be seen that the two equations (19) and (20) give
similar results. The corresponding strain rate expression to
equation (20) is given by

$$d(v_e/v_o)/d(t/t_o) = (\sigma_e/\sigma_o)^5 .$$ (21)

These equations are a particular form of the Rabotnov-Kachanov
equations in which the damage state variable ω is the normalised
form of the damage.

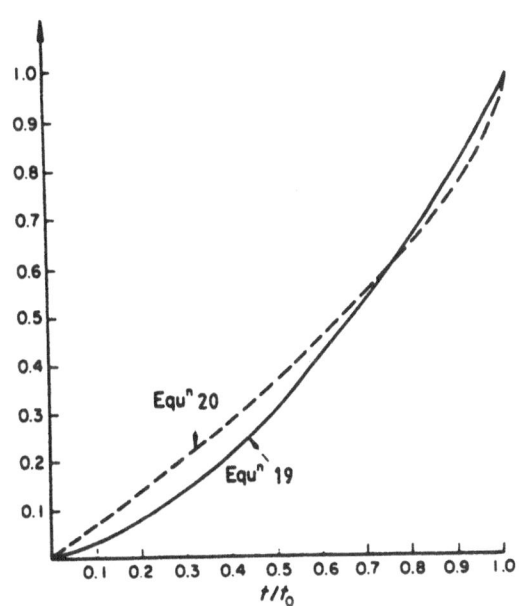

Fig. 3.   Comparison of predictions from damage growth
          laws for copper.

2.2.(b) The Dyson-McLean equations

Dyson and McLean (4) have performed constant uni-axial
tension and torsion tests on Nimonic 80A at 700°C In addition to
measuring tertiary strains and times to rupture, measurements
were also made of the number of voids per unit area of grain
boundary and of the total volume of voids.

For constant uni-axial stress tests carried out at stress
σ_o, the rupture time is t_o at which time the void density is n_o

and the total volume of the voids is $\bar{v}_o$. The constitutive
equations proposed by Dyson and McLean[4] give expressions for
the rate of void nucleation and of void growth. Damage is
expressed as the total volume of voids and the effect of damage
on the creep rate is also included. For multi-axial states of
stress the proposed rate equations may be written in the form

$$d(v_e/v_o)/dt = 3(\sigma_e/\sigma_o)^4(\bar{v}/\bar{v}_o)^{4/9}/t_o \; ;$$

$$d(n/n_o)/dt = 1/2(\sigma_1/\sigma_e)^2(v_o/v_e)^{1/2}d(v_e/v_o)/dt \; ;$$

$$d(\bar{v}/\bar{v}_o)/dt = (\sigma_1/\sigma_e)^{0.7}d(v_e/v_o)/dt \; ; \qquad (22a-c)$$

where $\qquad \bar{v}_o = 3\bar{v}_o/2n_o .$

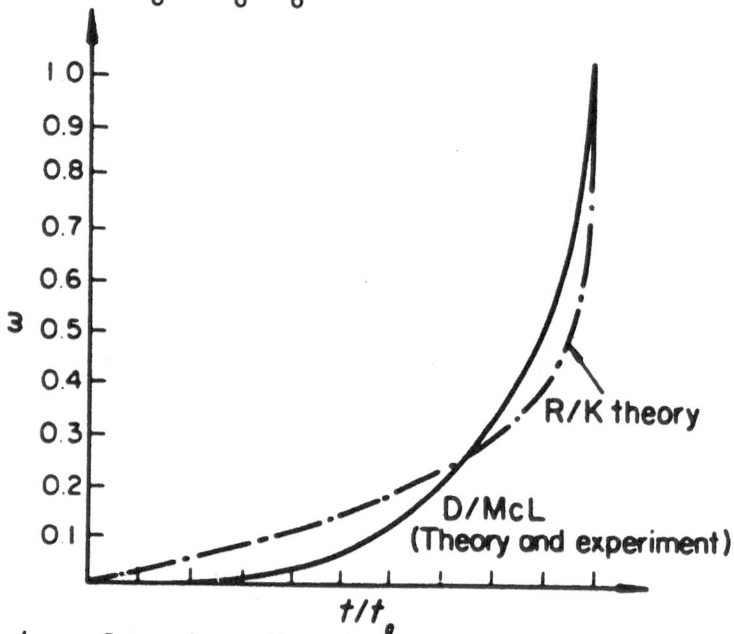

Fig. 4. Comparison of predictions from damage growth
laws for Nimonic 80 A.

The damage may be defined in terms in the total volume of
voids using an integral relation, and hence:

$$\omega = \frac{\bar{v}}{\bar{v}_o} = \frac{3}{4}\int_{v_o=o}^{v_1=v} (\sigma_1/\sigma_e)^2(v_o/v_1)d(v_1/v_o)\int_{v_1}^{v} (\sigma_1/\sigma_e)^{0.7}d(v/v_o). \quad (23)$$

For proportional loading, the ratio (σ_1/σ_e) is constant and the general damage equation (23) can be integrated and the constitutive equations (22) reduced to the single state variable form:

$$d(v/v_o)/d(t/t_o) = 3(\sigma_e/\sigma_o)^4 \omega^{4/3} \qquad (24a\text{-}b)$$

$$d\omega/d(t/t_o) = 9 \, \Delta^4(\sigma_{ij}/\sigma_o) \, \omega^{7/9} /2$$

where $\qquad \Delta(\sigma_{ij}/\sigma_o) = \{\sigma_1^{1.8} \, \sigma_e^{2.2}/\sigma_o^4\}^{1/4}$,

and the rupture condition is

$$\omega = \bar{v}/\bar{v}_o = \sigma_o/\sigma_1 \; ; \qquad (25)$$

in addition, it may be shown, for constant stress, that:

$$\frac{\bar{v}}{\bar{v}_o} = \left[\frac{\sigma_1}{\sigma_e}\right]^{2.7} \left[\frac{v_e}{v_o}\right]^{3/2} . \qquad (26)$$

For the uni-axial case, $\sigma = \sigma_o$, the variation of ω with time is given in Fig. 4. If now the damage rate equation is written in the Rabotnov-Kachanov form:

$$d\omega/d(t/t_o) = \Delta^4(\sigma_{ij}/\sigma_o)/5(1-\omega)^4 \; , \qquad (27)$$

and used with the rupture condition $\omega = 1$, it is found that this gives the damage-time graph shown in Fig. 4. The two equations (24) and (27) give similar results. The corresponding strain rate expression to equation (24a) is given by

$$d(v_e/v_o)/d(t/t_o) = (\sigma_e/\sigma_o)^4/(1-\omega)^4 . \qquad (28)$$

When $\Delta(\sigma_{ij}/\sigma_o) = 1$ the isochronous surface is obtained for the rupture time t_o, Fig. 2.

It would appear that for proportional stress or loading histories the phenomenological constitutive equations developed in Section 2.1.(b) using a single damage state damage variable satisfy the macroscopic observations and are also capable of physical interpretation. However, for non-proportional loading this approach indicates that it is necessary to provide rate equations for both nucleation and void growth and in these circumstances at least two state variables will be required. Fortunately, in many practical components, while stresses vary they do remain sensibly proportional. Nevertheless, problems will arise in which the stress fields have large rotations when the sequence of loading is applied. Growth and nucleation will occur at different rates in particular directions and it is to be expected therefore that the description of damage will be tensorial in form. It will also be necessary to understand how and if damage introduces important anisotropic effects.

3. CONTINUUM DAMAGE

In the previous section models have been developed which describe the evolution of creep damage, and its effects on creep deformation rates. Implicit in the approach is the requirement that at every material element contained within a homogeneously stress region the damage evolution and its effects manifest themselves uniformly. This requirement has been found to be satisfied in a wide range of experiments and has led to the use of the term CONTINUUM DAMAGE. The advantage of continuum damage is that, together with stress and strain field quantities, it can be used in continuum analysis to describe the time variation of the field parameters, and in particular the effect of damage evolution on stress redistribution and strain accumulation in components. Of particular importance in failure analysis is the ability to describe the initiation and growth of damage zones. The latter is of particular importance since recent work {18,19} has shown that the phenomena of notch strengthening and weakening, and of creep crack growth can be accurately predicted using creep continuum damage mechanics. The significance of the work is that it validates a method of design analysis which uses uni-axial creep curves, together with a knowledge of the multi-axial creep rupture criterion of the material, as input to conventional boundary value problem solution techniques to obtain stress, strain, failure predictions for engineering components working at high temperatures. Probably the most common, and readily accessible, boundary value problem solver is the Finite Element technique.

In the following section it will be shown how the finite element technique has been used, with uni-axial data and with multi-axial stress rupture criteria, to predict the behaviour of a range of engineering components, and how this provides an analysis tool for use in high-temperature component design.

4. NUMERICAL ANALYSIS

4.1. The Finite Element Method

In the following sections are presented the results of Finite Element analyses of a range of structural components. The computations have been performed using special purpose finite element systems; however, a detailed discussion of the systems is not given here and the reader is directed to individual references for numerical aspects. In all cases the finite element procedure commences with the solution of the elastic boundary value problem; this forms the boundary conditions for the solution of the initial value problem. In-elastic deformation and damage evolution takes place, which is described by equations (8) and (14); the field quantities (stress, strain and damage) are integrated numerically and material failure is described by the condition $\omega = 0.99$. In this way it is possible to trace the formation and spread of damage zones and to predict rupture lifetimes. Such results are now presented and discussed for a range of components.

4.2.(a) Continuum damage solutions for a uni-axially loaded tension plate containing a circular hole

Here consideration is given first to the situation in which a uniformly stressed uni-axial plate, containing a central

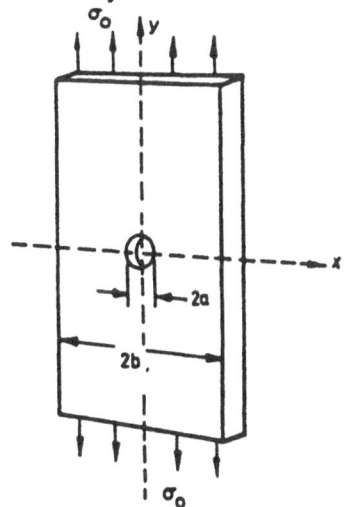

circular hole, undergoes steady load creep rupture. The problem has been studied both theoretically and experimentally for copper and an aluminium alloy. The plate under consideration is shown in Fig. 5. The results of experiments on copper, carried out at 250°C, and on an aluminium alloy, carried out at 210°C, show that when experimental lifetimes are plotted on the basis of the average stress acting on the minimum section, σ_N ($= \sigma_o/(1-a/b)$) then close agreement is obtained with uni-axial creep data.

Fig. 5. Uni-axially stressed tension plate containing a central circular hole.

The theoretical results were obtained from a finite element analysis (20). The multi-axial rupture behaviour of the materials was described by equation (14); a maximum effective stress, σ_e, criterion, $\alpha=0$, and a maximum principal tension stress criterion, $\alpha=1$, were assumed for the aluminium alloy and copper respectively. The computed lifetimes were found to agree well with experimental values.

Computed regions of damage are presented in Fig. 6 and compared with the result of a metallographic examination of a copper plate. Two zones of rupture have been computed; $\omega > 0.1$, which corresponds to moderately damaged material, and $\omega > 0.99$, which represents failed material. Very close agreement can be observed between the damage fields determined theoretically and experimentally.

Similar results have been obtained for aluminium alloy plates, and hence they are not discussed in detail here. However, one significant observation is that the results for the two materials do not reflect the fact that they satisfy different multi-axial rupture criteria. The reason for this is that the stress fields in the plate are essentially plane stress and are therefore confined to the tension-tension quadrant where the two rupture criteria almost coincide.

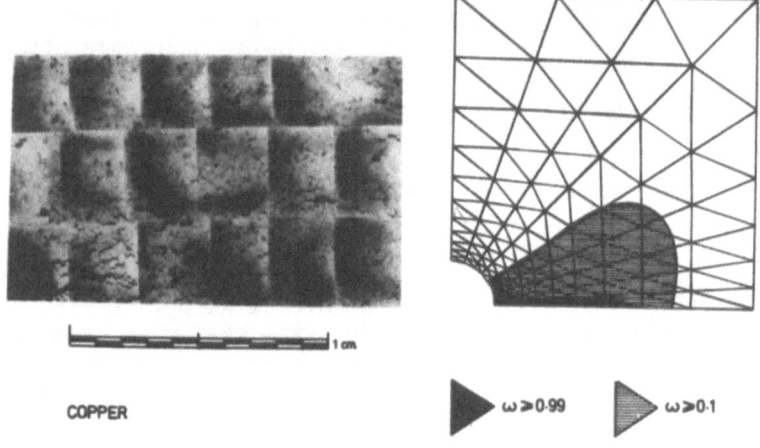

COPPER

$\omega \geqslant 0.99$ $\omega \geqslant 0.1$

Fig. 6.   Comparison of computed regions of damage with metallographic results.

The most striking result to emerge from this study is that the effect of the growth of continuous damage is to annul the localised high stresses found in the elastic and in the stationary state solutions and to redistribute stresses to values close to the net section values.

If this result is applicable to structures which contain more severe stress concentrators then it is of considerable technological significance since it greatly simplifies the level of complexity of design calculations.

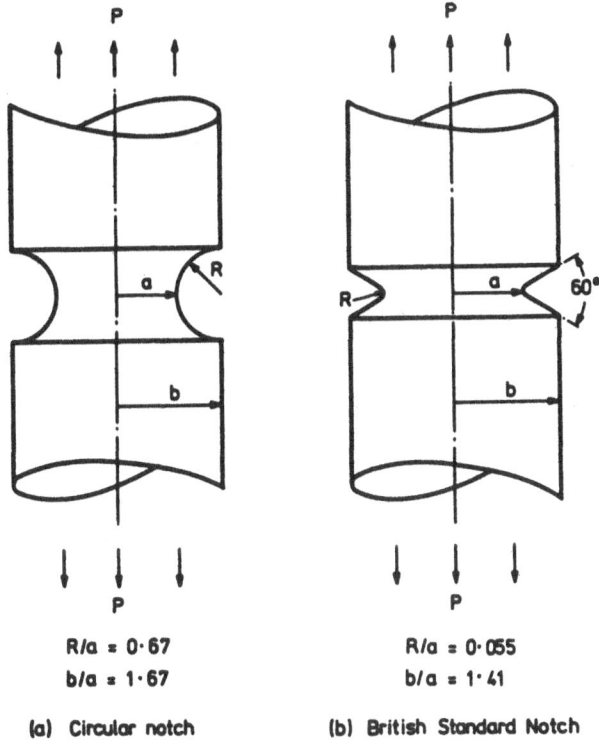

R/a = 0·67
b/a = 1·67

R/a = 0·055
b/a = 1·41

(a) Circular notch

(b) British Standard Notch

Fig. 7.   Circumferentially notched circular bars
under uniform tension.

4.2.(b) The influence of the multi-axial stress rupture criterion on the growth of continuum damage in axi-symmetrically notched bars

The structures to be considered in this section are the circular (21) and British Standard (22) notched bars shown in Fig.7. The elastic and stationary-state creep stress distributions across the minimum section (i.e. for $1 > r > 0$, where $\bar{r} = r/a$, at $z = 0$, and z is the axial coordinate measured from the centre of the notch) are significantly different.

Unlike the plane stress plates where σ_1 is similar in magnitude to σ_e, the stationary-state values of $\bar{\sigma}_1$ and σ_e deviate considerably. It may be expected therefore, that the rates of accumulation of creep damage, as determined by equation (14), and hence the structural lifetimes, will be dependent upon the form of the multi-axial stress rupture criterion or on the value of α in equation (14). This effect can be observed in Fig. 7 where the rupture lifetimes of circular notched bars and thin plates

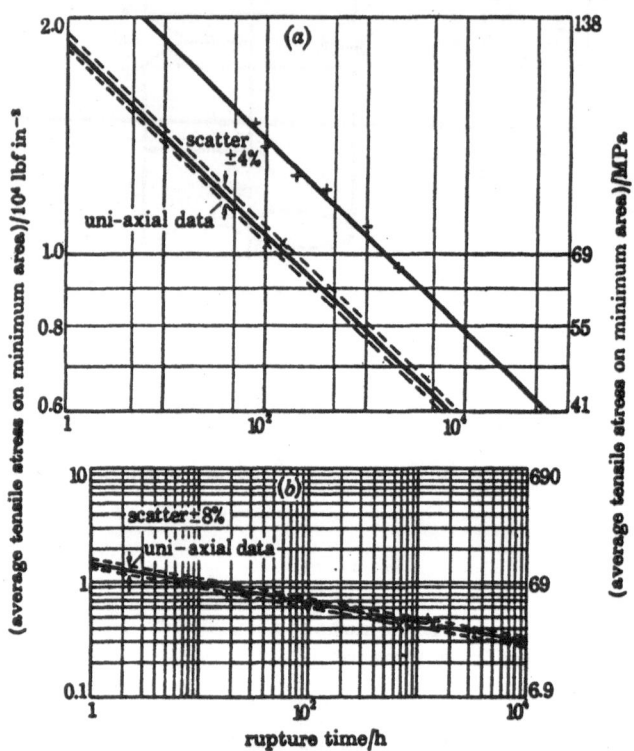

Fig. 8.   Comparison of rupture lifetimes for uni-axial plane stress plates containing circular holes and axisymmetric circular notched uni-axial tension specimens, cf. Fig. 7 (a) with uni-axial rupture data for (a) an aluminium alloy tested at 210°C and for (b) copper tested at 250°C.
+, notch specimens; X, plane stress plates.

containing circular holes are compared with uni-axial rupture data using the average stress acting on the minimum load bearing section. It can be seen in Fig. 8(b) that for copper the structural lifetimes can be determined by the net section stress and uni-axial data; for the aluminium alloys Fig. 8(a), the circular notched bars are stronger than uni-axial test specimens when judged on the basis of the net-section stress. This is not surprising since aluminium alloys satisfy a maximum σ_e stress rupture criterion, or $\alpha = 0$; examination of the stationary stress distribution shows that the average value of σ_e is supressed relative to the average value of the axial component of stress.

In discussions of notch rupture behaviour, it is convenient to introduce the concept of the normalised representative rupture stress. The performance of a notched or cracked specimen may be expressed in terms of the average stress which acts on the minimum section of the notch $\sigma_N = P/\pi a^2$, compared with the stress, σ_R, in a parallel sided uni-axial creep specimen which has the same lifetime as the notched bar. The normalised representative rupture stress Σ_R is defined by σ_R/σ_N. A specimen which requires a lower uni-axial stress σ_R, than the average stress σ_N is said to be notch strengthening ($\Sigma_R < 1$) and, conversely a specimen which requires a higher uni-axial stress σ_R, than the average stress σ_N is said to be notch weakening ($\Sigma_R > 1$). The results for copper bars, given in Fig. 8(b), show slight notch strengthening, $\Sigma_R = 0.95$, and those for the aluminium alloy bars show significant notch strengthening, $\Sigma_R = 9.74$. Both the circular and B.S. notches have been studied previously by using the finite element method of analysis (18). A discussion is now presented of the usage of the latter technique in a study of the behaviour of circular and B.S. notched bars manufactured from 316 stainless steel and tested under steady load conditions at 550°C. However, no multi-axial data is available at this temperature. Instead of performing complex multi-axial tests to determine α an alternative approach has been used. Normalised representative rupture stresses have been determined for both specimens using equations (8) and (14) but for a range of values of α, the results are presented in Fig. 9. Since the normalised experimental representative rupture stresses for the circular and the B.S. notched specimens are 0.858 and 0.942 respectively the value of α for 316 stainless steel at 550°C may be determined from Fig. 9. The theoretically determined values of the normalised representative rupture stress are 0.873 and 0.936 for the circular and B.S. notched bars respectively, for $\alpha = 0.75$. The fact that these values are close to the corresponding experimental values demonstrates that the theory of continuum damage is capable of predicting the behaviour of notched bars. The isochronous rupture locus which results is shown in Fig. 10 where it is compared with the results of tests (23) carried out on the same material at 600°C; reasonably good agreement may be observed.

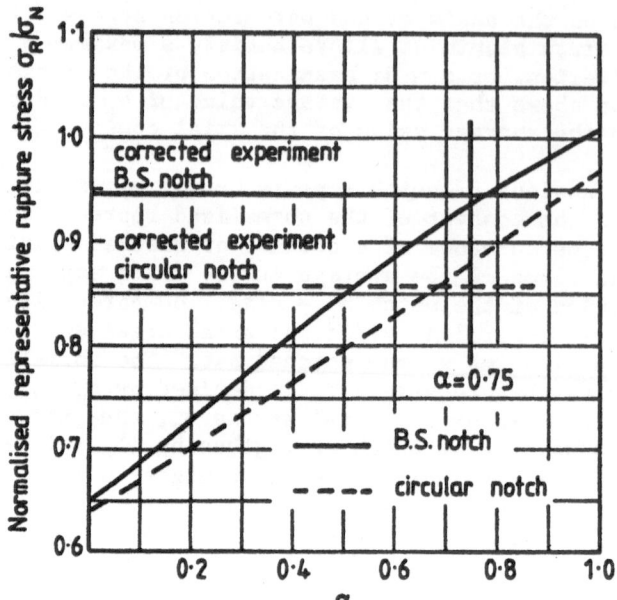

Fig. 9.   Variation of normalised representative rupture
           stress with α for 316 stainless steel.

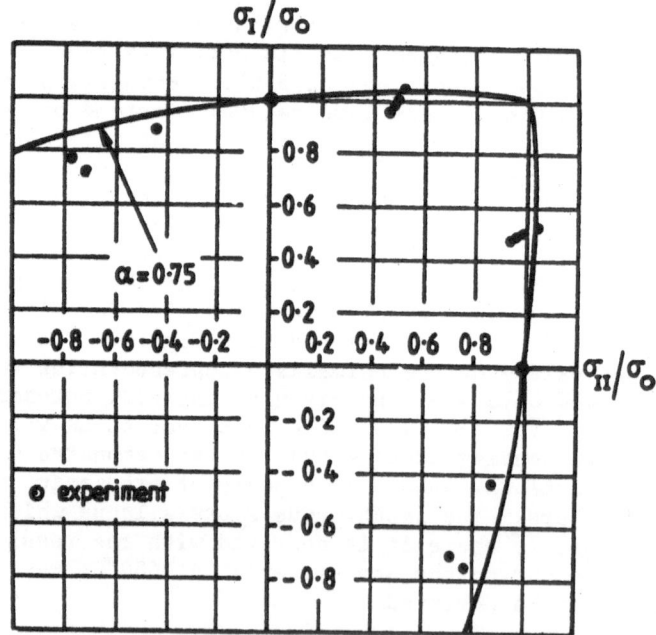

Fig. 10.   Comparison of isochronous rupture locus for α=0.75
            with experimental results obtained for 316
            stainless steel at 600°C, cf. ref. (23).

4.2.(c) Discussion

Two distinct features emerge on comparison of the results of circular and B.S. notches: firstly, that the circular notch shows characteristics of a specimen which uniformly damages in the region of the minimum section and in which failure takes place relatively uniformly towards the end of life, this behaviour is more pronounced in the copper and aluminium specimens studied by Hayhurst et al (18); secondly, that in the case of the B.S. notch the first crack forms early in the lifetime and propagates relatively slowly through the specimen. The latter situation is well removed from the type of behaviour found in the homogeneously stressed region of the circular notch; the extent of the regions of damage is confined to the close proximity of the crack and the behaviour is similar to that for the growth of a crack due to creep in a slightly damaged continuum. This observation suggests that the continuum damage process is probably the mechanism by which well-defined cracks grow by creep. This possibility is investigated in the next section.

4.3. The role of continuum damage in plane strain creep crack growth

Considerable effort has recently been devoted to the development of techniques to describe the growth of sharp cracks due to creep (24,25,26). The majority of the methods are ad hoc extensions to either linear elastic fracture mechanics (27), or post yield fracture mechanics (28) to high-temperature situations. Although the methods can be used to relate the results of experiments carried out on similar specimens of the same material, their range of applicability to different materials, temperatures and structural geometries is not clear (29). For example the available theories assume that discrete cracks grow in their original plane; it is well known that under certain conditions cracks grow by creep on planes which do not coincide with the original crack plane. An example of crack growth on a plane inclined at approximately 55° to the plane of the original crack is shown in Fig. 11 for an aluminum alloy.

The technological aim of a creep crack growth theory is to relate short-term performance of small laboratory specimens to that of large in-service components which contain cracks. Probably the only justifiable way of achieving this aim is to ensure that the governing physical mechanisms of the processes of creep crack growth are the same in both situations.

In this section the proposal is investigated that continuum damage mechanics theory can be used to describe the physical processes responsible for creep crack growth. The investigation reported considers both experimentally and theoretically the

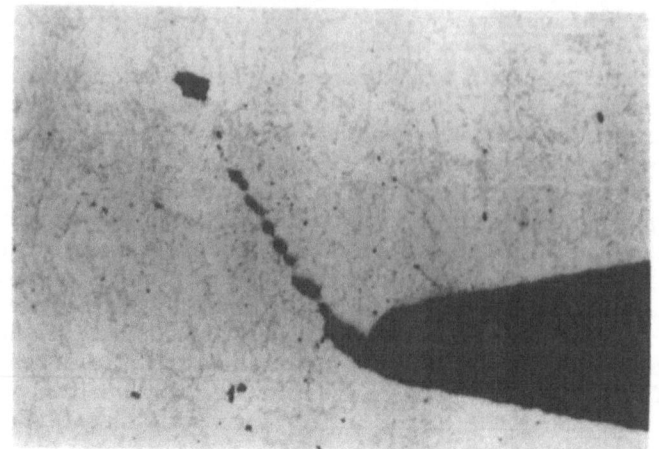

Fig. 11.   Crack tip shear damage, close to failure, in an
externally cracked aluminium alloy specimen tested
at 150°C under steady load, (Mag. x 48).

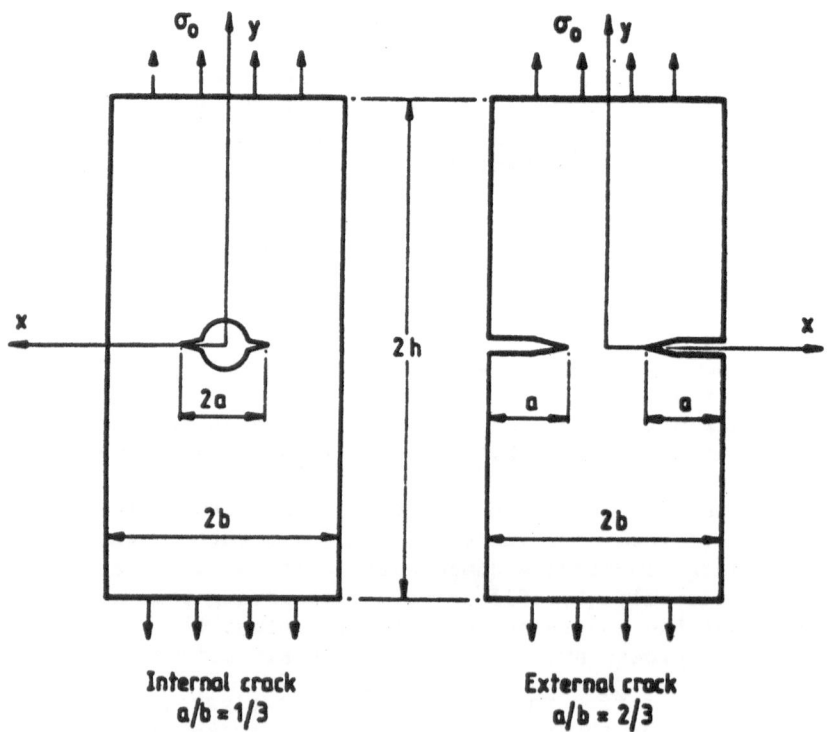

Fig. 12.   Definition of co-ordinate system for internally and
externally cracked specimens.

behaviour of the two specimens shown in Fig. 12. The selection
of the specimen geometries and the presentation of the
experimental results has been discussed in detail elsewhere
(19,30). The material considered is the 316 stainless steel,
discussed in an earlier section and tested at 550 ± 2°C.
Experimental values of the normalised representative rupture
stress given by $\Sigma_R = \sigma_R/\sigma_N$, where $\sigma_N = \sigma_o/(1-a/b)$, are 1.25 and
0.93 for the internally and externally cracked specimens
respectively. The internally and externally cracked specimens
undergo 25% weakening and 7% strengthening respectively.

In the theoretical study of the two specimens the finite
element procedures were used to quantify the growth of damage and
to allow the propagation of a zone of failed material. The
numerical results will now be presented and compared with the
results of metallographic examinations carried out on mid-
thickness planes of tested specimens.

Computed values of the normalised representative rupture
stress are 1.30 and 0.91 for the internally and externally
cracked specimens respectively. These results are in close
agreement with the experimental values; the level of agreement is
worst for the internally cracked geometry which predicts an
additional amount of weakening of 0.05.

Unlike the aluminium specimens tested by Hayhurst, Brown and
Morrison (19) the 316 stainless steel specimens showed no
evidence of the propagation of sharp cracks on initial loading.
However, initial geometry changes were observed which were shown
theoretically not to be significant. Theoretical predictions of
crack advancement, defined by $\omega > 0.999$, are shown in Fig. 13(a),
for the externally cracked specimen at 60% of life, and in
Fig. 11(a), for the internally cracked specimen almost at
failure; the predictions are compared with the distributions of
grain boundary damage shown in the micrographs presented in
Figs. 13(b) and 14(b). For both specimens the grain boundary
damage can be observed to have formed and linked up producing
cracked regions which are similar in shape and orientation to
those obtained theoretically.

4.4. Discussion

In this section it has been shown how the results of
stationary-state and of rupture calculations can be checked
against results obtained from experiments carried out on model
structures. In this way it is possible to check the design
methodology of combining constitutive equations with numerical
methods to predict the performance of structures. In addition it
has been demonstrated how such techniques can be used, together
with the theory of continuum damage mechanics, to predict the

rupture behaviour and the lifetimes of a broad range of structures, which includes cracked components.

a

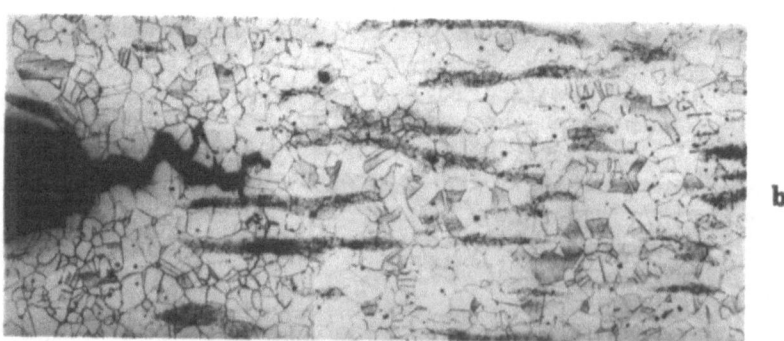

b

Fig. 13.   Comparison of (a) computed failure path $\omega = 0.999$, with (b) mid-thickness micrograph taken from the minimum section of an externally cracked specimen in 316 stainless steel at $t/t_f = 0.6$.

a

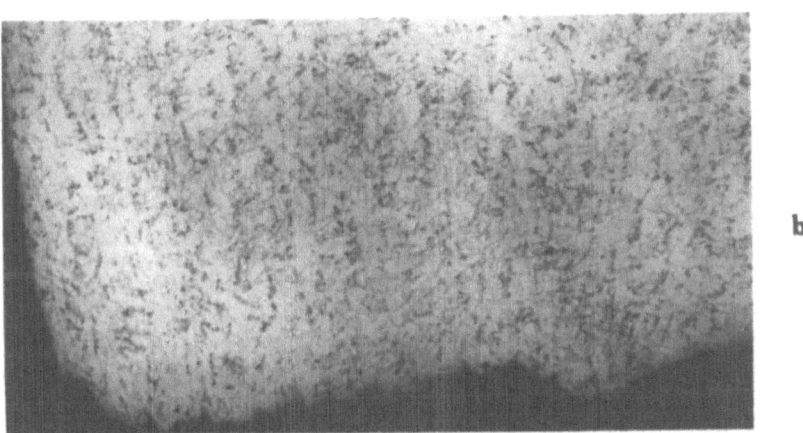

b

Fig. 14.   Comparison of (a) computed failure path ω = 0.999,
with (b) mid-thickness micrograph taken from the
minimum of an internally cracked specimen in 316
stainless steel at failure.

In the case of structures which contain modest stress concentrators, under plane stress conditions, it has been shown that the rôle of continuum damage is to cause considerable stress redistribution and to effectively blunt the region of the stress concentration.

In the axi-symmetrically notched bars, where complex stress-states are found the theory of continuum damage describes the two important features:
(i) stress redistribution due to damage, and
(ii) the multi-axial stress-rupture criterion of the material.
It has been shown that a description of both these features is necessary to predict the effects of notch strengthening and weakening frequently encountered in notch situations.

In structural components which contain initially sharp cracks it has been shown that the theory of continuum damage is capable of describing the high gradients of strain, stress and damage which are found in close proximity to cracks and of describing the multi-axial stress rupture behaviour of the material in such regions. The approach which has been described here has the advantage that, unlike other theories, it is not necessary to pre-judge the direction of crack growth and to assume crack-tip stress fields which neglect both localised and widespread creep damage.

5. FUTURE DEVELOPMENTS

5.1. Present position

The ability to model high temperature creep deformation and rupture using a single state variable theory has been clearly demonstrated; furthermore, the approach has been shown to be consistent with the physically based theories of the metals scientists. This approach is valid for steady stress and temperature conditions. However, few engineering components operate under such conditions and there is a need to develop models for non-proportional loading situations, and for situations in which temperature and stress change out-of-phase.

In addition, limited operation of components outside the shake-down regime is possible and adequate constitutive equations are needed to describe strain-time behaviour. Also, the interaction between creep and alternating plasticity, in certain classes of components, requires an accurate description of the effects of the combined damage processes and their influence on the strain rate behaviour.

In all cases, the need for constitutive equations which provide the input required to numerically solve the range of boundary – initial value problems are becoming more evident as computer technology and computer science develops.

5.2. Numerical Methods and Computer Software

Over the last two decades a significant development of computer software has taken place which allows the numerical solution, with accuracy and with speed, of a range of design problems which are geometrically and materially non-linear. The software is portable between machines and is capable of use by non-experts without long introductory/learning phases. Its development, for use on new machines and architectures, is taking place; and, it will inevetiably lead to its use on the more ambitious and industrially relevant problems.

The development of Computer Aided Design techniques, in particular pre- and post-processors for finite element applications, will improve the man-machine interface to such an extent that more complex multi-dimensional designer driven optimisations will be possible.

5.3. New Computer Hardware

The continuing trend of increased computational speed and of increased fast data storage, at decreasing cost, will generate a stronger desire by the designer to solve larger and more complex design problems. This desire will be stimulated by the requirement to verify the safety of high integrity engineering components and to optimise designs for better economic function. Examples of this can already be observed with developments using parallel processors. What is clear is that the designers ambitions will not be limited either by computer hardware or by computer software; but, by capability data and models to describe the physical processes controlling deformation and rupture in materials.

5.4. New directions

The way forward is to provide a physics based understanding of the process of deformation and rupture under a range of conditions of stress levels, of stress state, and of temperature, for the range of metallic and non-metallic materials now being used and introduced worldwide. The data must be in a form suitable for rapid access such that designers can carry out materials dependent optimisations in design. This inevitably requires the establishment of data banks and of expert systems which allow the appropriate quality and style of data access. However, before such systems can be satisfactorily used, laboratory data has to be generated; this activity is renowned for its cost and for the long times required to generate it.

A possible scenario may well be that development of low cost computer hardware and software takes place at a rate which is faster than the rate of generation of laboratory data; and, of the speed with which it is made accessible on international data banks. Consideration of this, together with the complex thermo-mechanical loading interactions that take place in engineering components, fabricated in metallic and in ceramic based materials, lead one to the conclusion that materials science could provide a limitation to the designers ability to use the full potential of computer technology to optimise the design process.

6. CONCLUSIONS

The paper has shown how mechanisms-based models for high-temperature behaviour of materials can be developed and used together with the theory of Creep Continuum damage, and with finite element based design techniques, to predict component performance from simple uni-axial laboratory tests. The approach has been shown to be valid for a range of engineering components including cracked members.

The impact on design methods of computer science and computer technology has been discussed and it is suggested that developments in these areas will provide a stimulus to change with the limitations being provided by a lack of: development in materials modelling; provision of necessary laboratory data; and, the establishment of appropriate data based systems.

7. REFERENCES

1. "The Theory of Creep", Kachanov, L.M., (English translation
    edited by Kennedy, A.J.), Chs. IX, X, National Lending
    Library, Boston, Spa (1960)

2. "Isothermal Creep Deformation and Rupture of Structures",
    Hayhurst, D.R., Ph.D. Thesis, Cambridge University (1970)

3. Leckie, F.A. and Hayhurst, D.R., Mech. Res. Comm., 2, 23 pp.
    (1975)

4. Dyson, B.F. and McLean, D., Met. Sci., 2, 37 pp. (1977)

5. Greenwood, G., Int. Congress on Metals, Cambridge, 1973.
    Microstructure and the Design of Alloys, 2, 91 pp. (1973)

6. Dyson, B.F., Canadian Met. Quart. 18, 31 pp. (1979)

7. Cocks, A.C.F. and Ashby, M.F., Prog. in Materials Science,
    27, 3-4, 189 pp. (1982)

8. "Mathematical Theory of Creep and Creep Rupture", Odqvist,
    F.K.G., (2nd Ed.), Ch.12, Oxford: Clarendon Press (1974)

9. "Complex-Stress Creep, Relaxation and Fracture of Metallic
    Alloys", Johnson, A.E., Henderson, J. and Khan, B., Ch.4,
    H.M.S.O., London (1962)

10. Hayhurst, D.R. and Leckie, F.A., Proceedings of IUTAM/ICM
    Symposium on Yielding, Damage and Failure of Anisotropic
    Solics. Villard-de-Lans, France, August 24-28 (1987)

11. Leckie, F.A. and Ponter, A.R.S., Ing. Arch., 43, 158 pp.
    (1974)

12. "Creep Problems in Structural Members", Rabotnov, Y.N.,
    Amsterdam: North Holland Publishing Company (1969)

13. Kachanov, L.M., IZV. Akad. Nauk. SSSR O.T.N. Teckh. Nauk. 8,
    26 pp. (1958)

14. Hayhurst, D.R., J. Appl. Mech., 1, 40, 244 pp. (1973)

15. Hayhurst, D.R., J. Mech. Phys. Solids, 20, 381 pp. (1972)

16. "Creep Fracture", Ashby, M.F. and Raj, M., Proceedings of
    the Conference on Mechanics and Physics of Metals,
    Cambridge: The Metals Society, Institute of Physics (1975)

17. Leckie, F.A. and Hayhurst, D.R., Acta Metallurgica, 25, 1059 pp. (1977)

18. Hayhurst, D.R., Dimmer, P.R. and Morrison, C.J., Phil. Trans. R. Soc. Lond. Q, 311, 103-129 pp. (1984)

19. Hayhurst, D.R., Brown, P.R. and Morrison, C.J., Phil. Trans. R. Soc. Lond. A, 311, 131-158 pp. (1984)

20. Hayhurst, D.R., Dimmer, P.R. and Chernuka, M.W., J. Mech. Phys. Solids, 23, 335 pp. (1975)

21. "Studies in Large Plastic Flow and Fracture", Bridgman, P.W., New York: McGraw-Hill (1952)

22. "Methods for Creep Rupture Testing of Metals", British Standards 1969 no. 3500

23. Chubb, E.J. and Bolton, C.J., Proc. Int. Conf. on Eng. Aspects of Creep, 15-19 Sept. 1980, Sheffield, 1, Paper C20180, 39 pp., London: I.Mech.E.

24. Neate, G.J. and Siverns, M.J., Conference on Creep and Fatigue in Elev. Tem. applics., Philadelphia: ASME/I. Mech.E. (1973)

25. "The Extension of a Macroscopic Crack at Elevated Temperature by the Growth and Coalescence of Microvoids", Riedel, H., (IUTAM Symposium, Leicester, U.K.), ed. Ponter, A.R.S. and Hayhurst, D.R., Berlin: Springer-Verlag (1981)

26. "Finite Element Analysis and Experimental Investigation on Creep Crack Propagation", Ohtani, R., (IUTAM Symposium, Leicester, U.K.), ed. Ponter, A.R.S. and Hayhurst, D.R., Berlin: Springer-Verlag (1981)

27. Webster, G.A., Proc. Conf. Mech. and Phys. of Fracture, January (Cambridge, England), paper 18, London: Institute of Physics (1975)

28. Rice, J.R., Jnl. Appl. Mech., 35, 379 pp. (1968)

29. Ainsworth, R.A., Int. Jnl. of Fracture, 18, 26 pp. (1982)

30. Hayhurst, D.R., Morrison, C.J. and Brown, P.R., Creep in Structures, 1980 (IUTAM Symposium, Leicester, U.K.) ed. Ponter, A.R.S. and Hayhurst, D.R., Berlin: Springer-Verlag, 564 pp. (1981)

5.3.2. Life Prediction and Residual Life Assessment

Professor G.W. Greenwood

School of Materials, University of Sheffield

Mappin Street, Sheffield S1 3JD, UK.

1. INTRODUCTION

The life of components operating under stress at elevated temperatures may be limited by excessive strain or by fracture and both of these may be influenced by environmental attack. Often a relationship has been noted between minimum strain rate $\dot{\varepsilon}_m$ and the time to fracture t_f, taking the form first proposed by Monkman and Grant (1956), with $\dot{\varepsilon}_m^{\beta} t_f \sim c$ where c is a constant, independent of stress or temperature over significant ranges, but influenced by the material noting its precise composition and microstructure. The further constant β is generally close to unity. This situation can be represented geometrically as illustrated in Fig. 1. The gradient of the line AB represents the minimum creep rate $\dot{\varepsilon}_{m1}$ leading to a creep life t_{f1}. Thus the line BC represents the product $\dot{\varepsilon}_{m1} t_{f1}$. The minimum creep rate $\dot{\varepsilon}_{m2}$ at some higher level of stress or temperature is represented by the gradient of the line DE and the creep life is now t_{f2}. In this case EF represents the product $\dot{\varepsilon}_{m2} t_{f2}$. When $\beta = 1$, then BC = EF.

On the above basis, creep lifetime prediction is readily made and, in principle, accelerated testing at relatively high stress levels may be undertaken to determine β and c for the material in question. However, there is increasing concern of the inadequacy of this approach since it may lead to non-conservative predictions through a tendency for the parameter c to decrease with a decrease in minimum strain rate. Taking the relation between $\dot{\varepsilon}_m$ and stress σ in the form of Norton's Law, with $\dot{\varepsilon}_m \alpha \sigma^n$, this is manifest at a 'knee' in plots of experimental data with axes log σ vs log t_f (Hayhurst and

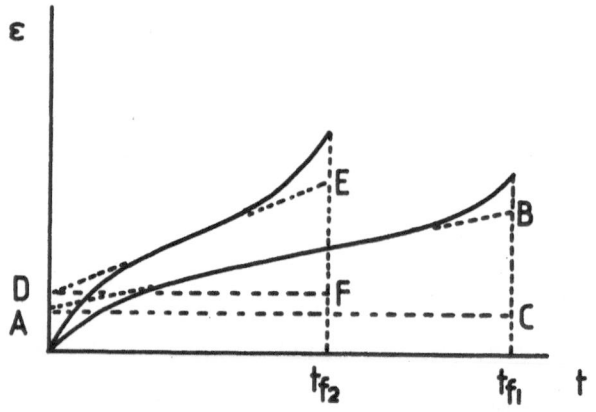

Fig. 1. Two creep curves are illustrated with the upper curve at the higher level of stress or temperature. The geometrical implication of the $\dot{\varepsilon}_m^\beta t_f/\varepsilon_f \sim c$, for the case where $\beta = 1$ is that BC = EF when the gradients of the lines AB and DE represent the respective minimum creep rates.

Lavender, 1985). The need to consider this feature is well recognised and has resulted in modified proposals. Dobes and Milicka (1976) pointed out that an equation of the form $\dot{\varepsilon}_m^\beta t_f/\varepsilon_f \sim c_1$ was more suited to the description of a wide range of experimental data, where ε_f is the strain at fracture and c_1 is a new constant. Although this latter equation retains a simple form, it is much more difficult to utilise in practical situations for it requires, in each instance, that the fracture strain is known. The implications of this go far deeper than is immediately apparent for they suggest that a unique relationship does not necessarily exist between the form of deformation and the mechanism of fracture. Atomistic theories of the fracture process have supported the implied separation of mechanisms of high temperature crack and cavity growth from those of the mode of overall deformation by identifying different regimes of behaviour (Beere, 1981).

Although the constants c and c_1 were considered to be not strongly dependent on temperature T because of its compensating effects on $\dot{\varepsilon}_m$ and t_f, the individual influence on each of these latter parameters of temperature on rupture life is highly significant (Woodford 1981). This has long been recognised and the Larson-Miller (1952) parameter P_{LM}, depending only on stress, has been written in the form $P_{LM} = T(C_2 + \log t_f)$ leading to a nearly linear relationship between $\log t_f$ and I/T at constant stress. The value of the constant C_2 taken as 20 by Larson and Miller, when t_f is measured in hours, has been questioned and modifications have been proposed. An example is the Manson-Haferd (1953) parameter P_{MH}, again a function of stress alonge, which may be written $P_{MH} = (\log t_f - \log t_a)/(T - T_a)$ where t_f and t_a are the creep lives at temperatures T and T_a respectively.

The latter procedure can be illustrated graphically as in Fig. 2. The logarithm of the experimentally determined creep life is plotted over a range of temperature between T_a and T_b where tests can be carried out relatively quickly at the same stress as that which the component would endure for a much longer time at some lower temperature in service.

The reciprocal of the gradient of this line corresponds to the parameter P_{MH}. To obtain an appropriate margin of safety in using this procedure, the line for the lower bound confidence limit PA is extrapolated down to the service temperature T_s to determine the creep life t_f.

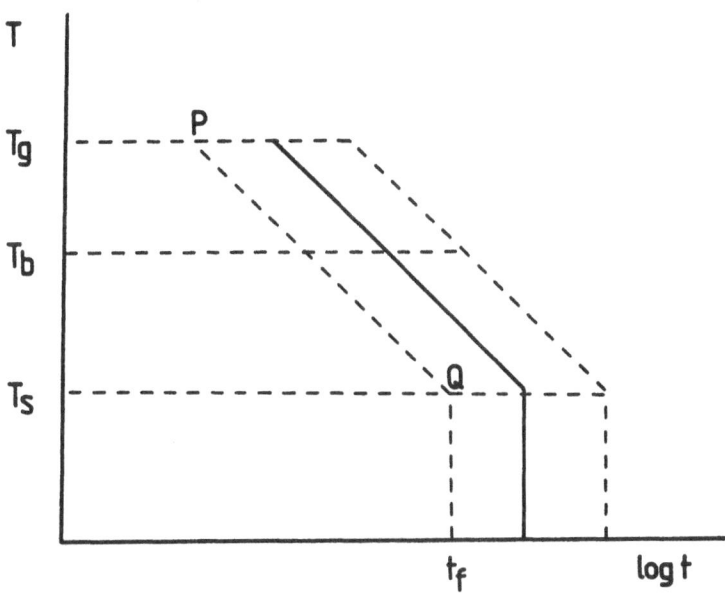

Fig. 2. The creep lifetimes are measured between temperatures T_a and T_b and the line is extrapolated down to the service temperature T_s so that the service lifetime may be predicted. In practice, a margin of safety is established by extrapolation of the lower sloping dotted line PQ, which corresponds to the lower bound of the conficence limit. A stress level equal to that sustained in service is used in all the tests.

2. DESCRIPTION OF THE CREEP CURVE

A somewhat different approach to the extrapolation of creep data has been proposed (Evans and Wilshire, 1985) which is based on the identification of a type of formula that can describe the complete strain-time relationship over the full creep life such as in the creep curves illustrated in Fig. 1. The proposed equation gives

$$\varepsilon = \theta_1\{1 - \exp(-\theta_2 t)\} + \theta_3\{\exp(\theta_4 t - 1)\}$$

where θ_1, θ_2, θ_3 and θ_4 are constants characteristic of the material but dependent upon the level of (uniaxial) stress and upon temperature. Here ε is the true strain and the creep experiments must be conducted at constant stress. The first term in squared brackets on the right hand side of the equation relates to the primary creep strain. Its form is essentially expirical though it may be tenuously related to dislocation mechanics. The second term represents the tertiary creep strain and is of a form that is broadly compatible with processes of deterioration that may arise through particle coarsening and by the nucleation and growth of cracks and cavities. The secondary range of creep is not explicitly described in this formulation though the equation may represent a significant region where the additive contribution of the two terms results in a nearly linear relation of true strain with time.

The experimental determination of the four constants at each required temperature and stress level and for each material, noting its precise composition and microstructure, presents a formidable task. Nevertheless, such experimental studies have been undertaken both for pure metals (Brown et al., 1987) and for commercially important 1 % Cr /2 Mo /4 % V steels (Hayhurst et al., 1985). Data processing by computer is essential to assess in a reasonable time the information from such studies and an appropriate formula mut be ascribed to each of the four θ parameters. These have been assumed to take the form $\log \theta_i = A_i + B_i T + C_i \sigma + D_i \sigma T$.

3. MICROMECHANISMS OF CREEP DAMAGE

An adequate understanding on a micro-scale of the detailed mechanisms leading to creep failure would form the most convincing basis of creep life prediction and for residual life assessment (Greenwood, 1978, 1987). However, although major progress has been made in extensive laboratory investigations, the wealth of information is complex and not yet approaching a situation where accurate predictions can generally be made. Nevertheless, many significant features have been identified of importance to designers of components (Penny and Marriott, 1971) and a much greater understanding of the creep behaviour of commercial materials is in prospect. Creep damage leading to fracture takes the form of cavity and crack formation which, for long creep lives almost invariably occurs on those grain

boundaries that are oriented nearly perpendicular to the
principal applied tensile stress. A schematic illustration of
cavity formation on a grain boundary under such a stress is
illustrated in Fig. 3. Failure occurs when cavities link up of
when cracks grow to cover a fraction (usually taken to be about

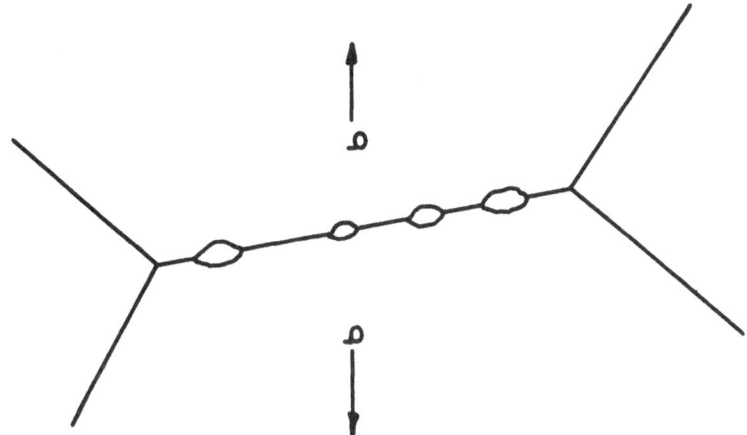

Fig. 3. Schematic illustration of cavities formed on a
grain boundary nearly perpendicular to the
principal tensile stress σ.

half) of the area of these grain boundaries (Gittus, 1975). The
fraction of seriously cavitated grain boundaries increases with
time as shown in Fig. 4 and a safe creep life may be assumed when the fraction of such boundaries reaches some specified value, typically about 0.5. For some low alloy steels a nearly linear relationship has been revealed on the variation of cavitated area with time, with consequent

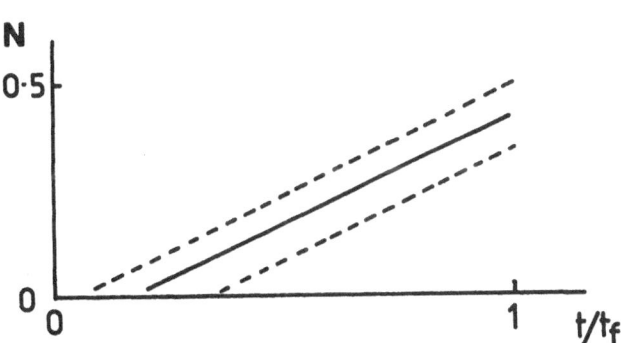

Fig. 4. The fraction N of seriously cavitated
grain boundaries is plotted against the
fraction of creep life t/t_f. Since
there is considerable scatter about the
mean line the upper bound confidence
limit shown by the upper dotted line can
be taken as an appropriate indication
of the safe creep life.

simplification of extrapolation procedures. However, there
is considerable scatter about the mean line and the value of an
upper bound confidence limit shown by the upper dotted line can
be taken as a measure of the remaining creep life.

Cavity nucleation still presents one of the most difficult
processes to understand. Clearly, particles in grain boundaries
of adequate size and with low cohesion with the matrix are likely
to prove the most common nucleation sites. Many experiments have
confirmed this role of particles but most recent studies (George
et al., 1987) have indicated the importance of precise
compositional control in iron and in steels of commercial
importance. Evidence is accumulating that in steels there is a
range of particle types that can be placed in order of their
propensity to nucleate cavities. Sulphide particles cause
easiest cavity nucleation though their deleterious effect can be
somewhat reduced by small addition of phosphorous. Oxide
particles can nucleate cavities when some sulphur is also
present, but carbon and phosphorous additions appear to reduce
nucleation on oxides. It has been noted that carbides do not
generally play a significant part in cavity nucleation.

Some of the difficulty in defining the exact role played by
particles lies in the observation that cavities can often be
nucleated continuously throughout creep life. Thus it is
considered that deformation processes, particularly grain
boundary sliding, play some part in the formation of cavities.
On this basis some assessment may be made of residual life but
clearly such investigation cannot be non-destructively and, at
least, it involves removal of some part of a component in a
critical region. It has also been noted that cavities may be
nucleated more easily in certain materials that have been subject
to prior deformation (Dyson et al, 1976). Several Nimonic alloys
are prone to such an effect which can substantially reduce their
creep life. Moreover, in these alloys the influence of
environment on creep life is also particularly strong.

Over many years conflicting proposals were strongly defended
about the mechanism of cavity growth. It is now increasingly
recognised (Beere 1981) that different mechanisms can operate,
some in series, some in parallel, depending upon the specific
material, its stress state and the temperature. At low stress
levels, the ability of grain boundaries to emit vacancies is
considered to be a feature of major importance. Some segregants
or particles at the grain boundaries may inhibit such action with
the result of substantially prolonging creep life. Cavity growth
by vacancy diffusion and condensation is amenable to theoretical
calculations (Cocks and Ashby, 1982) and this is perhaps the best
fundamentally understood feature. The volume to accommodate
growing cavities, however, may also require concurrent matrix

deformation and, if this does not occur, then cavity growth may be constrained (Dyson, 1979). In such calculations the effect of multiaxial stresses may be taken into account and this feature is likely to be of great importance in analysing the complex behaviour of components in service.

Although presenting more formidable theoretical challenges, considerable progress has been made in estimating rates of crack growth under creep conditions. Such cracks are frequently initiated through grain boundary sliding and the build up of stress concentrations where three grain boundaries meet (Gittus 1975). It seems clear that materials with coarse grains size, although they may have higher creep strength, may be more susceptible to grain boundary creep cracking (Greenwood, 1978). One of the most significant practical steps forward in enhancing the creep life of commercial materials has resulted from the realisation of the value of processing to give a sufficiently fine grain size which, in the case of ferritic materials, refers to the grain size of the prior austenite. This point is particularly significant in minimising creep damage at welds and in techniques for weld repair.

4. THE MACROSCOPIC APPROACH TO CREEP DAMAGE EVALUATION

In this approach (Kachanov, 1960), all the detailled features of cavity and crack development are incorporated into a single damage parameter w (Hayhurst, 1983) where the creep rate $\dot{\varepsilon}$ may be written $\dot{\varepsilon} = f(\sigma, w)$ and the rate of build up creep damage $dw/dt = g(\sigma, w)$. The problem now becomes one of determining the form of the functions f and g. For uniaxial stress, the creep rate normalised with respect to some chosen value $\dot{\varepsilon}_0$ at stress, σ_0 is assumed to be given by

$$\dot{\varepsilon}/\dot{\varepsilon}_0 = (\sigma/\sigma)^n K(t)/(1-w)^n$$

Similarly the rate of damage build up

$$\dot{w}/\dot{w}_0 = (\sigma/\sigma_0)^x/(1 + y)(1 - w)^y$$

where w_0, x and y are constants. When w = o and K(t) is constant, we have equation for Norton's law and $\dot{\varepsilon} \rightarrow \alpha$ as $w \rightarrow 1$. Thus, on integration, with limits w = o at t = o and w = 1 at the limit of creep life t_f, we obtain $t_f = 1/\dot{w}_0 (\sigma/\sigma_0)^x$. If w is taken as a fraction of the grain boundary area on which cracks have been produced and this is monitored as a function of the duration of creep, then this can form a further means of creep life estimation.

5. ESTIMATION OF REMAINING CREEP LIFE

There is a major advantage in using methods that do not require removal of material for destructive examination. This, however, implies close monitoring of operational parameters, involving strain, temperature and stress evaluation throughout the service period. It also requires the prior establishment of the form of behaviour of the material in question. With these requirements, considerable attention has been given to the validity of the proposed 'life fraction rule' (Robinson, 1952) which can be written in the form $\Sigma t_i/t_{fi} = 1$ where t_i is the time period endured under stress, temperature and environment which would result in a creep life t_{fi}.

There is substantial evidence that the life fraction rule provides a good description of the behaviour of a range of austenitic steels. For ferritic steels similar agreement has been noted when creep life variation has been brought about by temperature changes but predictions are much less concluded that the life fraction rule is useful in all instances in so far as temperature changes are concerned and it can also be applicable to stress changes for materials whose creep strength is not substantially altered by thermal exposure.

One of the most difficult problems to be faced lies in the reduction in creep life that can be caused by oxidation through atmospheric exposure during creep. This is particularly noted in experimental tests on small specimens. With larger specimens the effect is progressively reduced and this makes some estimation possible of the behaviour of the thicker material more typical of the wall thickness of vessels and pipe work used in practice.

In low alloy ferritic steels it has been recognised that failure results both from microstructural degradation which can cause progressive reduction in creep strength and from the development of intergranular creep cavitation. Both of these processes occur simultaneously and they are strongly influenced by chemical composition, prior heat treatment and microstructure, as well as by the stress and temperature conditions. The effects of microstructural degradation can be most simply assessed by measurements of hardness changes and these have proved a useful means of detection of cases where significant weakening has occurred.

Welds in ferritic steels have shown a particular propensity to failure and in these regions it is usually cavity development that presents the most serious problem. As mentioned earlier, welding techniques to minimise austenite grain size have proved a major advance in greatly reducing the widespread occurrence of this problem. Much has also been gained by detailed microstructural examination of material removed from weld regions after various periods of exposure under closely monitored conditions. By separate evaluation of the rates of cavity nucleation and of their growth, a systematic basis is becoming apparent which shows promise of further development. Such observations have been usefully compared with increasingly sophisticated models of cavity nucleation and growth, giving an opportunity for more soundly based estimates of residual creep life (Cane and Needham, 1983). At the same time such observations, although suffering the drawback of requiring destructive testing, are providing information about chemical compositional control of future power plant materials which can confidently be expected to have enhanced creep life. The use of cellulose acetate replicas has proved particularly useful (Neubauer and Wedel, 1983) in the optical examination of weldments for cavitational damage and cracking and it has been possible to examine relatively large areas non-destructively.

Although the occurrence of structural degradation can be ascertained by hardness measurements, it is only be detailed metallographic examination, generally by transmission electron microscopy, that the full details and implications can be inferred. The observation and measurements of rates of coarsening of carbide particles has been shown to give some indication of time of subjection to a specific operating temperature and this can assist life estimate by application of the life fraction rule. A further extension of this approach has become apparent through the recognition by microanalysis of the changes occuring in the chemical composition of carbides with exposure time at temperature.

6. REMAINING LIFE ASSESSMENT BASED ON ACCELERATED CREEP AND RUPTURE TESTING

Techniques based on accelerated creep rupture tests are increasingly acknowledged to present a reliable approach to creep life evaluation (Cane and Brear, 1986). Their main limitation lies in the limited number of samples that can be taken from critical regions.

There is the option of accelerating creep by stress or by temperature increase but experience indicates that testing at increased temperature at an unaltered stress level presents the most satisfactory basis. This view is substantiated by mechanistic considerations since each of the processes leading to creep damage is less sensitive to temperature than to stress changes. Even from relatively short (typically 1000 hr) post exposure accelerated creep test durations, reliable estimates may be made through application of the life fraction rule. It has been pointed out (Cane and Brear, 1986) that such post exposure testing is best carried out under inert atmosphere, so that undue oxidation does not occur on small specimens because of the higher temperature of test.

Further simplifications have been suggested to procedures for post exposure testing, since if the Monkman-Grant relation $\dot{\varepsilon}_m t_f$ = a constant, can be shown to apply, measurement of the creep rate at enhanced temperature T_a at which the creep life is t_a permits creep life t_f to be predicted at temperature T by making use of the Manson-Haferd parameter

$$P_{MH} = \{\log t_f - \log t_a\} / \{T - T_a\}.$$

This is a less time consuming approach and may sometimes be justified though it is inherently dependent on less well established features than those involved in the accelerated testing of creep rupture.

7. CONCLUDING NOTES

Creep life prediction is generally made more reliable by the monitoring of temperature and stress level throughout service together with measurement of strain accumulation. Every effort should be made to acquire such data for material serving in critical regions and particularly at welds. Some evaluation should also be made of oxygen penetration to ascertain whether it is likely to influence creep life.

Hardness measurements have proved a convenient way of assessing the extent of material degradation by particle coarsening and recovery mechanisms. Replication of the outer surfaces using cellulose accetate films has been shown to provide a means of finer scale defect detection than is possible with other techniques so that cavity size and number may be assessed. Further, replicas may be used for particle extraction and microanalysis thus allowing characterisation of microstructural changes.

The use of post exposure accelerated creep testing, at a similar stress level to that in service but at a higher temperature, together with an application of the life fraction rule, is widely acknowledged to provide the most reliable current practial method of residual life assessment.

More studies are required of the effects on creep life of transient conditions and of multiaxial stress application. The increasingly well developed microstructural interpretations of creep damage and their assimilation into macroscopic models, permitting numerical analysis, show promise of increasing application and of establishing more confidence in predictive methods.

8.   REFERENCES

(1)   "Cavities and Cracks in Creep and Fatigue", Beeré, W.B., Ed by Gittus J.H., Applied Science Publishers, London (1981)

(2)   Brown, S.G.R., Evans, R.W. and Wilshire, B., Mat Sci and Tech, 3 23 pp. (1987)

(3)   Int. ASME Conf. on "Advances in Life Prediction", Cane, B.J. and Needham, N.G., ASME, Albany, NY (1983)

(4)   Internat. Symposium on Creep-Resistant Metallic Materials, Cane B.J. and Brear, J.M., Czech Sci. and Tech. Soc, Brno, 274 pp. (1986)

(5)   Prog. Mater. Sci, Cocks, A.C.F. and Ashby, M.F., 27, 189 pp. (1982)

(6)   Met. Sci., Dobes, F. and Milicka, K., 10, 382 pp. (1976)

(7)   Proc. Roy. Soc., Dyson, B.F., Loveday, M.S. and Rodgers, M.J., London A349, 245 pp. (1976)

(8)   Can. Met. Quart., Dyson, B.F., 18, 31 pp. (1979)

(9)   "Creep of Metals and Alloys", Evans, R.W. and Wilshire, R.W., Inst. of Metals, London, (1985)

(10) Acta Metall, George, E.P., Li, P.L. and Pope, D.P., 35, 2471 and 2487 pp. (1987)

(11) "Creep Viscoelasticity and Creep Fracture", Gittus, J.H., Applied Science Publishers, London (1975)

(12) Phil. Trans. Roy. Soc., Greenwood, G.W., London, A, 288, 213 pp. (1978)

(13) Kovove Materialy, Greenwood, G.W., Bratislava 5.25, 569 pp. (1987)

(14) "Engineering Approaches to High Temperature Design", Hayhurst, D.R., Ed. by Wilshire B. and Owen, D.R.J., Pineridge Press, Swansea, UK, 85 pp. (1983)

(15) Int. J. Pres. Ves. and Piping, Hayhurst, D.R., Lavender, D.A., Worley N.G. and Salim, A., 20, 289 pp. (1985)

(16) Gos Izdat Fiz-Mat Lit, Kachanov, L.M., Moscow (1960)

(17) Trans ASME, Larson, F.R. and Miller, J. 174 (1952)

(18) NACA Report TN 2890, Manson, S.S. and Haferd, A.M., USA
(1953)

(19) Proc. Amer. Soc. Test Mater, Monkman, F.C. and Grant, N.J.,
56, 593 pp. (1956)

(20) Int. Conf. on "Advances in Life Prediction", Neubauer, B.
and Wedel, V., ASME, Albany, MY (1983)

(21) "Design for Creep", Penny, R.K. and Marriott, D., McGraw
Hill, London (1971)

(22) Trans ASME (E), Robinson, E.L., 74, 777 pp. (1977)

(23) "Creep Fracture of Engineering Materials and Structures",
Woodford, D.A., Proc. of Int. Conf. at Univ. Coll., Swansea
Ed. by Wilshire, B. and Owen, D.R.J., Pinderidge Press,
Swansea, UK (March 1981)

5.3.3. New approach to materials design: calculated phase

equilibria for composition and structural control

T.I. Barry and T.G. Chart

National Physical Laboratory

Teddington, Middlesex TW11 OLW, England

1. SUMMARY

In recent years great progress has been made in the application of thermodynamic methods to the development and use of materials. The principles of these methods are briefly described and the need for consistency in the representation and assessment of data is stressed. Examples of applications are given that relate to alloys, ceramics and coatings and to corrosion control.

2. INTRODUCTION

Thermodynamic analysis is employed as a tool in almost every aspect of the design, production and use of high temperature materials. This chapter covers alloy development, ceramics and corrosion control. Examples of the applications to specific problems are used to illustrate the methods and the sources of data.

Materials for use at high temperaratures are required to have various combinations of properties, including processibility, stability, strength, toughness, resistance to creep and corrosion and to have a product cost appropriate to the application. In order to achieve these properties, alloys may contain as many as ten or more elemental components. Ceramics too are variable in composition and their properties are particularly sensitive to the presence of additives and impurities which may accumulate in crystalline or glassy phases at grain boundaries. Moreover, by definition, high temperature materials operate in severe environments which may include reducing or oxidising gas streams containing numerous gaseous species of carbon; hydrogen, nitrogen, oxygen, sulphur, chlorine, alkalis and other elements. Molten salts may deposit on materials and act as fluxes for attack on protective oxide layers.

An important reason for employing thermodynamic methods is that they offer a means of systematising and modelling the behaviour of interacting phases in such complex systems, allowing the phase relations of multicomponent, multiphase systems to be modelled from those of their subsystems. It is seldom necessary to have data for systems of more than three or four components in order to make valid calculations on high order systems. For this reason the thermodynamic approach provides a method for guiding experiment. The reverse is also true, thermodynamic analysis is best undertaken in the context of a knowledge of the materials science of the problem and hence a combined thermodynamic and experimental approach will usually be desirable. This point will be seen to be relevant to all the examples described.

3. MODELLING AND REPRESENTATION OF DATA

Good experimental data are clearly necessary as the fundamental support of a thermodynamic database but they are seldom used directly in calculations of phase equilibria. The experimental data, usually for systems of 1 to 4 chemical elements at a time, must be assembled and critically assessed so as to provide sets of coefficients for mathematical equations (models). These coefficients then comprise the database used in calculations of chemical equilibria. The models selected should preferably be theoretically based and should require a small number of coefficients in order to describe the experimental data adequately. Furthermore, unless they prove inadequate, it is best to assess data for individual systems in terms of the current models and in such a way as to give numerical consistency with existing data for related systems. For example data for the solid solution with the halite structure between nickel oxide and wüstite should be consistent with existing data for the Fe-Ni, Fe-O and Ni-O systems and should preferably be undertaken at the same time as the assessment of data for the spinel and liquid oxide phases.

The models are needed to describe the mathematical dependence of the thermodynamic properties on temperature, pressure and composition for each phase. For pure substances and gases the temperature dependence is commonly provided by the following expressions for heat capacity or the Gibbs energy.

$$C_p = a + bT + cT^2 + dT^{-2} \tag{1}$$

$$G = A + BT + CT \ln(T) + DT^2 + ET^3 + FT^{-1} \qquad (2)$$

and since the enthalpy and entropy can be calculated from

$$H_T + H_0 + \int_{T_0}^{T} C_P \, dT \qquad (3)$$

$$S_T + S_0 + \int_{T_0}^{T} C_P/T \, dT \qquad (4)$$

The Gibbs energy can be determined by means of equation 5.

$$G = H - TS \qquad (5)$$

As implied by the terms H_0 and S_0 in equations 3 and 4, the Gibbs energy is not an absolute quantity but must be referred to both chemical and physical standard conditions in order to determine the coefficients A and B in equation 2.

The reference substances are almost always chosen to be the pure chemical elements in the phase in which they are stable at 298.15 K and 1 atm pressure. An exception is made for phosphorus for which the white rather than the red form is usually taken as the reference phase. Unfortunately various reference conditions (standard states) have been chosen. Both 0 K and 298.15 K are used as reference temperatures and 1 atm (101235 Pa) and 1 bar (10^5 Pa) are used as reference pressures. The reference pressure used in this Chapter is 101325 Pa and partial pressures are given as a fraction of this pressure. For components in solutions the standard states of interest for high temperature materials will usually be the pure substance in a defined phase as discussed further below. For solids the difference in reference pressures is trivial, but for gases the entropy difference of 0.1 J mol^{-1} K^{-1} is small enough to be overlooked but big enough to be significant. Details of the standard states used in various compilations are given by Barry {1}, but are in any case given in the compilations themselves {2-7}. Computer based databanks can be programmed to present data in various formats and to some extent with adjustable reference states, to match hard copy compilations.

In this chapter Gibbs energies of pure substances are expressed in the form G-H$_{ser}$ in which H$_{ser}$ is the sum of the enthalpy of the elements comprising the substance in their standard states at 298.15 K.

$$G - H_{ser} = \Delta_f H^o_{298} + (H^o_T - H^o_{298}) - T \, S^o_T \tag{6}$$

Where $\Delta_f H^o_{298}$ is the standard enthalpy of formation from the elements at 298.15 K, $(H^o_T - H^o_{298})$ is the enthalpy required to heat the substance from 298.15 K to T and S^o_T is the standard entropy of the substance at T.

4. SOLUTION PHASES

The Gibbs energy of a substance in solution in a particular phase contains additional terms to describe the ideal contribution to entropy and the non-ideal enthalpy and entropy of mixing.

$$G = \sum_i x_i \, G^o_i$$

$$+ \, RT \sum_i x_i \ln x_i$$

$$+ \sum_{i=1}^{m-1} \sum_{j=i+1}^{m} \sum_{h=0}^{r} x_i x_j (x_i - x_j)^h A_{ajh} \tag{7}$$

In this expression x_i and x_j are the mole fractions of components i and j, G^o_i is the Gibbs energy of pure component i, the term in $x_i \ln x_i$ represents the ideal contribution to the entropy of mixing, r is the order of the polynomial in compositions and the coefficients A_{ijh} form a polynomial in temperature that describes the Redlich-Kister/Muggianu {8} model of non-ideal mixing.

This is the most commonly used model but others {8} are frequently employed. The model is unable to describe without modification the behaviour of systems in which there is very strong interaction beween components; for example almost all metals have a strong affinity for sulphur. The effect is that the stable phases have crystal structures that tend to maximise the metal-sulphur interactions by putting the components on separate sublattices, with atoms having unlike nearest neighbours. Non-stoichiometry normally occurs as a result of the presence of vacancies on either sublattice. Again using sulphides as an example, solution can then be modelled by describing individual mixing of metals and vacancies on one sublattice and of sulphur and vacancies on the other.

Relatively little attention has been given to the thermodynamic modelling of the crystalline solution phases of importance in ceramic systems. Because the atoms on an individual sublattice are not usually nearest neighbours, interactions between them are usually relatively weak. However, the very large number of phases of importance in ceramics makes the building up of a body of critically assessed data using the consistent models and reference data a more complex task than is the case for alloy phases. Although relatively little work has been done on engineering ceramics {9,10}, literature relevant to conventional ceramics is being produced by geochemists. The review by Wood {11} gives a good indication of the areas covered.

Data for liquid oxides are also needed. The formation of liquid phases in ceramics is crucial to their production and ultimate performance as high temperature materials. Many of the important systems contain silica and it is encouraging that a number of groups have had success in the modelling of the behaviour of silicate melts. Some of these models may not adapt well to the description of multicomponent systems and no single model has yet become dominant {12-15}.

A term frequently used in the literature of alloy thermodynamics is lattice stability. For example, if the α-phase were chosen as the reference phase for pure A and the β-phase for pure B, the differences

$$G_A^{o\beta} - G_A^{o\alpha}$$

and
$$G_B^{o\alpha} - G_B^{o\beta}$$

are the lattice stabilities of the β phase for pure A and α phase for pure B respectively. A disadvantage of this method of expressing Gibbs energy is that any peculiarities in the data for the reference phase are incorporated in the data for all other phases. Moreover, the Gibbs energy of the reference phase becomes ill-defined outside its range of stability. The G-H$_{ser}$ formulation already described has much to recommend it because the data depend on the reference phase only at one temperature. In principle, the choice of reference for the pure components should have no effect on the data required to express the excess Gibbs energy of mixing. In practice small differences are difficult to avoid and hence data for mixing derived on the basis of one set of data should be checked by recalculation of the experimental data on which they were based before use in multicomponent calculations.

5. PARTIAL MOLAR PROPERTIES

Another important function is the chemical potential or partial molar Gibbs energy symbolised by G_i^α (the partial molar Gibbs energy or chemical potential of component i in phase α). The chemical potentials G_i^α, are related to the integral Gibbs energy, G^α, by the expressions

$$G^\alpha = \sum_i x_i \ G_i^\alpha \tag{8}$$

and
$$G_i^\alpha = \frac{\partial G^\alpha}{\partial n_i} \tag{9}$$

where
$$x_i + n_i/\sum_i n_i$$

From these relations it follows that

$$G_i^\alpha = (1 - x_i) \ \frac{\partial G^\alpha}{\partial x_i} + G^\alpha \tag{10}$$

6. THE CALCULATION OF EQUILIBRIUM

The emphasis on Gibbs energy in the previous sections arises because, for a system at constant pressure, the equilibrium distribution of phases and the distribution of elements between the phases is determined by the condition of minimum Gibbs energy (for a system at constant volume it is Helmholtz energy, A, that needs to be minimised where $A = G - Vdp$). An alternative way of expressing this condition is that the partial molar Gibbs energy (chemical potential) of each component is equal in all phases, $G_i^\alpha = G_i^\beta$. The equivalence of these two conditions is readily demonstrated graphically.

For the binary system A-B described in Figure 1, the Gibbs energies of the two solutions phases vary with composition as shown. It can be seen by comparing Equation 10 with Figure 1 that the chemical potentials of A and B at a particular composition in phase α are given by the intercepts corresponding to pure A and B of the common tangent to the integral Gibbs energy curve. For compositions with x_B less than x_1 the α phase is the more stable, whereas, with x_B greater than x_2 the β phase is more stable. For compositions with x_B lying between x_1 and x_2 a mixture of the two phases is stable, the α phase having the $x_B = x_1$ and the β phase $x_B = x_2$. Hence in this region the minimum Gibbs energy lies on the common tangent to the two curves and the chemical potentials of A and B are equal to G_A and G_B

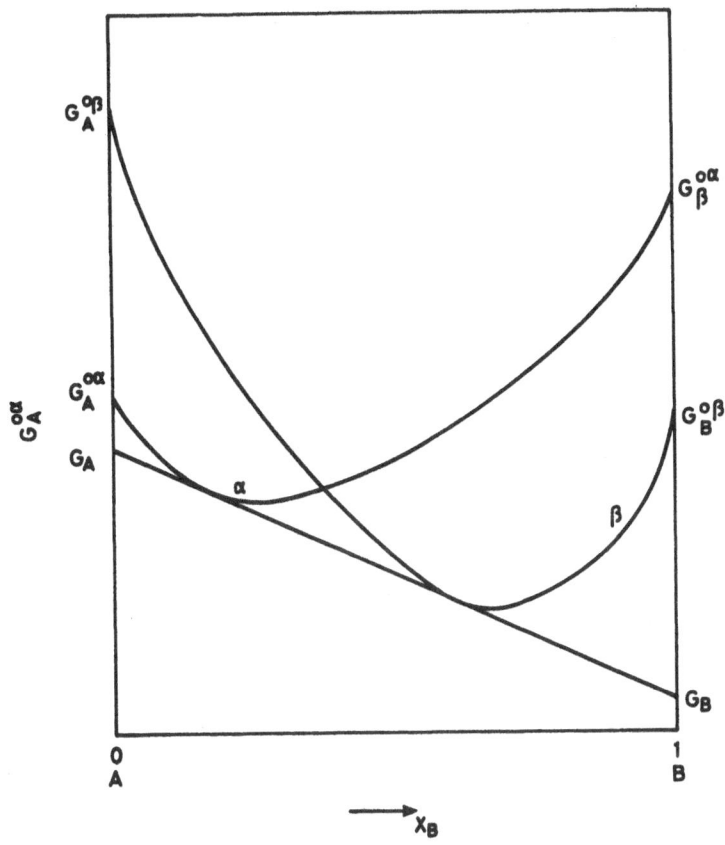

Fig. 1 Gibbs energy of the phases α and β at constant
        temperature and pressure. For the composition range
        $x_1 < x_B < x_2$ the α and β phases coexist. The integral
        Gibbs energy lies on the common tangent to the Gibbs
        energy curves at the composition of the mixture and the
        chemical potential of each of the components is equal in
        the two phases to the intercept of the common tangent on
        the corresponding axis.

respectively for both phases. It is important to note that the
Gibbs energy data for the α and β phases must relate to the same
reference states.

    As indicated by the above discussion, the principles of
equilibrium calculation are well established {16-20}. On the
other hand the reliable calculation of equilibrium in
multicomponent systems of disparate phase types places big
demands on the consistency of the data and the ability of the

software to cope with diverse problems, including:

- immiscibiity in phases,

- degenerate problems, where the effective number of components is less than the nominal number,

- cases where very small traces of components are important,

- the difficulty of making an initial selection when the number of phases is large,

- specification of the problem and its solution in terms that are meaningful for the application.

Unless otherwise stated, the diagrams presented here were prepared by means of MTDATA, the NPL Metallurgical and Thermochemical Databank, which incorporates the MULTIPHASE program for equilibrium calculations in high order systems {19,20}.

7. ALLOY DEVELOPMENT

Calculated phase equilibria have been used to assist in a wide range of problems arising from the development and use of high temperature alloys {21}. Examples include: solidification/melting behaviour relevant to segregation of impurities during casting, optimisation of compositions of prototype heat-resisting steels in which fcc → bcc and martensitic transformations are important, solid-state reactions involving the precipitation of embrittling phases (eg sigma), analyses of the possibilities of substituting for elements of strategic importance, (tramp elements) on recycling of scrap. Here three typical examples are given to illustrate the effectiveness of calculated phase equilibria in such applications.

8. HEAT RESISTING TITANIUM NITRIDING STEELS

Martensitic stainless steels are candidate materials for fast-breeder nuclear reactor fuel claddings because of their excellent resistance to void-swelling in comparison with austenitic steels. However, in their conventional hardened and tempered condition they have inadequate high temperature creep strength. One method of improving this situation is to produce a fine dispersion of titanium nitride in the matrix, the nitrogen being introduced via a gas-phase reaction.

The requirements are that the alloy contains sufficient chromium to provide oxidation resistance and sufficient titanium for effective nitride strengthening, is austenitic at the

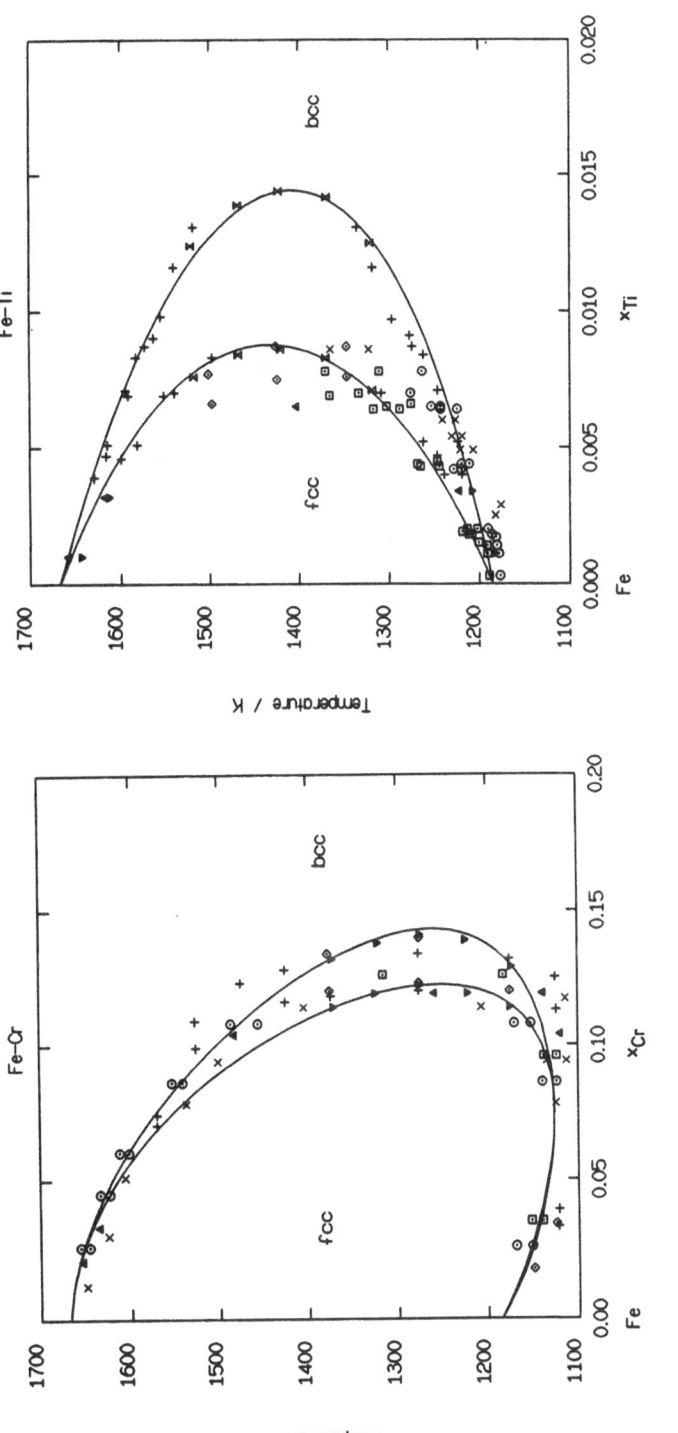

Fig. 2 Calculated partial phase diagrams (a) for the iron-chromium system and (b) for the iron-
titanium system showing details of the fcc/bcc equilibria (gamma-loops). Experimental data are
superimposed for comparison purposes.

nitriding temperature (1300–1500 K) and ferritic during service (650–1000 K). These constraints require an accurate knowledge of the phase equilibria since both chromium and titanium are bcc (ferrite) stabilisers, whereas high nickel contents lead to retention of the fcc (austenite) phase. Computer modelling is ideally suited to explore possible compositions.

Fig. 2(a) and (b) show calculated phase equilibria for the iron-chromium and iron-titanium systems relevant to this problem. The thermodynamic data employed {22,23}, which form the basis for multicomponent calculations, have been computer-optimised to take into account the experimental phase equilibria shown, and also the experimentally determined thermodynamic properties. Titanium markedly reduces the composition range over which the fcc phase is stable in comparison wih chromium. This is quantified in the calculated partial ternary isothermal section for 1300 K shown in Fig. 3, which is in close agreement with the limited available experimental data {24}. The amounts of the fcc and bcc phases present in a ternary alloy of composition 10 mol % Cr, 0.5 mol % Ti, balance Fe, are mapped as a function of temperature in Fig. 4. This type of diagram is

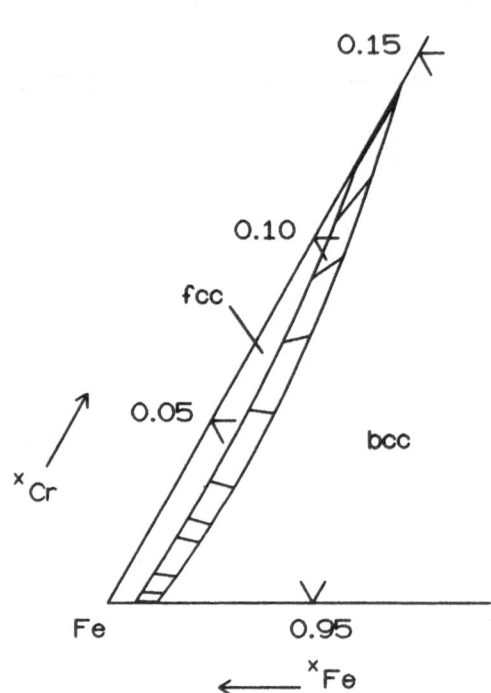

Fig. 3 Calculated partial isothermal section for the chromium-iron-titanium system for 1300 K showing the composition range over which the fcc phase is stable.

particularly useful to simulate the effect of moving through the gamma-loop in temperature with respect to a fixed composition. The composition of individual phases can be plotted if required.

A similar diagram for a six-component alloy additionally including carbon, silicon and nickel is shown in Fig. 5. This alloy, which contains 12 mol % Cr, is calculated to contain

90 mol % fcc at 1200 K but reverts to the bcc phase at temperatures below approximately 1000 K in accordance with requirement. A series of calculations in which the chromium, titanium and nickel contents were varied to produce optimum contents consistent with the required properties have led to an investigation of two prototype steels {25}. The calculated phase equilibria have been shown to be in almost exact agreement with the experimental results.

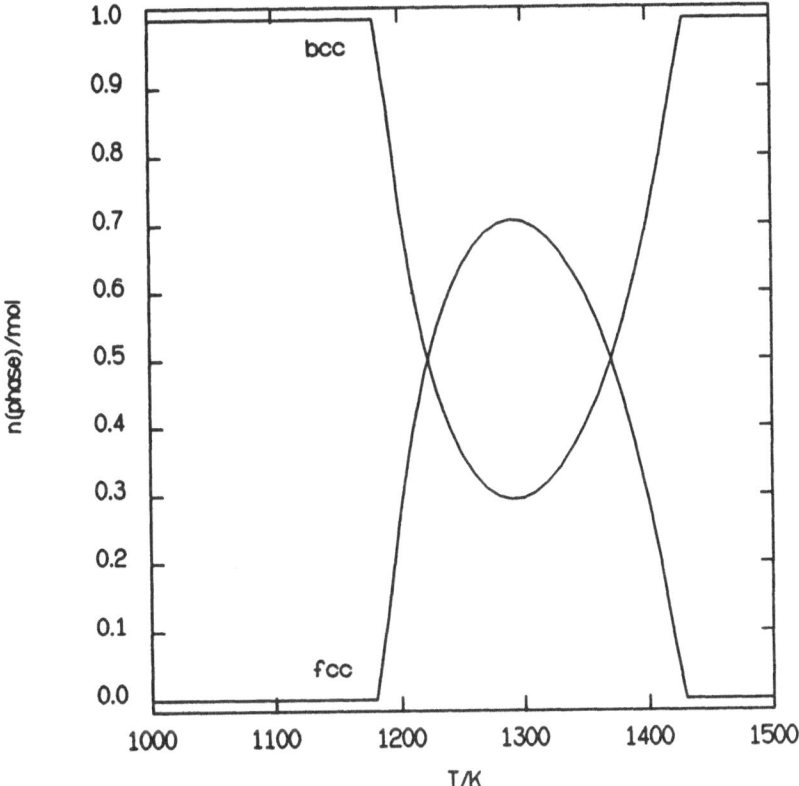

Fig. 4   Calculated amounts of the fcc and bcc phases present in a chromium-iron-titanium alloy of composition in mol % 10 Cr, 0.5 Ti, balance Fe as a function of temperature. This particular alloy is not completely austenitic at any temperature.

The equilibria involved in the above example are rather simple, the alloys concerned being essentially two-phase, fcc + bcc. The thermodynamic data employed, however, include all the possible competing phases including carbides, eg $M_{23}C_6$, Laves phases based on eg Cr_2Ti and Fe_2Ti, silicides and the embrittling sigma phase. The data allow the calculation of equilibria involving these phases and are just as applicable to nickel-based alloys as to plain-carbon and alloy steels.

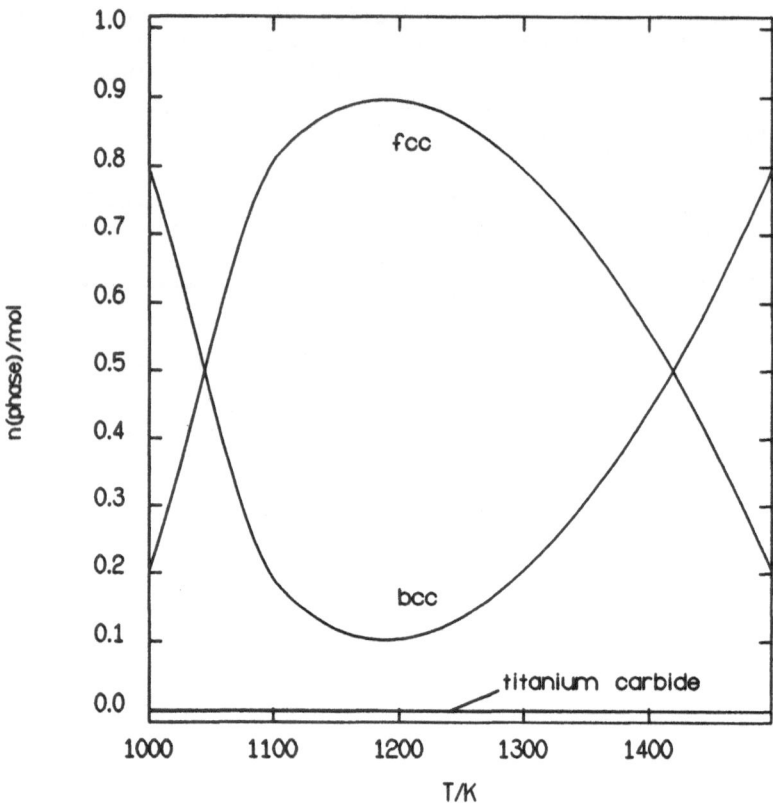

Fig. 5 Calculated amounts of the fcc, bcc and carbide phases present in a carbon-chromium-iron-nickel-silicon-titanium alloy of composition in mol % 0.05 C, 12 Cr, 2 Ni, 0.5 Si and 1 Ti, balance Fe, as a function of temperature. The carbon is present in the form of titanium carbide. Due to the small amount of carbon present and to the scale of the diagram this phase appears almost at the zero level.

9. PHASE STABILITY; SIGMA-PHASE EMBRITTLEMENT

The embrittlement of heat-resisting alloys due to precipitation of sigma-phase is well known and has been extensively investigated. A full understanding of this problem relies heavily on a knowledge of the relevant phase equilibria. However, published phase diagrams are limited primarily to ternary and some quaternary alloys, and accurate data are sparse, particularly for temperatures below about 900 K. It is now

possible to reliably calculate equilibria for quite complex
alloys, containing as many as eight to ten elements, providing
data for the relevant binary sub-systems are available.

A key sub-system is chromium-iron-nickel. Fig. 6 shows a
calculated isothermal section for this system at 1273 K,
validated by the superimposed experimental tie-line data of
Hasebe and Nishizawa {26}. The sigma-phase appears in this

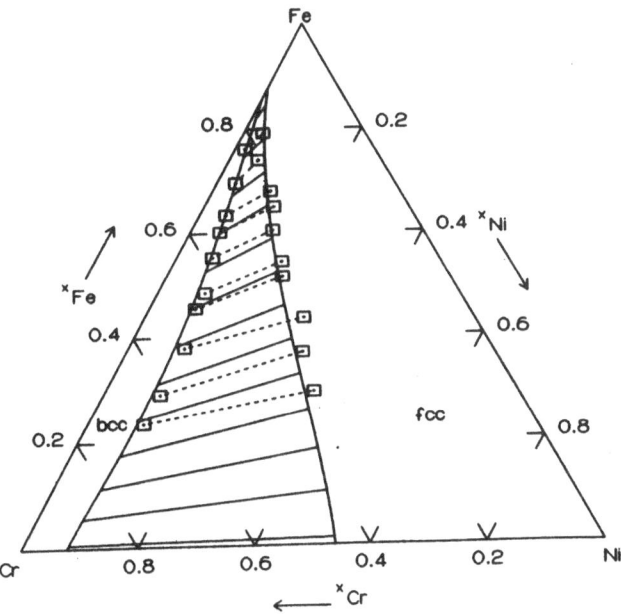

Fig. 6 Calculated and experimental isothermal section for the
chromium-iron-nickel system at 1273 K.

ternary system below about 1210 K, although it is not stable in
the binary chromium-iron system at temperatures higher than
1100 K {22}. Calculated equilibria for 1000 K are shown in
Fig. 7. These equilibria have been calculated using data from
the SGTE solution database {1,27} using a sublattice model for
the sigma-phase, which is more physically realistic in this case
than the Redlich Kister/Muggianu model, and is essential for
systems in which the intrusion of the sigma-phase is extensive eg
chromium-iron-molybdenum {28}. It should be noted, however, that
for relatively simple ternary and quaternary systems, including
chromium-iron-silicon and chromium-iron-nickel-silicon, the
simpler Redlich-Kister/Muggianu model has been applied very
successfully, {29,30} a particular aim of these calculations
being the development of fuel cladding-alloys for nuclear
reactors, and an understanding of void-swelling {30,31}. Phase

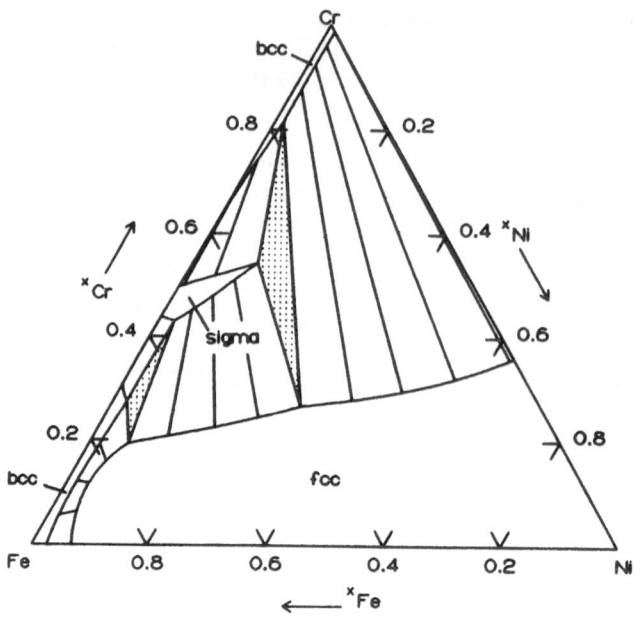

Figure 7   Calculated isothermal section for the chromium-iron-
            nickel system at 1000 K.

equilibria for a six-component system are shown in Fig. 8.   The
composition corresponds to a type 321 titanium-stabilised
stainless steel.   This diagram quantifies in a convenient manner
the proportions of the various phases present and indicates the
temperatures at which this particular alloy is sigma-prone.

10.   SOLIDIFICATION - MELTING BEHAVIOUR

The above types of phase diagram, when extended to
temperatures at which the liquid phase becomes stable, provide
solidification/melting temperatures, indicate compositions which
for a given temperature are completely liquid or solid, and
provide the proportions of the coexisting solid and liquid
phases.   This type of information has been applied to the
development of in-situ composites, {32,33} the modelling of
segregation of impurities during continuous casting {34,35} and
to liquid-phase sintering.

Directionally solidified gas turbine blades have been
considered to offer the prospect of achieving higher operating
temperatures when compared to conventionally produced materials.
However the search to find suitable candidate alloys requires

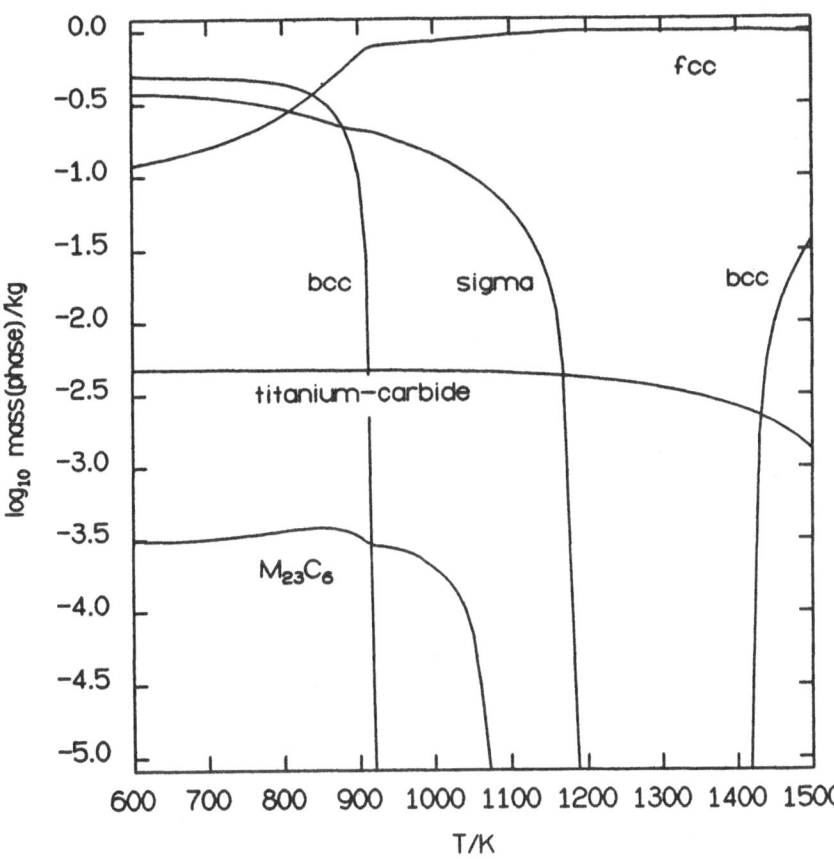

Fig. 8 Calculated equilibrium amounts of the phases present in a
type 321 titanium stabilised stainless steel of
composition in wt % 0.08 C, 18 Cr, 10 Ni, 1 Si and
0.4 Ti, balance Fe, as a function of temperature. Due to
the small amounts of carbides present, particularly
the $M_{23}C_6$ phase, a logarithmic ordinate has been used.

that phase diagrams for a large number of potential multicomponent systems be scrutinised, and generally these data are
unavailable. thermodynamic calculations provide a rapid and
cost-effective route to the required data. Chart et al. {32,33}
have calculated phase equilibria for a series of ternary and
quaternary cobalt-chromium and nickel-chromium bases systems
containing niobium, tantalum and zirconium, and predicted long
eutectic troughs between fcc-based solid solutions and solutions
based on Laves phases. Subsequent metallographic investigations
showed the calculated equilibria to be surprisingly accurate
{33}, especially when bearing in mind the paucity of available
thermodynamic data for some of the binary sub-systems.

Calculated multicomponent phase equilibria are being applied
successfully to the modelling of microsegregation of impurities
during continuous casting. This subject is of major concern
within the steel industry, affecting processing performance and
product properties {34}. A thermodynamic database and software
to allow the calculation of phase equilibria for alloy and
stainless steels containing up to twelve components is being
developed by Dinsdale et al. This forms part of a joint NPL/
British Steel Corporation project funded by ECSC. The partition
of the various elements between the various coexisting phases as
a function of both temperature and composition is required as
input to a finite difference computer model being developed by
Howe and Kirkwood {35}. The provision of these data by
thermodynamic calculation is the only practical method.

11. CORROSION AND COATINGS

In studies of corrosion it is well known that some oxides,
notably Cr_2O_3 and Al_2O_3, confer protection against further
attack, whereas others are less effective and the formation of
sulphides can be positively harmful because of the generally high
mobility of metal ions in sulphides. Because of these
possibilities, thermodynamic methods and especially predominance
area (or phase stability) diagrams are much used by corrosion
scientists to determine the effect of the environment on phase
stability {36,37}.

12. SULPHIDATION OF NICKEL

Fig. 9 shows a predominance area diagram for the Ni-O-S
system at 1000 K. Nickel is a major component of alloys used in
gas turbines. The conditions are normally oxidising, with
partial pressures of oxygen and sulphur dioxide of about 1.0 and
0.001 respectively. Superficially Fig. 9 appears to indicate
that nickel oxide will always form rather than a sulphide. In
practice, however, sulphides can form because, at the root of a
crack in the oxide scale, the oxygen potential will correspond
with that required for the coexistence of the nickel and nickel
oxide. Under these conditions any sulphur dioxide reaching the
nickel-nickel oxide interface is broken down by nickel to give a
mixture of nickel oxide and Ni_3S_2. Seen in this light the
independent variables of the diagram are the phases present an
the dependent variables are the potentials of $O_2(g)$ and $SO_2(g)$.

Several methods of drawing predominance area diagrams for
the simpler case of stoichiometric phases have been devised.
That developed by Barry {38} includes the possibility of
describing the simultaneous behaviour of more than one component
when they can form compounds with one another.

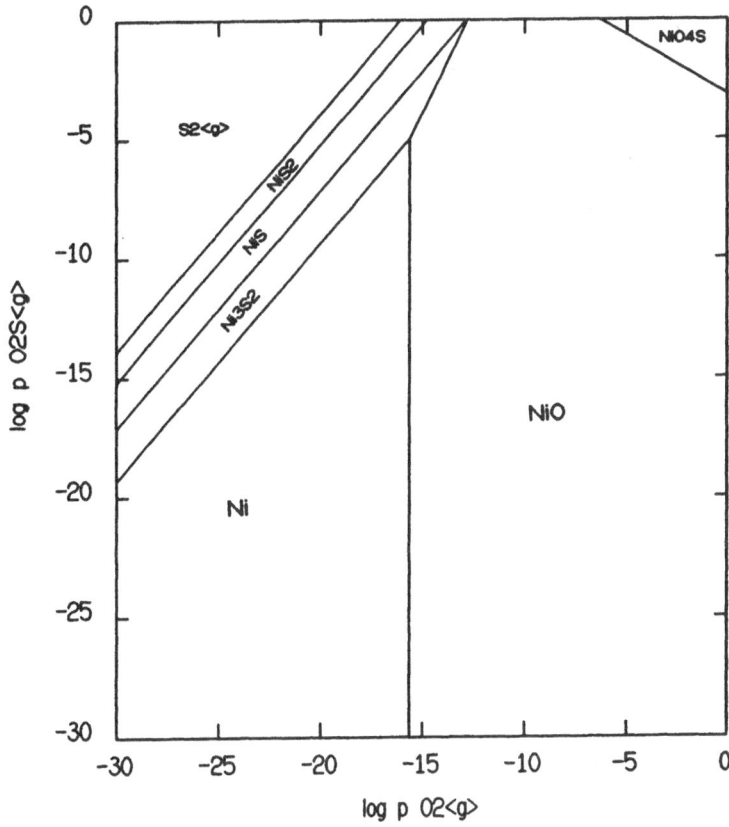

Fig. 9 Predominance area diagram for the Ni-O-S system at 100 K.
The region at the top left of the diagram is over-
constrained (because it corresponds to very high
pressures of S_2) and is automatically excluded from
further calculation.

13. MOLYBDENUM SILICIDE

Molybdenum silicide is used as a high temperature furnace
element material and has also been considered as a constituent of
corrosion resistant coatings. It withstands high temperatures
only because of the formation of a protective layer of glassy
silica on the surface. The predominance area diagram of Fig. 10
reveals a rather complex high temperature chemistry. For the
conditions specified, increasing the oxygen potential causes
$MoSi_2$ to be successively replaced by Mo_5Si_3, Mo_3Si, Mo, MoO_2 and
finally by gaseous molybdenum trioxides. More detailed
calculations are presented in Fig. 11, which shows the

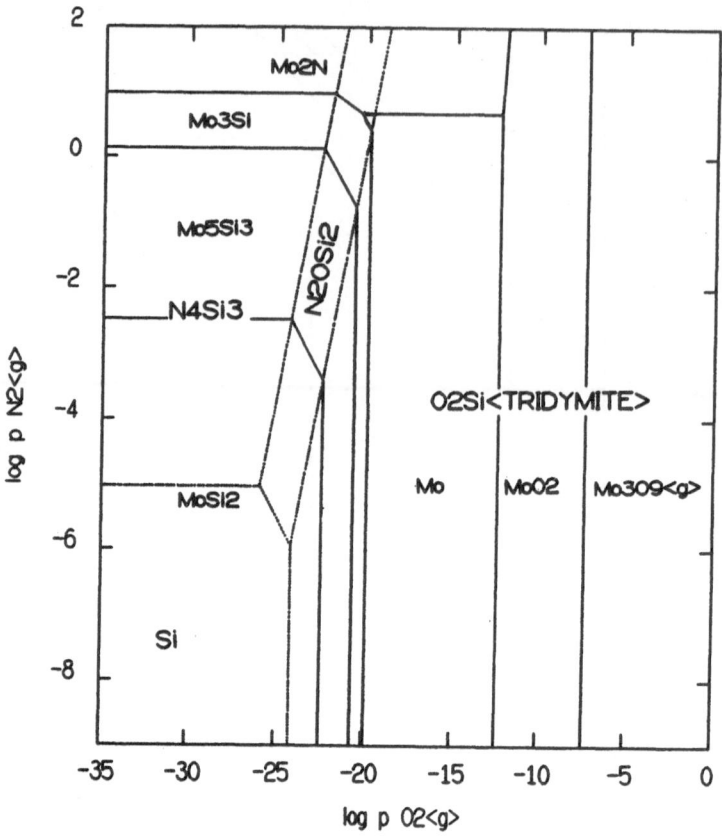

Fig. 10 Predominance area diagram for silicon and molybdenum
as a function of oxygen and nitrogen potentials for
1423 K. The gas volume is the ideal gas volume,
0.1168 m³, and the amounts of Si and Mo are 1 and
0.001 mol respectively. Because silicon is present in
the larger amount, its diagram (dotted lines, large
labels) is generated first. A molybdenum diagram (solid
lines, smaller labels) is then calculated for each area
of the silicon diagram, the component for silicon in each
calculation being the compound of silicon stable in that
particular area. The individual diagrams are then
automatically combined.

equilibrium amounts of all the significant products formed as a
result of progressive addition of oxygen to silicon and
molybdenum in the presence of nitrogen. Particularly noteworthy
is the formation of SiO(gas) at intermediate oxygen potentials.
This not only causes some loss of the protective silica but also

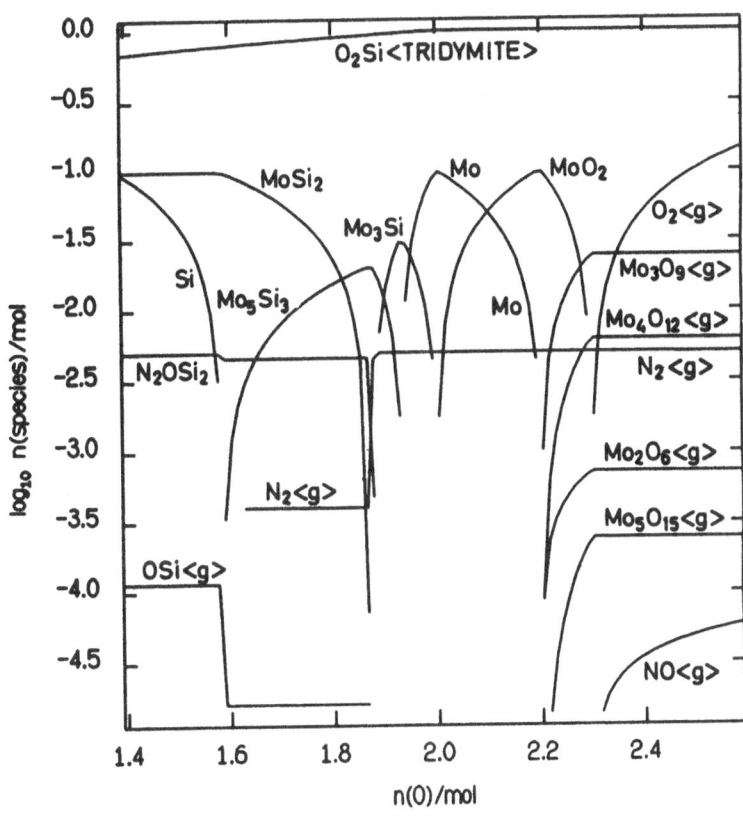

Fig. 11 Equilibrium calculation for the reaction of silicon,
molybdenum and nitrogen as a function of added oxygen. A
notable feature of the diagram is the formation of the
series of molybdenum compounds, $MoSi_2$, Mo_5Si_3, Mo_3Si, Mo,
MoO_2. The oxygen partial pressures determined for the
coexistence of pairs of these phases are identical with
those determined from the predominance area diagram of
Fig. 10.

permits recrystallisation of the glassy silica to tridymite or
cristobalite. These phases do not form a continuous oxide layer
and hence do not confer protection against further oxidation.
Moreover, they undergo very large changes in volume during rapid
changes of phase a low temperatures.

Thus the diagram quantifies the observation, well known to
users of silicon carbide and molybdenum disilicide furnace
elements, which rely for their oxidation resistance on the glassy

SiO_2 layer, that, paradoxically, moderately reducing conditions can lead to more rapid oxidation than oxidising conditions and eventually to catastrophic failure on thermal cycling. The same considerations can of course apply to coatings of other non-oxide engineering ceramics based on silicon, for example silicon nitride.

It should be mentioned that additional diagrams would be necessary to present the whole picture, since the chemistry is dependent on the proportions of molybdenum and silicon (which would vary through the layers between the $MoSi_2$ and the atmosphere) and the presence of additional components.

14. CORROSION OF IRON – CHROMIUM AND IRON – CHROMIUM – NICKEL ALLOYS

Depending on their composition or history the corrosion of iron-chromium alloys can cause the formation of two types of oxide scale: either a thin single phase layer with the corundum structure and a composition very close to Cr_2O_3 throughout its depth or a single phase spinel layer with two generally distinct composition ranges, that nearer the metal being close in composition to $FeCr_2O_4$ and the outer layer nearly pure magnetite, Fe_3O_4, which may be overlain with Fe_2O_3.

Fig. 12 shows an oxygen potential-composition diagram for the Fe-Cr-O system form the work of Rahmel, Schoor, Velasco-Tellew and Pelton {39}. How well does Fig. 12 explain observed phenomena? In order to describe the corrosion behaviour it is necessary to make the usual and reasonable assumption of local thermodynamic equilibrium, a condition of which is that the individual potentials of each of the components are locally equal and, in particular, equal at points where phases are in contact.

An alloy containing more than 15 mol% Cr will form a single ferritic phase at 1273 K. When the metal comes into contact with oxygen an oxide phase with the corundum structure will form with a composition very close to pure Cr_2O_3, the solubility of Fe_2O_3 at an oxygen potential of 10^{-22} atm being only about 10^{-12}. The diffusion rates of oxygen, oxide ions and chromium ions through this structure are very low and, although the diffusion coefficient for ferric ions is higher than that of chromium ions, the actual rate of diffusion of iron through the corundum structure will be very low indeed because of the low solubility of ferric ions at low oxygen potentials. Hence the corundum layer tends to be very thin and close to Cr_2O_3 in composition throughout its thickness.

If the chromium content of the alloy is originally less than about 12 mol% Cr or if the depletion of the chromium in the alloy

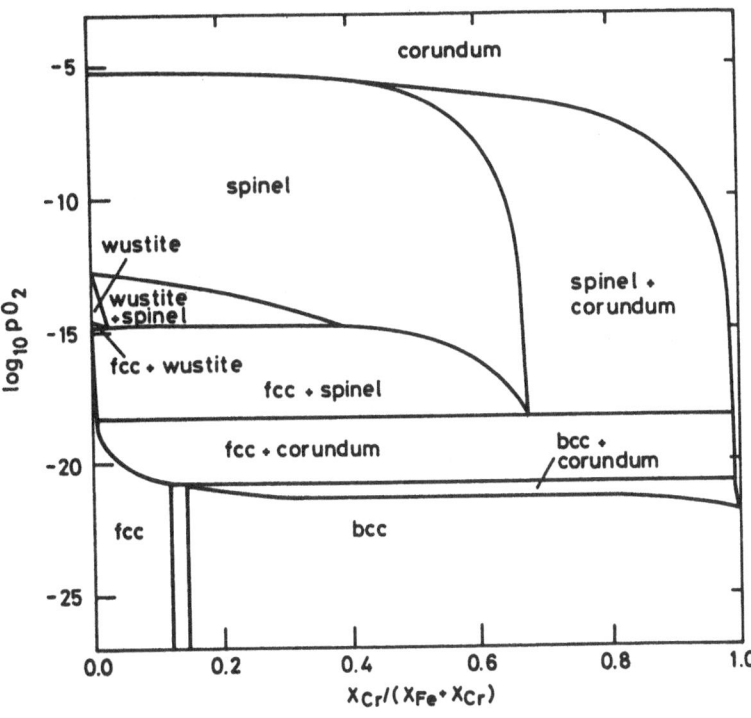

Fig. 12 Diagram showing the phases formed in the Fe-Cr-O
system as a function of oxygen potential and the chromium
content, expressed as a fraction of the total iron +
chromium. {39}

causes the concentration near the interface to fall below
12 mol%, the alloy phase stable near the surface will be
austenite rather than ferrite. If initial chromium depletion at
the interface is sufficient to cause the concentration of
chromium to fall less than about 0.2%, calculations show that the
stable oxide phase in contact with the metal will be spinel with
a composition close to $FeCr_2O_4$ rather than Cr_2O_3 with the
corundum structure and the oxygen potential at the interface will
rise above 10^{-18}. The spinel phase tolerates a wide range of
stoichiometry and hence its formation may be thermodynamically
and kinetically favoured in many circumstances. Thus, in alloys
containing even small amounts of additional components spinel may
form when the chromium concentration at the interface is greater
than 0.2% and the oxygen potential is less than 10^{-18}.

It cannot be predicted from the thermodynamic data alone
whether this chromium rich spinel grows by outward diffusion of

metal or inward diffusion of oxygen, though it is known that in practice inward diffusion of oxygen is responsible. The outer layers, however, must grow by metal ion diffusion, and, because of the slow rate of diffusion of Cr^{3+} relative to Fe^{3+} and Fe^{2+} ions, these layers are likely to approximate closely to magnetite in composition. The same considerations apply if grain boundary rather than lattice diffusion is responsible for ion transport. If the oxygen partial pressure of the atmosphere exceeds 10^{-5}, the diagram predicts that a layer with a composition close to Fe_2O_3 and the corundum (haematite) structure should overlay the spinel phase.

Thus the thermodynamic data are able to explain in a qualitative way two quite disparate types of behaviour and hence to enable the pattern of behaviour to be predicted as a function of thermal history and alloy composition.

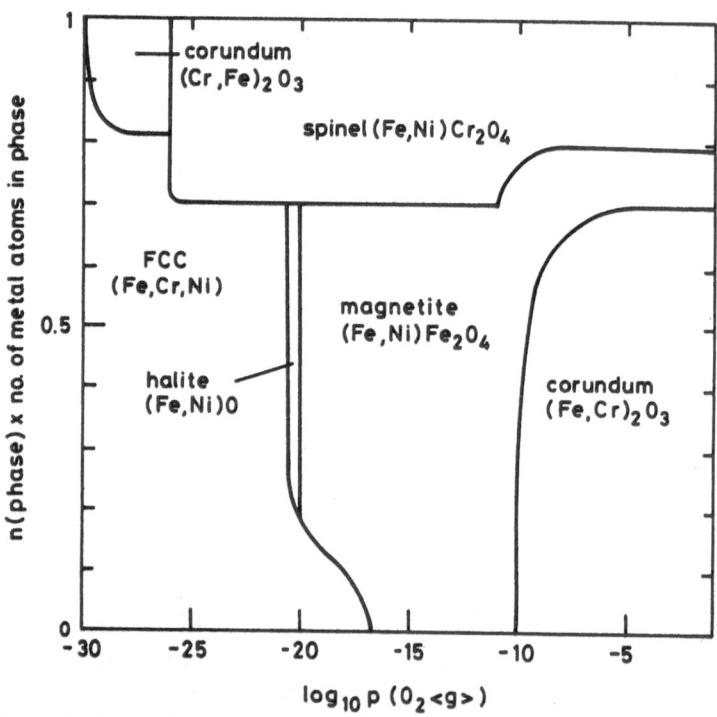

Fig. 13 The formation of phases at 1000 K as a function of oxygen partial pressure for the molar composition Fe 0.7, Cr 0.2, Ni 0.1. All the phases in a vertical section are in equilibrium. The calculation method also allows the compositions of the individual phases and the chemical potentials of the components to be determined.

The same types of calculations can be made for more complex alloys. Fig. 13 shows some of the results of a sample calculation using MULTIPHASE {19} for the oxidation of a iron-chromium-nickel alloy at 1000 K. In Fig. 13 the ordinate shows the amounts of the various phases in terms of the product of the molar amount of the phase times the number of metal atoms per formula unit, the total adding to one. Magnetite has been treated as a seperate phase from chrome spinel, because in practice there is usually a distinct separation between the two composition ranges. All the phases present in any vertical section are in equilibrium. The results can be plotted in many ways. In particular plots of the compositions of the individual phases show a characteristic pattern. As the oxygen potential increases, the chromium forms new phases and is depleted in the existing phases, whereas nickel becomes concentrated in the remnants of phases stable at lower oxygen partial pressures. This agrees with real behaviour: for example, islands of nearly pure nickel are often found embedded in iron chrome spinel. Quantitative prediction, for example of the overall corrosion process is possible in principle but requires models and diffusion data for the many kinetic processes involved. Local stresses may also play a role. Predictions based on a thermodynamic approach could prove very valuable in the understanding of corrosion in multicomponent alloys, the design of coatings and the formation of sulphides, chlorides and carbides during continuous or cyclic operation.

15. Hot salt corrosion

When gas turbines are operated in marine or other conditions involving exposure to sodium chloride and sulphur simultaneously, molten sulphates are likely to be deposited. In particular, they are likely to concentrate in crevices in the oxide scale, causing dissolution of the oxides, more rapid transport of the alloy components and hence accelerated corrosion {36}.

The very large number of components involved, including Ni, Cr, Al, Co, Na, K, Cl, S, O, N and H, and the lack of critically assessed data for the oxide and molten salt solution phases, makes this a difficult problem to tackle. Data and calculations relevant to molten salt attack on chromium oxides have been presented by Dinsdale and Barry {40}.

16. MOLTEN SALT SYSTEMS

In addition to their role in hot salt corrosion, molten salt systems are employed as electrolytes and fluxes important to a wide range of technologies including the refinement of alloys, brazing and the development of batteries. A knowledge of the phase equilibria for such systems is essential for a full

588

understanding of the reactions involved. It has been demonstrated over recent years that phase diagrams for multicomponent molten salt systems can be calculated from thermodynamic data for the binary sub-systems with particularly good reliability {41,42}.

Perry et al. {43} have studied the ternary system KCl-CaCl$_2$-ZnCl$_2$ in order more fully to understand reactions involved in the recovery and separation of actinide metals. Fig. 14 shows a calculated liquidus projection for this ternary system together with liquidus isotherms for selected temperatures. This ternary system is characterised by two extensive eutectic troughs emanating from the KCl-CaCl$_2$ binary system. Superimposed on the diagram are experimental data of Perry et al. {43}. These were made available subsequent to the calculation being performed.

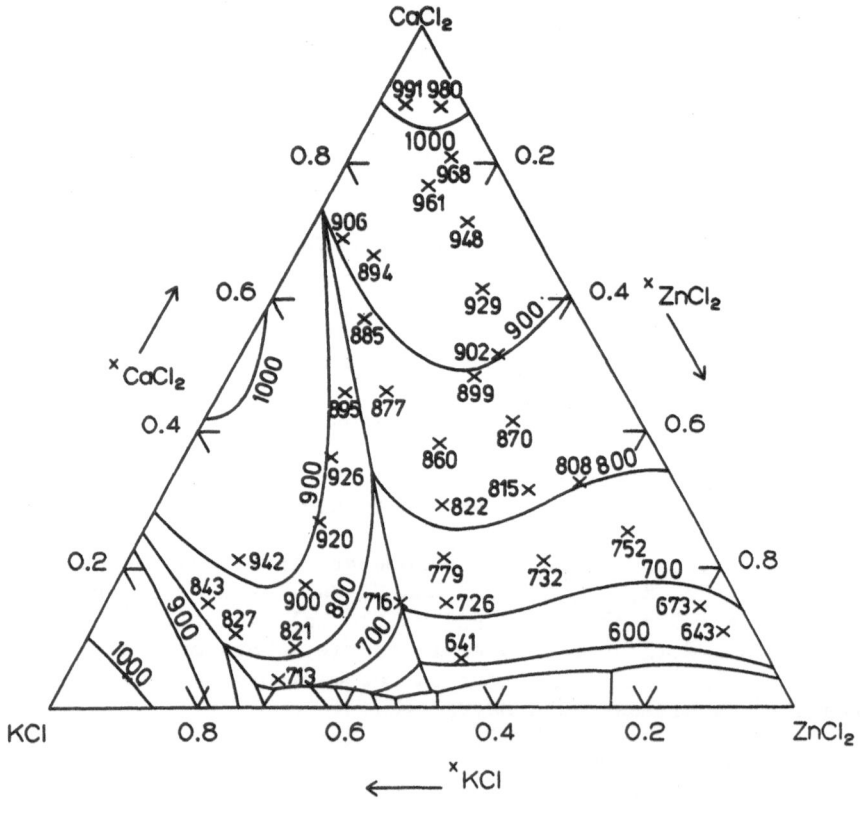

Fig. 14 Calculated liquidus projection and liquidus isotherms
for the KCl-CaCl$_2$ZnCl$_2$ system compared with experimental
melting temperatures. {43}

The agreement between calculation and experiment is remarkably good and it would not be possible to construct this diagram on the basis of the experimental data alone.

The above calculation was made using the relatively simple Redlich-Kister-Muggianu model of proven applicability to many alloy and molten salt systems. Calculations using the similar Kohler model {44} have been made by Skeaff et al. {42} for ternary systems involving chlorides of lithium, sodium, potassium, lead and zinc, for application to the electrowinning of lead and zinc. Again good agreement between calculation and experiment was observed. More sophisticated approaches are an advantage when cations and anions are dissimilar or of different valency. The Conformal Ionic Theory developed by Saboungi and Blander {45} has been applied with considerable success, eg to cryolite based systems by Lin et al. {46}.

17. CONCLUDING REMARKS

From even the few examples given it is evident that thermodynamic methods have much to contribute to the improved understanding of the development and use of high temperature materials. The current improvements in the mathematical modelling of the thermodynamic properties of stable and unstable phases are already encouraging developments in models for kinetic processes, including crystal nucleation, growth and transformation, and corrosion. Increased attention to ceramic properties can be expected, with applications in coatings and corrosion as well as ceramics in their own right.

REFERENCES

1. "Chemical Thermodynamics in Industry: Models and Computation", Barry, T.I., High temperature inorganic chemistry and metallurgy, ed. Barry, T.I., Blackwells Scientific Publications, Oxford (1985)

2. JANAF Thermochemical Tables, Third Edition: part 1, Al-Co, Part 2, Cr-Zr J. Phys. Chem. Ref. Data, 14, Chase Jr., M.W., Davies, C.A., Downey Jr., J.R., Frurip, D.J., McDonald, R.A. and Syverud, A.N., Suppl. 1 (1985)

3. "Thermodynamic Properties of Individual Substances", Gurvich, L.V., Veits, I.V., et al., Glushko, V.P., gen. ed., Nauka, Moscow, 4 vols. (1978-1982)

4. The NBS tables of chemical thermodynamic properties: Selected values for inorganic and C_1 and C_2 substances in SI units, Wagman, D.D., Evans, W.H., Parker, V.B., Schumm, R.H. Bailey, S.M., Churney, K.L. and Nutall, R.L., J. Phys. Chem. Ref. Data, 11, Suppl. 2 (1982)

5. Thermochemical Properties for Pure Substances, Barin, I. and Knacke, O., Springer Verlag, Berlin, 1973, Supplement, Barin, I., Knacke, O. and Kubashewski, O., (1977)

6. Thermodynamic Properties of the Elements and Oxides, Pankratz, L.B., US Dept. of the Interior, Bureau of Mines, Bulletin 672 (1982)

7. CODATA Thermodynamic Tables, Selections for some compounds of calcium and related mixtures: A prototype set of tables, Garvin, D., Parker B.V. and White Jr., H.J., Hemisphere Publishing Corporation, Washington (1987)

8. Ansara, I., Internat. Metallurgical Ref., $\underline{24}$, 20 pp. (1979)

9. Dörner, P., Gaukler, L.J., Krieg, H., Lukas, H-L., Petsow, G. and Weiss, J., CALPHAD, $\underline{3}$, 241-247 pp. (1979)

10. Sundman, B. and Agren, J., J. Phys. Chem. Solids, $\underline{42}$, 297 pp. (1981)

11. Wood, B.J., Reviews in Mineralogy Vol. 17, Thermodynamics of multicomponent systems containing several solid solutions, Eds. Charmichael, I.S.E., Eugster, H.P., Mineral. Soc. Amer. (1987)

12. Proc. Aust. Inst. Mining Metall., $\underline{254}$, Kapoor, M.L., Mehrotra, G.M. and Frohberg, G.M., 11 pp. (1975)

13. Second Internat. Symp. Metallurgical Slags and Fluxes, Gaye, H. and Wellfringer, J., Ed. Fine, H.A, and Gaskell, D.R., 1984, Metall. Soc., AIME, New York, 357-375 pp. (1984)

14. Hillert, M., Jansson, B., Sundman, B. and Agren, J., Metall. Trans., $\underline{16A}$, 261-6 pp. (1985)

15. Pelton, A.D. and Blander, M., Metall. Trans., $\underline{17B}$, 805 pp. (1986)

16. Thompson, W.T., Pelton, A.D. and Bale, C.W., CALPHAD, $\underline{7}$, 113 pp. (1983)

17. "Solgasmix", Eriksson, G. and Hack, K., CALPHAD, $\underline{8}$, 15 pp. (1984)

18. Sundman, B., Jansson, B. and Anderson, J.O., CALPHAD, $\underline{9}$, 153 pp. (1985)

19. MTDATA handbook: MULTIPHASE module, Dinsdale, A.T., Davies, R.H., Gisby, J.A. and Hodson, S.M., NPL (1987)

20. Hondson, S.M., Proc. Conf. 30th SIAM Anniversary Meeting, July (1982)

21. Barry, T.I. and Chart, T.G., EUR 9564 (1985)

22. Anderson, J.O. and Sundman B., CALPHAD, 11, 83-92 pp. (1987)

23. A critical assessment of thermodynamic data for the Fe-Ti system, Dinsdale, A.T., To be published.

24. Roe, W.P. and Fishel, W.P., Trans. ASM, 44, 1030-1046 pp. (1952)

25. Wilson, A.M., Gohil, D.D. and Laing, K., CALPHAD XV Conference, London, UK, July (1986)

26. Hasebe, M. and Nishizawa, T., Analysis and synthesis of phase diagrams of the Fe-Cr-Ni, Fe-Cu-Mn and Fe-Cu-Ni systems, ed. Carter, G.C., NBS Special Publication 496, U S Department of Commerce / National Bureau of Standards, 911-954 pp. (1978)

27. Ansara, I. and Sundman, B., The Scientific Group Thermodata Europe, ed. Glaeser, P.S., Elsevier Science Publishers, 154-158 pp. (1987)

28. Andersson, J.O. and Lange, N., TRITA-MC-0322, Royal Inst. Tech. Stockholm (1986)

29. Ansara, I., Chart, T.G., Chevalier, P.Y., Hack, K., McHugh, G., Rand, M.H. and Spencer, P.J., EUR 9657 (1985)

30. Chart, T.G., Putland, F.H. and Dinsdale, A.T., CALPHAD, 4, 27-46 pp. (1980)

31. Gittus, J. and Watkin, J.S., J. Nucl. Mats., 64, 300-302 pp. (1977)

32. Chart, T.G. and Putland, F.H., CALPHAD, 3, 9-18 pp (1979)

33. "Conference on In Situ Composites-III", Chart, T.G., Putland, F.H. and McLean, M., Prediction and assessment of new eutectic alloys in the Co-Cr-(Wr, Ta, or Nb) systems, ed. Walter, J.L., Gifliotti, M.F., Oliver, B.F. and Hibring, H., Ginn Custom Publishing, Lexington, Massachusetts, 441-450 pp. (1979)

34. Howe, A.A., CALPHAD XV Conference, London, UK, July (1986)

35. "Solidification Processing 1987", Howe, A.A. and Kirkwood, D.H., The numerical computation of microsegregation in alloys involving a peritectic reaction, The Institute of Metals, London, To be published.

36. Rapp, R.A., Materials Sci. and Eng. 87, 319 pp. (1987)

37. "High Temperature Corrosion", Shores, D.A., New perspectives on hot corrosion mechanisms, NACE-6, Ed. Rapp, R.A., Nat. Assoc. Corrosion Engineers, Houston, USA, 493 pp. (1983)

38. MTDATA handbook; COPLOT module, Barry, T.I., Davies, R.H., NPL (1987)

39. Rahmel, A., Schoor, M. Velasco-Tellez, A. and Pelton, A., Oxidation of Metals, 27, 199 pp. (1987)

40. Barry, T.I. and Dinsdale, A.T., Materials Science and Technology, 3, 501 (1987)

41. Lin, P.L., Pelton, A.D. and Bale, C.W., J. Amer. Ceram. Soc., 62, 414-422 pp. (1979)

42. Skeaff, J.M., Bale, C.W., Pelton, A.D. and Thompson, W.T., CANMET Report 79-23, Energy, Mines and Resources Canada (1979)

43. Perry, G.C., MacDonald, L.G. and Newstead, S., Thermochim. Acta, 68, 341-348 pp. (1983)

44. Kohler, F., Monatsh. Chem. 91, 738-740 pp. (1960)

45. Saboungi, M.L. and Blander, M., J. Chem. Phys., 63, 212-220 pp. (1975)

46. Lin, P.L., Pelton, A.D. and Saboungi, M.L., Metall. Trans. B, 13B, 61-69 pp. (1982)

5.3.4. Design Concepts for Ceramic Materials

J. Lamon

Battelle - Europe, Geneva Research Centre

7, route de Drize, CH-1227 Carouge/Geneva

1. INTRODUCTION

Design is a procedure, aimed first at defining a product, or a system for a given function, and then, ensuring that the product fits in with the function. The function induces various constraints (aesthetics, mechanics, economics, etc.). In engineering they can be quantitatively described. The product or the system must withstand these constraints to be able to properly assure its function. The response of material to service conditions is thus a key factor. Material selection is an integral part of design.

The principles of mechanics, dynamics, heat transfer, and so forth are well established for evaluating the service conditions. In contrast, new materials are appearing all the time. Therefore the designer has at present at his disposal a large range of materials : metals, ceramics, polymers, composites. These materials exhibit specific mechanical properties.

Metals are ductile, ceramics essentially linear elastic and brittle, composites have non-linear behaviour with brittle phases, etc. Until now the general trend consisted in applying the methodology formulated for metals to the newly emerging materials, since this methodology was available, and designers generally ignored the material.

The properties of metals are quite well-known. A design methodology based upon ductility has been used for a long time and is well established. All the metals and alloys have a certain amount of ductility that ensure that unnotched components yield before fracture, and that fracture when it occurs is of a tough ductile type. In designing with ductile materials a deterministic method is used.

A different situation prevails with ceramics. Engineering ceramics have characteristics that preclude the use of conventional design methodology employed for metals. Due to lack of ductility ceramics are highly sensitive to stress concentrations from microstructural flaws. They contain a lot of inherent flaws having random-like distribution. As a consequence failure is a probabilistic event. A simple deterministic approach no longer suffices for design purposes. A probabilistic approach is required.

Brittleness and flaw sensitivity are not eliminated in ceramics reinforced by whiskers or particles. A probabilistic approach is also required.

Ceramics reinforced with continuous fibers exhibit a decreased flaw sensitivity. However, strengths vary to such a degree that deterministic methods are not appropriate. Continuous fiber reinforced ceramics are yet unproven at high temperature, and design methodology has not been developed. Therefore, continuous fiber reinforced ceramics will not be addressed here.

Design with ceramics has mainly been empirical, relying upon the conventional methodology used for metals. There are many examples of ceramic component failures due to the fact that the conventional methodology which works for metals has been applied as such to ceramics.

From this point of view ceramics may appear as undesirable for a designer who is accustomed to working with metals. However they display advantageous properties (in particular for high temperature applications) that can be fully capitalized provided a sound methodology is employed. Furthermore, one may consider that the actual utilization of ceramics is held up due to the use of an inappropriate design methodology.

2. DESIGNING WITH METALLIC MATERIALS : CONVENTIONAL METHODOLOGY

Designing with metallic materials is not currently a major problem. The properties of metals are generally well-known. Designers are accustomed to working with them.

The diagram of Fig. 1, proposed by Ashby and Jones (1) is a good description of design methodology. It shows the main steps of a design procedure, starting with the definition of service conditions and ending with the production of components.

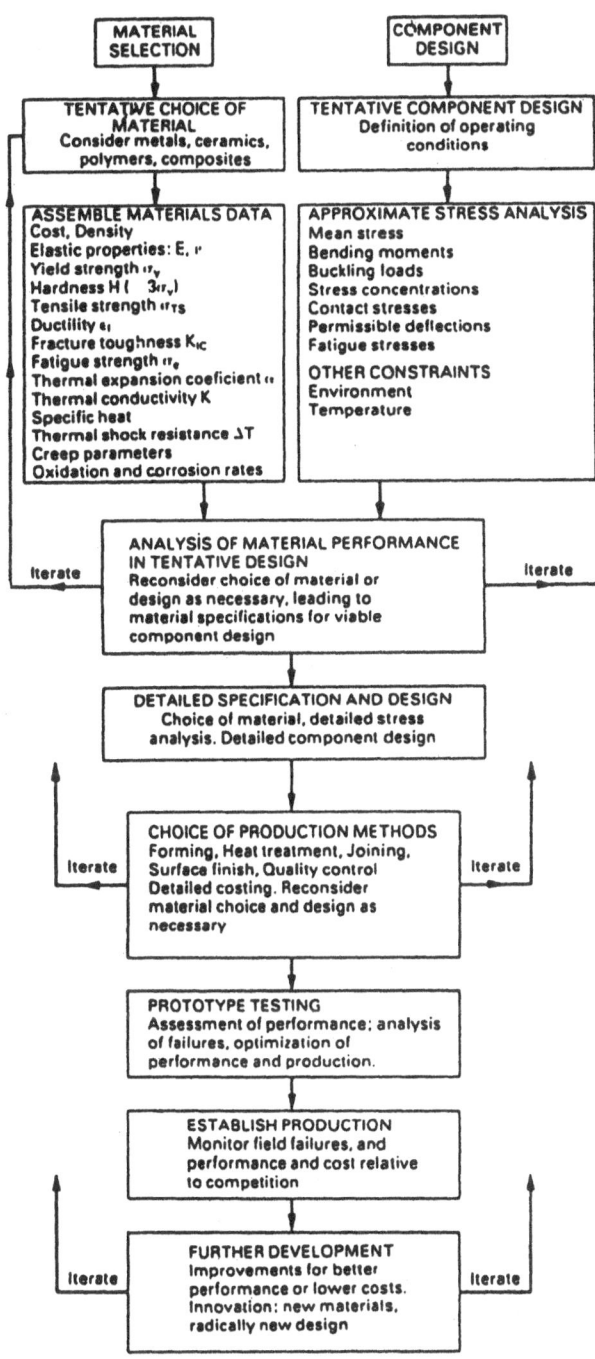

Fig. 1   Design methodology proposed by
Asby and Jones (1)

Once the operating conditions have been specified in terms of applied loads (mechanical as well as thermal loads, heat transfer, loading rates, etc.) using conventional tools (principles of mechanics, dynamics, heat transfer, ..) the first important step is the <u>material selection</u>. This selection may be guided by specific properties required for a given function such as for example low friction, light weight, insulation, wear resistance, heat resistance, corrosion resistance. Once the materials which best meet these prioritized requirements are selected as candidates, further selection is then essentially based upon their ability to withstand the mechanical stresses and deformations induced by service conditions.

The question is thus to determine which materials will not cause component failure. The answer to this question requires a stress analysis, which converts the applied conditions into stress levels and strains. The stress levels are compared with the characteristic material strength. The characteristic material strength can be the yield stress (static loads), the creep rupture stress or the fatigue endurance limit depending on the temperature and the loading situation. The strains are compared to the allowable maximum dimensions when each constituent of a structure is analyzed seperately. They are compared to strain limits when high temperature conditions prevail.

Material selection thus requires extrapolation of the properties measured on small test specimens to large components with complex geometry. In the case of metals this operation is easy, because properties are not significantly sensitive to scale effects and methods are well established.

In particular fracture mechanics provide the essential relationship between the three conditions for fracture : the defect, the stress and the material (Fig. 2). We can go around the triangle in any way we like. For example, we can fix the design of a structure by the applied stress or strain. The fracture toughness requirements for the material can then be specified from the size of defect that can confidently be expected to be found in the structure. Alternatively, for a given material, we can determine the size of the defects that can be allowed to remain in the structure. Or, in a similar way, the fracture toughness can define a minimum wall thickness in order that brittle failure be avoided.

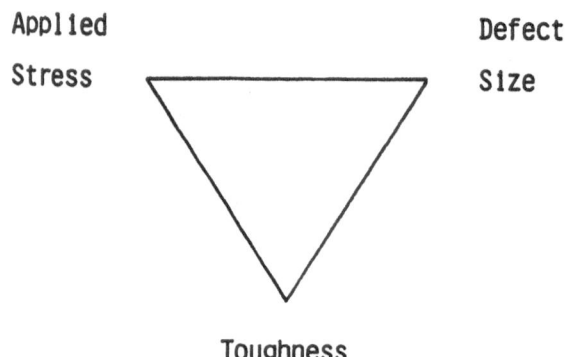

Applied Stress · Defect Size · Toughness

Fig. 2. Deterministic fracture mechanics: The interrelated variables of fracture

This approach to material selection as well as to assessing the acceptability of cracks, applies only when the defects are larger than the microstructure, and are present in a limited number, in other words when they constitute an anomaly. This approach is therefore applicable to dense materials (such as metals) but not to ceramics where the presence of a multitude of minute flaws is a very common situation.

After material selection, it is time to evaluate the component performances to determine whether this component is viable. If the component can bear the loads, moments, etc., then the design can proceed. If the component cannot perform adequately then the following iterative process takes place :

- either a new material is chosen, or
- the component design is changed (or both), or
- the operating conditions are improved.

Several iterations may be required until the problem is overcome. Subsequently the actual design may have to be refined to check that the tentative design and the material selection lead to a safe solution.

Production costs have then to be considered. Depending upon the price of the products a second major interation may be required for the choice of material or component design, allowing a low cost production path.

Finally a prototype is produced and tested. If this is satisfactory, full-scale production is established. But the designer's role does not end at this point. Continuous analysis of the performance of a component usually reveals weakness or ways in which it could be improved or made more cheaply. There is always scope for further innovation : for a radically new design, or for a radical change in the material selected, for example using ceramics.

3. DESIGNING WITH ADVANCED CERAMICS : ALTERNATIVE TECHNOLOGY

For ceramics, the sequence of operations in the design procedure as well as the material selection process are basically the same as for metals. A first selection based upon a few prioritized parameters (corrosion resistance, high temperature capability, thermal shock resistance, wear resistance, etc..) allows a number of possible materials to be narrowed down to a few. Then material selection is based upon material strength and reliability.

However, there are some problems, in particular in the methodology for strength correlations and reliability assessment. This question will be discussed later, but it is worth considering first some other parameters that are crucial in ceramic design.

3.1. The Choice of Production Methods

There are a number of factors other than cost which determine the choice of fabrication methods.

The properties of ceramics are dependent upon the fabrication technique. For example silicon nitride exhibits lower strength and toughness when processed by reaction bonding (RBSN) instead of pressureless sintering technique (SSN) (Table 1). RBSN may also exhibit higher sensitivity to creep than SSN, and be less reliable. However, RBSN presents a very interesting near net shape capability avoiding the need to perform very expensive and difficult machining operations.

Even with the same processing technique, ceramics do not exhibit unique properties. As illustrated in Table 1, properties are greatly depend upon the supplier.

Table 1 - Mechanical Properties of Silicon Nitrides

Material	Mean 25°C Bend strength (MPa)	
		Hot-pressed Si_3N_4 (2)
Norton NC-132 (1% MgO)	710	
Norton NCX-34 (8% Y_2O_3)	873	
Harbison-Walker (10% Ceria)	529	
Kyocera SN-3 (4% MgO, 5% Al_2O_3)	516	
Ceradyne Ceralloy 147A (1% MgO)	600	
Ceradyne Ceralloy 147Y (15% Y_2O_3)	605	
Ceradyne Ceralloy 147Y (8% Y_2O_3)	573	
Fiber Materials, Inc. (4% MgO)	460	
Toshiba (4% Y_2O_3, 3% Al_2O_3)	728	
Toshiba (3% Y_2O_3, 3% Al_2O_3, SiO_2)	576	
Westinghouse (4% Y_2O_3, SiO_2)	627	
NASA/AVCO/Norton (10% ZrO_2)	628	
Battelle HIP (5% Y_2O_3)	620	
Kyocera SN-205 (5% MgO, 9% Al_2O_3)	260	Sintered Si_3N_4 (2)
Kyocera SN-201 (4% MgO, 7% Al_2O_3)	342	
GTE Sylvania (6% Y_2O_3)	537	
AiResearch (8% Y_2O_3, 4% Al_2O_3)	547	
Rocketdyne SN-50 (6% Y_2O_3, 4% Al_2O_3)	356	
Rocketdyne SN-104 (14% Y_2O_3, 7% SiO_2)	356	
		Reaction bonded Si_3N_4 (3)
Unknown	138	
Unknown	193	
Ford RBSN	~ 290	
Norton NC350	~ 297	

The selection of production methods as well as of the supplier who determines material properties must be an integral part of the material selection step.

Although certain fabrication techniques lead to high properties, they are not appropriate for production. For example, hot pressing which gives the most resistant grade of Si_3N_4, cannot be used for the manufacture of components having a complex shape. Furthermore, engine components for which the use of ceramics is being considered have the form of cylinders, plates or irregular shapes such as turbocharger "hot ends", gas turbines combustors, vanes and shroud rings. For such irregular shapes, the most suitable techniques are slip casting and injection moulding.

The fabrication technique must also be selected with respect to component surface finish, and dimensional tolerances. Ceramics are very hard materials. Therefore, machining is a difficult and very expensive operation. Most ceramic components, particularly those with complex shapes and close dimensional tolerances cannot generally be formed and densified to their specified shape and dimensions. If tolerances of better than ± 1 % are required, the sintered components must be machined using diamond tools, a process which is very expensive and time consuming. Moreover, machining may be a fracture-initiating operation. Machining creates a lot of microcracks at the surface of components which may subsequently operate as sources of failure drastically affecting component life.

It is therefore imperative to select production methods which eliminate machining of sintered parts. For instance this can be achieved by developing a forming method for manufacturing high-precision products.

3.2. <u>Cost</u>

Today the cost may be very high for certain ceramic parts and the cost of these parts in high-volume production is still the subject of much debate. However ceramics are still in their phase of development, and the market is continuously increasing. Workers in Japan predict growth rates for structural ceramics of between 25 % and 40 % for the next decade. As processes are further developed and volume increased, the cost of ceramic components should decrease. One company has thus claimed to produce in medium volume (100 000/yr) a structural silicon nitride diesel engine component at a cost reported to be competitive with that of equivalent part in nickel-based superalloy (4).

3.3. Ceramic Joining

Because of inherent brittleness and high notch sensitivity, ceramics cannot be joined to ceramics or metals in the same way as metals. Bonding is often preferable to mechanical attachment because contact fracture problems are averted. Mechanical attachment is generally achieved over limited areas of contact, subject to normal and to transverse loads. Consequently, stress concentrations develop around the contact. The relative fracture resistance of bonded systems has not been widely investigated. However, it has been established that several techniques provide useful information (5). Fig. 3 shows the properties dictating the resistance of bonded systems.

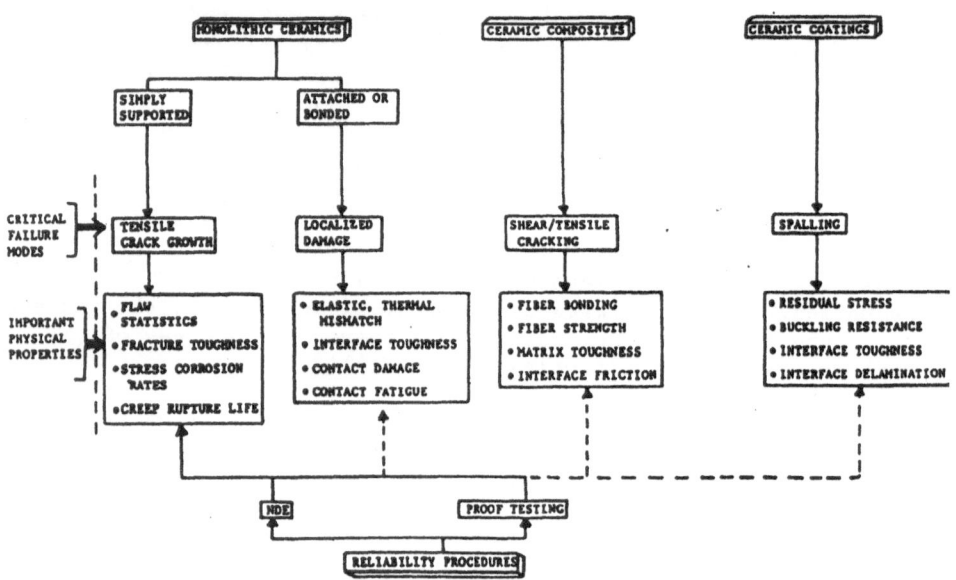

Fig. 3. A summary of property requirements needed for engineering design purposes with various types of ceramics (5)

3.4. Material Selection and Assessment of Reliability

Although the component size, shape, surface finish requirements, cost of manufacture and attachment to other parts determine the method of fabrication and the design of component, the underlying priority is the strength and the reliability of the ceramic component.

3.4.1. Ceramic Properties

Ceramics are highly sensitive to stress concentrations, a result of their inherent brittleness. Furthermore, due to low toughness, they are intolerent of even small flaws (< 100 μm), whereas, in contrast aluminium and alloys may tolerate cracks of around 1 mm. It is impossible to prevent the occurrence of all such small flaws during processing. As a consequence ceramics always contain a large amount of very small flaws (pores, voids, inclusions, agglomerates, etc.). The flaws have a certain distribution in size, nature and orientation.

It is impossible to locate the critical flaw among the large numbers of flaws pre-existing in ceramics. This renders the fracture of ceramics a probabilistic event.

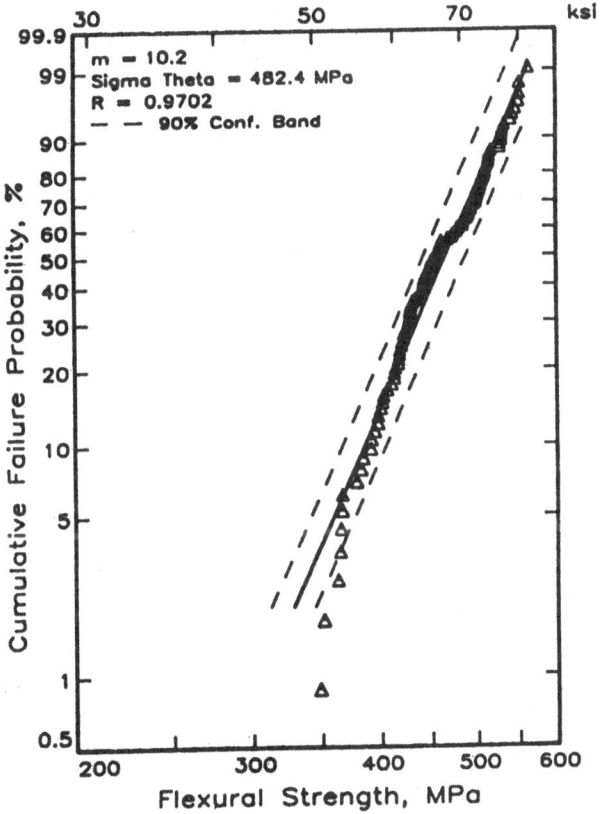

The presence of microstructural flaw populations has several important implications in ceramic design. The flaws are always responsible for failure. Moreover, the failure strength is not characterized by a unique value. Several identical specimens made out of the same material, having the same dimensions and tested in identical conditions fail at different stress levels Fig. 4. Moreover, strength is not an intrinsic property. It depends upon several parameters such as the specimen size, the stress state and the nature of flaw

Fig. 4. Flexural strength distribution for cold pressed and sintered alpha silicon carbide at 25°C (6)

602

populations that induce failure (6-10). Therefore the data measured on small test specimens do not represent the component strength.

Methodologies based upon a statistical description of failure should allow the use of laboratory data to infer the strength and the reliability of a structure under more complicated stress states encountered in service. They allow the use of strength as a fracture criterion although it is not intrinsic and they provide a safety factor defined in terms of failure probability.

Fig. 5 represents these aspects of probabilistic fracture mechanics. It provides the relationship between the three criteria for failure of ceramics : the flaw populations, the stress and the reliability. As with the deterministic approach triangle, we can go around the triangle in any way we like. For example, we can fix the applied stress in

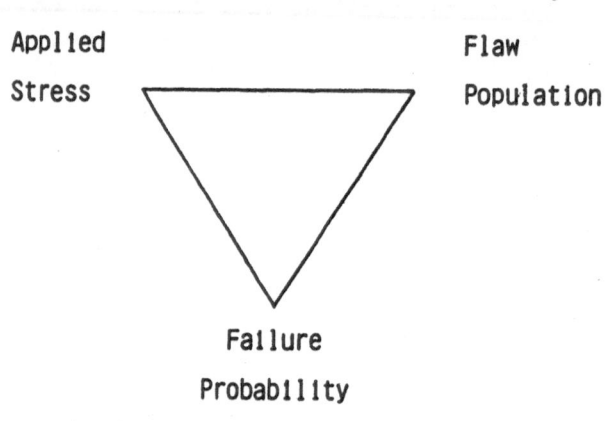

Fig. 5. Probabilistic fracture mechanics:
The interrelated variables of
fracture

the structure and derive for a given material component reliability. Or, for a given reliability, derive the characteristics of the appropriate material and of the permitted population of flaws.

To design reliable ceramic components, the adequate initial mechanical resistance of components is the first goal to be achieved. However, in high temperature environments, component failure may occur before the desired service life is reached, as a result of growth of the pre-existing flaws or new flaws which may appear. For high temperature applications, the second design issue is therefore to ensure a suitable service life.

3.4.2. Metallic Design Concept

Historically, design with ceramics has been empirical. This led to many failures in the past. The designs evolved from these failures. With structural ceramics, a metallic design based concept has been mainly used until now. This design procedure consists in replacing metal components piecemeal rather than developing an entire ceramic system. This is particularly true for such applications as engines. Fig. 6 shows the different steps of the procedure which consists in manufacturing a ceramic component having exactly the geometry of the metallic part, and then in testing this component. Generally testing leads to failure. The following step uses stress analysis (by finite element method) and fractographic inspection of broken component to remedy the failure. A new component with improved geometry is then made and tested. Failure is again analyzed as previously and this iterative procedure is carried out until the failure is overcome. In most cases this stage is never reached, which is a matter of design

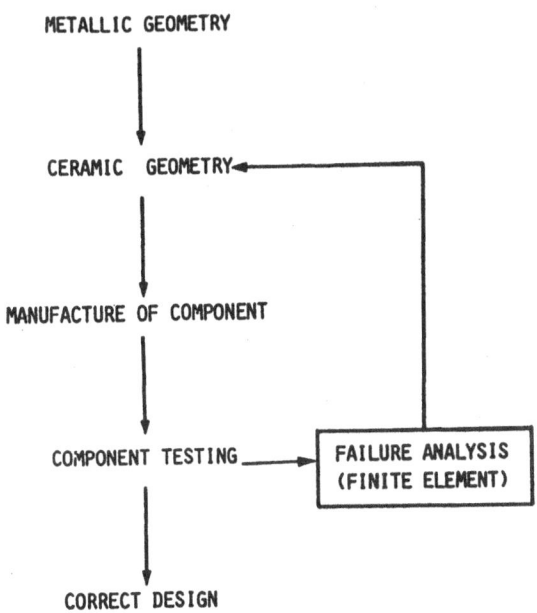

Fig. 6. Metallic design concept: The methodology initially used for the design of ceramic components

procedure instead of material capability. This - time and money - consuming empirical procedure neglects the main features of ceramic behaviour. In particular, the results of stress analysis cannot be analyzed as such and compared to material properties, due to the above-mentioned scale effects. The calculated stresses apply to small elements which are generally several orders smaller than the test specimens used for measuring material strength. Therefore, a single stress analysis is irrelevant to decide whether the material is appropriate. It just provides qualitative information on the location of highly stressed regions. However, due to the specific microstructure of ceramics, the high stress regions are not necessarily critical. A probabilistic approach is more appropriate to identify the high risk of failure regions in the component.

3.4.3. Ceramic Design Concept

Design methodology has evolved, and a sound methodology based upon a probabilistic approach is now more accepted by

ceramic users. However there are still only limited attempts by
industry to effectively employ this concept (see for instance
ref. (11-14)).

The basic idea in ceramic design is that component design is
guided by reliability. Reliability calculation is based on
weakest link theories for failure. The well-known Weibull theory
of fracture statistics is generally employed. However it has
been found inadequate in critical applications of structural
ceramics ((10) and refs. therein). The logical solution is to
use probabilistic fracture mechanics where linear elastic
fracture mechanics concepts are combined with alternative
statistical models (see for instance refs. (7-10)).

For complex geometries and loading conditions experienced by
components, a computer is necessary. A recent development of an
advanced computer code for ceramic design based upon
probabilistic fracture mechanics has been reported (15).

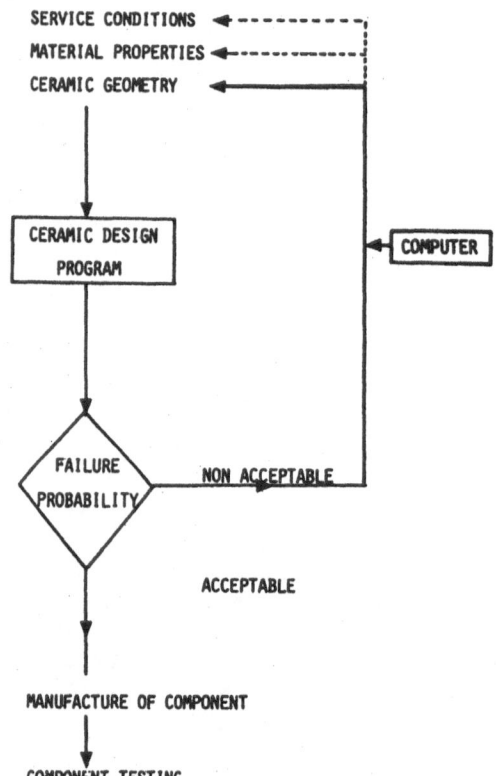

Fig. 7 illustrates
the modern design
procedure for ceramics.
The ceramic design
computer code determines
the reliability of
components. The
obtained reliability
then permits the
diagnosis of design, and
will guide the decision
to be made. If the
reliability is judged
insufficient, geometry
will be changed first,
until an acceptable
reliability is reached.
When an acceptable value
is obtained the decision
for component
manufacture can be made,
provided the above-
mentioned criteria are
also satisfied. If an
acceptable value cannot
be obtained with the
actual material, a new
material has to be
sought. At this step,
the program determines
the required properties.

Fig. 7. Ceramic design concept: The
modern methodology for
ceramics

Specifications are expressed as a function of the testing conditions, for an easy comparison with materials data base.

Depending upon the reliability obtained with the new material, the decision to manufacture can be made or new iterations have to take place. The situation may arise that no existing ceramic can lead to an acceptable reliability. At this

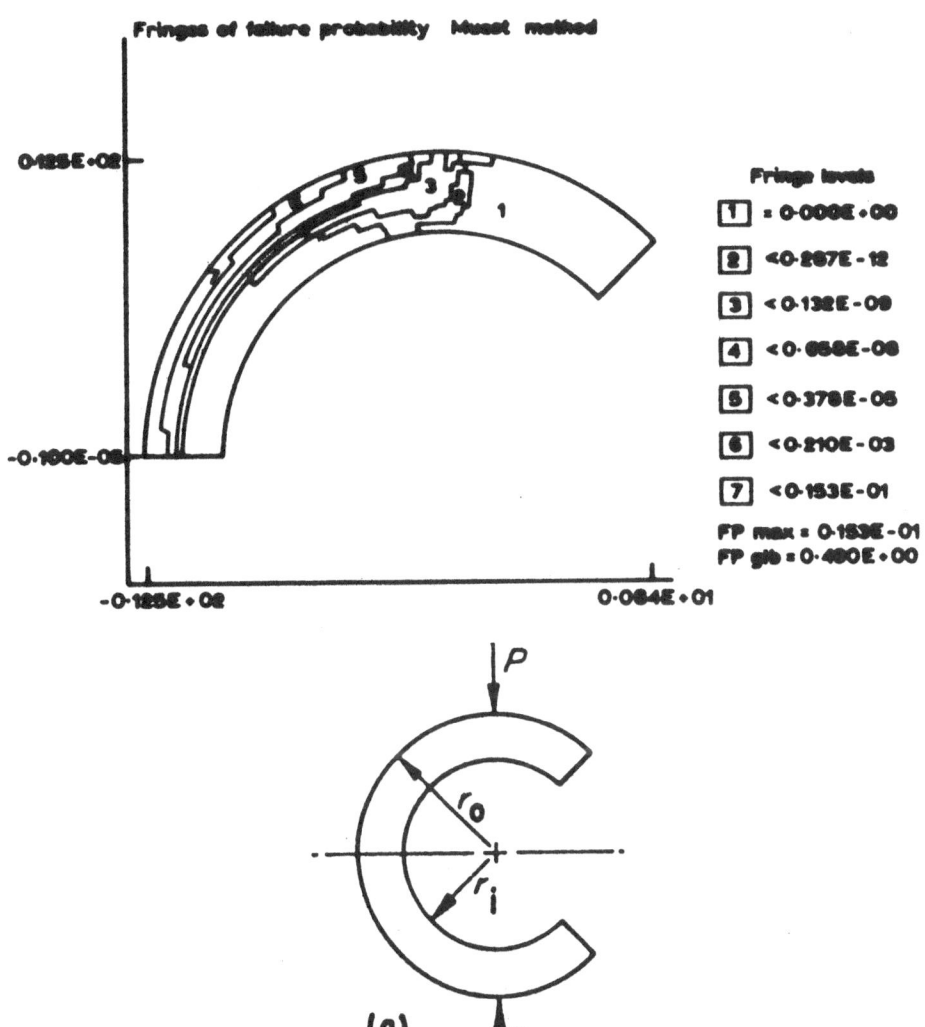

Fig. 8. Diagram showing a reliability plot obtained for a SiC C-ring specimen subjected to a diametral compression load of 250N. Reliability was calculated using a computer code developed by Battelle-Geneva based upon the multiaxial elemental strength model (8,9)

stage, the operating conditions have to be redefined.

As previously with metals, the present design procedure does not require early manufacture of a prototype. A prototype is produced and tested only at the final stage.

Fig. 8 shows an example of reliability map calculated for a SiC C-ring diametral compression specimen using the computer code developed by Battelle-Geneva. The calculation provides two useful data : first the component reliability (denoted FP glb in Fig. 8) and the distribution of high risk of failure regions.

The first parameter measures the quality of design. In the present case, a 49 % failure probability seems to be too high. However, the acceptability of this value should be judged in respect to other criteria such as the cost. The failure probabilities map then shows where the geometry must be changed. Geometry improvement may be aimed at or guided by the elimination of high risk of failure regions.

3.4.4. Lifetime prediction

The second issue of ceramic design is now service life prediction as delayed fracture may result from the damage experienced by the components in high temperature environment.

Based upon metallic experience, most designers tend to consider that delayed fracture is caused by the slow propagation of a unique crack. Therefore they incorporate a Paris type equation $V = AK_1^n$ in the failure probability expressions.

The problem of performance prediction is generally more complicated at high temperature. The phenomenon of delayed failure has not yet been satisfactorily characterized. Slow crack growth may be the final stage of failure. However, it is not necessarily the predominant mechanism and damage is not necessarily characterized by the above equation. A lot of work tends to show that creep may make a decisive contribution to delayed failure. Thus, for correct prediction of ceramic reliability, it is more appropriate to identify the damage process and use the pertinent data. Data are lacking in this area and further studies are required.

Fig. 9 represents the complete methodology for ceramics, including the initial mechanical capability of component and time effects induced by high temperatures. A large amount of work still has to be performed in the area of high temperature failure of ceramics. The high temperature mechanical behaviour is also fundamentally different from the fatigue behaviour of metals. Therefore emphasis has to be placed upon identification of the damage mechanisms responsible for delayed failure and upon characterization of this damage for lifetime predictions. As a consequence the dotted lines of Fig. 9 will be turned into solid lines.

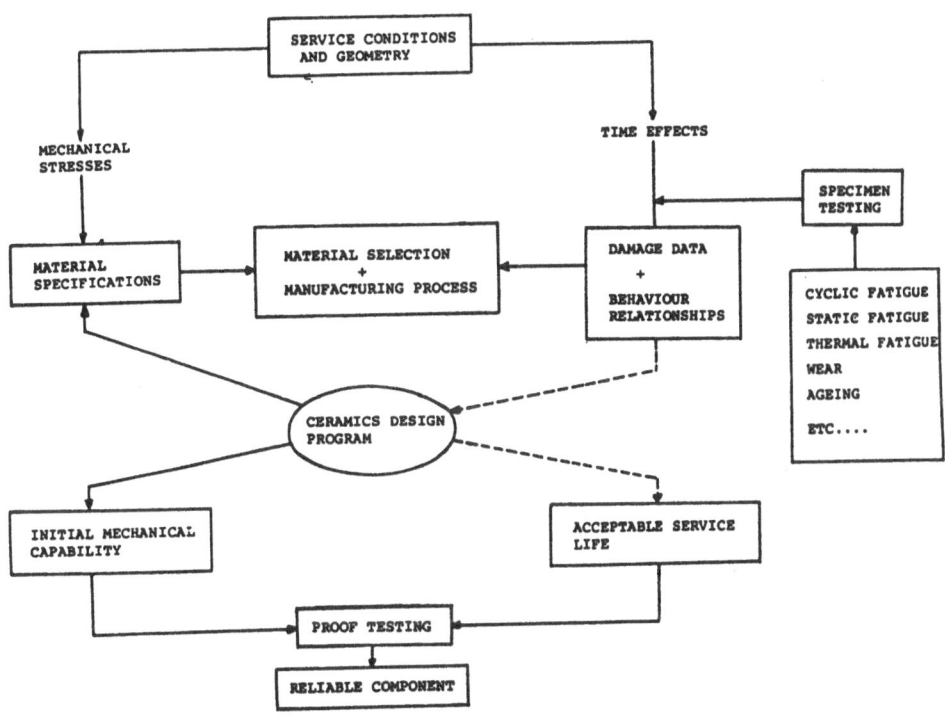

Fig. 9. Complete methodology for ceramic design including initial mechanical capability as well as fatigue effects

4. CONCLUSIONS

Metals have been used for a long time in structural applications. Designers are accustomed to working with them. Methodologies for material selection and reliability evaluation are well established and routinely used.

A different situation arises concerning engineering ceramics. Ceramics are a newer material to structural applications under critical conditions. Roughly the sequence of operations in the design procedure is the same as with metals. However some problems arise concerning material selection and reliability assessment.

A sound design methodology for ceramics is emerging. This methodology is based upon probabilistic fracture mechanics instead of the deterministic approach used for metals. From a designer's point of view, a major problem with ceramics is that they contain a multitude of minute flaws. Location of the critical defect is therefore unpredictable and undetectable. Design of ceramics is thus guided by the component reliability. A design is good when reliability is high.

Design is not only a matter of method, but also a matter of material. The designer has to know the material he is working with. The emergence of engineering ceramics required a sound methodology. The increasing use of other new materials will similarly require appropriate methodologies.

Although sound design methodology is under development there is much work to be done to establish a complete design methodology including high temperature effects. It is true that there is an important lack of data and understanding on the behaviour of ceramics at high temperature. A material data base that can be easily used by the designer is still not available and current data bases are incomplete. Moreover certain data are questionable as the testing conditions are not specified. This implies an effort for the standardisation of the testing methods with respect to design requirements.

5. REFERENCES

(1)   "An Introduction to Microstructures, Processing and Design", Ashby, M.F. and Jones, D.R.H., Engineering Materials 2, Int. Series on Materials Science and Technology, Vol. 39, Pergamon Press, (1986).

(2)   Ceramics Source 1986, The American Ceramic Society, Vol. 1 (1985).

(3)   Bennett, A., Materials Science and Technology, September, (1986) (2), pp. 895-899.

(4)   Hartsock, D.L. and McLean, A.F., Ceramic Bulletin of the Am. Ceram. Soc. 63(2), pp. 266-270 (1984).

(5)   "Engineering Property Requirements for High Performance Ceramics", Evans, A.G., Materials Science and Engineering, 71 (1985), pp. 3-21.

(6)   Seshadri, S.G. and Srinivasan, M., J. Eng. Mat. Tech. 105 (1983), pp. 219-223.

(7)   Lamon, J., Rev. Int. Hautes Températures et Réfractaires, 22 (1985), pp. 115-127.

(8) Lamon, J. and Evans, A.G., J. Am. Ceram. Soc. 66 (3), pp. 177-182 (1983).

(9) Lamon, J., ASME Gas Turbine Conference and Exhibit, Houston (USA) (1985), ASME-85-GT-151.

(10) Lamon, J., J. Am. Ceram. Soc. 71, pp. 106-112 (1988).

(11) Lindgren, L.C., Heitman, P.W., and Thrasher, S.R., SAE Technical Paper Series, 831520, October 3-6, (1983).

(12) Trantina, G.G., de Lorenzi, M.G., Journal of Engineering for Power, October (1977), pp. 559-566.

(13) Swank, L.R. and Williams, R.M., Ceramic Bulletin 60 (8), pp. 830-834 (1981).

(14) Matsui, M., Soma, T., Ishida, Y. and Oda, I., SAE Technical Paper Series, 860443, February 24-28 (1986).

(15) "Ceramic Reliability: Statistical Analysis of Multiaxial Failure using Weibull Approach and the Multiaxial Elemental Strength Model, Lamon, J., to be published by the American Society of Mechanical Engineers.

5.3.5. Structural Design Methods and Rules for Metallic

Components Operating at High Temperatures

Dr. F. Schubert

KFA Jülich GmbH

1. INTRODUCTION AND DEFINITIONS

The successful operation of high temperature components is dependent upon the amount of information concerning materials specification, materials behaviour and data as well as the dimensioning and construction rules which are applied by the designer. The structural design of a component comprises

- materials evaluation
- dimensioning and geometrical layout
- structural analysis and analysis of the operational behaviour
- routines for testing and inspection.

The basic principles of structural design are to allow safe operation of a component; to ensure integrity and continuous functioning: and to avoid any damaging incident or catastrophic failure during the whole period of the defined operation of a componenet and during the total service life of a plant. There are different national rules and codes (1 .. 26) for pressure vessels and high temperature components in fossil fuelled power plants, nuclear power plants and petrochemical equipment, as well as the manufacturer's specifications for components of steam and gas turbines (27). The design rules lay down the essential basic principles for the guidance of the designer, the fabricator, the user and the inspection authorities. They try to meet the best balance between safety requirements and commercial and economical factors.

The design of high temperature components is more complicated than that for components operating at ambient temperature, because various time dependent degradation processes and failure modes have to be considered. An appropriate selection of materials and design methods should prevent failure by any one of the following modes:

- creep rupture due to long term loading;
- creep-fatigue failure due to load cycles in the operating and upset conditions;
- gross disortion due to incremental collapse and ratchetting caused by thermally induced cyclic loading;
- loss of function due to excessive deformation;
- creep induced structural instabilities, such as creep buckling;
- failure due to environmental attack;
- spontaneous fracture due to critical stress concentration around a flaw or to unstable crack growth under cyclic or static loading.

The design consideration takes into account all loading conditions such as:

- primary stressed (dead weight, internal, external pressure, stress controlled fatigue);
- secondary stresses or thermally induced strains (low cycle fatigue, ratchetting, constraints);
- peak stresses (local stress and/or strain exhaustions)

The majority of existing design rules with their origins in the rules used for ambient temperatures, propose the limitation of stress, indicating that there is a balance between stress and strain, "the design by rules". A limited number of design codes or design rules provide the use of inelastic analysis which takes in account that there is a time-dependent stress-strain history, giving a balance between stress and strain rate. It is not the stress that is limited, but the remaining strain, the "design by analysis".

Rules for fossil fired power plants do not reflect different categories of stresses. For limitation of stress often the secondary and peak-stresses are incorporated as additional stresses by defining the design stresses. In nuclear codes, the stresses are classified according to their potential to cause failure and to their different safety margins.

The aim of this survey is to give an introduction to the methodology for handling the critical points of high temperature design and analysis and to emphasize the research and development work in that field.

2. MATERIAL EVALUATION

The selection of material for creep exposed components puts the main emphasis on the following properties:

- thermal expansion, E-Modulus, creep resistance, stress-strain behaviour, creep-fatigue resistance, corrosion resistance, microstructural stability.

For some typical applications, some further points have to be considered: weldability, coatability, fatigue behaviour of a given geometry and surface condition and fracture mechanics criteria. For the evaluation of the real stress-strain-time behaviour in a section of a component, constitutive equations are required for the calculations of both the stress and strain-distributions.

The evaluation of new steels or alloys for high temperature components has to provide answers to all the above properties. The slow pace of experimental test, means that the qualification of a material for high temperature application takes generally a decade.

Today, the main emphasis is put on creep, creep fatigue and constitutive equations, and, in those cases where corrosion is the life limiting mechanism, corrosion behaviour and/or development of corrosion resistant coatings. The development of materials is not a task for code-work. The definition of specifications and thereby the guarantee of a given scatter band is a code concern.

3. DIMENSIONING

3.1. Limitation of stresses

The procedure for determination of the minimum wall thickness by rules can be handled satisfactorily against primary stresses. Incorporation of thermally induced stresses, should ensure that a component has a good chance of surviving for the planned operation time. Typical rules for the dimensioning of the wall of a cylindrical component under internal pressure or subject to a pressure difference across the wall are compared in Table 1.

The required wall diameter is then defined by the dimensioning pressure, temperature and time and safety margins, e.g. a margin for weldments, a safety margin for wall thickness tolerance and, in some cases, a safety margin for corrosion. The allowable stress level is derived from creep-rupture data.

Tab. 1: Determination rules for wall thickness of tubes subject to pressure difference across the walls

CODE	EQUATION FOR DIMENSIONING $\frac{Da \cdot p}{s} =$	FACTOR FOR WELDMENT	OPERATION TIME	ALLOWABLE STRESS
AD-Merkblätter	$2\,\sigma_{allow}\,v + p$	$v = 0.85 - 1.00$ $= 0.80$ (no $R_{m,t}$-values)	100 000 h	$\overline{R}_{m,t}/1.5$
TRD	$(2\,\sigma_{allow} - p)\,v + 2p$	$v = 0.80 - 1.00$	100 000 h 200 000 h	$\overset{\vee}{R}_{m,t}/1.0$ $R_{m,t}/1.11$ (reducer fittings)
HTR-Auslegungskriterien	$2\,\sigma_{allow} + p$	--	$t = t_{dimensioning}$	$\overset{\vee}{R}_{m,t}/1.5$ $R_p\,1.0\,t/1.0$
BS 5500	$2\,\sigma_{allow}\,v + p$	$v = 1.0$	100 000 h 200 000 h	Appendix k
ASME Section I Power Boilers	$2\,\sigma_{allow}\,v + 2\,p\,y$	$v = 1.0$ (machined) $v = 0.9$ (unmachined)	100 000 h	
ASME SECTION III Nuclear Power Plant Components ASME Section VII, 2	$2\,\sigma_{allow} + p$	--	100 000 h	$\overline{R}_p\,1.0\,t/1.0$ $\overline{R}_{m,t}/1.5$ $\overline{R}_{m,t}/1.25$
ASME SECTION VIII, 1 Pressure Vessels	$2\,\sigma_{allow}\,v + 0.8\,p$	$v = 1.0$ (defined) $v = 0.7$ (not yet defined)	100 000 h	
API - RP 530	$2\,\sigma_{allow}\,v + p$	$v = ?$	20 000 h 200 000 h	$\overset{\vee}{R}_{m,t}/1.0$
Remarks	distribution factor for plastic strain	$y = f\,(T) = 0.4 - 0.7$		‾ mean value v minimum value

614

In codes for nuclear plants for temperatures above the intersection of S_m and S_t, Fig. 1, the stresses due to loading are compared with the stress intensities defined as

S_m = time independent stress intensity
S_m = reference design stress intensity
S_t^o = time dependent stress intensity.

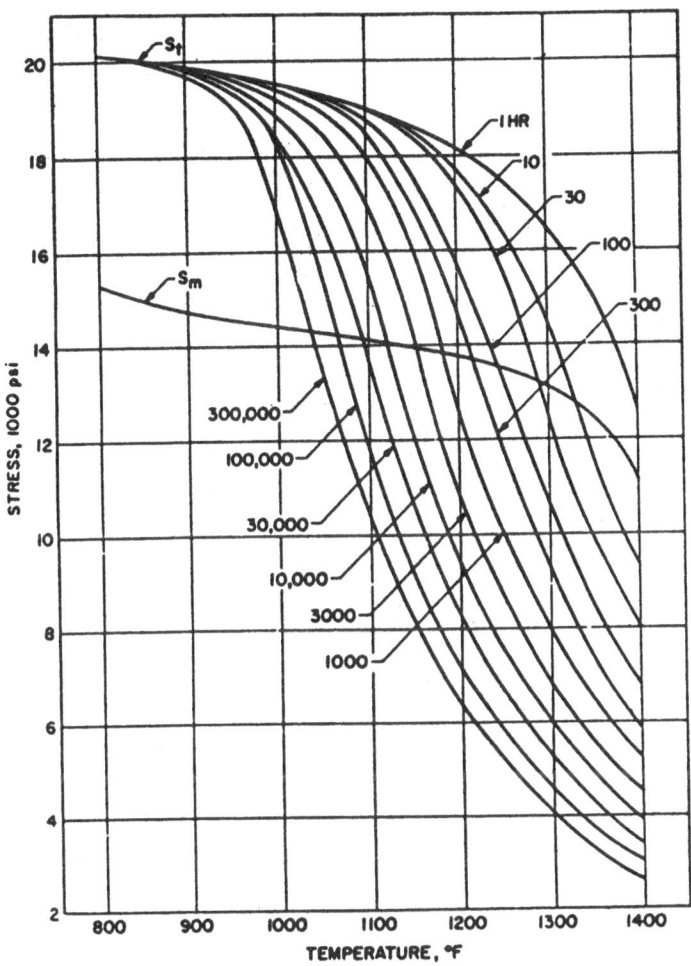

Fig. I-14.3C S_{mt} — Ni-Fe-Cr (Alloy 800H)

Fig. 1. Stress limits for time independent (S_m) and time dependent (S_t) intensity for dimensioning according to ASME code case N 47 (11).

The time dependent stress intensity S_t limit (S_t may be compared with S_t = K/S) is determined by

S_t =   minimum of R_p, & %, t, $T_{/1}$

or minimum of R_m, t, $T_{/1.5}$

For high temperature components sometimes the onset of tertiary creep and the creep rupture strength at a temperature 15°C higher than the maximum operation temperature are considered.

The rules given for wall thickness determination (Table 1) are very similar; there are only a few minor modifications with respect to the allowable operation time, to the criteria to derive the allowable stresses and to the differences in the safety margin for wall thickness tolerance and for corrosion.

The primary stress for high temperature components used in the dimensioning procedure is not always easy to determine, if the primary stress is defined as that fraction of the total stress which remains after small scale permanent deformation, and which causes continued creep. Therefore, within a structure, any stress field which balances the volumetric forces and loads applied to the surface is an upper bound of the primary stress. The primary stress itself, however, must be derived from the multiaxial loading condition within a component. The stress intensity value, resulting from the application of mechanical loads such as internal pressure, weight, centrifugal forces, reaction of supports for other components, must be determined from the stress tensor from which the principal stresses must be derived. Therefore, some of the structural design codes prefer a classification of stress categories (11, 25, 22).

Whereas design rules for non-nuclear plants provide for the use of the main stress for definition of primary stress, the codes for nuclear plants try to operate with membrane stresses and bending stresses for pressurized components. For the calculation of primary stress, the primary membrane stress and the primary bending stress are defined, from which the design stress can be derived using various deviatoric stress theories.

For the dimensioning procedure, it seems that the thermal loadings are not treated or not adequately considered. This raise the question whether the procedure of dimensioning by rules can sufficiently assure the integrity of a component.

Therefore, the design rules for conventional plants propose an additional procedure for components with cyclic loadings and, the codes for nuclear power plants prefer methods of inelastic analysis. These procedures are not included ("express verbis") in the dimensioning rules. A well trained designer, however, will consider those points.

3.2. Additional procedure for components with cyclic loadings

Additional procedures are proposed for components with cyclic loading (e.g. in the AD-Merkblätter S1 and S2). The basis is "inspection during the operating time" and "determination of the effect of loading history". All components which have to be treated according the pressure vessel rules (4) must be regularly inspected during the service life e.g.: - every two years, by external inspection, - every five years, by internal inspection, - every ten years, by pressure tests.

The consumption of allowable life time, in German rules defined as exhaustion e, is the linear accumulation of the exhaustion due to creep e_z and fatigue e_w.

$$e = e_z + e_w \tag{2}$$

The exhaustion values are:

$$e_w = \Sigma \, (N_k/N_{Ak}) \cdot S_L \leq 1.0 \tag{3}$$

with

N_k = number of loading cycles with the stress range of $2_{a, \, k}$

N_{Ak} = number of cycles to crack initiation with the stress range of $2_{a, \, k}$ with $T = 0.75 \, T_{max} \, a + 0.25 \, T_{min}$

S_L = 10 safety margin for cyclic number

$$e_z = \Sigma \, (t_j/T_{mj}) \cdot S \leq 1.0 \tag{4}$$

with
t_j/T_{mj} = creep exhaustion fractions
S = 1.5 (for forged and rolled steel products).

If the exhaustion value "e" has reached, or even exceeded, a given limit, non-destructive testing of the affected component regions must be conducted. By this concept the remaining uncertainties are excluded by the inspection routines.

The uncertainties in using design by rules may be illustrated by the following comments:

- not all possible loading conditions have been considered;
- the tolerances of temperature and wall thickness exceed the degree which is controlled by the safety margins;

- under upset conditions the component may be exposed to loading conditions that have not been predicted;
- non-destructive test routines cannot be carried out in certain component areas;
- the hypothesis for calculation of the accumulated exhaustion may not be sufficiently accurate.

To avoid mistakes in design of nuclear power plant components all possible failure modes must be excluded either by constructional methods and/or by analysis, as it is exercised for high temperature exposed metallic components of fast breeder reactors (FBR ASME Code Case N 47 (11), RCR-MR (25)) and high temperature gas cooled reactors (Fachkreis "HTR-Auslegungskriterien" (22)) AGR, HTR, HTGR, where the inclusion of an analysis of all postulated operational and upset events is proposed, which are classified in categories of different levels of severity, even at an early stage of developing the design concept.

The French code characterizes the levels by naming the types of damage which have to be avoided.

Level A: protection against:
- immediate or time-dependent excessive deformation
- immediate or time-dependent plastic instability
- time-dependent fracture
- elastic or elasto-plastic instability, immediate or time-dependent
- progressive deformation
- fatigue

Level B: is not defined in RCC-MC, the avoidance of failure is covered by the requirements of level A

Level C: protection against the following types of damage:
- plastic instability, immediate or time dependent
- time-dependent fracture
- elastic or elasto-plastic, immediate or time-dependent instability.

If the deformation of level C loading accelerates the strain limits of level A, it is necessary to ensure by inspection and analysis that the function of the component will be satisfactory. Level D criteria are to protect the components against the same failure as level C, however, with lower safety margins and accepting that a return to operation will not always possible, i.e. the plant would be written off.

4. DETERMINATION OF ALLOWABLE SERVICE LIFE WITH LIFE FRACTION RULES

The definition of allowable operation time is usually proposed by life time fraction rules. At the present time, the separate estimation of creep and fatigue exhaustion seems to be the only practicable method.

With the design curves for creep and fatigue the sum of consumption of allowable service time is calculated for each stress and strain range level (Fig. 2) (28).

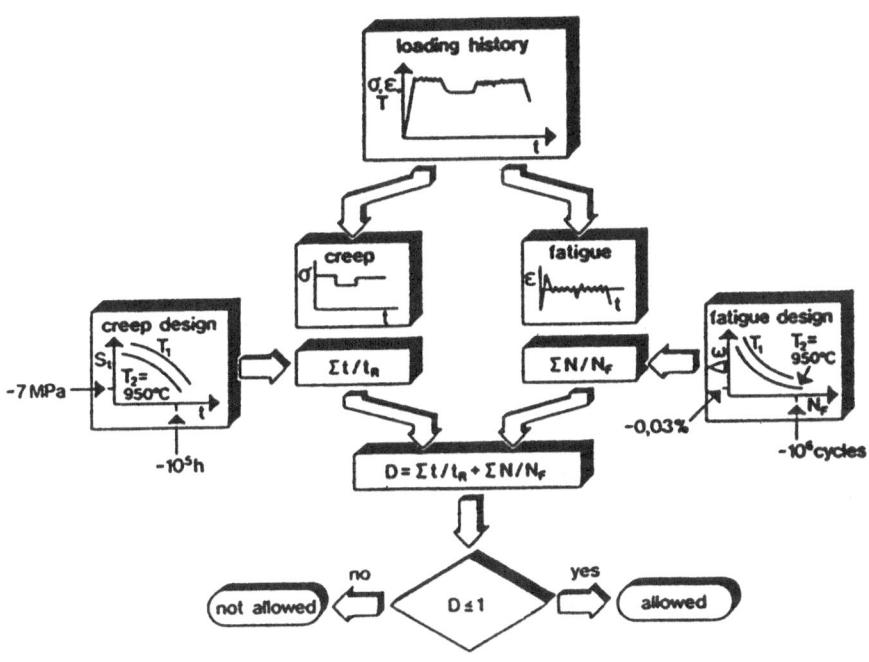

Fig. 2.   Design against creep/fatigue interaction at elevated temperatures, according to ASME Code Case N-47

$$\Sigma\, t_i/t_{io} + \Sigma\, N_j/N_{jo} = D \leq 1 \tag{5}$$

t_i = duration under the stress $_i$ at the temperature T_i
t_{io} = allowable opertion time (S_t^i) at $_i$ and T_i

N_j = number of cycles at strain range $_i$
N_{jo} = number of allowable cycles at strain range $_i$

The designer can be guided by ASME Code Case N-47 which is applicable for materials up to 800°C. The Code Case optionally envisages an elastic examination method or an examination of the results of an inelastic analysis for the creep damage part of the accumulation rule.

For inelastic analysis the strain limits are

- 1% for membrane strains averaged over the wall thickness,
- 2% for membrane and bending strains,
- 5% for local strains.

The strain criterion does not differentiate between remaining strains due to plasticity and those due to creep. The strain limits for weldments are set at 50% of the above values. The basis for this factor is not very well established, and it should be improved on the basis of experimental results.

The design fatigue strain range is obtained from temperature dependent design fatigue curves, which are determined from the average values of the fatigue of low cycle fatigue life with the safety factors: $\nu = 2$ for the strain amplitude at a given number or cycles to failure or $N_f = 20$ for the number of cycles to failure for a given strain amplitude.

The allowable life-time of a creep-fatigue exposed component is determined by a linear accumulation rule, given in Fig. 2. For Alloy 800 H, the total creep fatigue damage factor in ASME Code Case N-47 is $D = 1$. For AISI 304 and 316 austenitic stainless steels, however, the factor is approximated by a bilinear function equal to or less than unity. Results of X10NiCrAlTi 32 20 (similiar to Alloy 800 H) and NiCr22Co12Mo (Similiar to Alloy 617) LCF-tests /29/ with hold times show that for S10NiCrAlTi 32 20 and NiCr22Co12Mo the creep fatigue interaction factor D should also be less than unity for the temperature range of 700 to 850°C.

Using low cycle fatigue tests with different holding time to proof life time fraction rules raises the question, how to account for the exhaustion due to creep. The calculation of relaxation curves for the very first step of holding time as well as the treatment of holding-period in the compression mode are not yet fully understood. Here is a task for future research work.

To simplify the superposition of primary stress and thermally induced alternating strains and stresses, design curves were derived from the experimental results of low cycle fatigue tests, with hold times either in the tension or compression region. However, also in this case, there seems to be no uniform concept in the interpretation of what is consumption of allowable life-time and what is damage (30).

The use of the linear accumulation rules of Miner and Robinson, works only acceptably when the interaction between fatigue and creep damage is restricted to the actual event of failure.

Methods such as "strain range" partitioning and the frequency-modified Manson-Coffin Rule (31) are better applicable for fatigue failure modes. Chaboche (32) has chosen a phenomenological expression based on continuum mechanics which separates creep and fatigue, which seemd to work acceptably even for a multiaxial loading situation.

The limited number of multiaxial fatigue tests, however, are not yet enough to propose a treatment of a multiaxial fatigue-creep-life prediction rule. This must be a subject for further research work.

5. CONTROLLING EXCESSIVE STRAINS

As pointed out earlier, many high temperature components are subjected not only to the primary stress controlled load but also the thermal shock and to thermal fatigue, which lead to a strain-controlled stress distribution in the wall of a component. If these superimposed temperature transients change cyclically then structural deformation behaviour can develop by plastic effects or creep-relaxation in such a manner that strain increments in one direction add up in each cycle. This behaviour is called "ratchetting" or "creep ratchetting".

Depending on the primary to secondary stress ratio, the creep strain under ratchetting conditions can be much higher than that due to creep under primary stress only related to the highest cycle temperature. To guarantee that the creep strain under creep ratchetting conditions is not higher than the limits given for creep strain before the ASME Code Case N-47 (11) specifies ratchetting tests based on elastic methods for the design of components.

5.1. Elastic creep ratchetting analysis

The analysis begins with the definition of the load and temperature cycle to which the component will be exposed. Then the stresses in the components are calculated elastically at each point in the cycle. Parameters X and Y are then defined:

$$X = \frac{\left(P_L + \frac{P_b}{K_t}\right)_{max}}{S_y} \text{ and } Y = \frac{Q_{r\ max}}{S_y} \quad (4)$$

S_y = yield strength

$Q_{r\ max}$ = maximum secondary stress range

P_L = membrane stress

P_L = bending stress, K_t = shape factor of real component

$\left(P_L + \dfrac{P_b}{K_t}\right)_{max}$ = Maximum of possible combinations of membrane and bending stress

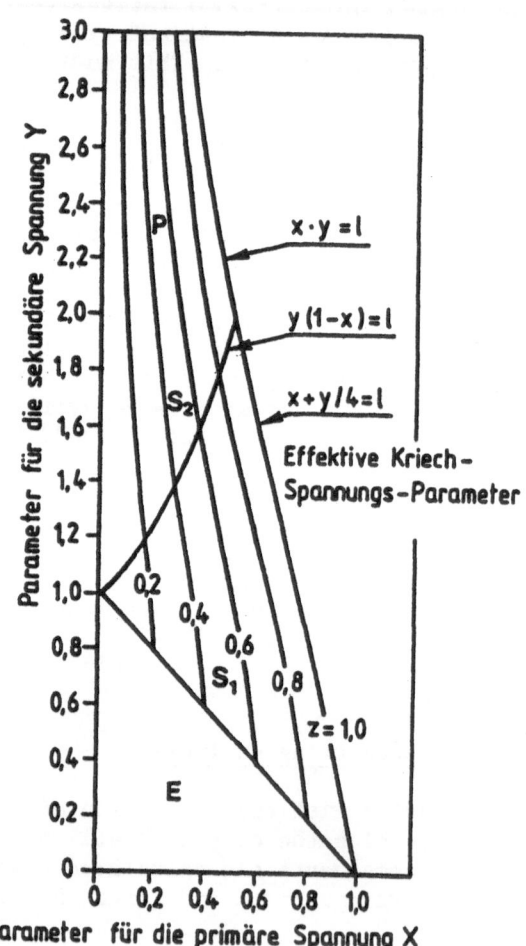

Fig. 3.　O' Donnel-Porowski modified Bree-Diagram (11, 36, 37, 38)

Sy is the average of the instantaneous yield points at the
max./min. wall averaged temperature during a cycle. Tests 1 and
2 ensure that the stresses are always less than the yield stress.
Shake-down will happen so that there can be no progressive
deformation. Test 3 makes use of the modified "Bree diagramm".
The axes are in terms of the parameter X and Y.

Shake-down is within the areas marked E, S1 and S2. Outside
these regions, the diagram provides an estimation of the total
strain which has accumulated in the component. The lines of
constant Z are used to define an effective stress σ_c.

The creep ratchetting strain is determined by multiplying the
effective stress σ_c by 1.2 T. The procedure is limited to
temperatures in the range 500 to 650°C. The French Code (24)
considers also purely elastic and plastic behaviour to define the
progressive deformation and ratchetting if creep can be
neglected.

In the case of creep relevance, this type of simple
definition is no longer adequate, the designer has to deal with
the increase in deformation by creep when there is a cyclic
deformation such as thermal deformation.

5.2. Inelastic creep ratchetting analysis

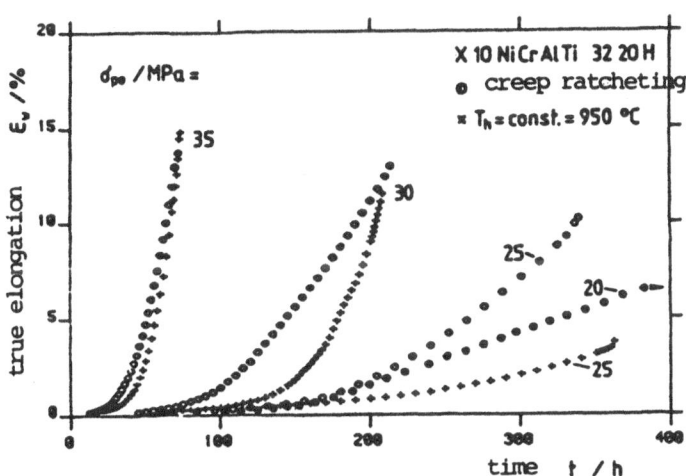

Fig. 4.   Comparison of creep strain versus time behaviour with
and without creep ratchetting load (37)

The calculation of creep ratchetting strain can be made
based on a simple two-shell sandwich model. A tube wall is
divided into two shells, with temperature cycling taking place,
for example, in one shell representing the inner tube wall. It

is postulated that the shells representing the tube wall are rigidly connected. A further approximation is with a finite element calculation.

A synthesis from experimental work and calculation may help to overcome this problem. In a recent investigation (36, 37) tubes held at 950°C were primary loaded in the axial direction and a thermal gradient was imposed over the wall. In Fig. 4 the results obtained with and without ratchetting are compared.

By a modified finite element technique it has been possible to compute ratchetting rates which agree with experiment.

The task of further R and D-work, is to derive simplified inelastic methods.

6. CONTROLLING STRUCTURAL INSTABILITIES

6.1. Shape instabilities

Geometrical instabilities may occur if a sudden change in the balance of loading condition occurs. In principle several kinds of instability are possible: buckling or creep buckling due to excessive deformation or bifurcation buckling.

The danger of kinking and buckling must be avoided. Further stability problems arise due the technical geometrical tolerances in producing semi-finished products, e.g. tubes or ring forgings always have a slight eccentricity, ovality (deviation from the perfect circular shell geometry). The mathematical treatment of buckling problems requires a calculation, in which geometrical and material dependent non-linearities are integrated. Bifurcation buckling is a kind of "Eigen value" calculation.

For high temperature components in which creep cannot be ignored the buckling problem is further dependent on creep behaviour of the material, the relevant creep law.

The structural design codes solve the buckling damage problem by limiting either the allowable duration of the upset condition or by a stress (or a pressure) safety margin. A theoretical treatment of creep buckling is very complicated, even using Hoff's simplification.

The calculations either by Hoff's method or by a finite element method allow the buckling phenomenon to be described only within an order of magnitude, due to the scatter in the constants n and k of Norton's creep law (Fig. 5). The required assurance against buckling is provided by the conservative stress and time limits.

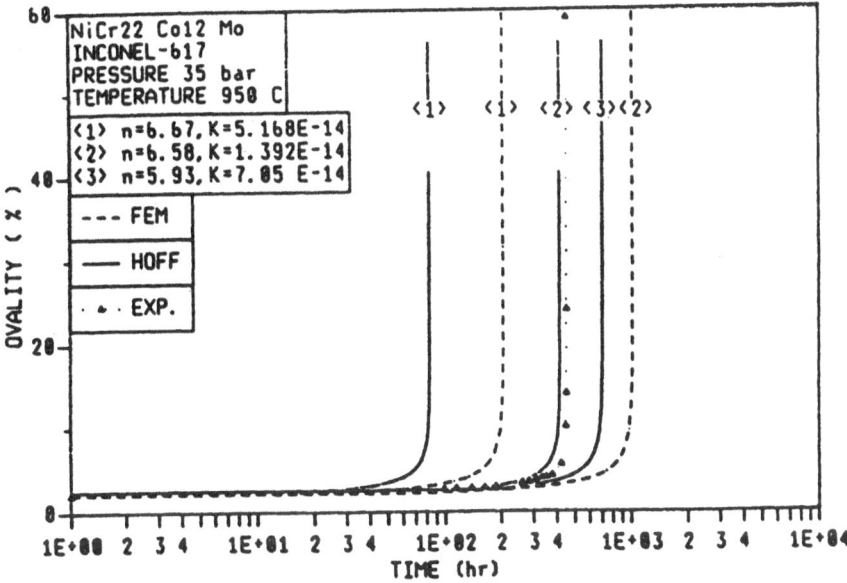

Fig. 5.   Comparison of a creep buckling curve with those
calculated by Hoff method and a finite element method,
taking the scatter of K and n values in Nortons creep
law into account (40).

In the "HTR-Auslegungskriterien" (22) it is therefore
proposed that the buckling collapse time be calculated with the
highest values for k and n.

6.2. Components with flaws

Nuclear components of light water reactors with flaws or
postulated flaws, according ASME Code Case Appendix G 10 and KTA
3101.3, may be treated by fracture mechanics methods for safety
assessment. In non-nuclear components, fracture mechanics
methods are used to indicate how long a component can withstand
stable crack growth before a critical crack situation leads to
spontaneous fracture. A design against fracture mechanics limit
factors is not in use for components operating at elevated or
high temperatures, e.g. above ca. 300°C for pressure vessel
steels.

For the use of PM-Superalloy discs for aircraft engines,
together with a new concept for determining the allowable
operational cycles, a fracture mechanics evaluation is under
consideration. For high temperature gas cooled reactors, as well
as for steam turbine rotors (41), basic work on high temperature

fracture mechanics is in progress (42).
In high temperature applications, both fatigue and creep crack
growth, as well as their synergistic influences are being
investigated. Until now, no commonly used concept for high
temperature has been agreed, therefore no structural design code
provides fracture mechanics for high temperature applications.

There are, however, some results indicating that for
austenitic steels and nickel-base superalloys, fatigue crack
growth can be correlated well with the stress intensity cyclic
amplitude ΔK_I (42) (Fig. 6).

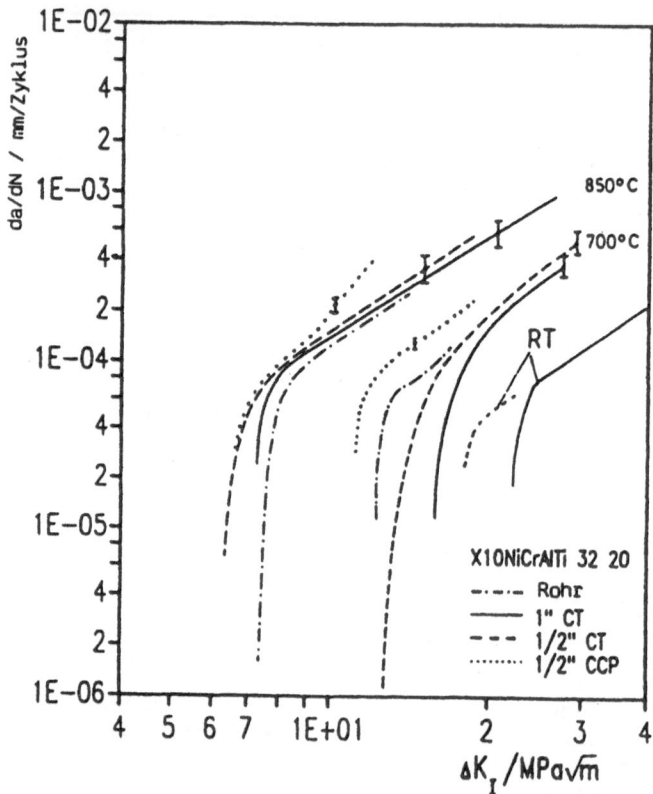

Fig. 6.　　da/dN (crack growth per cycle) vs. ΔK_I-curves for
　　　　　different specimen-types and circumferential notched
　　　　　tube (diameter 120 mm, wall thickness 10 mm) of
　　　　　X10NiCrAlTi 32 20 (42).

Fatigue crack growth at 700 and 800°C can be described by the
linear elastic ΔK_I concept. The assessment of components can be
done on the basis of the Paris law.

The situation concerning creep crack growth has not yet reached this clarification. The use of C*-concept is deemed possible. Nevertheless, a design of high temperature components with fracture mechanics values is far from becoming a commonly adapted tool.

7. FINAL REMARKS

The commonly used structural design codes and proposals for design rules for components exposed at temperatures where creep must be limited, define in the first step the allowable stresses and compare them with time dependent creep design values. This restriction is very successful if the loading is stress controlled (primary stress). The treatment of thermally induced, strain controlled fatigue loading using limitation of stresses does not give satisfactory results, as it leads to either over or underestimation of thermal transients. Therefore the trend is to the use of inelastic analysis methods from the very beginning of the design process. Time dependent inelastic analysis, however, is not only time and cost consuming, but has its own inherent problems. The problems are:

- Constitutive equations:

The constitutive equations should describe the three dimensional stress/strain behaviour. In the work for the HTR design rules beside the analytical approach given in (22), the Norton creep law in the three-dimensional formulation is used. Further, the applicability of the ORNL-model as well as the Interatom-model for HTR components is under consideration.

The problem with all three kinds of approach is the scatter of the materials constants in the mathematical equations. Further the treatment of plastic and creep deformation by the same constitutive equation brings difficulties, because the yield surface is also dependent of the strain which is in turn dependent upon the deformation rate. In the so called "unified models" a great number of materials parameters must be introduced, which render the use for inelastic analysis more difficult.

- Anisotropy in the semi-finished products:

Comparision of creep behaviour of specimens of rod material with axially loaded tube specimens shows the importance of the introduction of anisotropic behaviour in a component by the manufacturing procedure. This is even more important for those components which have a highly anisotropic structure such as unidirectionally solidified turbine blades or ODS-material.

- Weldments:

The joint itself is very difficult to treat by inelastic analysis due to differences in stress-strain behaviour within the different areas of the joint: filler metal, heat affected zone, and base metal. The limitation of the strain in the weldment reflects this fact.

- Flaws in the components:

Fracture mechanics analysis is required to demonstrate that within the next inspection period or the total service time, crack growth will not lead to a catastrophic fracture. High temperature fracture mechanics where both, fatigue and creep crack growth must be examined is at present under intensive examination.

- Gaps in the knowledge of expected loading history:

In addition to the uncertainties and scatter in material and product properties, the uncertainties in the loading history prevent an exact calculation and prediction of the real life-time of a component. The calculated stress distribution itself is dependent on the particular constitutive equation used.

The problems with the inelastic analysis of high temperature components mentioned above are the focus of current activities and efforts are being made to achieve a better understanding of the basis for design codes and for the evaluation of simplified methods. The future work must be carried out as a joint effort of specialists in metallurgy, materials mechanics and structural analysis.

ACKNOWLEDGEMENT:

The author would like to thank all participants of the German working group: " Fachkreis HTR-Auslegungskriterien", without whom this publication would not have been possible. Thanks are also due to Mr. P.J. Ennis for critical reading and correction of the manuscript.

8.   LITERATURE

(1) "Design of High Temperature Metallic Components", ed. Hurst, R.C., Elsevier applied Science Publishers, London, New York (1984)

629

(2) Bundesminister für Wirtschaft (BMW): Verordnung über Druckbehält, Druckgasgehälter und Füllanlagen (Drukbehälteverordnung - DruckbehV) und Allgemeine Verwaltungsvorschrift (fordert Einhaltung der TRB - /3/), BMW, Bonn, Febr. 1980

(3) Bundesminister für Arbied und Sozialordnung (BMA): Technische Regeln Druckbehälter (fordert Einhaltung der Anforderungen der AD-Merkblätter - /4/), Hauptverband der gewerblichen Berugsgenossenschaften e. V., Zentralstelle für Unfallverhütung und Arbeitsmedizin, Bonn, Stand 1985

(4) Arbeitsgemeinschaft Druckbehälter (AD): AD-Merkblätter, Vereinigung der Technischen Überwachungsvereine e. V., Essen, Stand 1985

(5) Bundesminister für Wirtschaft (BMW): Verordnung über Dampfkesselanlagen (Dampfkesselverordnung - DampfkV) und allgemeine Verwaltungsvorschrift (fordert Einhaltung der TRD - /6/), BMW, Bonn, Febr. 1980

(6) Deutscher Dampfkesselausschuss (DDA): Technische Regeln für Dampfkessel (TRD), Vereinigung der Technischen Überwachungsvereine e. V., Essen, Stand 1985

(7) American Society of Mechanical Engineers (ASME): ASME Boiler and Pressure Vessel Code, Section I, Rules for Construction of Power Boilers, ASME, New York, 1983 and Addenda 1985

(8) American Society of Mechanical Engineers (ASME): ASME Boiler and Pressure Vessel Code, Section III, Rules for Construction of Nuclear Power Plant Components, Division 1, ASME, New York, 1983 and Addenda 1985

(9) American Society of Mechanicel Engineers (ASME): ASME Boiler and Pressure Vessel Code, Section III, Rules for Construction of Pressure Vessels, Division 1, ASME, New York 1983 and Addenda 1985

(10) American Society of Mechanical Engineers (ASME): ASME Boiler and Pressure Vessel Code, Section VIII, Rules for Construction of Pressure Vessels, Division 2, New York, 1983 and Addenda 1985

(11) American Society of Mechanical Engineers (ASME): ASME Boiler and Pressure Vessel Code, 1983, Code Cases, Nuclear Components: Case N-47-21, Components in Elevated Temperature Service, Section III, Division 1, ASME, New York, 1981

(12) American Society of Mechanical Engineers (ASME): ASME Boiler and Pressure Vessel Code, 1983, Code Cases, Nuclear Components: Case N-201-1, Class CS, Components in Elevated Temperature ASME, New York, 1982

(13) Amercian Society of Mechanical Engineers (ASME): ASME Boiler and Pressure Vessel Code, 1983, Code Cases, Nuclear Components: Case N-253-2, Construction of Class 2 or Class 3 Components for Elevated Temperature Service, Section III, Division 1, ASME, New York, 1983

(14) American Society of Mechanical Engineers (ASME): ASME Boiler and Pressure Vessel Code, 1983, Code Cases, Nuclear Components: Case N-290-1, Expansion Joints in Class 1 Liquid Metal Piping, Section III, Division 1, ASME, New York, 1983

(15) American Society of Mechanical Engineers (ASME): ASME Boiler and Pressure Vessel Code, 1983, Code Cases, Boilers and Pressure Vessels, 1473-1, Short Time High Temperature service, Section VIII, Division 2, ASME, New York, 1981

(16) American Society of Mechanical Engineers (ASME): ASME Boiler and Pressure Vessel Code, 1983, Code Cases, Boilers and Pressure Vessels, 1489, Elevated Temperature Desing, Section VIII, Division 2, ASME, New York, 1981

(17) American Society of Mechanical Engineers (ASME): ASME Boiler and Pressure Vessel Code, 1983, Code Cases, Boilers and Pressure Vessels, 1933, UNS N10276 Alloy for Service Temperatures up to 1250°F, Section VIII, Division 1, ASME, New York, 1983

(18) American Petroleum Institute (API): API Recommended Practice 530: Recommended Practice for Calculation of Heater Tube Thickness in Petroleum Refineries, API, Washington, 1978

(19) Britisch Standards Institution (BSI): BS 5500, Specification for Unifired Fusion Weldes Pressure Vessels, BSI, London, 1985

(20) British Standards Institution (BSI): BS 3915, Carbon and Low Alloy Steel, Pressure Vessels for Primary Circuits of Nuclear Reactors, BSI, London, 1983

(21) "Erarbeitung von Grundlagen wu einem Regelwerk über die Auslegung von HTR-Koponenten für Anwendungstemperaturen oberhalb 800°C", Jül-Spez. März 1984

(22) "Auslegungstemperaturen für hochtemeraturbelastete metallische sowie des Spannbeton-Raktordruckbehälter zukünftiger HTR-Anlagen", German Working Group, Statusbericht, Jül-Spez-347, Febr. 1986

(23) British Standards Institution (1975), BS 806 Specification for Ferrous Piper and Piping Installation for and in Connection with large Boilers

(24) "High Temperature Components Design in Chemical/ Petrochemical Industry", P. H. G. Holl, in (1)

(25) "Design and Construction Rules for Mechanical Components of FBR Nuclear Islands", AFCEN: RCC-MR (1985)

(26) "Status of Structural Design Code for Metallic High Temperature Gas Cooled Reactor Components", Nickel, H. and Schubert, F., in Special Issue on High Temperature Gas Cooled Reactor Materials, Nuclear Technology (1984)

(27) "Entwicklung eines strukturabhängigen Werkstoffmodells für komplexe Hochtemperaturbeanspruchungen", Information obtained within the Research Programme, (begin: April 1987) by BBC, KWU, MTU and others

(28) "Constitutive Equations for the Description of Creep and Creep Rupture Behaviour of Metallic Mateirals at Temperatures Above 800°C", Penkalla, H.J., Over, H.H. and Schubert, F., aus /32/, S. 685 (1984)

(29) "Special Issue on High Temperature Gas Cooled Reactor Materials", Nuclear Technology, ed. Post, R.G., Wirtz, K., Nickel, H., Rittenhouse, P.L. and Kondo, T.

(30) "Life Time Prediction Models", Fischmeister, H.F., Danzer, R. and Buchmayer, B., COST 501, Liège 1986

(31) "The Development and Application of SRP as a Tool in the Treatment of high Temperature Metal Fatigue", Manson, S.S., AydRD CP-24 (1978)

(32) "Méchaniques des materiaux Solides", Lemaitre, J. and Chaboche, J.L., Dunod, 1985

(33) "Life-time and Creep Ratcheting Calculation of Two Typical HTR-components", Bieniussa, K., Breitbach, G., Over, H.H., Penkalla, H.J., Schubert, F. and Seehafer, H.J.

2nd int. seminar on "Standard and Structural Analysis in Elevated Temperature Application for Reactor Technology, Venice, Oct, 1986

(34) "Phenomenological Consideration on Creep and Fatigue at High Temperature Application to the Life Time Calculation of HTR-Components" Over, H.H., Penkalla, H.J., Breitbach, G. and Bieniussa, K., SMIRT-Post Conference (Paris), 1985

(35) "Necking and Rupture of Rods Subjected to Constant Tensile loads", Hoff, J. Appl. Mech. 20, 105, 1959

(36) "Experimentally Verified Creep Ratcheting Analysis", Breitbach, G., Over, H.H. and Schubert, F., Post-SMIRT-Conference, Paris (1987)

(37) "Influence of Cyclic Temperature Transients on Creep Behaviour of Heat Exhanger Tubes", Zottmaier, R., Over, H.H., Schubert, F. and Nickel, H., Nucl. Eng. and Design (1986)

(38) "Elastic-Plastic Behaviour of Thin Tubes Subjected to Internal Pressure and Intermittent High Heat Fluxes with Application to Fast Nuclear Reactor Fuel Elements" Bree, J., J. of Strain Analysis, Vol. 2, No. 3., S. 266, 1967

(39) Trans. 6th Int. Conf. on Structural Mechanics in Reactor Technology, O'Donnell, W.J. and Porowski, J.S., Paris, 1981

(40) "Experimental and Theoretical Investigations of Creep Buckling on NiCr22Co12Mo Tubes at 950°C", Schubert, F., Ahmed, K., Breitbach, G. and Nickel, H., 13. MPA-Seminar, Stuttgart 1987

(41) "Evaluation of the Creep Crack Growth by Means of a Crack Tip/Far Field Concept", Ewald, J., Maile, K. and Tscheuschner, R., 13. MPA-Seminar, Stuttgart 1987

(42) "Fatigue and Creep Crack Growth in Methane Reformer Tubes at Temperatures above 700°C", Rödig, M., Kienzler, R., Nickel, H. and Schubert, F., 13. MPA-Seminar, Stuttgart 1987

6. FUTURE TRENDS IN HIGH TEMPERATURE TECHNOLOGY

AND THE IMPLICATIONS FOR MATERIALS R & D

Dr. T.B. Gibbons

N.P.L., Div. of Materials Applications

Teddington, Middlesex, U.K.

INTRODUCTION

The rapid development of metallic materials for high
temperature applications that occurred during the 1960's and
early '70s has given way in the last decade or so to a
consolidation phase with the emphasis on optimisation and
detailed improvement in alloys. Innovative developments in
materials processing technology (melting, casting, fabrication)
has enabled the upward trend in temperature capability and load-
carrying capacity to be maintained and considerable scope for
further developments in this area remains through the use of
computer-aided techniques. Also there has been an increase in
the effort directed towards a better understanding of the
behaviour of alloys in service as a basis for more effective
usage and more efficient design and this activity will
undoubtedly develop and intensify in the decade ahead.

Consistent with this shift of emphasis away from alloy
development and particularly with regard to materials for gas
turbine applications, the alloy user rather than the producer
has provided the main driving force for improvement. A good
illustration of this trend has been the establishment of
directional solidification technology for gas turbine blading, a
process invented by an engine manufacturer and subsequently
licensed to various investment casting plants. The necessary
alterations in the chemical composition to fine-tune the most
promising alloys to optimise the performance in the directionally
solidified form was also led by the aero-engine producers. In
this context it is generally agreed that the days of alloy
development in the hope that a user will be found, are over. The

costs involved are simply too great for any one company to risk a totally speculative investment of this kind. As a result there is a greater tendency nowadays for groups of companies, viz. alloy producer, fabricator, potential user, to work collaboratively towards a new development with goals and target parameters provided by the potential user. This is more efficient in principle and avoids the necessity for any one organisation to carry all the risks. An important consequence of this type of evolution is that the alloy producers have lost much of the pre-eminent position held some years ago. Several renowned research centres have closed down or been greatly reduced and in national and international committees concerned with aspects of materials usage (eg standards committees etc.) the voice of the user is frequently heard more loudly. Furthermore there is an increasing tendency for collaboration in pre-competitive research and most organisations now recognise the value of working together with appropriate partners. The best example of the new spirit of collaboration which has developed in the last fifteen years or so, at least on the European scene, is the COST 500 series of projects, dealing with a wide range of materials problems and involving alloy manufacturers and users in closely integrated research programmes. A critical factor has been the recognition that work-sharing through a well structured collaboration can result in a reduction in research costs and large amounts of data can be collected more efficiently and in a considerably shorter timescale than would be possible if individual organisations worked in isolation.

In the next ten years or so it is highly probable that these trends will develop to meet the new challenges faced by the materials engineer working in the power engineering industry and elsewhere on the design, operation and maintenance of high temperature plant. Developments will be paced by the needs of the power engineering industry to reach new levels of efficiency by operating at higher temperatures and pressures with no loss in durability or availability of individual power generating units and by the requirement to meet the more stringent constraints on environmental pollution. One of the key issues in this time-frame will be assessment of the realistic potential for the reliable use of ceramic materials for critical high temperature components.

In the following sections some of the likely developments in the immediate future are considered in the context of trends in the evolution of the relevant technologies. The potential for progress in the field of information technology and computer manipulation of data and in the important area of improved standards for materials property measurement are also discussed. Some ideas on developments in the organisation and management of research are outlined.

DEVELOPMENT DIRECTIONS IN POWER ENGINEERING TECHNOLOGY

Since the oil crisis of the 70's the main driving force for development in power engineering has been associated with the need for greater efficiency in the use of oil and other fossil fuels. Aero-engines in particular have been developed with improved specific fuel consumption as a main objective and sold with guarantees on levels of fuel economy. With the stabilisation of oil prices, concern with fuel economy has eased but has been replaced by other concerns such as durability and availability of plant and equipment and improved emission control, particularly in power producing installations, to minimise pollution and protect the quality of life in industrial environments.

Currently aero-engines, even for military applications, are supplied against a requirement that specifies guaranteed lifetimes for individual components. This has focussed attention on the need for improved design procedures, particularly for non-isotropic materials, greater reliability in coatings both as a protection against corrosion attack, and as thermal barriers, and a more highly developed NDE technology based on condition monitoring. Industrial engine builders are increasingly adopting the technologies of the aero-space application as illustrated by the introduction of directionally solidified blades and the use of high strength Ni-base alloys for discs. Also the wider use of coatings to protect against corrosion attack and thermal barriers to reduce maximum temperature levels in critical areas can be anticipated as industrial developments for the 1990's.

For power generation equipment the emphasis will be on emission control, high levels of availability in a flexible operating environment (eg two-shifting etc.), and a construction policy which minimises capital investment. The need for improved emission control will sustain a strong interest in novel coal-burning technologies particularly those involving fluidised bed combustion and gasification. Current pilot and demonstrator systems operating in various parts of the world have confirmed the technical validity of the concepts involved. However the operating environments, for heat exchangers for example, are particularly aggressive and pose challenging problems for the materials engineer in the context of the long-term reliability and overall availability of these systems. In conventional power plant the trend in development will be towards higher steam temperatures, to give increased efficiency along with improvements in combustion systems (eg low NO_x burners) and flue gas desulphurisation (FGD) for emission control. However some of the more promising steels for application in higher steam temperatures may be limited in terms of available ingot size for critical components such as intermediate pressure rotors but this may not necessarily be a disadvantage. Somewhat smaller generating units will offer flexibility in total system response to variability in demand as well as lower capital costs.

Internal combustion engines will continue to develop towards greater power outputs per unit of swept volume, reduced weight and increased thermal efficiency to reduce fuel consumption. The selective use of ceramic materials both monolithic and in the form of coatings for critical parts will continue to give benefit in reduced heat loss, lower weight and increased wear resistance. In this context it is interesting to note that the main advantage in the use of ceramics for turbo-chargers by Japanese automobile manufacturers lies in the reduced mass of the ceramic component and hence the reduced inertial forces, rather than in any advantage in temperature capability. The more traditional materials for diesel engines, eg cast irons, have changed little even though power output from individual units has increased significantly. As a result, thermal fatigue problems have arisen in certain high performance engines and consequently some development in cast irons can be expected with the aim of providing improved high temperature capability. Novel surface treatment techniques involving for example the use of laser methods, will be used to increase the durability of critical parts such as valve seats and facings.

A key factor underlying these developments towards more efficient power generating systems will be the availability of improved design methodologies which take more account of the detailed effects of the service environment, ie stress, temperature and chemistry of the working fluid, on component performance. The closer relationships that have developed in the last decade between design and materials engineers will encourage the evolution of less-conservative design procedures. In many cases these will be derived from an improved mechanistic understanding of materials behaviour in service conditions and will be assisted by a better knowledge of stress distributions in complex components, which will become available both by the application of computer-based analysis techniques and from the monitoring of component performance in service. Currently few design procedures incorporate any allowance for the effect of attack by the chemical environment on the behaviour of components, in for example, combustion atmospheres. However, progress in understanding the complex interactions which occur between deformation and fracture mechanisms and metal wastage processes will provide a basis for the introduction of information of this type in the prediction of component behaviour. An important factor in the development of improved design methodologies will be the availability of facilities for the testing of model components to assess the validity of novel approaches to design. Facilities of this type are not widely

available and some consideration should be given to the future
needs in this area within Europe.

INFORMATION TECHNOLOGY AND DATA PROCESSING

Rapid developments in the exploitation of computer based
systems, giving access to databanks, for design applications and
in process modelling and control, will have important
implications for the high temperature materials industry in the
next 5 - 10 years. Hitherto the main medium of communication in
materials engineering has been by publication of papers in the
technical press, but this has expanded to such an extent that few
can maintain the necessary level of awareness in more than
selected areas of individual specialisation. Improvements in
electronic communication will encourage an increased tendency for
data transfer by computer between collaborating groups, as an
alternative to publication of data in the scientific press for
general dissemination. In this way databanks of carefully
selected and validated information will become available for
specialised applications in component design, lifetime prediction
in complex plant, process control etc. Larger, more widely
accessible databanks will also be assembled and will provide a
basis for alloy selection and for detailed intercomparison of
candidate materials for specific applications. The extent to
which units from these larger databanks will provide input to
more specialised systems will depend on whether the "pedigree" of
the data can be established as a basis for user confidence.

A related activity concerns the use of computer based
modelling both to predict material performance in service
conditions and in the development of "right first time"
processing of high value components such as forgings and castings
for the aerospace industry.

In the prediction of the performance of advanced materials,
well validated databases will be used, which will be constructed
with the specific end-use in mind, and the controlling factor in
the successful exploitation of systems of this type may prove to
be the availability of reliable and relevant constitutive laws to
represent the material behaviour adequately. In a well-
integrated materials engineering system the databases will
interact with the design methodologies described in the previous
section. Computer techniques will also be used in the estimation
of remanent life of plant where the development of a "global" and
a "local" strategy can be anticipated. The global approach will
involve a "real-time" computer system where temperatures and
stresses will be logged continuously at critical locations in the
plant and a continual assessment of remanent life will be
provided using, in the simplest form, a Miner's Rule type of
approach or one of the more advanced techniques, such as the

CRISPEN system now being developed at NPL. The local approach will become important when a given percentage of life has been used up and will involve a detailed assessment of the material condition at points such as welds, bends, etc. where, from stress and temperature considerations, the most damaging situations are expected. At these points, replication techniques will be used or specimens of material will be removed and tested destructively to give a more precise picture of the extent of damage. Systems like these are already being used in chemical and other types of plant and new installations will be designed to operate on the basis of this type of "condition monitoring" procedure.

In the use of computer based modelling for materials processing, software packages already exist for forging and casting of aerospace parts but there is an acknowledged lack of good input data. Thus, in the modelling of, for example, forging processes, data on thermal conductivity is a critical requirement but this type of information is not readily available in the temperature range of interest and for alloys of commercial relevance. Similarly a shortage of information on the flow characteristics of liquid metal presents a limitation on the accuracy of the modelling of casting processes. However, the main challenge lies in the ability to incorporate data on properties and microstructure into this type of approach so that processing parameters can be controlled to give the desired microstructure in the finished component. Progress has undoubtedly been made in this direction, particularly in the forging of aerospace components but further significant headway will be made in the next 5 - 10 years.

STANDARDS AND PRENORMATIVE RESEARCH

The need for more efficient, highly cost-effective designs in modern plant, the revolution in design procedures brought about by the introduction of computer-based methods, and the greater awareness of the operating conditions experienced by components in service, have resulted in a growing demand for better and increasingly sophisticated measurement methods to assess materials performance. These developments have implications for future trends in the standardisation process as applied to materials measurement methods and specifications and for the evolution of accreditation systems for test laboratories. Some aspects of these issues will be briefly considered in this section.

The importance of documentary standards, particularly those relating to the measurement of materials performance to provide input data for the modern design methodologies has focussed attention on the need for a new route to standardisation.

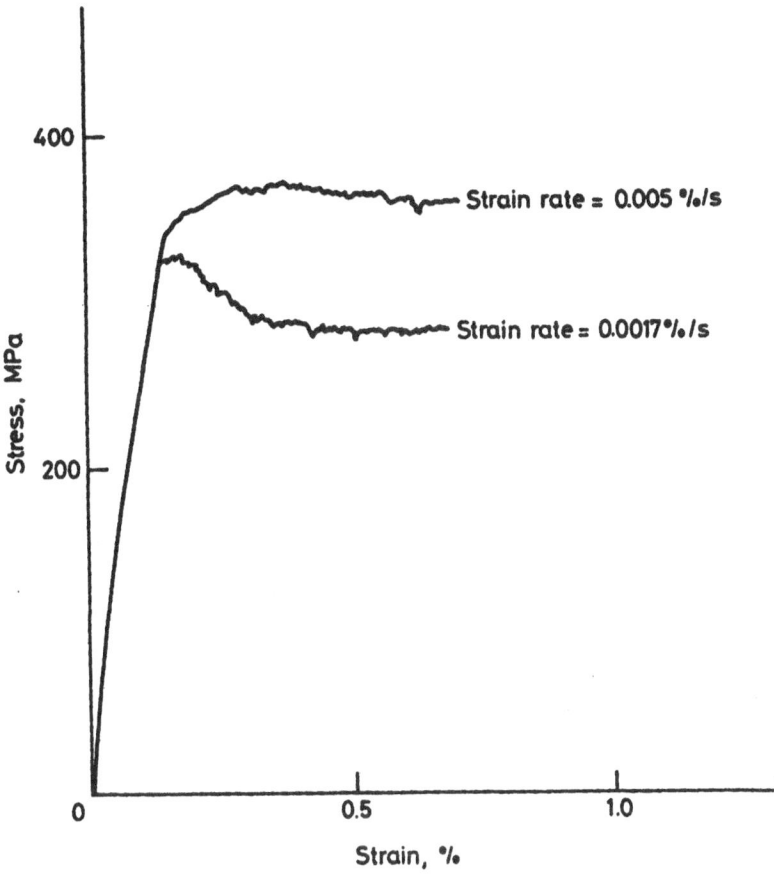

Fig. 1   Stress-strain curve for Nimonic 75 at 725°C showing 30%
         difference in 0.2% "proof stress" for the range of strain
         rate permitted by BS 3688.

Hitherto national standards for key performance measures such as tensile strength, creep strength, etc. have been produced largely on the basis of consensus in national and international standards committees and Fig. 1 illustrates one of the problems that can result. Intercomparison of the various national standards in these areas shows certain anomalies and detailed differences and harmonisation at the international level tends inevitably to lead to adoption of the "lowest common denominator". This is clearly an unsatisfactory situation and the aim of harmonisation ought to be to raise the general level of standards thereby contributing to increased product quality.

For high technology industries in a fiercely competitive international environment it is recognised that improved methods of measuring materials performance are required which will relate more precisely to the real operating conditions and which will be based on a knowledge of the science of materials behaviour. The key to this more rational route to the development of materials measurement procedures lies in the provision of a reliable database of information on materials behaviour which can be used to set appropriate limits of control on critical test parameters. Research programmes aimed at providing the technical information from which new and improved measurement methods will be developed, ie prenormative research, will become an increasingly important feature of the generic research activity in Europe. The availability of research information from other sources in national and international programmes will provide further input to the prenormative database.

However for this type of approach to succeed new structures will have to be established which will encourage the transfer of technical information into the sphere of standardisation. Appropriate technical committees, possibly along the lines of The American Society for Testing and Materials (ASTM) or the Japanese Industrial Standards Committee, will have to be established to define and satisfy the standardisation needs for new and emerging technologies in materials development and usage such as engineering ceramics. The aim will be to try to avoid many of the difficulties that occurred in the area of metallic materials, particularly with steels, by encouraging collaboration between users and producers at an early stage and by providing the technical underpinning to support the standards-making process. A model for the type of structure required is provided by the High Temperature Materials Testing Committee (HTMTC) in the UK which acts as a pressure group to produce state-of-the-art publications and codes of practice for critical applications in high temperature materials. These codes of practice are produced according to the requirements of the national standards body and, until fully accepted, serve as pre-standards which can be used with confidence by materials suppliers and users. The

international collaborative Project VAMAS (Versailles Agreement
on Advanced Materials and Standards) performs a similar function
for a selected number of material topics within the major OECD
countries.

The development of the type of structure envisaged here,
along with the appropriate technical back-up, will be an
important contribution to an improvement in the infrastructure
for materials property measurement and an invaluable asset in
Europe in the context of the Single Market requirements for
harmonisation of standards.

The availability of appropriate documentary standards is the
key to a successful accreditation system for test laboratories
since accreditation requires the demonstration of an ability to
comply with the requirements of standard procedures when carrying
out performance testing of materials. The demands of design
engineers for better quality data (reduced scatter bands) and
also the need to specify the pedigree of data for inclusion in
materials databanks has encouraged a positive response to
accreditation and some effort to improve on existing procedures
can be expected in the next few years. Currently accreditation
schemes are essentially passive and largely involve demonstration
of compliance with standard procedures, traceability of
calibration and an appropriate system of management and
accountability. Developments from this position are likely to
involve a more active type of evaluation to confirm that
measurements can be made to a given level of reliability and
accuracy and will encourage the introduction of proficiency
testing. Proficiency testing implies a source of reference
materials and in this regard, the activities of the Community
Bureau of Reference (BCR) to support the development of reference
materials for mechanical testing will, by providing a supply of
transfer standards, encourage improvement and harmonisation of
test procedures.

RESEARCH ORGANISATION AND MANAGEMENT

The rapid growth in international collaboration has been the
most significant organisational development in materials research
in Europe in the last decade. The flourishing COST projects
which have grown from the original COST 50 on materials for gas
turbines to the seven major Projects in the 500 series covering a
broad range of materials topics (Table 1) exemplify the
successful development of the principles of concerted action.
National and international collaboration within Europe has
developed as a result of the widespread recognition that few
organisations have in-house the broad range of skills, expertise
and equipment to tackle some of the highly complex problems of
materials behaviour which must be solved in order to improve and

advance current technologies. Reductions in research budgets and limited recruitment of skilled staff in many European research centres have been additional factors in influencing the growth of collaborative ventures.

TABLE 1   COST concerted-action projects in materials

Project No.	Title
501	Advanced materials for Power Engineering Components
502	Corrosion in the Construction Industry
503	Powder Metallurgy
504	Advanced Casting and Solidification Technology
505	Materials for Steam Turbines
506	Industrial Application of Light Alloys
507	Thermochemical Data for Light Alloys

The great advantage of concerted-action is that the collaboration proceeds on the basis of work-sharing with appropriate safeguards to ensure equal contributions from each participant and no funds are transferred across national frontiers. Hitherto this arrangement has worked well in the general area of pre-competitive research but extension of the principle to the evaluation of the performance of model components and to demonstrator projects to confirm the validity of advanced concepts is desirable and may well become feasible in the next few years, provided suitable funding arrangements can be developed. This approach will apply especially when it is recognised that the major competitors lie outside Europe and also where the collaborating organisations involve materials supplier, component manufacturer and plant user, along with perhaps specialised research centres. "Vertical" integration of this type avoids potential conflicts of interest between competing companies. However a strong formalised management structure will be required to ensure success rather than the largely volunteer effort typical of the present COST Projects on materials research.

In parallel with these developments in concerted-action activities there has been a very substantial growth in the cost-shared materials research programmes funded by the European Commission. These have increased both in terms of the scope of the individual programmes and of the amount of funding available to support research effort in Community countries. Thus the new BRITE-EURAM Project due to begin in early 1989 will have a budget of ca. 420 Mecu available over the four year time span envisaged for the activity.

These various research initiatives along with the programmes of the laboratories of the Joint Research Centre will undoubtedly contribute to a strengthening of the technical base for materials in Europe. However there is also little doubt that, without improved coordination between the various agencies involved, the full overall benefit to European industry will not be realised.

Indeed the challenge for research management in Europe in the years to 1992 and beyond will be to establish a coherent strategic plan for materials research, set in the context of an overall strategy for the future direction of European high-technology industry to exploit and develop market opportunities in the face of overseas competition. An agreed industrial strategy would provide the necessary framework for the definition of priority requirements in materials technology which would form the basis of collaborative research programmes, including demonstrator projects, targetted to the needs of key industrial sectors. Such a complex and ambitious task will not be readily accomplished but must be tackled if the full potential of European technical innovation is to be realised. Success will require the participation of individuals with a high degree of technical competence, vision and commitment and free from the constraints of limited-horizons and parochial self-interest imposed by national and international bureaucracies. Models for an approach of this type already exist in the national collaborative activities initiated for advanced technologies in Japan and the USA.

It is gratifying to note that the European Aerospace Industry which has long experience of collaborative projects and a track record of technical and commercial success, has recently undertaken a strategic assessment of this type in the Euromart Study on aeronautical research and technology in Europe.

7. Conclusions

The conclusions and recommendations for future R&D priorities summarised in this chapter have been drawn from the individual authors contributions and from the analysis of the editors of the individual subject area, which is based upon all of the contributions to the book, pertinent to the particular materials area, and from their own expertise.

The conclusions retain the order and numbering of the authors contributions, to facilitate cross-referencing.

7. CONCLUSIONS

2. MATERIALS POTENTIAL

2.1. Alloys

2.1.1. High Temperature Steels for Power Plant

There are a number of R&D programmes in Europe which should continue to be supported with the primary aim of giving improved critical components such as super heater and reheater piping exposed to products of combustion, headers and tubing, turbine casings, rotors and bolting, and welds for conventional electricity generating plant. The benefits sought are higher steam pressures and temperatures, thinner sections to reduce thermal stresses, greater reliability and longer plant life. Highlighted areas include:

For low alloy ferritic steels; to find the tolerance levels for embrittling impurities and develop refining techniques to mass produce pure steels. Trials will be needed for high purity super clean LP rotors of NiCrMoV steel.
For 9 to 12 Cr steels; to establish T91, explore its long term properties, weld properties including susceptibility to type IV cracking, develop production techniques and demonstrate components in service. To develop modified 12CrMoV rotor steels and demonstrate in-service performance within five years in Europe and to develop stronger 9-12 Cr steels for the longer term and do plant trials.

For austenitic steels. To develop stronger alloys in case the stronger ferritics prove unsatisfactory. To develop high chromium austenitics for very high temperature corrosive environments for superheaters and reheaters with full in-service trials, and to evaluate the properties of weldments and transition joints in all developed materials. The stronger ferritics will present particularly severe welding problems.

2.1.2. High Temperature Steels for Petrochemical Processes

Alloy tubes with compositions in the range 20-28 Cr, 20-48 Ni, 0.4C wt% have been available to the petrochemical industry since 1950. They rely for strength on a distribution of fine stable carbides. Improvements in creep strength have been provided in later generations of these alloys by adding niobium or tungsten to give primary and secondary carbides more resistant to coarsening. Lower carbon versions are also provided for parts subject to bending stresses. More recently small additions of titanium, zirconium and certain rare earth elements have been added to give further increase in strength and in-service ductility. None of these alloys can be hot worked and so they are used in the cast condition. Straight tubes are spun cast, and can in principle be made as duplex tubes. There are immediate needs for design codes, standards, specifications and testing method validation. These do not exist at present, and for their generation would require an impartial data bank of accredited materials properties. This could speed up the introduction of new materials.

Current design procedures are not related to likely modes of failure or to actual lives in plant. Very long term steady state creep data are of little relevance to the conditions of exposure in typical plant. Well instrumented industrially relevant test facilities are required particularly if there is any increased emphasis placed on plant reliability. It is also timely to set up a joint programme on coke deposition.

2.1.3. High Temperature, Oxidation Resistant FeCrAl Alloys

The FeCrAl range of ferritic steels has the highest survival temperature in oxidising, sulphidising and intermittent reducing atmospheres of all commercial alloys. They have very poor creep strength, however, and hence are restricted to low stress applications. Further R&D is required to produce a solid solution strengthened or precipitation hardened series of these alloys perhaps by addition of niobium or molybdenum. In the very high temperature range the most likely method for strengthening has to be the use of oxide dispersions. Composites using alumina are available but their price is prohibitive for many applications.

The FeCrAl alloys require careful grain size control to avoid high brittle-ductile transition temperatures. Since the steels are ferritic up to the melting point, the use of phase transformations to refine the grain size is not an option. Also as with other ferritic high chromium steels, they suffer from 475°C embrittlement a phenomenon which is still not fully understood. Further research on the physical metallurgy of these potentially useful alloys and their response to process variables is clearly required.

A small yttrium addition has many useful functions but of particular value are its contributions to grain size control and to the adhesion of protective oxide films. Other rare earth additions may give further improvements. Some exploratory work is required here.

2.1.4. Superalloys base: Nickel, Cobalt, Iron and Chromium

Conventional superalloys as pioneered by the aerospace

industry for use as jet engine components have probably reached the end of their development potential.

Future developments of materials capabilities will therefore come from ODS alloys, and metal matrix and ceramic matrix composites. These are described in following separate sub-sections.

Another range of improved materials properties will come from new and improved processing techniques to give freedom from flaws and more refined structures. For discs in particular, the preparation of flaw free components with a high resistance to crack propagation is a priority item, coupled with life prediction techniques. Composite blade and disc technology will involve much R&D on joining techniques.

The use of corrosion resistant coatings and thermal barrier coatings will become more widespread and require much research effort.

2.1.5. Oxide dispersion Strengthened Alloys

The potential of the ODS concept is by no means fully realised. Application to nickel-base alloys has been commercially successful, using recently developed mechanical-alloying (MA) processing. It has taken a long time to develop and guarantee the exact complex micro structure needed to give competitive properties. It has been necessary to include a very strong solid solution matrix for strength at intermediate temperatures, an elongated grain structure in the tensile stress direction and decorated narrow grains tranverse to the stress axis, to provide crack arresting barriers, in addition to the fine dispersion of insoluble particles to give high temperature strength.

There is considerable scope for generic alloy development to improve properties and extend the strengthening mechanisms to other systems (e.g. titanium). Fabrication and joining procedures will present major development targets.

2.1.6. Refractory Metals and Alloys

The BCC refractory metals, with melting points of 2500°C plus offer the potential of many applications at high temperature. The metals Nb and Mo for example show high specific strengths at high temperatures making them attractive candidate turbine materials for use above the limits set for nickel-base superalloys.

There is a wide range of applications of refractory BCCs at

very high temperatures in non-oxidising atmospheres including
certain molten salts and glasses. It is important that engineers
should know more about these metals. However, the exploitation
of these materials at high temperature in oxidising environments
cannot be considered until a solution has been found to their two
major drawbacks:

1. a very high reactivity with oxygen even at intermediate
   temperatures
2. a tendency to embrittle by pick-up of hydrogen, nitrogen and
   oxygen from the air.

Coating techniques have long been tested without success.
The chance of finding a universal solution to these problems is
very small, but by concentrating on one specific application, the
likelihood of success is greatly increased.

There is serious interest in exploiting the high strength to
weight ratio of niobium as a disk material for use at
intermediate temperatures. This will require much R&D effort to
realise early hopes.

2.1.7. Titanium

Titanium as a relatively recently exploited material offers
considerable scope for further development. While the major
volume applications will be at lower temperatures, the metal
shows considerable application as a turbine material. The
following specific developments are likely to repay the costs of
R&D in Europe:

- development of high quality complex castings for aero-engine
  components, particularly hollow turbine blades.
- development of effective coatings to resist oxidation, and
  pick-up of oxygen, nitrogen and hydrogen during long term
  exposure to high temperatures.

Much longer term developments include:

- development of improved structure, by rapid solidification
  powder metallurgy, oxide or fibre dispersion etc. to improve
  mechanical properties at high temperature provided the
  oxidation problem is first solved.

In view of the importance of aero engine developments in
Europe there is a parallel need to keep at the forefront of
materials development for that application. Because development
costs of a new product or process are very high, collaboration on
a European scale could be of great benefit in allowing cost
sharing and in speeding up the rate of progress.

2.1.8. Platinum Group Metals

The high cost of the Platinum Group Metals, restricts their use to areas where their properties are unique and essential for the application. These applications divide broadly into two areas using the properties of PGMs i) to act as catalysts and ii) to show outstanding oxidation resistance at very high temperatures either singly or alloyed with other PGMs, combined with high strength particularly when dispersed with zirconia (ZGS alloys).

General problem areas restricting the development of the metals and warranting research priority are:-

- purity. The metals are strongly susceptible to contamination influencing especially, workability, ductility, strength and structural properties. In particular silicon contamination of platinum leads to formation of intergranular low melting silicides. Improvement of purification processes, and understanding of interaction with containment materials will benefit both properties and fabricability.
- diffusion across interfaces at very high temperature seriously limits the operating life of PGMs in components. The development of diffusion barriers especially for the couples:-
      PGM/refractory metal
      PGM/base metal
      PGM/ceramic
  will greatly improve the economic value of PGM applications.
- joining technologies. Diffusion across joints at high temperature leads to Kirkendall voiding. Research into matching of filler materials and into novel welding processes e.g. laser welding, would repay costs.

2.1.9. Intermetallic Compounds for High Temperature Use

Research began over 20 years ago on intermetallic compounds in the hope of developing materials to supercede the conventional superalloys particularly for use in high temperature aerospace applications.

Compounds based on Ni_3Al and NiAl have been investigated most thoroughly but even the properties hoped for by further development have been overtaken by those of the latest conventional superalloys. Fundamental difficulties come from their inherent brittleness and from difficulty of fabrication to complex shapes. Service at the highest temperatures of a jet engine does not look very hopeful at present, but some basic work is continuing.

More recently intermetallics based on Ti_3Al have attracted

interest for high modulus/weight ratios combined with oxidation
resistance. They might compete with titanium alloys in the range
400° to 700°C for low stress applications. Here again
improvements in ductility and fracture toughness are essential
for their success, without sacrificing their attractive
properties. Experience indicates that to be a very difficult
task.

Some of the more refractory intermetallics are sufficiently
brittle to be compared directly with the more advanced refractory
engineering ceramics. While some ceramics have lower density and
better resistance to oxidation and creep, the intermetallics can
offer better thermal shock resistance, but whether they could
avoid catastrophic damage from ingested debris during cold start
up remains a critical question. They might however find useful
application as components of composites.

Control of composition and addition of boron have both been
found to give stronger grain boundaries, and very recently
published work has indicated the importance of choosing
compositions which freeze in the disordered state and then become
ordered by a solid state reaction. The resulting grain structure
contains stacking faults which may confer additional toughness.
There is clearly scope for basic work on long range ordered
alloys.

2.2. Coatings

2.2.1. Claddings and Co-extruded Tubes

The technology for the development of duplex tubes where the
one component of the tube is chosen mainly for its high creep
strength while the other mainly for corrosion resistance is well
established. Such tubes have already been in service for up to
20 years.

The next major steps in the successful development of co-
extruded tube materials in Europe will be driven by the
requirement of advance cycle steam plant for electricity
generation and will require:

- the optimisation of the composite billet production route for
  selected combinations of alloys to improve yield and thereby
  reduce unit costs.

- the corrosion assessment, including plant trials, of improved
  alloys with chromium contents in the range 30 % - 40 % and
  nickel of 30 % - 50 % balance iron.

- the determination of long term degradation modes of various

materials combinations and the validation of two stage
composite welding processes.

- establishment of new applications of co-extruded tubes over a
  wider range of industrial processes. A wider study would
  identify further materials combinations with potential for
  commercial exploitation but requiring further R&D prior to
  large scale applications.

2.2.2. Coatings for Turbine Applications

The requirements for surface coatings applied to strong
substrate alloys to enhance resistance to environmental attack,
become increasingly more demanding as operating temperatures go
up and fuels and atmospheres become more aggressive. Also
problems arise at lower temperatures for turbines likely to
ingest seawater or burn low grade fuel. There are a number of
areas of importance for R&D where there is substantial scope for
improvement.

Metallic Coatings

For metallic coatings to give added oxidation resistance it
is essential to supplement the largely empirical developments by
fundamental studies of scale growth in realistic environments
particularly on aluminium rich coatings as influenced by small
additions of other elements (Pt, Si, Y etc.).

Existing coatings reduce the resistance of components to
thermal fatigue crack initiation, the damage mechanism must be
studied and a method of prevention found. In spite of the wide
range of coating techniques available on a laboratory scale,
there is need for industrial scale development of the better
alternatives.

More advanced titanium alloys already have mechanical
properties which fit them for service at temperatures well above
those where pick up of oxygen becomes unacceptable. Finding a
satisfactory coating presents particular difficulties but the
rewards are potentially very great.

For next generation gas turbines entirely new materials will
have to be used for the new higher temperature regions. Many of
these will rely on coatings, the best example being graphite
fibre-carbon matrix composites which have zero resistance to
oxidation above quite modest temperatures. For graphite
protection it is necessary to prevent internal oxidation using
phosphates or boron, to provide a protective surface film and use
a viscous glassy material to seal the cracks in the coating.

Ceramic coatings

Thermal barrier coatings currently represent the best opportunity for increasing engine performance and durability in the near-term future. Following a rapid and rather empirical development, the primary need is now to consolidate experience and to build up fundamental information on which development can be based. Essential to this goal will be:

- the understanding of coating degradation modes, (rôle of microstructure, microcracking, porosity, residual stresses, oxidation of bond layer, ceramic/metal interfaces etc.).

- modelling of the composite, substrate/bond layer/coating, in order to provide guidelines to optimise coating formulation and deposition. The provision of a test rig in Europe would allow testing under service conditions, to give realistic evaluation of newly developed barrier coatings.

In addition there is enormous scope for extending the application range of these coatings to more corrosive environments (molten salts, contamined fuels/atmospheres) and to higher temperatures, for which it will be necessary to identify new coating compositions and new processes for deposition, especially for complex geometries.

Coating techniques developed here have many potential applications in other industries.

2.3. Composites

2.3.1. Metal Matrix Composites MMCs

The MMC materials are made up primarily from tough alloy matrices strengthened by:-
a. equiaxed refractory particles (e.g. tungsten carbide cutting tips)
b. short fibre whisker refractories (e.e. SiC in Al or Ti)
c. long parallel fibres (including metal wires) -
   e.g. Steel wires in Al
        SiC in titanium
        W wires in nickel or steels.
In all cases diffusion barriers are essential to prevent chemical interaction or embrittlement of the fibres. Except for aluminium matrices fabrication must be well below the melting point i.e. using powder techniques. In each case the aim is to achieve usable creep strength at temperatures above those possible for conventional alloys.

Future development requires improved fibres of lower cost,

alloy matrices tailored for compatibility with the fibre and of the same thermal expansion coefficient, improved barrier layers and development of processing techniques. For long fibres there is the possibility of exploiting preformed laminates, woven structures and mixed fibre hybrids.

The longer term looks to the promise of better fibres and use of intermetallic compound matrices.

2.3.2. Ceramic - Matrix Composites

The main interest here is in short single crystal fibres (whiskers) or continuous fibres in ceramic matrices. Materials include alumina, zirconia, silicon carbide, silicon nitride and graphite for use up to 1200°C (or for graphite fibre in graphite matrix up to well above 1600°C). The benefits conferred by whisker and fibre re-inforcement of monolithic ceramics are enhancement of toughness at low temperatures, reliability, improved Weibull moduli, and creep strength. The potential for improvement in this new and rapidly growing field, is very considerable. Immediate problems are :- for whiskers, high toxicity and cost of small scale production of whiskers and of composites.
- for fibres, lack of high temperature stability, especially with respect to oxidation/corrosion, and insufficient strength to allow high temperature pressing. Research priorities must include:

1. Provision of alternative/improved whiskers, fibres and filaments with low health hazard and better strength degradation resistance at high temperatures (e.g. via higher purity, controlled stoichiometry etc.) combined with low cost manufacturing technology.

2. Surface treatment and/or coating of whiskers, fibres & filaments to enhance control of matrix/reinforcement bond strength and toughness e.g. pyrolytic coatings of SiC or C or multiple SiC/C/SiC to improve energy absorption in fracture by multiple interface shear/pull-out and also to enhance stability at high temperatures. Study of matrix/reinforcement interface behaviour is also required in conjunction with microstructural design.

3. Lower cost processing/production technology and the potential of energy efficient composites based on low temperature processing e.g. macro defect free cement and SiC fibres in poly-sialate-siloxo-matrix.

4. The potential of pre-stressed composites e.g. SiC reinforced RBSN as well as transformation or mechanical toughening in

particulate or dual phase composites.

5. Matrices tailored to the required high temperature performance
   and suitable for composite manufacture, e.g. glass, glass
   ceramics, oxide, nitride and carbide ceramics.

2.4. Ceramics

2.4.1. Engineering Ceramics and Pyrolytic Materials

Engineering ceramics must be structurally stable and,
ideally, resistant to oxidation up to very high temperatures but
it is the requirement for very high tensile strengths above the
temperature range served by super alloys which distinguishes them
from modern high purity refractories. The main problem is to
achieve fracture toughness in the low temperature range. Sharp
edged pores on grain boundaries or foreign particles in the
powder mix can act as crack nucleators, in a matrix which almost
inevitably has poor resistance to crack propagation. For the
turbine applications there are also problems of thermal shock and
impact of ingested matter. Some engineering ceramics are also of
interest for electrical properties, e.g. for use as very high
temperature electrical heater elements operating in air ($MoSi_2$)
or as solid electrolytes in fuel cells.

Current research is targetted to eliminate crack nucleating
features by process control or use of vapour deposition
techniques, and by the introduction of crack arresting features
into the microstructure. There is considerable scope for
improvement of engineering ceramics and some orientations for
future R&D are given below.

1. Improved and constant strength and toughness together with
   strength retention for long times at high temperatures by
   improved purity and freedom from deleterious residual
   sinter-aid phases. Study of transformation toughening and
   whisker, fibre, or filament reinforcement, is essential.

2. Purer and more easily sinterable powders together with
   fugative sintering aids. Powders of consistent particle size
   distribution optimised to allow processing to fine grained,
   flaw free, ceramics of more consistent, reproducible and
   predictable properties.

3. Reduced materials and production costs combined with cost
   effective and reliable methods of non-destructive evaluation.

4. Pyrolytic materials for surface coating of porous or
   corrodable ceramics and to form free standing shapes of
   acceptable strength and fracture properties, by control of

processing parameters to produce pure, high strength, fine grained, equiaxed structures which could also serve as matrices for composites.

5. Reliable ambient and high temperature data on selected ceramics to provide an adequate design data base and a framework for materials' specifications.

6. Studies in the short term on silicon nitride, sialons and silicon carbide, alumina and zirconia. In the longer term, effort should be devoted to binary and ternary systems plus alternative oxides, carbides, nitrides and silicides.

2.4.2. Refractories

The steel industry and other metal industries are demanding improved high purity refractories and special items for new metallurgical processes e.g. continuous casting at a time when, owing to the dramatic decline in demand for refractories, the producers can least afford it. Examples of developments required during the next 10 years are given briefly below. Key properties include creep strength, thermal conductivity, resistance to thermal shock, and corrosion resistance.

1. The development of high quality refractory shapes, monolithics and increasingly advanced ceramics (e.g. Si_3N_4 and SiC), and to establish the benefits of use of synthetic raw materials together with areas for application of advanced/technical ceramics and attendent cost benefits.

2. Research on chemically bonded brick materials for the steel industry and on the relative economic and technical benefits of electro-fused MgO as compared with additions of antioxidants (Al, Mg, Si) in relation to slag resistance. The mechanism by which antioxidants may sometimes result in decreased slag resistance also requires elucidation.

3. Development of low cement castable monolithics for heavily cooled linings. These must be thermally stable and capable of being installed in thin layers. Development of mortars to be employed with Si_3N_4, SiC and resin bonded bricks.

4. Industrial scale production of low cost fibres for insulation, free from health hazards in manufacture, installation and use and of higher purity to reduce contamination of furnace processing environments.

5. High quality refractories/ceramics for heat-exchangers, incinerator linings and gasifier filters.

3. MATERIALS PRODUCTION

The emphasis of Chapter 3 is on the R&D requirements for improving ways of making the feed materials through to the final fabricated component of high temperature materials. Many of the advanced manufacturing technologies and quality assurance procedures are pioneered by the needs of aerospace engineering. They are adopted where appropriate elsewhere once their worth has been demonstrated.

3.1. Alloy Production

The need is to improve quality and reliability but also to reduce costs of manufacture. As in other industries this will come from automation, instrumentation for process monitoring at all stages and tighter control over raw materials, refractories, temperatures at all stages and computer aided design and manufacture. Developments in computers and relevant software are inevitable, any powerful new NDT technique will be welcomed particularly those which indicate the presence of unwanted inclusions at an early stage of manufacture e.g. in the melt or in as-produced powder. Alongside this are needed techniques to remove such defects, e.g. by melt filtration or by water elutriation of powder.

While dramatic improvements in alloys are not expected there are new processes which justify further development to manufacture products in existing alloys. Vacuum arc remelting can be replaced by electro slag refining to give a finer structure. This in turn may be replaced by a drip melting and casting process to give even finer structure if present difficulties can be resolved. Even more rapid rates of solidification may be introduced in the longer term not only to refine structures but also to retain more alloying elements in solution in the higher temperature range thus preventing the formation of coarse embrittling precipitates. The chief drawback of consumable arc melting or electroslag melting is that inclusions and insoluble particles or lumps do not have any chance to float out or sink to the bottom of the melt so that they can be removed. The electron beam cold-hearth remelting process should provide a new method for obtaining super clean quality in ingots.

With water or liquid metal cooled copper moulds a "skull" of frozen metal acts essentially as the crucible thus avoiding contamination by the crucible. Such a process is particularly useful for remelting secondary scrap. The same technique is being developed for melting titanium alloys. Improvement in scrap recycling would be a major factor in reducing cost providing the product has an acceptable quality.

There is still scope for reducing the incidence of porosity in castings by careful control of minor alloying elements and impurities. For this work the factors involved in feeding interdendritic cavities must be studied.

Many of the recently introduced superalloys can be forged only in a very narrow range of temperatures. Fast acting automated presses are therefore required to give near constant temperature working due to the heat evolved by the processes of deformation.

Production by the powder route using rapidly solidified atomised droplets makes it possible to use alloy compositions which would be coarse and brittle if slowly cooled. The chief problems as with all powder processes are the accidental pick up of foreign particles and prevention of surface contamination. Screening to remove unwanted material is not satisfactory. Prevention of contamination in the first place is essential using "cleaner" powder preparation methods such as the plasma melted rotating electrode process currently being developed.

Ability to make a component to near final shape in one process would not only reduce cost but also reduce the amount of scrap for recycling. Any method likely to achieve this should be investigated. Such processes include hot isostatic pressing, the Osprey process, thixo- and rheocasting.

Improvements to be investigated in sintering include surface treatment of powder in halogen atmospheres and partial liquid phase sintering. Ceramic mould and core materials for precision casting require further study and improvement. Further improvement in cutting tips would be welcome since superalloys are difficult to machine.

There is also scope for more work on surface treatments to confer resistance to crack initiation or to oxidation. Ion implantation and laser surface treatment techniques are still in their infancy and promise major development.

3.1.3. The Role of Computers in Production of High Temperature Materials

The increasing introduction of computers into manufacturing processes for, design, modelling, materials selection, production routines, flow control and process planning will have a major impact on production technology in the next decade.

3.2. Ceramics and Refractories

3.2.1. Ceramics and Pyrolytic Materials

Cold Consolidation and Sintering

1. Improved powders of controlled particle and agglomerate size distribution are required together with R&D in sintering shrinkage to improve reproducibility and reduce machining.

2. Improved organic additions are necessary for powder consolidation by pressing or moulding to give better lubrication and resistance to pressure release damage and which are easily removed by thermal treatment and do not produce defects.

3. Better slips, moulds and mould coatings are required for slip casting. Injection moulding requirements are better blending of constituents, alternative binders and improved burn-out schedules.

4. Processing requirements include enhanced throughput to reduce unit costs and 'fast' firing to minimise unit energy costs plus improved low thermal capacity furnace technology, high purity efficient fibre insulation and alternative methods of heating.

Hot Pressing

1. Die materials of improved life compared with carbon are required plus lower cost machining techniques for finishing near-net-shape hot pressed components.

2. Further investigation of superplasticity in ceramics is necessary to permit superplastic forming.

3. Investigation is required of the technical benefits and cost penalties of hot isostatic pressing in comparison with die hot pressing as well as critical appraisal of HIP sintering and a comparison of these with competing processes.

Reaction Bonding

1. R&D is required on improved materials with greater reproducibility and on process optimisation to attain better properties and reduced costs.

2. Development of surface impregnation and pyrolytic coating processes is required to improve the oxidation/corrosion resistance of porous ceramics/refractories and to confer

better erosion resistance or tribological properties.

3. R&D is required on the potential of reaction bonding for producing more complex ceramics (e.g. $Si_3N_4 - ZrO_2$) and on the use of reaction bonding to produce pre-fired refractory shapes and monolithic refractories.

4. Further investigation of thermal spray deposition processes is desireable for production of hollow components and filament wound ceramic matrix composites by the reaction bonding route.

3.2.2. Powder Production

1. Low cost powders of controlled particle size, shape and size distribution with acceptable agglomerate size and distribution together with high purity are an essential requirement. Lower cost powder precursors are equally essential.

2. Investigation is required of sol-gel, co-precipitation and other routes capable of producing powders or precursor particulates of controlled size and shape and also allowing preparation of mixed powders of controlled purity, homogeneity and sintering characteristics.

3. Factors controlling sinterability of powders and the possibility of surface 'activation' to improve sintering characteristics require research and development.

3.3. Composites

3.3.1. Metal Matrix Composites (MMC)

Fabrication of tungsten carbide cutting tips has reached an advanced stage of development to give a material of high hardness up to 650°C. The combination of tungsten carbides or other refractory carbides with a cobalt alloy matrix is ideal, the combination is compatible up to very high temperature and wetting is readily achieved. These advantages are absent in almost all other MMCs of current interest. Poor wetting requires treatment of the reinforcing component to modify the surface, or use of pressure to cause infiltration. In addition matrices such a aluminium or titanium whether molten or as powder inevitably form surface oxide films which interfere with infiltration and wetting. Fabrication techniques must be developed to circumvent these difficulties.

Some detailed R&D requirements are as follows:

1. Investigation of particulate reinforcement of matrices other than aluminium is necessary as well as development of

alternative aluminium alloy matrices for short fibre composites and the extension of the working routes to higher melting point matrices. The liquid phase sintering plus forging route also warrants investigation.

2. R&D on thermal spray forming and electrocodeposition as manufacturing techniques for particulate composites is justified as is arc weld deposition of particulate composite material to form coatings.

3. Further study is required of alternative matrices for fibre composites including steel and titanium.

4. Development of low cost fabrication technology for long multi- or mono filament fibre composites and hybrid composites as well as three dimensional reinforcement is essential.

5. There is a requirement for improved fibres compatible with various matrices and having good thermal stability as well as reasonable cost and assured sources of supply. Special coating techniques to reduce matrix-fibre interaction will be increasingly important.

3.3.2. Ceramic Matrix Composites (CMC)

Ceramics reinforced with SiC whiskers can be fabricated by conventional processing techniques. Future development needs - improved and lower cost whiskers, efficient mixing to control fibre distribution, and methods for making complex shapes reliably and cheaply. Both empirical and fundamental scientific investigations are essential.

Long fibre preform composites are made by infiltration of a fibre pre-form by chemical vapour deposition, or liquid impregnation + pyrolysis (for SiC, Si_3N_4, or carbon) or sol impregnation + sintering, or mixtures of these methods sequentially. Future needs are to obtain higher fibre loadings, to develop better lower cost fibres and matrices and to bring relevant branches of science to bear on the processes involved in composite production.

There is a range of long fibre composites made by slurry impregnation of continous fibres which are cut stacked, heated to remove binder and finally hot pressed. Glass and glass-ceramic matrices are particularly suited to such a sequence. For more refractory materials liquid phase sintering under pressure has been successful for graphite fibre reinforced Si_3N_4. Coated filament winding is an alternative route. There is much to be done in this area of fabrication technology, and full application of relevant basic sciences is essential including inorganic and

organic chemistry, colloid science, solid state chemistry and
fluid mechanics also in the applied sciences are textile
engineering, chemical and mechanical engineering. Very stringent
quality control is also vital, since these material have very
poor inherent tolerance of flaws. More sensitive NDT techniques
are needed.

3.4. Joining

3.4.1. Joining of Metallic Materials

Even for conventional fusion welding of materials in current
usage there is scope for improvement in welding techniques by use
of automation and instrumented monitoring of temperature
distributions. Further improvements include narrow gap weldments
to cut down the volume of weld metal combined with high power
input techniques to reduce structural degradation in heat
effected zones. There is still not enough known about properties
of welds relevant to long term in-service performance.

If higher steam temperatures or pressures are achieved with
high chromium ferritics the welds will need to have improved and
matching properties. The new steels will present quite severe
welding problems, requiring considerable in-depth study.

Many of the new materials including ODS metals and MMCs
completely lose their strength and structure when fusion welded.
"Low temperature" joining processes will therefore have to be
tried and evaluated e.g. diffusion hot pressure bonding and
friction welding with a variety of relative motions. NDT methods
for these bonds will need to be developed. The strengths at high
temperatures and toughness will have to be measured for a variety
of welding conditions. The feasibility of brazing or very high
power input welding techniques could also be explored.

3.4.2. Joining of Advanced Ceramics

Many potential applications for high temperature ceramics
depend critically for success on the development of a method for
joining either to other ceramics or to metals. Conventional
techniques using local heating or fusion are ruled out because
they cause cracking and decomposition or unacceptable structural
degradation. It is also vital to choose materials of matched
thermal expansion to avoid high internal stresses and thermal
fatigue of joints. In this context the use of brazing in
conjunction with interlayers of graded thermal expansion
coefficient requires further investigation for temperatures up to
about 1100°C for metal/ceramic joints. Brazing is used primarily
to join oxide ceramics but for higher temperature service
uniaxial or hydrostatic hot pressure (diffusion) bonding is

promising. Surface modification techniques (e.g. ion implantation etc.) should be investigated to promote diffusion bonding. Liquid phase and reaction bonding are also worthy of effort.

There is much scope for R&D of the evaluation of joints and standardisation of testing methods. There is a dearth of data for input into models for computing temperature and stress distributions and failure mechanisms. There is also a need for well specified quality control techniques sensitive to very small flaws.

There is also a need for some fundamental innovative thinking. Ceramics are different from metals, and will require specific solutions for joining particularly for the more severe applications, typically at high temperatures and stresses, where basic and applied R&D is vital to understanding of joint behaviour.

4. MATERIALS CONSTRAINTS IN THE HIGH TEMPERATURE INDUSTRIAL TECHNOLOGIES

4.1. Fossil Energy

4.1.1. Combustion Technologies

The primary aims are: 1. to reduce the cost of electricity production, 2. to use non-renewable energy sources more efficiently and, 3. to reduce pollution of the environment from noxious gases, dusts and heat. Improved HTMs are critical requirements for achieving these objectives. In many cases the three requirements can be simultaneously satisfied by a well planned combustion scheme e.g. 1. staged low temperature combustion + full flow filtration of dust + gas turbine topping + steam raising, and 2. more efficient heat to electricity conversion to reduce thermal pollution.

Staged combustion for NOx control and/or use of fluidised beds involves submitting economiser/evaporator and superheater tubes to the products of incomplete combustion and/or erosion, requiring materials of improved corrosion resistance particularly at the lower temperatures where existing materials fail to form and maintain protective oxide films.

In conventional pulverised coal burners, tubes downstream of the combustion zone are corroded by deposited molten alkali sulphates and chlorides. Crude and residual oils containing vanadium can cause even more severe corrosion of austenitic steels and reduce permitted steam temperatures by 30°C. Coextruded or coated tubes could be of value here. The temperatures for gas turbines are modest but protection of aerfoil surfaces from type II hot corrosion is required. For second generation fluidised bed systems full flow positive filtration of hot gas will require major improvements in ceramic filter materials and design. For indirect fired gas turbines heat exchangers operating to give turbine entry temperatures well above 750°C are needed, perhaps fabricated from ODS alloys, or fibre toughened ceramics.

There are now several routes competing for recognition for obtaining increased efficiency of coal burning to generate electricity including gasification + gas turbines or fuel cells both offering load-following capabilities; or hot topping using gas turbines or using MHD, both more suitable for base load operation.

4.1.2. Steam Cycle Power Plant

This subsection is concerned with all steam plant components not exposed to the combustion gases, i.e. headers, main pipes, turbines and casings. Current steam parameters 540° to 570°C, 16 to 24 MPa are limited by corrosion and creep properties of materials. Doubling steam pressure and raising superheat by 100°C would increase efficiency by about 10% provided reliable materials were developed. For plant not on base load operation, thermal fatigue resistance is required in addition to higher creep strength. More stringent NDE would be required to detect flaws which could grow to a critical size in service together with in-service monitoring. No established materials could be expected to survive under conditions imposed in plant with higher than current steam parameters.

4.1.3. Coal Gasification

Gasification is a first stage in use of coal as a feedstock for the chemical industry, for making liquid fuel, for generating power using gas turbines, or for reducing environmental pollution from coal burning in steam plant. A favoured scheme gasifies coal completely by high temperature reaction with oxygen and steam to give a gas of high calorific value consisting mainly of $CO+H_2$. Advanced systems could be built using existing refractories and alloys since similar conditions are faced in iron and steelmaking plant. More efficient and reliable plant could be envisaged if certain lines of materials development are successful. Specified items include:

1. R&D to define safe operating limits for heat exchanger materials in raw product gas from fluidised bed combustors.

2. Alloys and coatings to allow superheated steam temperatures of

565°C at 160bar in raw product gas heat exchangers; these should have good oxidation/sulphidation resistance and high integrity coatings are also required. The possibility of Type II hot corrosion by molten salts would probably impair the value of alumina forming coatings.

3. Improved ceramics for raw product gas heat-exchangers including high integrity silicon carbide. Better refractories and ceramics are required for linings including high chromia, dense, refractories. The effect of coal ash composition or slag attack must also be studied. Investigation is also needed of the use of fluxes to lower the slag melting points of European coal ash derived slags.

4. Development of on-line validation/monitoring techniques for cracking, corrosion, etc. of metallic, refractory and ceramic materials.

4.1.4. Fuel Cells

Since in fuel cells the chemical energy of fossil fuel is converted directly to electrical energy without intervention of a heat engine much higher conversion efficiencies are possible and high temperatures are not essential. Alkaline cells operate at 80°C and phosphoric acid cells at 200°C. The high temperature cells include the molten carbonate fuel cell at 650°C and the solid oxide cell at 1000°C.

For the Carbonate Cell the materials needs include:

1. Cathode materials superior to NiO-a conducting ceramic to avoid dissolution and metal deposition in the eletrolyte. Stronger more stable high specific surface area anodes, nickel based or to replace nickel, such as alternative metals or conducting ceramics.

2. Corrosion resistant separator plate materials which are stable in molten carbonate under reducing or oxidising conditions including non-wettable ceramics, conducting or insulating for construction of separators or for coatings on separator plates.

3. Improved non-wetting catalyst support structures for internal reforming, embracing nickel with ceramic support materials.

For Solid Oxide Fuel Cells future needs include:

1. Materials of enhanced durability resistant to strength degradation and having stable pore and grain structures to provide thermomechanically compatible structures both for

single component and laminated multilayer structures.

2. Reduction of operating temperatures to below the present 1000°C. Materials with improved oxygen-ion conductivity at lower temperatures are required, possibly modified ZrO_2 with additions of 3d transition metals which may also improve sintering characteristics. Whilst this approach may improve endurance, long term performance and hopefully initial electrochemical efficiency, alternative approaches may be justified to develop active catalytic area including electrode material with mixed electronic and ionic conductivity; also reduced polarisation, and internal and contact resistance.

3. R&D on the influence of temperature, pressure, sulphur and other contaminants on cell behaviour and materials. Effort is also needed on the influence of fabrication process parameters on the properties of ceramic materials to enable optimum selection of materials and reproducible manufacture.

4.1.5. Magneto Hydrodynamic Energy Conversion (MHD)

Channel materials are the most critical for current attention and priority should be directed in the first instance to the development of very high temperature materials. Specific areas of particular importance are; controlled sintering of powder materials to develop microstructure for higher HT strength, controlled doping of electrode materials for HT electrical conduction, brazing and soldering to fasten electrodes and insulators to channel mainframe, coating techniques for channel component manufacture and repair, development multilayer, metal-ceramic structures to control thermal and electrical transport in the electrode and for fastening. Apart from these very high temperature materials problems, those of slag handling, seed recovery and NOx + SOx emission must be tackled.

Service life testing and evaluation facilities are critical to industrial acceptance.

From the point of view of national energy strategies the potential of MHD is poorly exploited. MHD provides a topping unit with no moving parts. It could be retrofilled to 50 existing conventional power stations in Europe to give 50% increase in power output, and efficiency increase from 32 - 35% to 40 - 45% (in 15 year old plant). A joint retrofit MHD plant, 250 MWth, is proposed as a European joint development based in Italy.

4.2. Petrochemical Plant

The manufacture of polymer materials from petroleum source materials through to final consumer products is subject to keen international competition both in price and delivery terms. The provisions of reliable high temperature materials at a suitable cost plays an important part in the achievement of competitive status. The development of alloys to enable processes to run at higher temperatures and pressures is a continuing vital activity of the metallurgical industry in Europe. Parallel improvements in process control and on-line NDT is an essential activity in maintaining reliability of HTM products.

The absence of internationally agreed standards for quality, composition and design stress allowables is a big disadvantage at the present time. Greater interaction between European suppliers and users of HTMs in the present petrochemical industry would be a distinct advantage. The aim here, initially, would be to establish an accredited data base for the various alloys, determined by well defined test procedures. Then allowable stresses could be perhaps less conservative than those currently used. Constant strain rate testing may give more relevant data.

The primary processes of cracking and reforming require pressure retaining components operating in the range 900°C to 1200°C exposed to naphtha and steam inside and flames outside. The latest materials offered by suppliers claim to give doubled creep strength by using minor alloying additions but these are difficult to check by chemical analysis. A validated mechanical test method is required.

For the new materials new weld fillers are needed matched in thermal expansion, creep strength and modulus. Further weld development is required to avoid the need for tedious life testing.

Directional solidification of cast products has a very big effect on the very anisotropic elastic moduli of austenitic materials. More must be known about this if designers are to introduce low cycle fatigue into their design and life assessments of plant. Thermal expansion coefficients must also be measured.

Coatings should be evaluated to improve corrosion resistance and to reduce carbon deposition.

Long term exposure of materials must not impair weldability since periodic repairs are necessary or the operator may want to improve the arrangement of the plant to increase efficiency.

4.3. Nuclear Energy

Fission

1. The principle research on high temperature materials problems in fission reactors in the next decade relate to the UK CO_2 cooled Advanced Gas Cooled Reactors and European helium cooled High Temperature Reactors and sodium cooled Fast Reactors.

   Accident analysis may require data about materials properties well above normal operating temperatures. Thus for PWR loss of coolant, accident analysis requires high temperature data for fuel elements and core structures.

2. For helium gas turbine cycles, blade cooling is not feasible and traces of impurity can cause internal oxidation and carburisation since protective films are not formed. Coatings of noble metal or alloys free from Al and Ti may be required.

3. Thermal reactors use the worlds limited uranium reserves very wastefully. At least fifty times the energy could be extracted using fast (breeder) reactors to convert the 238 U which consititutes 99.3% of natural uranium to fissile material. Sometime before fossil fuel is exhausted an economic reliable fast reactor must be developed. The presently favoured sodium cooled fast reactor has high capital cost and in Europe a "back to the drawing board" concerted rethink is being instigated.

   The HTM problems include void swelling of core materials – currently 9 - 12 Cr steels are being developed – thermal shock damage in the sodium circuits which must be prevented by appropriate design and operational control, and last but not least the provision of superheaters and evaporators etc. with sodium at 550°C and high pressure steam, guaranteed to be free from any kind of leak for 25 years of base load or load-following operation.

   The merits of low core rated gas-cooled fast reactors should not be entirely overlooked. Fast reactor R&D is already well coordinated on a European basis.

Fusion

The next stage experimental machine for attempting the controlled release of energy by fusing light atoms ITER is being designed on a world basis. One published estimate of the cost of the machine is that it will be more than $ 2 billion (A serious next stage inertial confinement machine would cost about the same amount). Decision whether to build could be made in the early

1990s. Problems of high temperature materials may be avoided by running at low temperatures since the main purpose of the experiment is to release energy in a "long burn". A thin inner surface layer of the torus lining is exposed to the plasma and refractory materials are undergoing tests in existing machines. Carbon composites, ceramics, coatings, and tungsten are all of interest.

The next even more difficult stage of development will be both to use the energy released, and to breed tritium. The coolant circuit materials will suffer fast neutron displacement damage and transmution, possibly exposed to corrosive materials, high pressures and temperatures. There will also be a number of stress transients additional to those imposed by the conventional end of the power generation plant, including quiescent periods between each burn, and major transients from loss of magnetic field. The consequences of even very small leaks in the coolant circuits could be serious. These problems however lie well outside the ten year forward look of this book.

4.4. Engines

4.4.1. Aero Gas Turbines

The next generation of military engines need to have higher power to weight ratios and increased fuel efficiency, necessitating higher fan and overall pressure ratios and higher turbine entry temperatures, TETs appoaching 2100°C given by stoichiometric burning. Direct operating costs (DOC) could be reduced in designs using fewer components.

For civil engines lower DOC is the paramount requirement. For current turbo fans DOC is made up of 50% fuel cost, other factors being purchase cost, maintenance costs and replacing of components. A small, 5%, fuel saving could come from adopting the proposed engine core improvements mentioned above for military engines. A much bigger, up to 20% improvement in fuel efficiency could be obtained from larger more expensive engines with higher bypass ratios, but these are attractive only if fuel costs double.

The higher rotational speeds and higher compression ratio per stage in future engines will impose higher stresses on rotating parts, and increased TET will impose higher temperatures. Higher strength/weight ratio materials will be essential with adequate creep strength, erosion protection and good high and low cycle fatigue resistance and adequate damping capacity.

New or improved materials currently under study are as follows:

1. Materials for use up to 800°C. Perhaps Ti_3Al intermetallic, or continuous fibre Ti composites with a corrosion resistant coating. At present titanium alloys are limited for long term exposure to 550°C until a suitable coating can be found.

2. Glass – glass ceramic SiC fibre composites for use up to 1000°C of lower weight and better fire resistance than Ti alloys. For military engines with TETs of 1400°C to 2100°C blade cooling to temperature limits set by nickel base alloys is not economic and materials in the HP turbine region will need to operate above 1400°C. Only materials such as SiC fibre – SiC matrix have the slightest chance of success here.

SiC fibres as currently available degrade at 1200°C. Graphite fibre in carbon suitably protected has been proved capable of surviving for a few hundred hours up to 1600°C in simple shapes such as for nozzles and heat shields on space vehicles. The new composite high temperature materials have a long way to go and some fundamental difficulties have to be overcome before they will look attractive to the engine builder. If and when they do arrive entirely new design fabrication and quality control procedures will be needed.

4.4.2. Marine Gas Turbines

Custom built or adapted aero gas turbines are used to propel warships, hovercraft and hydrofoils, also for pumping of gas or oil in bulk carriers, electric power generation and fire fighting duties. Gas turbines are also used for a range of duties on offshore exploration and oil production platforms. Marine turbines tend to run at modest temperatures and power ratings but do have to contend with ingested salt and impurities in low grade fuel. Blade materials and coatings with improved resistance to Types I + II hot corrosion are therefore required. A combustion test rig is needed to reproduce realistic conditions for corrosion testing of new materials and coatings.

4.4.3. Turbines for Motor Vehicles

A turbine powered car was demonstrated in 1950. Turbines offer reliability, low cost, responsive performance, clean exhaust, compactness, lower weight, and ability to use a wide range of fuels. High TET and use of the hot exhaust gas to heat inlet air are essential to give competitive efficiency. A favoured method is to use rotary regenerators made of thermal shock resistant refractories. Development of materials for the rest of the system will follow closely aero gas turbine progress, but the rotary regenerator presents special problems requiring long term testing. Joints, seals and bearings for high temperature service must also be developed. There is the additional need for low cost components to eliminate expensive fabrication and coating procedures. Demonstration engine projects are essential for making real progress. Improved NDE, proof testing and design methods for ceramic components are also necessary.

4.4.4. Reciprocating Engines

For the "high temperature" moving parts the requirements are, low coefficient of thermal expansion, creep resistance, and high strength to weight ratio. Improvements of aluminium alloys can be obtained by dispersing ceramic fibres in them.

In the longer term more of the developing tougher ceramics might find application as components or coatings to endure high temperature and corrosive products in burnt low grade fuel. Other potential benefits of ceramics might be to provide thermal insulation of the combustion chambers, to provide wear resisting surfaces, to allow weight saving compared with metals, and to be used as sensors monitoring engine parameters.

For ceramic materials R&D needs may be summarised to include:-

- Tougher ceramics,
- Better strength and reliability in ceramics made by low cost routes,
- Better adhesion for coatings,
- Research on optimising and understanding the wear performance of ceramics for engine applications,
- Development of materials with improved resistance to fuel impurities,
- Better understanding of engineering methods for fixing ceramics to metals
- Low cost effective methods for non-destructive evaluation of components.

4.5. Refractories for Iron and Steel Production

For new processes and particularly for certain critical components new advanced refractories and production trials are required. Iron and steel making industries still are the major customers for refractories and therefore will have the greatest influence on new materials developments. With the emphasis on high quality and composition control of metals there is the corresponding need for pure refractories for any lining to which the metal is exposed. Chemically purified feed materials are therefore needed in place of direct use of mineral rocks, sands and clays. A second requirement is avoidance of bits of refractory material in castings from moulds, ladles, tundishes and filters. Thus the refractory must resist spalling and flaking. Finally there are some critical limiting requirements in new processes for pouring spouts, burner nozzles, tubes, nozzles, sliding gates and rings for continuous casting, radiant tubes and skid buttons for which very advanced and expensive materials are needed.

Recuperators are very high temperature heat exchangers operating above the temperatures where metals can be considered either due to excessive creep or melting, or due to lack of oxidation resistance. There is work in a number of countries to find a suitable ceramic material and recuperator design but after a long time there is no positive achievement and the prognosis is not good. Regenerators are still the practical option but require on-off operation. Fecralloy with oxide dispersion strengthening was seriously considered at one time for temperatures up to 1250°C.

The blast furnace provides 10 zones of different temperatures and chemical and physical environments for the refractory lining, and therefore is a good testing ground for most types of refractory brick that might be needed elsewhere.

There are requirements for silica bricks of improved thermal conductivity in coke ovens to reduce wall temperatures and better bonding systems for SiC grains in bricks for blast furnace linings. Development is also required to adapt magnesium-graphite bricks to the specific requirements of torpedo car linings.

4.6. Processing of Superalloys

The market for HT materials for use in manufacturing superalloy components is too small to justify specific development programmes, and benefit must be gained from exploitation of materials developed for other purposes. Desired improvements in refractories include removal of unwanted impurities, reduced wettability, more compact material, bricks of complex shapes to avoid the use of ceramics, ramming powders of matched shrinkage rates, porous plugs of better wear resistance, and atomisation nozzles of improved wear resistance to reduce contamination of powders.

Development of satisfactory filters to remove solid particles from the melt would be particularly welcome. Better low cost isostatic forging die materials would also be welcome.

Improved insulating materials are needed, also lower cost die and tool materials plus further improvement of ceramic cutting tool materials.

4.7. Sensors for High Temperature Applications

The next ten years is likely to see the introduction of computer control of industrial processes using on-line direct input from monitoring instruments. Many of the sensors will have to operate at high temperatures for very long times and in various aggressive environments. Process variables to be measured include
1. temperature using thermocouples, thermistors, radiation pyrometers and differential extensometers.
2. Gas and liquid composition using solid electrolytes eg thoria for oxygen level in sodium, doped β alumina for hydrogen in gases.
3. Pressure and acceleration sensors e.g. piezo electric materials. Major applications for successful instruments abound in mass markets for control of NOx and particle emission from car exhausts, and control of industrial and domestic static combustion systems related to fuel economy.

Some detailed materials requirements are as follows:-

1. Improved thermocouple materials to provide enhanced life and greater resistance to aggressive environments plus reduced costs. Attention must be directed to thermocouple elements, insulation and sheathing including metal/insulant/contaminant interactions. There is scope for study of novel thermocouples such as C/SiC prepared by pyrolytic deposition techniques.

2. Improved transparancies for radiation pyrometers and data are needed on the emissivity of refractories and the effect of surface finish. Special optical fibres are required for IR transmission.

3. Innovatory anisotropic materials are required for extensometry and improved selective sensors are needed for gas analysis (e.g. carbide/nitride electrodes) plus coating technology for deposition of non-metallic electrodes. Gas sensors are required which show a linear rather than a logarithmic response.

4.8. Furnace Materials

In an ideal modern furnace the throughput would be limited only by the heat transfer properties of individual items of the work load and their ability to survive thermal stress. To achieve that may require higher temperature heat sources: radiant tubes, metallic and ceramic electrical resistors, higher flame temperature using preheated air and/or oxygen enrichment, in turn requiring more refractory burner nozzle and furnace linings, and large fans to distribute the heat, operating at the full gas temperatures.

Another separate requirement is for efficient shift or intermittent operation. Furnace construction then has to be of low thermal capacity i.e. light weight with a thin inner lining of thermal shock resistant refractory stable in the furnace atmospheres and exposed to the maximum temperatures, backed by high temperature (above 1400°C, ceramic fibre) insulating layers.

There are specific problems associated with the use of controlled atmospheres. Maintaining a seal at the points of entry of burners and heating elements is difficult, and for electric heating insulation must be maintained.

Systems for supporting the work load and conveying it through the furnace also present the need for a special range of desirable properties including strength toughness, stability, low thermal capacity, corrosion resistance etc.

There is plenty of scope for using improved high temperature metals, ODS alloys, tough ceramics, short and long refractory fibres, cermets.

5. OPTIMISATION OF COMPONENTS

5.1. Testing Metrology

Properties may be measured merely to indicate that a given batch of material has been correctly processed and gives values lying within the acceptable run of the mill scatter band. Tests must be simple, relevant, preferably non destructive or requiring only small specimens, but sufficiently sensitive to reject unsuitable material. For design and generic safety analysis quite different properties usually have to be measured. The results are accumulated in data banks, which can then be used to generate standards such as the new European Standards to become mandatory in Europe in 1992 and to supercede national standards and ISO. There is room for considerable improvement in the specification of testing procedures, the calibration of equipment and the accrediting of individual laboratories and operators on a regular basis in Europe so that uniformity can be achieved.

New Standards or Codes of High Temperature Testing Practice are required in the following areas:

a. Creep testing of internally pressurised tubes.

b. Creep testing of circumferentially notched specimens used to generate triaxial stress state creep data.
c. Torsion creep testing.
d. Instrumented impact testing.
e. Hot hardness.
f. Wear testing.
g. Dynamic calibration of (i) Load cells (ii) Extensometers used for fatigue testing.
h. Measurement of crack growth under creep and fatigue conditions
i. Methods of validating testing machine software.
j. Determination of Young's modulus.
k. Thermal expansion measurements.

Measuring high temperature properties presents a number of problems, in gripping the specimen, controlling temperature for long periods e.g. 1 year, avoiding drift of pyrometers, and measuring strain continuously. Capability for up to 1600°C should soon be provided, and up to 2000°C in 10 years time. There are also problems in measuring relevant low temperature properties of ceramics and fibre composites.

5.2. Non-Destructive Evaluation

Each of the classes of material used for high temperature service or under development for such service will require considerable adaptation of the NDE techniques which have been developed and optimised mainly for examining structural steels for service at room temperature.

The main high temperature materials areas to covered are as follows. Refractory metals and alloys, mainly nickel and titanium based. Ceramics and ceramic composites. Thermal barrier and tribological coatings. Ceramic particles and fibre reinforced metals. Carbon fibre reinforced carbon. Inspection of bonds and joints involving any of these materials.

All of these materials raise special NDT problems at present unsolved.

In view of the wide diversity of the problems raised by the different classes of material, research priorities should be ordered as follows :

a. Rapid analysis of ultrasonic characteristics of nickel and titanium based materials.

b. New methods, including thermal, eddy current and radiation back-scatter techniques for the assessment of thermal barrier coatings.

c. Rapid high resolution scanning methods for the detection of defects in high strength ceramics. This should include both radiographic and ultrasonic techniques.

d. Methods for assessing unfired ceramic mouldings.

e. Fundamental studies of such materials as metal matrix and composites and carbon fibre reinforced carbons.

The development of a "new" material without an adequate NDE technique can be a barrier to its confident adoption in critical applications.

5.3. Design of Materials for Components

5.3.1. Modelling Deformation and Rupture Behaviour

At the present state of development computations are an adjunct to component testing, to help in planning tests and in understanding test results. Since computations have become much more powerful and computing costs are decreasing rapidly, while at the same time making and testing components is very expensive, the use of computations is bound to increase. Computer hardware and software are available or are rapidly becoming available for dealing with HTMs. Over the next ten years undoubtedly progress will be limited by lack of well accredited data of the kind needed for creep computations. For failure predictions there are two major problems
1. the time for onset of tertiary creep shows a very wide scatter, and
2. failure often occurs at the site of a structural flaw.
Both these observations indicate the need for a soundly based statistical treatment.

Materials exposed to high temperatures and corrosive environments may suffer structural changes which alter their mechanical properties in service. The particular changes taking place at any instant depend not only on the temperature and stress imposed but also on the structural changes caused by the previous stress-temperature history of the specimen. Thus computation of in service stress - temperature - time - strain response of a structure involves at least a series of finite element computations at time steps small enough to achieve stability i.e. to cope sensibly with the fastest deforming element. Sequential stress strain distributions can be computed either by using empirical equations obtained from stress strain measurements on specimens of simple shape or by modelling all the microstructural changes which occur under stress including grain growth, changes in dislocation distributions, nucleation of voids and growth of voids. Using the former methods it is possible to predict whether a stress concentrating feature will become diffuse or will propagate as a crack for the simple case of constant temperature and applied load. For structures free from stress concentrations much more complex situations can be done with a super-computers, using physical models and where stress reversal and interactive changes occur simultaneously over a wide temperature range. Failure prediction is however much more difficult.

5.3.2. Life Prediction and Residual Life Assessment

The development of constitutive relationships for life prediction and residual life assessment of structural steels subject to failure by accumulation of creep damage has received considerable attention over the last decade. The increasingly well developed microstructural interpretations of creep damage and their assimilation into macroscopic models, permitting numerical analysis, show promise of increasing application and of establishing more confidence in predictive methods.

The reliability of methodologies, however, depends crucially upon the comprehensiveness of the available information on the behaviour of materials in service. Structural changes can be monitored by hardness measurement and replication to detect fine scale defects, and to extract particles for examination. A method of measuring oxygen penetration is needed also. If exposed material is available for accelerated creep testing useful estimates of remnant life can be made. The effects of multiaxial stresses and temperature + stress transients on crack growth require further study.

To monitor in-service behaviour, surveillance specimens of all components and welds, should be exposed in relevant places in the plant where they can experience appropriate stress-temperature and environmental variations. Such specimens can be examined at intervals and give a running prediction of remnant life. Similar specimens exposed under laboratory conditions can be used to demonstrate the effects of variation in plant operating conditions. Monitoring of plant temperatures and strain should be done throughout the service life.

5.3.3. Calculated Phase Equilibria for Composition and Structural Control

When the appropriate thermodynamic data are available it is possible to model the behaviour of multiphase systems for alloys containing ten or more elements from knowledge of subsystems containing only three or four elements. The flow of ions through multilayer oxide/sulphide films on alloys can also be modelled from knowledge of solubility and diffusion rates in the various films. Experimental data must be assessed and used to produce coefficients for the thermodynamic equations taking care to generate consistent sets. Examples where calculations are used in alloy development include segregation during solidification, optimising composition of steel with respect to stability of fcc, bcc, martensite and sigma phase, substitution of strategic elements, and effects of scrap recycling.

A typical oxidation model for a nickel-iron-chromium alloy involves depletion of the metal surface in preferrentially oxidised species, and two oxide layers of different composition. Many kinetic processes are involved in evolution of the oxide structure and quantitative modelling would require diffusion data for all the elements of interest in each of the layers. Breakdown of films due to internal stresses would require different treatment, and empirical experimental data input. The modelling of molten salt attack has also made some progress.

Computer modelling of alloys and of their performance in service contributes to the understanding of high temperature materials and supplements empirical experimentation. It does however require input data of a different kind from that usually gathered. Modelling of the various rate processes relevant to metallurgy and ceramics technology is a new development likely to make great progress over the next decade. There are many applications to ceramics and coating technology yet to be explored.

5.3.4. Design Concepts for Ceramic Materials

The well-established design concepts for metals in structural components are not appropriate to ceramics. A new design methodology for ceramics, based on probabilistic fracture mechanics is emerging. Since the critical flaw size for ceramics is too small to be detected, design is guided by failure probability (reliability).

Much work is needed to establish a complete design methodology, including high temperature behaviour.

There is a serious lack of data and understanding on the behaviour of ceramics especially at high temperature. A material data base that can be easily used by the designer is still not available and current data bases are incomplete. Data and behavioural information in certain areas are almost completely absent, e.g. under dynamic loading (fatigue) conditions, or thermal stress conditions. High temperature failure modes have received little study. Moreover ceramic performance data is extremely sensitive to testing conditions and much of the data produced to data are questionable as the testing conditions are not specified. This implies an effort for the standardisation of the testing methods with respect to design requirements.

5.3.5. Design Methods and Rules for Metallic Components

The basic principles of structural design are to allow safe operation of a component; to ensure integrity and continuous functioning: and to avoid any damaging incident or catastrophic failure during the whole period of the defined operation of a component and during the total service life of a plant. There are different national rules and codes for pressure vessels and high temperature components in fossil fuelled power plants, nuclear power plants and petrochemical equipment, as well as the manufacturer's specifications for components of steam and gas turbines. The design rules lay down the essential basic principles for the guidance of the designer, the fabricator, the user and the inspection authorities. They try to achieve the best compromise between safety requirements and commercial and economic factors.

In some areas, current design codes are adequate to restrict component loading to safe levels (e.g. stress-controlled creep). Codes applying to more complex systems as are often found at high temperature are empirical and less than satisfactory.

An appropriate selection of materials and design methods should prevent failure by any one of the following modes:

- creep rupture due to long term loading;
- creep-fatigue failure due to load cycles in the operating and upset conditions;
- gross disortion due to incremental collapse and ratchetting caused by thermally induced cyclic loading;
- loss of function due to excessive deformation;
- creep induced structural instabilities, such as creep buckling;
- failure due to environmental attack;
- spontaneous fracture due to critical stress concentration around a flaw or due to unstable crack growth under cyclic or static loading.

For flow free specimens the allowable operation time is based on a linear sum of creep strains and fatigue cycles of various stress amplitudes to which the component is expected to be exposed each divided by the strain or number of cycles expected to cause failure, and then modified by a safety factor.

The treatment of creep failure during stress cycling and creep fatigue interactions are topics for future research, together with that of multiaxial stressing. Simplified inelastic methods are required to compute creep ratchetting during thermal cycling.

For components with flaws high temperature fracture mechanics has to be developed and much basic work has to be done, including fatigue crack growth.

A problem common to all design techniques is the wide scatter of measured materials constants in the mathematical equations and the combination of rapid plastic deformation with creep deformation.

Materials with highly directional properties need design techniques to model high temperature performance. Inelastic analysis of welds is difficult since the different parts of the weld have different physical and mechanical properties. The codes deal with this simply by halving the strain and cycle allowables.

Added to uncertainties in materials properties is the uncertainty of the operating schedules and parameters to which components will be submitted in service. This makes regular inspection a vital part of plant operation.